Sustainability Science

Michael P. Weinstein • R. Eugene Turner
Editors

Sustainability Science

The Emerging Paradigm and the Urban Environment

Springer

Editors
Michael P. Weinstein
PSEG Institute for Sustainability Studies
Montclair State University
Montclair, NJ, USA

R. Eugene Turner
Department of Oceanography
and Coastal Sciences
Louisiana State University
Baton Rouge, LA, USA

ISBN 978-1-4614-3187-9 e-ISBN 978-1-4614-3188-6
DOI 10.1007/978-1-4614-3188-6
Springer New York Dordrecht Heidelberg London

Library of Congress Control Number: 2012937794

Printed on acid-free paper

Springer is part of Springer Science+Business Media (www.springer.com)

Photo by Mike Peters/Montclair State University

*This book is dedicated to Robert W. Kates,
a pioneer in Sustainability Science who
inspired a generation of systems thinkers.*

Preface

Balancing human needs with the ability of ecosystems to provide the goods and services that we all depend on is a fundamental formula for the global sustainability transition (Fig. 1). Equilibrium can be attained either by increasing these goods and services or by reducing our consumption of them, or in today's world, both!

Furthermore, demographic shifts and new patterns of settlement have placed unprecedented pressures on human well-being, ecosystem functions, and the interactions between them. Society has yet to adequately address the challenges of diminishing resources, i.e., by facing challenges that make sustainability more feasible technologically, and simultaneously more difficult politically and economically. First, there has been a dramatic growth in per capita domestic product in many regions of the globe and an increased ability to meet human needs. Second, despite recent successes in decreasing harmful consumption per unit value of product, worldwide consumption of energy and other natural resources in industrialized nations continues to accelerate (Kates and Parris 2003; Brown et al. 2011).

Authorities worldwide have called for the prioritization of uses in order to minimize conflicts, protect resources, and ensure that all uses are compatible with sustainability goals. The public interest is addressed through recommendations to balance long- and short-term strategies with greater decentralization of governance to regional and local levels. Ecosystem-based management has been widely advocated as a central organizing principle for addressing land-use impacts holistically and reconciling multiple use conflicts at different geographic scales. Nevertheless, academicians, governance organizations, decision-makers, and the general public have yet to confront one very real issue:

> Where multiple desirable but competing objectives exist, it is not possible to maximize each…[and] in any system with multiple competing objectives, it will not be possible to meet every one.
>
> United States Commission on Ocean Policy 2004

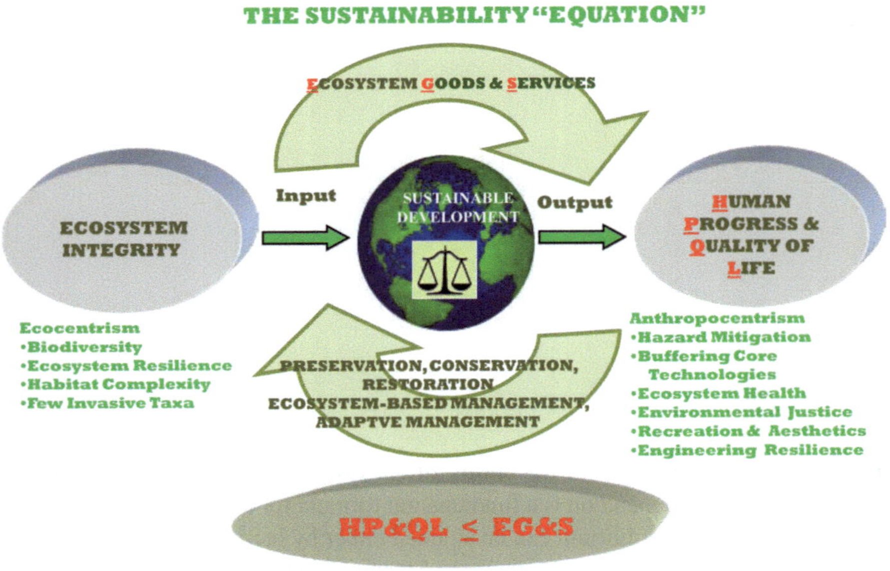

Fig. 1 The "Sustainability Equation" balancing human needs with ecosystem integrity

Any solution to the emerging conflicts arising on the path to long-term sustainability will, in part, require the integration of the biophysical and social sciences into a new transdisciplinary science that we refer to as "sustainability science,"[1] continued development and refinement of a number of new approaches and concepts including a systems approach to problem-solving, social learning, resolution of the "paradox of the dual mandate,"[2] and enhanced incorporation of human dimensions into resource management.

There is a growing awareness that the intractability of environmental problems can be explained in part by the social context in which they arise. When perceptions of a problem vary broadly, and when there is uncertainty in the scientific assumptions and outcomes that underlie the process, then a consensus is difficult to achieve. Under such circumstances, tensions can arise among stakeholders (Fig. 2), even

[1] There are many definitions of sustainability science; the National Academy of Sciences through its Proceedings offer the following: "…an emerging field of [transdisciplinary] research dealing with the interactions between natural and social systems … how those interactions affect the challenge of sustainability: meeting the needs of present and future generations while substantially reducing poverty and conserving the planet's life support systems."

[2] Whereas complexity, interdependence, high levels of uncertainty, unpredictability, and dynamism characterize natural systems—traits that prevent competitive dominance by any one species—human-dominated systems require predictability and stability to ensure uninterrupted provision of resources for human use. The paradox of the dual mandate arises from the need to reconcile society's desire to preserve, restore, and rehabilitate natural ecosystems while at the same time ensuring the provision of reliable, predictable, and stable supplies of goods and services at a time of escalating demand (Roe and van Eeten 2001; Berkes 2006).

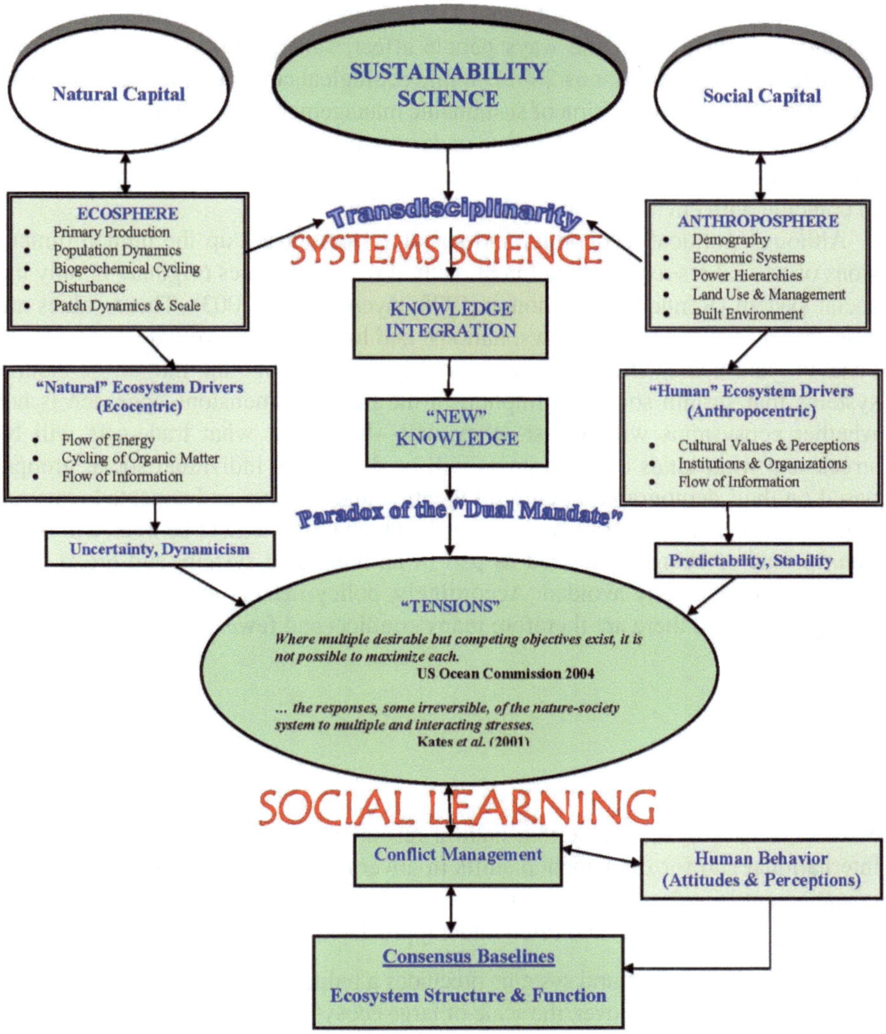

Fig. 2 (modified from Weinstein 2009) A conceptual framework for achieving the sustainability transition through the integration of the natural and social sciences in a transdisciplinary, systems approach to managing natural and social capital. Developing "new knowledge", changing human behavior and perceptions through a lens of social learning, and achieving consensus on how to effectively manage ecosystems for sustainability will be consummated through the emerging discipline of Sustainability Science and Conflict Management. Resolution of the "paradox of the dual mandate"—reconciling society's desire to preserve, restore, and rehabilitate natural ecosystems while at the same time ensuring the provision of reliable, predictable, and stable supplies of goods and services—will be a key component of any future success

when all are committed to sustainable development (Weinstein et al. 2007). This understanding of the social character of environmental problems has focused the attention of researchers, stakeholders, and policy-makers on the important role of governance, participation, and collaborative decision-making in better managing, if not solving, environmental problems. The human dimensions of natural resource management incorporate the ways people affect, value, utilize, and benefit from ecosystems (Salz and Loomis 2005). While ecological considerations are essential, the successful implementation of sustainable management depends on, and is driven by, societal values. We need a better understanding of the human-induced causes and social drivers of environmental change and how human behavior can be made to coincide with environmental and social priorities.

Although political, economic, and social systems make up the human dimensions of natural resource management, natural resource values originate in only the social system (Kennedy and Thomas 1995; Ayensu et al. 2003). These values are manifested as environmental laws, national and local budgets, volunteerism, voting behavior, and management decisions and largely determine the fate of the natural systems that sustain societies. Implicit in the human dimensions approach is not whether ecosystems will persist—they will—but rather what trade-offs will be struck and what kinds of ecosystems will be desired by individual social groups, based on their demographics, cultural identity, and existing and expected resource requirements. The present scenario is one in which issues tend to be treated in isolation, instead of being considered as part of an integrated system, and broad-scale decisions are generally avoided. Accordingly, policy-makers may too easily avoid the trade-offs and there are therefore many conflicts and few solutions.

Difficult Choices

Most citizens now recognize that natural resources are not inexhaustible, and an international call for fundamental shifts in governance, political will, and resource management is underway. The challenges we face in the move towards global sustainability are substantial and often underappreciated:

1. The complexity of natural systems precludes a reductionist experimental approach to management. Moreover, the scale of large ecosystems make controlled and replicated experiments virtually impossible. Consequently, our "imperfect science" and the effects of natural variability and uncertainty lead to an inability to reach consensus and accurately predict the environmental consequences of our actions. We are often left with a wide range of opinions on the issues (Ludwig et al. 1993).
2. With acquired wealth comes political and social power that is often used to promote further unlimited exploitation of natural resources (Ludwig et al. 1993).
3. Traditional demography and economics do not incorporate sufficient appreciation of environmental principles. Furthermore, ecologists tend to disregard human influence and instead concentrate on ecosystem function and dynamics. Numerous authors have suggested that the failure to agree on a collective vision of how to attain sustainability lies in the limitations and disconnects among disciplines (Kaufman and Cleveland 1995; Holling 2000; Clark and Dickson 2003; McMichael et al. 2003; Naveh 2005).

4. Anthropocentrism and the "we versus them" mentality stemming from the "arrogance of humanism" is a concept that expresses humankind's faith in its technology to manage nature so that all can prosper (Ehrenfeld 1981). In anthropocentric terms, humans have the "right" to control the natural world for the benefit of humanity. Even a cursory examination of the published literature reveals the sometimes large divide between ecocentrists and anthropocentrists, scholars and practitioners, functionalists and compositionalists (Callicott et al. 1999), environmental organizations and industry, commercial and recreational fishermen, public and government, etc. (Weinstein and Reed 2005). Thus, the ultimate compromises and sacrifices required—a distasteful concept to many, and possibly the root cause of the "we versus them" mentality that pervades sustainability management—will be necessary to accommodate human needs. Thomas Friedman (2007) stated this idea succinctly: "if you think we can deal with these huge problems without asking [the American] people to do anything hard, you're a fool or a fraud." Successfully balancing the demands of competing uses is perhaps the greatest challenge we face.

In the end, the successful transition to sustainability rests on a complex infrastructure that translates science-based information into public policy. This, in turn, elicits effective responses from society at large (Baird 2005). It is the performance and long-term capacity of this diverse array of entities (including scientific and educational institutions) from global to local scales that will ultimately determine the tempo and mode of the transition. Our fate rests in societal action involving all stakeholders, consensus building, and accepting the compromises and sacrifices that will ensure environmental and social justice for all. We hope that this book will contribute towards those goals, and quickly!

This book is organized into five thematic sections and an Epilogue; a summary of which precedes each compilation of chapters: Part I. Managing the Earth's Life Support Systems: The Emergence of Sustainability Science and Transdisciplinarity; Part II. Balancing Ecology and Economy: Natural Capital and Quality of Life; Part III. From Science to Policy: Managing the Commons, Social Learning, and Social Responsibility; Part IV. The Ecology of Cities; Part V. Restoring and Rehabilitating Ecosystems: Return from the Precipice; and Epilogue: The Challenge of Sustainability—Lessons from an Evolutionary Perspective. Key topics address emerging research and policy in (a) sustainability science, (b) the ecology *of* cities, (c) landscape ecology—scale, spatial patterns, organizational levels, and ecological processes, and (d) related topics in resource exploitation and management, ecosystem health and habitat restoration, the valuation of natural and social capital, habitat and biodiversity conservation, social learning, ecosystem-based management, and integrated watershed-coastal zone management.

Montclair, NJ, USA — Michael P. Weinstein
Baton Rouge, LA, USA — R. Eugene Turner

References

Ayensu E, van R Claasen D, Collins M et al (2003) International ecosystem assessment. Science 286:685–689

Baird RC (2005) The human dimension in ecosystem management: institutional performance of the Sea Grant paradigm. In: Hennessey T, Sutinen J (eds). Sustaining large marine ecosystems: the human dimension. Elsevier, New York

Berkes F (2006) From community-based resource management to complex systems: the scale issue and marine commons. Ecol Soc 11:45

Brown JH, Burnside WR, Davidson AD et al (2011) Energetic limits to economic growth. Bioscience 61:19–26

Callicott JB, Crowder LR, Mumford K (1999) Current normative concepts in conservation. Conserv Biol 13:22–25

Clark WC, Dickson NM (2003) Sustainability science: the emerging research program. P Natl Acad Sci 100:8059–8061

Ehrenfeld DW (1981) The arrogance of humanism. Oxford University Press, New York

Friedman TL (2007) The power of green. New York Times, 15 April

Holling CS (2000) Theories for sustainable futures. Conserv Ecol 4:7–13

Kates RW, Parris TM (2003) Long-term treands and a sustainability transition. P Natl Acad Sci 100:8062–8067

Kaufman RK, Cleveland CJ (1995) Measuring sustainability: needed—an interdisciplinary approach to an interdisciplinary concept. Ecol Econ 15:109–112

Kennedy JJ, Thomas JW (1995) Managing natural resources as social value. In: Knight RL, Bates SF (eds). A new century for natural resource management. Island Press, Washington

Ludwig D, Hillborn R, Walters C (1993) Uncertainty, resource exploitation, and conservation: lessons from history. Science 260:17–18

McMichael AJ, Butler CD, Folke C (2003) New visions for addressing sustainability. Science 302:1919–1029

Naveh Z (2005) Epilogue: towards a transdisciplinary science of ecological and cultural landscape restoration. Rest Ecol 13:228–234

Roe E, van Eeten M (2001) Threshold-based resource management: a framework for comprehensive ecosystem management. Environ Manage 27:195–214

Salz RJ, Loomis DK (2005) Human dimensions of coastal restoration. In: Thayer GW, McTigue TA, Salz RJ et al (eds). Science based restoration monitoring of coastal habitats. NOAA, Silver Spring

US Commission on Ocean Policy (2004) An ocean blueprint for the 21st century. Final report. US Commission on Ocean Policy, Washington, DC

Weinstein MP (2009) The road ahead: the sustainability transition and coastal research. Estuar Coasts 32:1044–1053

Weinstein MP, Reed DJ (2005) Sustainable coastal development: the dual mandate and a recommendation for "commerce managed areas". Rest Ecol 13:174–182

Weinstein MP, Baird RC, Conover DO et al (2007) Managing coastal resources in the 21st century. Front Ecol Environ 5:43–48

Acknowledgments

Major funding for the precedent to this book – *The International Symposium on Sustainability Science: The Emerging Paradigm and the Urban Environment* – was provided by the PSEG Foundation, and the PSEG Public Affairs and Sustainability Practice. Additional co-sponsors whom we gratefully acknowledge include the USEPA, Norris McLaughlin & Marcus, P.A., S/L/A/M Collaborative, Sodexo, NJ Department of Environmental Protection, PS & S, and NJ Natural Gas.

Contents

Contributors

James Aronson Centre d'Ecologie Fonctionnelle et Evolutive (CNRS-UMR 5175), Montpellier, France

Ronald C. Baird Rhode Island Sea Grant, University of Rhode Island, Narragansett, RI, USA

Guillaume Bernard Groupement d'Intérêt Publique pour la Réhabilitation de l'Etang de Berre (GIPREB), Berre-l'Etang, France

Jean-Michel Bocognano Grand Port Maritime de Marseille, Marseille, France

Richard Burroughs Department of Marine Affairs, University of Rhode Island, Kingston, RI, USA

Florian Claeys École Normale Supérieure, Département de Biologie, Paris, France

Robert Costanza Institute for Sustainable Solutions, Portland State University, Portland, OR, USA

Michael G. Frisk School of Marine and Atmospheric Sciences, Stony Brook University, Stony Brook, NY, USA

Rudolf de Groot Environmental Systems Analysis Group, University of Wageningen, Wageningen, The Netherlands

Alan D. Hecht Office of Research and Development, US EPA, Washington, DC, USA

Eric Higgs School of Environmental Studies, University of Victoria, Victoria, BC, Canada

Robert W. Kates Senior Research Associate, Kennedy School, Harvard University, Trenton, ME, USA

Christopher Kennedy Department of Civil Engineering, University of Toronto, Toronto, ON, Canada

Judith A. Layzer Department of Urban Studies and Planning, MIT, Cambridge, MA, USA

Robin Leichenko Department of Geography, Rutgers University, Piscataway, NJ, USA

Simon Levin Department of Ecology and Evolutionary Biology, Guyot Hall, Princeton University, Princeton, NJ, USA

Steven Y. Litvin Hopkins Marine Station, Stanford University, Pacific Grove, CA, USA

Bonnie J. McCay Department of Human Ecology, Rutgers University, New Brunswick, NJ, USA

Philippe Picon Groupement d'Intérêt Publique pour la Réhabilitation de l'Etang de Berre (GIPREB), Berre-l'Etang, France

Michael R. Redclift Department of Geography, Kings College London, London, UK

William E. Rees School of Community and Regional Planning, University of British Columbia, Vancouver, BC, Canada

Thomas K. Rudel Department of Human Ecology, Rutgers University, New Brunswick, NJ, USA

Daniel Simberloff Department of Ecology and Evolutionary Biology, University of Tennessee, Knoxville, TN, USA

John D. Sterman MIT Sloan School of Management, Cambridge, MA, USA

Jac. A.A. Swart Science & Society Group, Energy and Sustainability Research Institute Groningen, Groningen, Netherlands

Robert W. Taylor Earth and Environmental Studies, Montclair State University, Montclair, NJ, USA

R. Eugene Turner Department of Oceanography and Coastal Sciences, Louisiana State University, Baton Rouge, LA, USA

Henny J. van der Windt Science & Society Group, Energy and Sustainability Research Institute Groningen, Groningen, Netherlands

Michael P. Weinstein PSEG Institute for Sustainability Studies, Montclair State University, Montclair, NJ, USA

Vanja Westerberg Ecole nationale supérieure agronomique de Montpellier (SupAgro), Laboratoire Montpelliérain d'Économie Théorique et Appliquée (LAMETA), Montpellier, Cedex 1, France

Jianguo (Jingle) Wu School of Life Sciences and Global Institute of Sustainability, Arizona State University, Tempe, AZ, USA

Part I
Managing the Earth's Life Support Systems: The Emergence of Sustainability Science and Transdisciplinarity

By focusing on the science–policy interface through a "systems" lens, sustainability science addresses the fundamental character of interactions between nature and society and society's capacity to guide those interactions along sustainable trajectories (Kates et al. 2001). The underlying principles of this nascent field suggest, moreover, that a sustainable biosphere is not only necessary but economically feasible, socially just, and ecologically sound. It targets the need to break down artificial and outdated disciplinary gaps between the natural and social sciences through the creation of new transdisciplinary knowledge and its practical application to decision-making. New applications of "use inspired" science that incorporate different perspectives of society become more relevant and, in turn, contribute to more transparent and democratic processes of governance (Gibbons et al. 1994; Nowotny et al. 2001).

Conflict mitigation, consensus building, and trade-offs in the form of sacrifice and compromise will become the norm for sustainable management of coupled human-environment systems because growing demands on resources can no longer be met by access to unexploited sources. An integrated systems approach is required, taking into account conflicting goals and interlinkages among environmental issues (Ayensu et al. 2003; Naveh 2005), as well as the geographic scales of both the issues and political jurisdictions. Success will depend on the ability to create new paradigms that will resolve the growing tensions among the involved communities. More effort at the interface between science and society is needed in order to make the transition from the centralized, top-down approach of government institutions to more decentralized, regional, and local approaches to resource management (Bruckmeier 2005).

References

Ayensu E, van R Claasen D, Collins M et al (2003) International ecosystem assessment. Science 286:685–689

Bruckmeier K (2005) Interdisciplinary conflict analysis and conflict mitigation in local resource management. Ambio 34:65–73

Gibbons M, Limoges C, Nowotny H et al (1994) The new production of knowledge: the dynamics of science and research in contemporary societies. Sage, London

Kates RW, Clark WC, Corell R et al (2001) Sustainability science. Science 292:641–642

Naveh Z (2005) Epilogue: towards a transdisciplinary science of ecological and cultural landscape restoration. Restor Ecol 13:228–234

Nowotny H, Scott P, Gibbons M (2001) Re-thinking science: knowledge and the public in an age of uncertainty. Polity Press, Cambridge

From the Unity of Nature to Sustainability Science: Ideas and Practice

Robert W. Kates

Abstract The ideas of sustainability science are at least two centuries old, but only a decade in practice. This introductory paper reviews some of the key concepts underlying sustainability science beginning with Alexander von Humboldt and the unity of nature, discusses the basic foundation of the science, and illuminates the three major tasks of sustainability science: fundamental research on use-directed problems; nurturing the next generation of sustainability scientists; and moving knowledge into action.

Keywords Sustainability science • New knowledge • Use-directed research • Student learning

Ideas of Sustainability Science

I have selected some of the major ideas that contributed to the development of sustainability science from a much larger set, beginning first with Alexander von Humboldt's dream of understanding the unity of nature. This was followed by George Perkins Marsh's vision of nature as modified by human action. Then much later, the International Union for the Conservation of Nature (IUCN) linked nature and human development, which led to the World Commission on Environment and Development, and culminated in the US National Academy of Science (NAS) report of *Our Common Journey* and the call for a sustainability science.

R.W. Kates (✉)
Senior Research Associate, Kennedy School, Harvard University, 33 Popple Point, Trenton, ME 04605, USA
e-mail: rwkates@gmail.com

M.P. Weinstein and R.E. Turner (eds.), *Sustainability Science: The Emerging Paradigm and the Urban Environment*, DOI 10.1007/978-1-4614-3188-6_1,

The Unity of Nature

Let me begin with Alexander von Humboldt's dream. Humboldt, then 29 years old, set out his dream in a letter to friends in 1799 as he awaited his sailing from Spain to Venezuela and the beginning of a 5-year exploration of the Orinoco River and the Andes mountains: "In a few hours we sail round Cape Finisterre. I shall collect plants and fossils and make astronomic observations. But that's not the main purpose of my expedition—I shall try to find out how the forces of nature interact upon one another and how the geographic environment influences plant and animal life, In other words, I must find out about the unity of nature." (Alexander von Humboldt; as quoted in Nicolson 1995).

He would pursue that goal until the final posthumous publication of Volume 5 of the *Kosmos* in 1862. But his dream was not to be shared widely, for by then the Academy had discovered another more powerful approach to understanding nature, but not its unity. To pursue this new approach of reductionism, specialization increased, disciplines were born, and graduate degrees were invented.

Nature Modified by Human Action

Beyond the unity of nature, a second great idea was that of a nature modified by human action. George Perkins Marsh, the remarkable Vermonter wrote *Man and Nature* in 1862 and revised it as *Earth Modified by Human Nature* in 1874 (Marsh 1965[1862, 1874]) documenting for the first time, the destructive impacts of human activity on the biosphere. A more detailed examination of human activity took place in 1956 (Thomas 1956) but 30 years later a systematic review moved beyond modification and found the earth transformed (Turner et al. 1990). Along the way, Vernadsky had integrated human knowledge with the biosphere in a "noosphere" (Vernadsky 1998[1926]) and Rachel Carson helped initiate the modern environmental movement with her *Silent Spring* (Carson 1962).

Nature Linked to Human Development

The next great idea about an earth already transformed by human action was to link nature or the environment to development, particularly human development. Thus the idea of sustainable development was born, emerging in the early 1980s from scientific perspectives on the interdependence of society and environment that were fostered by the International Union for the Conservation of Nature and Natural Resources (1980) in the *World Conservation Strategy*. It gained considerable political attention through the publication by the Brundtland led World Commission on Environment and Development (1983–1987) report, *Our Common Future* (World Commission on Environment and Development (WCED) 1987) and the subsequent

United Nations Conference on Environment and Development held in Rio de Janeiro in 1992. There were no scientists on the World Commission and little science present in Rio. Ten years later at the Johannesburg world summit, there was some scientific presence, in part because work on sustainability science had already begun.

NAS-NRC Board on Sustainable Development 1995–99

Beginning in 1995, I served as vice-chair along with William Clark of Harvard University on the National Academy of Sciences Board on Sustainable Development when we began a 5-year effort to reconnect science and technology to sustainable development. We sought to make the concept of sustainable development both manageable and measurable by focusing on a minimal sustainability transition over two relatively foreseeable generations. Using one of the major social science findings of the time, i.e., the demographic transition, we envisioned a world population of about nine billion in 50 years. To decide what constituted a sustainability transition we chose three normative goals that had emerged at the top of priority-setting negotiations of international conferences and summits: meeting the human needs of the nine billion, preserving the life support systems of the planet, and reducing hunger and poverty. We argued for acting on what we already know and creating a sustainability science for what we needed to know (National Research Council-Board on Sustainable Development [NRC-BSD] 1999).

Friibergh Workshop on Sustainability Science

But after finishing our report and presenting it to a meeting of the World Academies of Sciences (Interacademy Panel on International Issues 2000) which embraced the notion of a sustainability science, we also realized that there was much to learn on how to do sustainability science. So with assistance from Sweden we convened a small international workshop in Friibergh to identify the core questions and methodologies of sustainability science (Kates et al. 2001). This meeting was followed by a series of regional meetings in Africa, Asia, Latin America, and North America (International Council for Science [ICSU] 2002).

Our discussions at Friibergh and in the subsequent regional meetings revealed profound differences in problems and perspectives between scientists based in developed countries and those in developing countries. Scientists in developed countries focused primarily on global issues, whereas their colleagues in developing countries were concerned primarily with local issues. The two groups were separated by a variety of economic, digital, and capacity divides (Fig. 1). Scientists in the "north" worried about the effects of affluence and consumption, climate changes and its causes, and undertook theory-driven research. Scientists in the "south" worried about the effects of poverty and under-consumption, and the impacts of climate change, and they undertook action-driven research. Scientists in the north

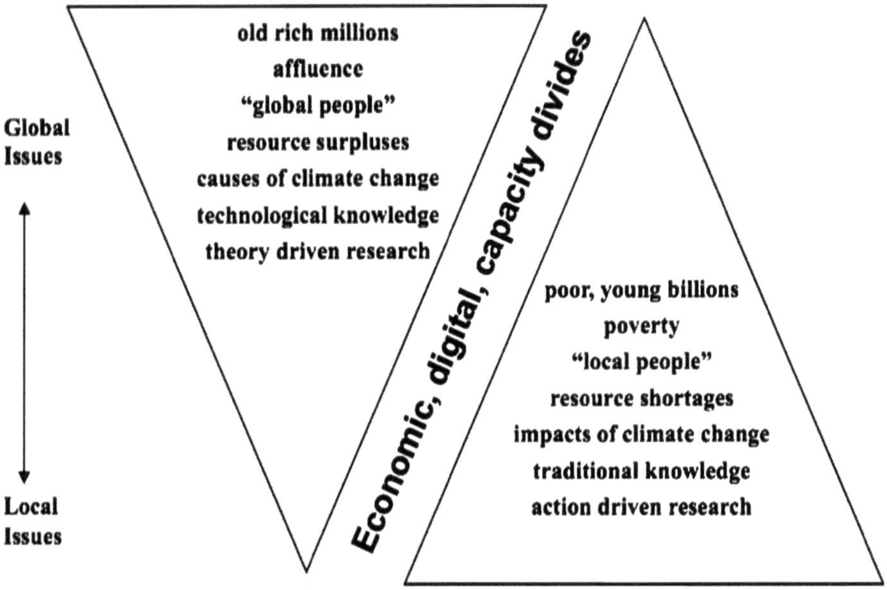

Fig. 1 Differences in problems and perspectives in developed and developing countries. Scientists based in developed countries focus primarily on global issues, worry about consumption, climate changes and its causes, and undertake theory-driven research with needed tools and funds. Scientists based in developing countries focus primarily on local issues, worry about poverty and under-consumption, the impacts of climate change, and undertake action-driven research, often short of facilities and funding (based on Figure 1 in Kates et al. 2001)

took for granted broadband Internet access and many sources of funding. Scientists in the south tried to cope with interrupted electricity, worked at multiple jobs to support themselves, and had few funding sources.

Such differences notwithstanding, the workshops also reflected broad agreement that science and technology have an enormous potential to make important contributions to a sustainability transition. Realizing that potential, however, will require that serious efforts be made to promote science for sustainability. It will have to be more than the status quo such as simply renaming work we are already doing or claiming that specific work limited to either environmental science or development studies is sustainability science. And sustainability science is not just an extension of existing research agendas (e.g., that of earth systems science) or action agendas (e.g., that of climate change) to include the various goals of sustainability.

Sustainability Science

The science and technology needed to develop sustainability is essentially integrative of the natural, social, and engineering sciences; seeks to bridge the communities engaged in promoting environmental conservation, human health, and economic development; and brings together the worlds of knowledge and action.

Our discussions also revealed agreement that much of the science and technology developed to support sustainability will be regional and place-based, and focused at intermediate scales where multiple stressors intersect to threaten or degrade human-environment systems. In a sense, sustainable development differs in every place as human needs and life support systems vary, and as hunger and poverty are smaller or larger. It is at these intermediate scales that the complexity of coupled human-environment systems is more readily comprehensible, where innovation and management happen, and where significant transitions toward sustainability may have already begun.

We also agreed in our workshops that sustainability science addresses fundamental questions of scale, nonlinear processes and complexity, and the unity of nature and society. How is the universal related with the particular, the whole with its parts, and the global with the local? How can knowledge of the component parts explain the properties of larger systems, and are such properties knowable? How do the earth, its living biota, and our human species work together?

Thus our global conversations concluded that the science and technology needs for sustainability will have fundamental and applied characteristics. These needs will be addressed with cutting-edge questions regarding nature-society dynamics, while recognizing the need to address sustainability concerns in a problem-solving mode, and to apply what we already know in science-based action programs. Stokes' (1997) quadrant model of scientific research is instructive with his 2×2 view of research contrasting the quest for fundamental understanding with the quest for utility (Fig. 2). His model has a "Neils Bohr" quadrant that is high in fundamental understanding and little immediate utility, and a contrasting "Thomas Edison" quadrant that has pure applied focus. The sustainability science quadrant of use-inspired fundamental research, in contrast, is exemplified by the discoveries of Louis Pasteur.

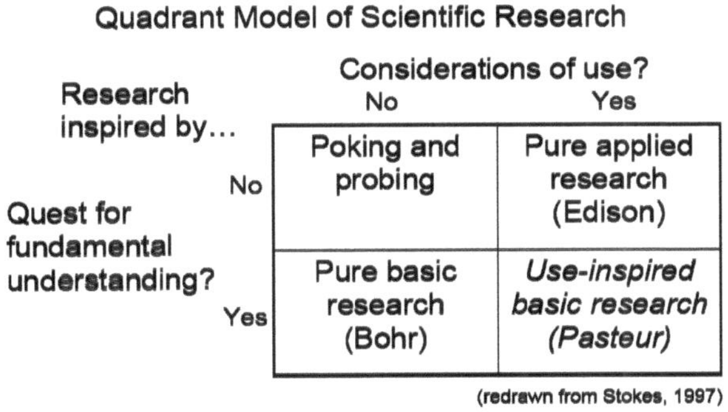

Fig. 2 Quadrant model of scientific research. Stokes' (1997) two by two view of research contrasts the quest for fundamental understanding with the quest for utility

For a succinct definition of sustainability science, I prefer that of the Proceedings of the National Academy of Sciences (PNAS 2010) as: "...an emerging field of research dealing with the interactions between natural and social systems, and with how those interactions affect the challenge of sustainability: meeting the needs of present and future generations while substantially reducing poverty and conserving the planet's life support systems." Variants of this definition are widely accepted, but as sustainability science spread around the world, different emphases emerged, as in Europe (Jäger 2009; European Commission 2009), Japan (Komiyama et al. 2010), or the US (Clark 2007; Matson 2009).

Major Tasks of Sustainability Science

There were three major tasks for sustainability science in its first decade; (1) fill the Pasteur quadrant (Fig. 2) and do fundamental research on use-directed problems; (2) nurture the next generation of sustainability scientists; and (3) move new knowledge into action. I will illustrate the progress currently being made for each task with three examples.

Fundamental Research on Use-Inspired Problems

Over this last decade, core fundamental research questions and themes have been identified. Central to these has been the study of coupled human-environment systems. Existing models for many of the components of human-environment systems have been evaluated for their suitability for integrated sustainability assessments. And a growing body of fundamental research on use-inspired problems has been published.

Core Research Questions and Themes

There have been two major efforts to articulate core research questions for sustainability science. At Friibergh, seven core questions were identified and these survived a set of follow-up regional meetings in Africa, Asia, Latin America, and North America (Kates et al. 2001; International Council for Science [ICSU] 2002). Then at the initiative of the US National Science Foundation, a second conference "Towards a Science of Sustainability" was convened in 2009 at the Airlie Center in Warrenton, Virginia. There were twice as many conferees at the Airlie conference as at the original Friibergh workshop, but unlike that workshop, most attendees were from the United States. The conference identified six sets of major thematic research areas (Clark and Levin 2010). These core questions and themes are compared in Table 1. Four of the core questions and research themes were almost identical in both Friibergh and Airlie efforts, but some were expanded in the Airlie version. Three of the Friibergh core questions (in italics) were not specifically singled out at the Airlie Conference, while two new themes (also in italics) were added about the trade-offs between natural and human systems and rigorous evaluation of sustainability trajectories were added.

Table 1 The Friibergh international workshop of 2000 (Kates et al. 2001) identified seven core research questions for sustainability science. A decade later, the US Airlie House conference, *Towards a Science of Sustainability* (Clark and Levin 2010), selected six major research themes. Those that differ between the efforts are in *italics*. H-E; Human Environment

Sustainability science research questions and themes	
Friibergh 2000 (Kates et al. 2001) Core questions	Airlie House 2009 (Clark and Levin 2010) Major themes
Integrative H-E models	H-E theory and models
Long-term trends	Long-term trends and transitions
Vulnerability or resilience	H-E systems adaptability
Incentive structures	Guidance of H-E systems
Limits or boundaries	*H-E trade-offs*
Monitoring and reporting	*Evaluation of sustainability trajectories*
Better integrated activities	

Research questions are now better defined as a result of these two attempts to articulate the overlapping questions and themes. The view of the participants at the Airlie conference was, however, that major progress has been made only in the two themes concerned with the original core set of research questions identified as "Long-term trends" and "Vulnerability or resilience." Selected progress has been made on "Limits or boundaries" (primarily about climate), in "Incentive structures" (primarily for common resources or conservation), in "Monitoring and reporting" (primarily from space), and in "Better integrated activities" (primarily from interdisciplinary efforts). The two new themes on coupled Human-Environment (H-E) system trade-offs and sustainability trajectories seem well justified. The results of both the Airlie and Friibergh meeting were the identification of the basic need for better theory and models to bridge the gap between those expert in modeling approaches but not in H-E systems and those empirical scientists knowledgeable about H-E systems but not modeling complexity. Only in climate modeling has there been a significant improvement in the merger of theory and models.

Elaborating Human-Environment (H-E) Systems

At the heart of sustainability science are the closely coupled human-environment systems that are more easily described with box and arrow models than detailed with numbers and equations. Although I use these all the time, I suspect that box and arrow diagrams are more useful to their authors than to prospective viewers. Nevertheless, I will use two recently described box and arrows diagrams from Dasgupta et al. (forthcoming) to illustrate several points (Figs. 3 and 4). The basic structure of interacting H-E systems has been used for many years (e.g., Burton et al. 1978) although labeled variously as "nature-society," or "socio-ecological," as well as "human environment." There has been much effort over the past decade that

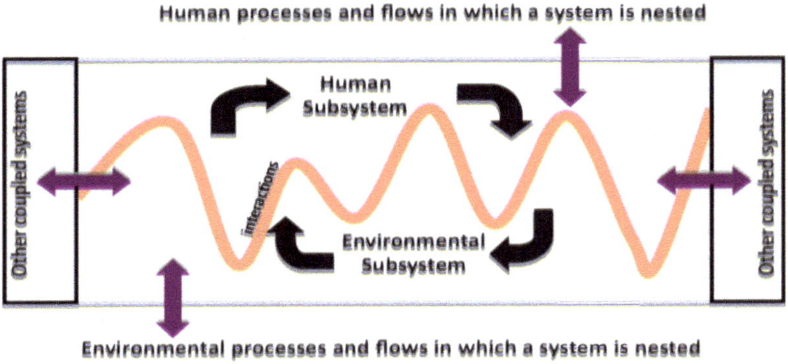

Fig. 3 The coupled Human-Environment system. The basic overview of interacting H-E systems have been labeled variously as nature-society, socio-ecological, or human environment and contain a human subsystem, environmental subsystem, and other external systems that impact these subsystems. *Source*: Dasgupta et al. forthcoming

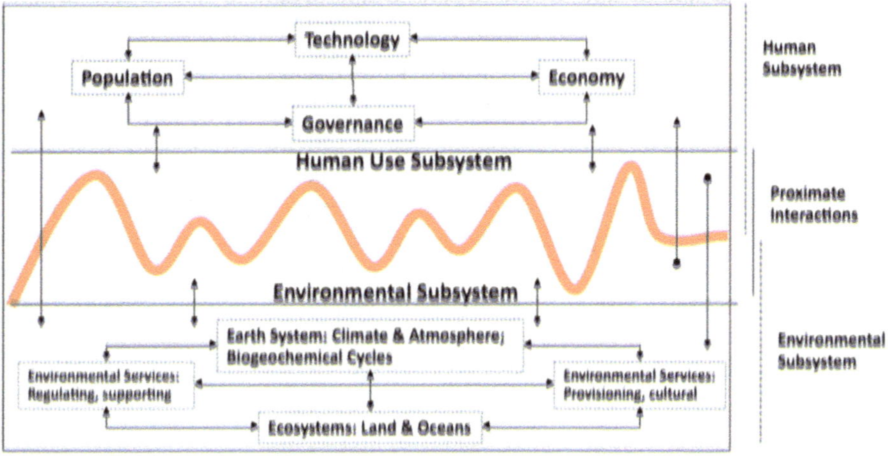

Fig. 4 Inner workings of coupled Human-Environment system. Each of the subsystems has been expanded with the human subsystem including population, technology, governance, and economy and the environmental subsystem including a biogeophysical earth system, ecosystems, and ecosystem services. *Source*: Dasgupta et al. forthcoming

has focused on detailing the interactions (wavy lines) and exogenous flows from other coupled systems (Fig. 3), and on elaborating the human and environmental subsystems (Fig. 4).

There has been an emerging consensus that the important elements of human subsystems are population, technology, governance, and economy (Fig. 4). But these four elements are still studied primarily as separate entities and usually introduced into models as external inputs or as scenarios. A major addition to the conceptual model of environmental subsystem has been idea of ecosystem services (Daily et al. 1997). This idea is now widely used as the major output of the environmental subsystem. Judging from the literature of the last decade, much of coupled H-E system study focuses on the environmental subsystem with less attention to the human subsystem (see Tables 2 and 3).

Table 2 Models evaluated for integrated sustainability assessment by MATISSE project. The models are listed by acronym. See Lotze-Campen (2008) for the full titles. Source of the table is Lotze-Campen (2007)

Category	Models (examples)
Biophysical models	
Climate	GCM (HadCM, ECHAM): EMIC (CLIMBER, MAGICC/SCENGEN)
Hydrology	WaterGAP, SWIM, IRM-ABM
Biogeochemistry	LPJ, VECODE, 4C, WOFOST, ACCESS
Socioeconomic models	
General economy	GE (GTAP, WorldScan, GEM-CCGT, GEM-E3, SNI-AGE): Macroeconometric (E3ME, NEMESIS, QUEST-II, GINFORS)
Partial economy sectors	Energy (POLES, PRIMES, MARKAL); Agriculture (WATSIM, IMPACT, CAPRI, RAUMIS); Transport (TREMOVE)
Demography	PHOENIX, IIASA Population Project
Public health	MIASMA, PHSF, TARGET
Integrated models	
Land use change	FARM, AgLU, MAgPIE, CLUE, SFARMOD, CORMAS
Qualitative systems analysis	SYNDROMES, QSA-SCENE
Integrated assessment	IMAGE, ICLIPS, FUND, MIND, DEMETER, RICE-FEEM, GENIE, IMPACT-WATER
Scenario building and planning	QUEST, POLESTAR, THRESHOLD-21, FAIR

Table 3 Sustainability science research by topic as reflected in titles to PNAS articles. The titles of the 232 PNAS papers for the years 2003–2010 listed under sustainability science (PNAS 2010) were classified under 16 topics, using nine describing human needs and seven describing life support systems used to prepare a *Reader in Sustainability Science and Technology* (see below; Kates 2010)

	Topics number	%
Human well-being		
Population	3	1
Health and well-being	13	5
Poverty/affluence	11	4
Habitation and transportation	7	2
Peace and security	0	0
Energy and materials	20	7
Food and fiber	44	16
Water and sanitation	2	1
Disasters	8	3
Total	108	38
Life support systems		
Climate and stratospheric ozone	60	21
Land	35	12
Atmosphere	6	2
Water	9	3
Oceans and fisheries	20	7
Biodiversity	25	9
Ecosystem services	18	6
Total	173	62
Grand total	281	100
PNAS papers 2003–2010	232	

Modeling tools already exist for many of the components of human-environment systems described above. A major European effort (2005–2008) on Methods and Tools for Integrated Sustainability Assessment (MATISSE) focused on methods and tools for integrated sustainability assessment (Lotze-Campen 2008). This assessment included comparative evaluations of dozens of available biophysical, socioeconomic, and integrated models (Table 2). Overall, the MATISSE assessment concluded that integrated sustainability assessments were in their infancy (Lotze-Campen 2008).

Published Research

The growing sets of sustainability science research results have many new sources for publication that include traditional journals (e.g., *Global Environmental Change* or *Proceedings of the National Academy of Sciences [PNAS])*, new journals (e.g., *Sustainability Science*), Internet journals (e.g., *Ecology and Society*), review volumes (e.g., the *Annual Review of Environment and Resources*), and pieces written for the general public (e.g., *Environment*).

During the last decade, the *Proceedings of the National Academy of Sciences* established a Sustainability Science section in which public access is readily available (PNAS 2010). Many of the best of research results appear there (only about 1 in 5 submissions to PNAS are accepted). There were 232 articles in that section from 2003 to 2010, and many of them appeared in special features which characterized the range of current research in sustainability science. The topics covered include features on marine reserves, ecosystem services, poverty, health, governance, food security, climate change, and land use change.

Results of a more detailed analysis of these same 232 titles (Table 3) classified the papers into 16 topics; nine described human needs and seven described life support systems. Where titles seemed to contain multiple topics, they were allotted to all of the topics, and so the number of papers listed in Table 3 is larger than the actual number of papers. Twelve were unclassifiable by title.

I concluded, based on these data, that the first eight years of Sustainability Science Research seem to have been devoted primarily to environment (life support systems 62%) rather than development (human needs 32%) and were concentrated in only six topics: climate (21%), agriculture (16%), land use, primarily forests (12%), biodiversity (9%), energy and materials (7%), and oceans and fisheries (7%).

Bias toward environmental science is as clear as the underrepresentation of the varied fields of development sciences that include health and human development, economic and social development, governance, and the multitude of technologies that make development possible. A second bias is that despite considerable involvement of developing countries, the current agenda (as represented in these papers) reflects the priorities of what are primarily environmental scientists from the developed countries. There is a similar bias toward global aspects, despite the emphasis in sustainability science on regional and place-based studies.

Nurturing the Next Generations

Sustainable development is a century-long goal and the best sustainability scientists are probably still unborn. Thus the second of our tasks is that of nurturing the next generation of sustainability scientists addressed here by three current examples: (1) educational opportunities at Arizona State University (ASU); (2) a recently published *Readings in Sustainability Science and Technology*; and (3) the first distributed graduate seminar in sustainability science.

ASU School of Sustainability

President Michael Crow has been refiguring ASU into a new type of American University since 2002 by featuring among many things an emphasis on sustainability.

A sustainability effort is within all of its various schools that has the intent to introduce sustainability principles to all of its 72,000 students. Furthermore, there is also a specific School of Sustainability which had enrolled, as of September 2010, 581 undergraduate majors and 84 graduate students, including 58 PhD candidates. A minor in sustainability is available to all undergraduates at ASU.

The school of sustainability represents one major approach to sustainability science education—the creation of new sustainability science degrees. At least ten PhD degree programs and many more Master's programs are available around the world either in sustainability science or sustainable development. It is much too early to know how such degrees will be received in academia as an entré to graduate work or academic employment. Thus, an alternate approach is to create teaching programs with sustainability nomenclature within existing disciplines, programs, or centers. At ASU, for example, undergraduate concentrations and degree programs are available in the Business School, the Institute of Design and the Arts, the Engineering school, the Law School, and the College of Liberal Arts and Sciences. But as environmental disciplines or programs are often the hosts for such concentrations and degrees, the result may be to add to the environmental bias discussed above.

To help students bridge the gap between textbooks and careers or graduate education, ASU encourages and supports numerous opportunities for undergraduates to participate in internships, workshops, and service and all students are required to take and successfully complete an applied/capstone learning or research experience. Further, the School of Sustainability partners with community businesses and other organizations to provide undergraduate and graduate internships. An internship coordinator helps students match their interests and capabilities with outside internship hosts and monitors these to ensure that the internship experience counts as credit toward fulfilling degree requirements (Buizer 2011, Personal communication, e-mail, 1/24/2011).

Readings in Sustainability Science and Technology

There is need for new sets of educational materials as opportunities continue to increase in sustainability science. The *Readings in Sustainability Science and Technology* are the result of a 4-year effort and available as the Harvard, Center for International Development Working Paper 213 (Kates 2010), and available on the Internet. It is written for teachers of advanced undergraduate students and beginning graduate students, and it suggests a three-part architecture for courses organized around major domains of sustainability science. The first part presents an overview of sustainable development that begins with the history of sustainable development and the dual goals of sustainable development—the promotion of human development and well-being while protecting the earth's life support systems. Readings are provided for the current status, long-term trends, and the impacts of nine essentials for human well-being and seven essential life support systems. It concludes with a description of the interactions of human society and the life support systems sketched—simply, realistically, and imaginatively.

The second part of the reader focuses on the "what, why, and how" of sustainability science and technology. For "what to do" in sustainability science, it examines three essential qualities of the emerging science: its use or needs orientation, a focus on human-environment systems, and the goal of integrated understanding. As to why to do sustainability science, it considers normative values and the science of identifying and analyzing values and attitudes. As to "why do sustainability science," it examines the current practice of the science, the analyses undertaken and the distinctive methods and models used.

The reader ends by linking knowledge systems and action using examples of both global and local solutions to meeting the needs of human well-being and the earth's life support systems. It specifically identifies three critical needs that constitute grand challenges: poverty, climate change, and peace and security.

In all, there are some 93 readings to choose from for course material supported by a mini-text of 60 pages to place them in context. In its current form, the Reader has been reviewed extensively and many suggestions for additional topics are incorporated. Because it is an electronic reader, it can be frequently updated and users are requested to send commentary and suggestions for new or replacement readings to: sustsci_reader@hks.harvard.edu.

Sustainability Science Distributed Graduate Seminar

In the fall semester of 2010, seven universities (Arizona State, Cornell, Florida International, Harvard, Minnesota, the National University of Mexico [UNAM], and Princeton) took part in an Internet-based distributed graduate seminar attended by 120 students. The majority of the lecturers were authors of draft chapters from a book to be completed in 2012: Sustainability science: an introduction for researchers which served as the primary readings for the course and were supported by readings from the reader described above. Students for each session and at a particular school took the initiative in a commentary that was later opened to all. Course materials and video recordings of the sessions can be currently downloaded from their website (https://groups.nceas.ucsb.edu/sustainability-science). Overall, student and faculty viewed the course as an exciting success.

Moving Knowledge into Action

Perhaps the most challenging task of sustainability science is moving knowledge into action. I will begin with tracking the global and local agenda of needed action, report on major opportunities that act on what we already know how to do, and conclude with an exciting state-wide effort that focuses on solutions to sustainability problems.

Knowledge into Action Challenges

There has been a convergence of agreement over the past decade about what constitutes the central challenges of sustainable development, which is essentially an agenda of priority areas (Table 4). The list of these challenges begins with Dr. Bruntland and the World Commission on Environment and Development (WCED) which identified six challenges for sustainable development: (1) population and human resources; (2) food security; (3) species and ecosystems; (4) energy; (5) industry; and (6) urban areas (WCED 1987). The National Academy of Sciences Board on Sustainable Development grouped two together and reduced these challenges to five (NRC-BSD 1999). Four of these five were kept by the Secretary-General of the UN as an achievable agenda for the 2002 Johannesburg successor conference to Rio, and to which he added "water" while dropping population (Annan 2002). My own Reader keeps all of these but expands the agenda to specifically include three grand challenges for sustainable development: poverty, climate change, and peace and security (Kates 2010).

Table 4 Putting knowledge into action: global agendas. An agenda of priority areas or central challenges of sustainable development can be constructed from the four major documents beginning with *Our Common Future* from the World Commission on Environment and Development (WCED 1987), *Our Common Journey* from the NRC-BSD (1999), the *Achievable Agenda* of the Secretary-General of the UN (Annan 2002), and the *Reader on Sustainability Science and Technology* (Kates 2010)

Our common future (WCED 1987)	Our common journey (NRC-BSD1999)	Achievable agenda (Annan 2002)	Readings in sustainability science and technology (Kates 2010)
Population and human resources	Human population		Population
		Health	Health and well-being
Food security	Agriculture	Agriculture	Agriculture and food security
Species/ecosystems	Living resources	Biodiversity	Biodiversity ecosystem services
Energy industry	Energy industry	Energy	Energy materials
Urban	Cities		Urban growth
		Water	Water and sanitation
			Poverty
			Climate change
			Peace/security

It is not clear how this set of challenges will emerge in 2012 when the United Nations again convenes a world conference in Rio de Janeiro. Judging from the preliminary documentation (United Nations and Sustainable Development 2010), there will be a review of progress which will probably touch on many of these challenges. But the themes of the next Rio conference are not on challenges but on means of addressing them, specifically economic mechanisms labeled collectively as the "green economy" and the effectiveness of the complex of institutions that address aspects of sustainable development.

Acting on What We Already Know

As noted in Table 4, we have had a consistent agenda of sustainability challenges and problems for more than two decades. Sustainability science does not, therefore, start from a blank slate; indeed we already know ways of moving into a sustainability transition over the next two generations. We know, for example, how to reduce the expected nine billion population by as much as 10% through encouraging current downward trends in fertility by addressing unmet contraceptive needs, by educating women, and by delaying marriage. Africa can produce much more food just by increasing its use of fertilizer. The rate of improvement in energy intensity can be met by using technologies and behaviors that are readily available. The rapidly growing cities in developing countries can work more efficiently, provide housing and employment, and become increasingly green. We have learned much about how to restore degraded ecosystems. Some progress has been made in all of these areas over the last decade, but in none have these been sufficient to compensate for the growth in fossil fuel use or the unplanned expansion of existing cities.

There is also the process of localization. Because the challenges of sustainable development are place-based problems, sustainability scientists need to join hands with practitioners from local communities, industry, government, and civil society to address their specifics. This is beginning to happen, but there are still too few examples of solutions offered or problems solved. The growing body of research may at best be only slowly yielding solutions for important global and local problems in the priority action areas of population, settlements, agriculture, energy and materials, and living resources identified in the NAS report (NRC-BSD 1999). Progress in all of these areas has been slow in this first decade of needed action. Finally, and perhaps most importantly, is the need to overcome the normal academic penchant to emphasize current uncertainty and thus the need to do more research, and also to fail to acknowledge how much is already known about the crucial needs for a transition into sustainability.

Maine Sustainability Solutions Initiative

Thus sustainability scientists are better at research than in finding and implementing solutions to local and regional sustainability problems. A major initiative is underway in my home state of Maine to go beyond a purely research agenda to address Maine's sustainability problems and opportunities. It is a 5-year effort funded jointly by the State and the EPSCoR program of the National Science Foundation and currently involves over 100 faculty and 200 students in 10 institutions of higher education in the State. The sustainability problems being addressed are those posed by the convergence of four important and long-term changes in landscape: (1) there is major change in forest resource management (87% of Maine lands are in forest) moving from paper company ownership to diverse ownership and development;

(2) there is significant new urbanization in southern Maine; (3) climate change has already affected the State and will increase in the future; and (4) renewable energy, particularly in hydropower, biofuels, and wind, may make the State a significant exporter of energy. The major state research institutions have a current portfolio of 20 projects that focus on problems and needed solutions where these four trends intersect. There is also research on the research itself. We need to address the efficacy of organizational innovations that facilitate interdisciplinary effort. Three other projects, therefore, study the process of how to move knowledge into action.

Some of the problems encountered within this large research program in Maine mirror problems like the bias to environmental problems found in science projects working at the global scale. Academic researchers also have difficulty addressing big problems in their local expression and in identifying and finding solutions. Most begin with what they already study and have difficulty to move beyond what they know to address bigger questions or to integrate their work with others. The Maine project recognizes these problems and some progress is being made in dealing with all of them.

From the Unity of Nature to Sustainability Science

Having reviewed both the ideas and practice of sustainability science, we can now return to the opening theme of the paper—the yet unborn future sustainability scientist Alexis V. Humble. She too has a dream:

> In a few hours, we launch from Cape Canaveral the first of the next generation of Sustainability Earth Observation Satellites (SEOS). We shall collect full spectrum data on populations of people, plants, and places; the atmosphere and oceans; the movements of energy, materials, and information; of warfare and welfare, and of environment and development. But that's not the main purpose of our science. We shall try to find out how the forces of nature and society interact upon one another and how the geographic environment and the complex system of life can sustain itself.

In other words, we must find out about the unity of nature that we humans are both a part of, and apart from.

References

Annan K (2002) Towards sustainable development. United Nations, New York

Burton I, Kates RW, White GF (1978) The environment as hazard. Oxford University Press, New York

Carson R (1962) Silent spring. Houghton-Mifflin, Boston

Clark WC (2007) Sustainability science: a room of its own. Proc Natl Acad Sci USA 104:1737

Clark WC, Levin SA (2010) Toward a science of sustainability. Report of conference toward a science of sustainability. Airlie Center, Warrenton

Daily GC et al (1997) Ecosystem services: benefits supplied to human societies by natural ecosystems. Issues Ecol 2:1–18

Dasgupta P, Clark WC, Bongaarts J et al (forthcoming) Sustainability science: an introduction for researchers

European Commission, Research Directorate General for Environment (2009) Report from the workshop on sustainability science. Brussels, 28 Oct 2009 http://ec.europa.eu/.../report_workshop_sustainability_science_october_2009.pdf. Accessed 12 January 2011

International Council for Science (2002) Science and technology for sustainable development: report from the synthesis meeting sponsored by the initiative on science and technology for sustainability. TWAS and ICSU series on science for sustainable development 9. ICSU, Paris

Interacademy Panel on International Issues (2000) Transition to sustainability in the 21st century. Interacademy Panel, Trieste http://www.interacademies.net/?id=3568. Accessed 12 January 2011

International Union for the Conservation of Nature and Natural Resources (1980) World conservation strategy: living resource conservation for sustainable development. http://data.iucn.org/dbtw-wpd/edocs/WCS-004.pdf. Accessed 12 January 2011

Jäger J (2009) Sustainability science in Europe. Background paper prepared for European commission's DG for research. http://ec.europa.eu/research/sd/pdf/workshop2009/background_paper_sust_science_workshop_october_2009.pdf. Accessed 12 January 2011

Kates RW ed (2010) Readings in sustainability science and technology. CID working paper No. 213. Center for International Development, Harvard University, Cambridge. http://www.hks.harvard.edu/centers/cid/publications/faculty-working-papers/cid-working-paper-no.-213. Accessed 12 January 2011

Kates RW, Clark W, Corell R et al (2001) Sustainability science. Science 292:641–642

Komiyama H, Takeuchi K, Shiroyama H et al (eds) (2010) Sustainability science: a multidisciplinary approach. Brookings Institution Press, Washington, DC

Lotze-Campen H (2007) Review of experience with existing models and their suitability for integrated sustainability assessment (ISA). MATISSE working paper No. 16. http://www.matisse-project.net. Accessed 12 January 2011

Lotze-Campen H (2008) The role of modeling tools in Integrated Sustainability Assessment (ISA). Int J Innovation Sustain Develop 3:70–92

Marsh GP (1965) [1864] Man and nature: or physical geography as modified by human action. Belknap Press, Cambridge

Matson PA (2009) The sustainability transition. Issues Sci Technol Summer; http://www.issues.org/25.4/matson.html#. Accessed 12 January 2011

National Research Council, Board on Sustainable Development (1999) Our common journey: a transition toward sustainability. National Academy Press, Washington

Nicolson M (1995) Historical introduction In: Wilson J (ed) Personal narrative of a journey to the equinoctial regions of the new continent by Alexander von Humboldt. Penguin Books, London

PNAS (2010). Sustainability science. http://www.pnas.org/site/misc/sustainability.shtml. Accessed 12 January 2011

Stokes DE (1997) Pasteur's quadrant: basic science and technological innovation. Brookings Institution Press, Washington

Thomas WL Jr (ed) (1956) Man's role in changing the face of the earth. University of Chicago Press, Chicago

Turner BL II et al (eds) (1990) The earth as transformed by human action. Cambridge University Press, Cambridge

United Nations, Preparatory Committee for the United Nations Conference on Sustainable Development (2010) Report of the secretary-general: progress to date and remaining gaps in the implementation of the outcomes of the major summits in the area of sustainable development, as well as an analysis of the themes of the conference. A/CONF.216/PC/2. United Nations, New York

Vernadsky VI (1998)[1926] Biosphere: complete annotated edition. Copernicus Books, New York

World Commission on Environment and Development (WCED) (1987) Our common future. Oxford University Press, New York

Sustaining Sustainability: Creating a Systems Science in a Fragmented Academy and Polarized World

John D. Sterman

Abstract From climate change, deforestation, and depletion of fossil fuels to overexploited fisheries, species extinction, and poisons in our food and water, our society is unsustainable and it is getting worse fast. Many advocate that overcoming these problems requires the development of systems thinking. We have long been told that the unsustainability of our society arises because we treat the world as unlimited and problems unconnected when we live on a finite "spaceship Earth" in which "there is no away" and "everything is connected to everything else." The challenge lies in moving from slogans to specific tools and processes that help us understand complexity, design better policies, facilitate individual and organizational learning, and catalyze the technical, economic, social, political, and personal changes we need to create a sustainable society. Here I outline a design for a systems science of sustainability that rises to this challenge. Where the dynamics of complex systems are conditioned by multiple feedbacks, time delays, accumulations, and nonlinearities, our mental models generally ignore these elements of dynamic complexity; where the consequences of our actions spill out across time and space and across disciplinary boundaries, our universities, corporations, and governments are organized in silos that focus on the short term and fragment knowledge. I describe how sustainability research, teaching, and engagement with the policy process can be organized to provide scientifically grounded, reliable knowledge that crosses disciplinary boundaries, that engages multiple stakeholders, that grapples with unavoidable issues of ethics, values, and purpose, and that leads to action.

Keywords Sustainability • System dynamics • Limits to growth • Ecological footprint • Overconsumption • Mental models

---

J.D. Sterman (✉)
MIT Sloan School of Management, E62-436,100 Main Street, Cambridge, MA 02142, USA
e-mail: jsterman@mit.edu

M.P. Weinstein and R.E. Turner (eds.), *Sustainability Science: The Emerging Paradigm and the Urban Environment*, DOI 10.1007/978-1-4614-3188-6_2,

The Challenge

First, the bad news: our civilization is unsustainable and it's getting worse fast. Humanity is overwhelmingly dependent on nonrenewable resources, especially fossil fuels, and the resulting greenhouse gas (GHG) emissions are rapidly changing the climate (IPCC 2007). Most of the world's fisheries are overexploited, and world capture fishery production is falling (FAO 2008). Extinction rates "exceed normal background rates by two to three orders of magnitude" with one-fifth of tracked species "classified as Threatened" (Hoffman et al. 2010). The food we eat, water we drink, and products we consume expose us to carcinogens and endocrine disruptors (e.g., US EPA http://www.epa.gov/iris). Humanity's total ecological footprint exceeds the global carrying capacity (Wackernagel et al. 2002). We have exceeded sustainable planetary boundaries for vital elements of the ecosystems upon which our lives depend, including GHGs, nitrogen, and biodiversity loss (Rockström et al. 2009). And the demands we place on those ecosystems are growing: world population, which reached 7 billion in 2011, is projected to exceed 9.3 billion by 2050 and 10.1 billion by 2100 (United Nations 2011). Real Gross World Product (GWP) is growing at an average rate of 3.5% y^{-1} (World Bank 2010). Billions in developing nations legitimately seek to rise out of poverty and live like those in the developed world, with the housing, refrigerators, air conditioners, flat screen TVs, cars, jet travel, vacations, and consumption that lifestyle entails, while those in the developed world seek even greater consumption than they enjoy today.

The good news? After decades of false starts, "sustainability" is becoming mainstream. Most large corporations have programs promoting corporate social responsibility and environmental stewardship. Universities, including business schools, offer sustainability courses and programs. Scholarly papers and journals devoted to sustainability are growing. Store shelves in developed nations stock more and more eco-friendly products. EnergyStar, LEED, fair trade, and other certification programs abound. Toyota has sold over one million Priuses™ in the USA, and consumers can choose among efficient hybrid, plug in hybrid, and electric vehicles offered by a growing number of carmakers. Firms from GE to Walmart have committed themselves to sustainability, and even oil companies trumpet their devotion to the environment. To paraphrase fictional Wall Street profiteer Gordon Gecko, "green is good."

But is the burgeoning sustainability movement itself sustainable? And do current approaches to sustainability actually make a difference to the sustainability of human society? The answer to both questions is no. Despite notable successes and many important contributions, the current sustainability movement, in business, public policy, and education, is neither effective nor itself sustainable. Why not, and what can be done?

Here I argue that the most efforts by firms, individuals and governments in the name of sustainability are directed at symptoms of unsustainability rather than causes. These include policies to reduce waste, cut energy and material use, reduce GHG emissions, promote green products and local consumption, and so on. Many of these activities are necessary to create a more sustainable society and economy. But they are not sufficient. They fail to address the underlying source of the unsustainable world we have created. I argue that the focus on symptoms and low-leverage policies reflects a widespread failure of systems thinking.

We have long been told that the unsustainability of our society arises because we treat the world as unlimited and problems unconnected when we live on a finite "spaceship Earth" in which "there is no away" and "everything is connected to everything else." Many sustainability advocates argue that overcoming these problems requires the development of systems thinking (e.g., Suzuki 2007; McKibben 2010; Senge et al. 2008). If the world's peoples developed a more holistic appreciation of the intricate interconnections binding us to one another and to nature, it is argued, we would internalize social and environmental externalities, consider the welfare of future generations in making decisions today, and act in consonance with our collective long-term best interests. I agree. The challenge lies in moving from slogans about systems to specific tools and processes that help us understand complexity, design better policies, facilitate individual and organizational learning, and catalyze the technical, economic, social, political, and personal changes we need to create a sustainable society. Here I outline a design for a systems science of sustainability that rises to this challenge. First, I describe the characteristics of complex systems that lead to policy resistance—the tendency for our attempts to solve problems to be defeated by unintended reactions of the system to these interventions. Policy resistance arises from the gap between the complexity of the systems in which we live and the often simplistic and erroneous mental models of those systems that guide our decisions and behavior, from the short time horizons we consider, and from the fragmentation of knowledge into disciplinary silos. I illustrate some of the specific tools of systems thinking with a variety of examples. The goal is to design sustainability research, teaching, and engagement with the policy process that generates scientifically grounded, reliable knowledge that crosses disciplinary boundaries, that engages multiple stakeholders, that grapples with the unavoidable issues of ethics, values, and purpose, and that leads to action.

Characteristics of Complex Systems

Policy Resistance

Thoughtful leaders throughout society increasingly suspect that the policies we implement to address difficult challenges have not only failed to solve the persistent problems we face, but are in fact causing them. All too often, well-intentioned programs create unanticipated "side effects." The result is *policy resistance*, the tendency for an intervention to be defeated by the system's response to the intervention itself (Forrester 1969, 1971a; Sterman 2000). Forest fire prevention and suppression policies work in the short run, but as a consequence of initial success, the fuel burden builds, increasing the incidence and severity of fires. Buying vehicles with better gas mileage in response to high gasoline prices reduces the demand for petroleum, lowering gas prices and undermining the demand for more efficient cars. Powerful pumps help farmers access deep aquifers in arid regions, but speed the drop in the water table, reducing water availability. In these and many other cases, our best efforts to solve problems often make them worse (Table 1).

Table 1 Examples of policy resistance

- Road building programs designed to reduce congestion have increased traffic, delays, and pollution (Sterman 2000)
- Low tar and nicotine cigarettes actually increase intake of carcinogens, carbon monoxide, and other toxics as smokers compensate for the low nicotine content by smoking more cigarettes per day, by taking longer, more frequent drags, and by holding the smoke in their lungs longer (Tengs et al. 2005)
- Health plan policies "limiting what drugs can be prescribed—intended to prevent the unnecessary use of expensive drugs—[are] having the unintended effect of raising medical costs" (Horn et al. 1996)
- Antilock brakes and other automotive safety devices cause some people to drive more aggressively, partially offsetting their benefits (Wilde 2002)
- Forest fire suppression causes greater tree density and fuel accumulation, leading to larger, hotter, and more dangerous fires, often consuming trees that previously survived smaller fires unharmed (US Forest Service 2003)
- Flood control efforts such as levee and dam construction have led to more severe floods by preventing the natural dissipation of excess water in flood plains. The cost of flood damage has increased as flood plains were populated in the belief they were safe (Sterman 2000)
- The impacts of large dams "are more negative than positive and, in many cases, have led to irreversible loss of species and ecosystems" (World Commission on Dams 2000)
- Antibiotics have stimulated the evolution of drug-resistant pathogens, including multiple-resistant strains of TB, *S. aureus*, and sexually transmitted diseases (Fong and Drlica 2003)
- Pesticides and herbicides have stimulated the evolution of resistant pests, killed off natural predators, and accumulated up the food chain to poison fish, birds, and, in some cases, humans (Palumbi 2001)
- Despite dramatic gains in income per capita and widespread use of labor-saving technology, Americans have less leisure today than 50 years ago and are no happier (Layard 2005; Kahneman et al. 1999)

Policy resistance arises from a narrow, reductionist worldview. We have been trained to view our situation as the result of forces outside ourselves, forces largely unpredictable and uncontrollable. Consider the "unanticipated events" and "side effects" so often invoked to explain policy failure. Political leaders blame recession on corporate fraud or terrorism. Managers blame bankruptcy on events outside their organizations and (they want us to believe) outside their control. But there are no side effects—just *effects*. Those we expected or that prove beneficial we call the main effects and claim credit. Those that undercut our policies and cause harm we claim to be side effects, hoping to excuse the failure of our intervention. "Side effects" are not a feature of reality but a sign that the boundaries of our mental models are too narrow, our time horizons too short.

Policy resistance also arises from a mismatch between the characteristics of complex systems (Table 2) and the simplistic mental models we use to make decisions.

Table 2 Policy resistance arises because systems are

- *Constantly changing*: Heraclitus said, “all is change.” What appears to be unchanging is, over a longer time horizon, seen to vary. Change occurs at many time scales, and these different scales sometimes interact. A star evolves over billions of years as it burns its hydrogen fuel, but can explode as a supernova in seconds. Bull markets can rise for years, then crash in a matter of hours
- *Tightly coupled*: the actors in the system interact strongly with one another and with the natural world. Everything is connected to everything else. “You can’t do just one thing”
- *Governed by feedback*: because of the tight couplings among actors, our actions feed back on themselves. Our decisions alter the state of the world, causing changes in nature and triggering others to act, thus giving rise to a new situation, which then influences our next decisions
- *Nonlinear*: effect is rarely proportional to cause, and what happens locally in a system (near the current operating point) often does not apply in distant regions (other states of the system). Nonlinearity often arises from basic physics: bacteria in a river can convert sewage into harmless byproducts until the sewage load becomes so large that dissolved oxygen is depleted, at which point the fish die and anaerobic bacteria produce toxic hydrogen sulfide. Nonlinearity also arises as multiple factors interact in decision-making: pressure from the boss for greater achievement increases your motivation and effort—up to the point where you perceive the goal to be impossible. Frustration then dominates motivation—and you give up or get a new boss
- *History dependent*: many actions are irreversible: you cannot unscramble an egg (the second law of thermodynamics). Stocks and flows (accumulations) and long time delays often mean doing and undoing have fundamentally different time constants: during the 50 years of the Cold War arms race the nuclear nations created more than 250 tons of weapons-grade plutonium (^{239}Pu). The half-life of ^{239}Pu is about 24,000 years
- *Self-organizing*: the dynamics of systems arise spontaneously from their internal structure. Often, small, random perturbations are amplified and molded by the feedback structure, generating patterns in space and time. The stripes on a zebra, the rhythmic contraction of your heart, and persistent cycles in predator–prey populations and the real estate market all emerge spontaneously from the feedbacks among the agents and elements of the system
- *Adaptive and evolving*: the capabilities and behaviors of the agents in complex systems change over time. Evolution leads to selection and proliferation of some agents while others become extinct. People adapt in response to experience, learning new ways to achieve their goals in the face of obstacles. Learning is not always beneficial, however, but often superstitious and parochial, maximizing local, short-term objectives at the expense of long-term fitness and well-being
- *Characterized by trade-offs*: time delays in feedback channels mean the long-run response of a system to an intervention is often different from its short-run response. Low leverage policies often generate transitory improvement before the problem grows worse, while high leverage policies often cause worse-before-better behavior
- *Counterintuitive*: in complex systems cause and effect are distant in time and space, while we tend to look for causes near the events we seek to explain. Our attention is drawn to the symptoms of difficulty rather than the underlying cause. High leverage policies are often not obvious
- *Policy resistant*: the complexity of the systems in which we are embedded overwhelms our ability to understand them. As a result, many seemingly obvious solutions to problems fail or actually worsen the situation

Where the consequences of our actions spill out across space and time, we tend to focus on the local and short term. Where complex systems are dynamic, tightly coupled, governed by feedback, nonlinear, self-organizing, adaptive, and evolving, our mental models tend to be static and narrow. We ignore interconnections and the delayed and distal impacts of our decisions. We divide the world into silos, whether a firm, with separate and often competing fiefdoms of sales, production, finance, research, and so on; governments with separate departments of energy, interior, agriculture, transportation, and so on; or universities with separate departments and disciplines.

Much of the debate around sustainability frames the issue as conflict between the economy and the environment, as if these were distinct domains competing against one another: growth vs. social justice, jobs vs. nature, logging vs. spotted owls, polar bears vs. drilling in the Arctic National Wildlife Refuge. But these boundaries are not features of reality. They are mental constructs. Boundaries are "invisible fences in the mind" (Sterman 2002), the result of the mental models we create, the categories into which we place people. Without underestimating the differing interests that arise in a heterogeneous population, I argue that framing sustainability as a zero-sum game of contending objectives reflects a narrow and deeply dysfunctional mental model. The economy, society, and environment are not separate domains to be traded off against one another. The economy is embedded in a social and political context, which in turn is embedded in ecosystems upon which all life depends. The interests of business, society, and the environment are therefore fundamentally aligned: We cannot have healthy firms, a healthy economy and healthy people if growth and the pursuit of profit destroys the environment, and we cannot have a healthy environment if people live in poverty, ill-fed, without decent housing, health care, education, or economic opportunity. Environmentalists tend to stress the first half of this mutual dependency: Destroy the environment and we destroy both society and the economy. But it is equally true that the health of the environment depends on a society and economy that secures people's human rights and fulfills people's needs. Where there is poverty, hunger, conflict, and war, there the environment suffers. Creating an effective science of sustainability and building the public understanding required for action requires us to develop the skill to recognize the boundaries of our mental models and then expand them so that we become aware of and take responsibility for the feedbacks created by our decisions.

Feedback: (Almost) Nothing Is Exogenous

Contrary to the open-loop mental model so prevalent in people's thinking, the world reacts to our interventions (Fig. 1). There is feedback: Our actions alter the environment and therefore the decisions we make tomorrow. Our actions may trigger so-called side effects we did not anticipate. Other agents, seeking to achieve *their* goals, act to restore the balance we have upset; their actions also generate intended and unintended consequences.

Fig. 1 Sources of policy resistance. The boundary of the decision-makers' mental model is represented by the thin lines, showing the basic feedback loop through which we seek to bring the state of the system in line with our goals. Policy resistance arises when we fail to account for the so-called side effects of our actions, the responses of other agents in the system, human and natural (and the unanticipated consequences of these), the ways in which experience shapes our goals, and the time delays often present in these feedbacks

It is hard to underestimate the power of the feedback view. Indeed, almost nothing is exogenous. Ask people to name processes that strongly affect human welfare but over which we have no control, and many people cite the weather, echoing Mark Twain's famous quip that "everybody talks about the weather, but nobody does anything about it." But today even the weather is endogenous. We shape the weather around the globe, from global warming to urban heat islands, the Antarctic ozone hole to the "Asian brown cloud."

Human influence over the weather is now so great that it extends even to the chance of rain on the weekend. Cerveny and Balling (1998) showed that there is a 7-day cycle in the concentration of aerosol pollutants around the eastern seaboard of the USA. Pollution from autos and industry builds up throughout the workweek and dissipates over the weekend. They further show that the probability of tropical cyclones around the eastern seaboard also varies with a 7-day cycle. Since there are no natural 7-day cycles, they suggest that the weekly forcing by pollutant aerosols affects cloud formation and hence the probability of rain. Their data show that the chance of rain is highest on the weekend, while on average the nicest day is Monday, when few are free to enjoy the out of doors. Weekly cycles in temperature, cloud cover, and other meteorological variables have now been documented in many regions of the world (Forster and Solomon 2003; Bäumer and Vogel 2007; Laux and Kunstmann 2008). Few people understand that driving that SUV to work helps spoil their weekend plans.

In similar fashion, we are unaware of the majority of the feedback effects of our actions. Instead, we see most of our experience as a kind of weather: something that happens to us but over which we have no control. Failure to recognize the feedbacks in which we are embedded, the ways in which we shape the situation in which we find ourselves, leads to policy resistance as we persistently react to the symptoms of difficulty, intervening at low leverage points and triggering delayed and distal, but powerful, feedbacks. The problem intensifies, and we react by pulling those same policy levers with renewed vigor, at the least wasting our talents and energy, and all too often, triggering an unrecognized vicious cycle that carries us farther and farther from our goals. Pumping water from deep aquifers for irrigation causes the water table to fall, requiring more powerful and costly pumps. To offset the rising costs, governments often subsidize electric power for the farmers, increasing pump use and speeding the drop in the water table and need for subsidies.

Policy resistance breeds a sense of futility about our ability to make a difference. One of the challenges in building a more sustainable world is helping us to see ourselves as part of a larger system in which our actions feed back to shape the world in ways large and small, desired and undesired. The greater challenge is to do so in a way that empowers us to take action instead of reinforcing the belief that we are the helpless victims of systems we can neither comprehend nor change, mere leaves tossed about by storm systems of inscrutable complexity and scope.

All dynamics arise from the interaction of just two types of feedback loops, reinforcing (or positive) and balancing (or negative) loops (Fig. 2).[1] Reinforcing feedbacks tend to amplify whatever is happening in the system: the larger a population, the greater the number of births, further increasing the population. The greater a nation's investment in capital plant, equipment, and infrastructure, the larger its gross domestic product (GDP) becomes, increasing the resources available for investment still further. The higher the concentrations of GHGs such as carbon dioxide and methane in the atmosphere, the warmer the earth becomes; as higher temperatures melt permafrost, bacteria metabolize previously frozen organic matter, releasing still more CO_2 and methane and leading to still more warming. These self-reinforcing feedbacks are all processes that generate their own growth, leading, respectively, to population and economic growth and the potential for runaway climate change.

Balancing feedbacks counteract and oppose change. The larger a population relative to the carrying capacity of its environment, the lower the net birth rate will be, slowing population growth. The more oil we discover, the less remains to be discovered in the future. High levels of air and water pollution harm human health, leading to political pressure for action and, eventually, regulations to limit pollutant concentrations. These loops all describe self-correcting processes that seek balance and equilibrium.

Research on mental models shows few incorporate any feedback loops. Axelrod (1976) found virtually no feedback processes in the cognitive maps of political leaders. Dörner (1980, 1996) found that people tend to think in single-strand

---

[1] The scientific literature generally uses the terms "positive" and "negative" to denote self-reinforcing and self-correcting feedbacks. However, laypeople persistently conflate "positive feedback" with "good" and "negative feedback" with "bad," as in "my boss gave me negative feedback on my performance." However, either type of feedback can be good or bad, depending on how the loop is operating and on one's values. The positive feedback of compound interest on credit card debt is "bad" if you are the debtor, but "good" for the card issuer. To avoid the confusion, I use the terms reinforcing and balancing rather than positive and negative.

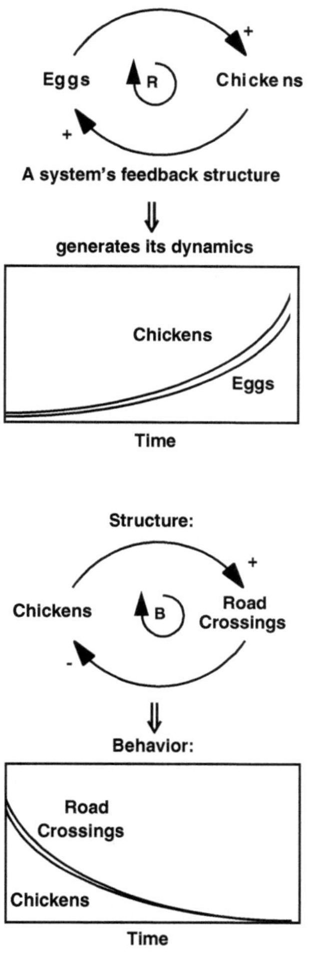

Reinforcing feedback: Reinforcing loops are self-reinforcing. In this case, more chickens lay more eggs, which hatch and add to the chicken population, leading to still more eggs, and so on. A Causal Loop Diagram (Sterman 2000) captures the feedbacks in a system. The arrows indicate the causal relationships. The + signs at the arrowheads indicate that the effect is positively related to the cause: an increase in the chicken population causes the number of eggs laid each day to rise above what it would have been (and vice versa: a decrease in the chicken population causes egg laying to fall below what it would have been). The loop is self-reinforcing, hence the loop polarity identifier **R**. If this loop were the only one operating, the chicken and egg population would both grow exponentially.

Of course, no real quantity can grow forever. There must be limits to growth. These limits are created by balancing feedbacks.

Balancing feedback: Balancing loops are self-correcting. They counteract change. As the chicken population grows, various self-correcting loops will act to balance the chicken population with its carrying capacity. One classic feedback is shown here: The more chickens, the more road crossings they will attempt. If there is any traffic, more road crossings will lead to fewer chickens (hence the — [negative] polarity for the link from road crossings to chickens). An increase in the chicken population causes more risky road crossings, which then bring the chicken population back down. The **B** in the center of a loop denotes a balancing feedback. If the road-crossing loop was the only one operating (say because the farmer sells all the eggs), the number of chickens would gradually decline until none remained.

All systems, no matter how complex, consist of networks of reinforcing and balancing feedbacks, and all dynamics arise from the interaction of these loops with one another.

Fig. 2 Reinforcing and balancing feedback loops

causal series and had difficulty in systems with side effects and multiple causal pathways, much less feedback loops. Booth Sweeney and Sterman (2007) found limited recognition of feedback processes among both middle school students and their teachers. People tend to assume each effect has a single cause and often cease their search for explanations when the first sufficient cause is found (see, e.g., Plous 1993).

Compounding the lack of feedback in people's mental models, research also shows that people do not understand the behavior of even the simplest feedback systems. Wagenaar (1978) and Wagenaar and Sagaria (1975) studied people's ability to understand exponential growth processes. They found people tend to extrapolate linearly instead of exponentially, assuming a quantity increases by the same *absolute* amount per time period, while exponential growth *doubles* the quantity in a fixed period of time. When the growth rate and forecast horizon are small, linear extrapolation is a reasonable approximation to exponential growth. However, as the growth rate increases or the forecast horizon lengthens, the errors become huge. I often demonstrate this phenomenon in my classes with the "paper folding" task, starting with a sheet of copy paper, which I show the students is about 0.1 mm thick:

> "Consider an ordinary sheet of paper like this one. Fold it in half. Fold the sheet in half again. The paper is still less than half a millimeter thick. If you were to fold the paper 40 more times, how thick would it be?
>
> Do not use a calculator. We are interested in your intuitive judgment. Along with your estimate, give your 95% upper and lower confidence bounds for your estimates (i.e., a range of estimates you are 95% sure includes the right answer. Your 95% confidence bound means you believe there is only a 5% chance the correct answer falls outside the upper and lower bounds you give)."

	Lower bound (95% sure it is between lower and upper bound)	Your estimate	Upper bound (95% sure it is between lower and upper bound)
42 folds			

Typical of results with a wide range of audiences, the median estimate in a sample of 95 graduate students at the MIT Sloan School of Management was 0.05 m (less than 2 in.), and the mean, skewed by a few who offered higher numbers, was 134 km (≈83 miles). The correct answer? Each fold doubles the thickness of the paper. After 42 doublings the thickness has increased by a factor of $2^{42} \approx 4.4$ trillion, from 0.1 mm to 440,000 km (≈273,000 miles), farther than the distance from the earth to the moon. The mean response is only 0.03% of the correct value. None of the confidence bounds included the correct value—not only do we fail to understand exponential growth, but we are grossly overconfident in our judgments (see, e.g., Plous 1993). Some students provided the correct formula, but still failed to grasp its implications, such as the student who wrote, correctly, that the paper would be "0.1 mm*2^{42} thick," but whose upper confidence limit was 1.2 km, less than three-quarters of a mile.

These misperceptions of reinforcing feedback and exponential growth are powerful barriers to understanding the sustainability challenge. For most of our existence as a species on the planet, humans were small in number and population growth slow. People had limited power to deplete resources and poison the environment. To be sure, humans could and did despoil local environments (e.g., Diamond 2005), but the consequences, though severe for the affected populations, remained local; in most cases people could move to regions where resources were abundant

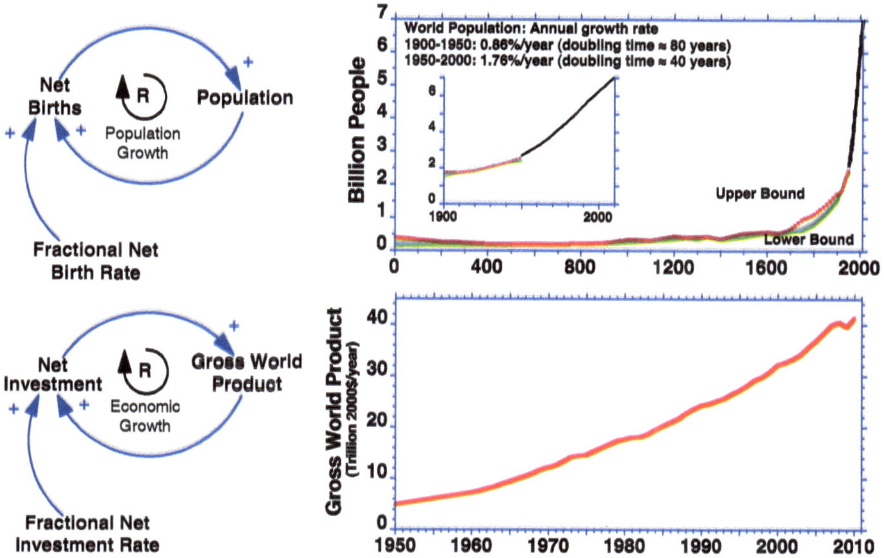

Fig. 3 Reinforcing feedbacks underlying population and economic growth. *Source:* population: UN (2011); GWP: 1970–2010: World Bank, http://databank.worldbank.org; 1950–1970: Worldwatch Institute, Signposts 2001

(though such migrations often triggered conflict and war). But exponential growth explodes very quickly. World population reached one billion sometime in the early 1800s, roughly 100,000 years after modern humans evolved, but it took only about a century to reach two billion. Only about 40 years were required to add the next billion. World population by the end of 2011 was roughly seven billion (Fig. 3) and it took only about a dozen years to add that last billion. More important, the impact of that population is growing even faster. Real Gross World Product (GWP) grew from 1950 to 2009 at an average rate of 3.5% y^{-1} (Fig. 3), doubling every 20 years. At that rate, in a century, the real value of goods and services produced worldwide would grow by a factor of 32.

Such astounding growth in population and material throughput cannot continue indefinitely on a finite planet. The question is not if growth will cease, but when and how. Yet corporations seek ever-growing sales, governments strive for ever-greater economic growth, and individuals desire ever-higher incomes. Many believe that the goal of environmental policy is to enable "sustainable growth," an impossibility. Material growth in a finite world must eventually cease; by definition it cannot be sustained (for excellent discussion, see e.g., Meadows et al. 2004; Daly 1991; Daly and Townsend 1993). Yet, as I write this (August, 2011), Google returns over seven million results for the term "sustainable growth" and hundreds of organizations, including firms and well-intentioned environmental groups, promote "sustainable growth" and other oxymoronic constructs that reinforce the idea that endless growth is not only possible but a worthy goal. Thus, "DuPont has a mission of sustainable growth, which we define as the creation of shareholder and societal value while we reduce our environmental footprint along the value chains in which we operate." The Clinton Foundation offers its "Sustainable Growth Initiative." And the World Environmental Organization lists the "100 top sustainable growth sites."[2]

Nonlinearity

The interactions among the feedbacks in complex systems are typically nonlinear. Consider the interaction of a population with the carrying capacity of its environment (Fig. 4). The population could be yeast in a Champagne cask or the population of earth. The larger the population, the more net births, forming the reinforcing population growth feedback R1 and leading to exponential growth—as long as the fractional net birth rate is constant. However, every organism grows in the context of its carrying capacity. The carrying capacity is the size of the population the habitat of that species can support. It is determined by the resources available in the environment and the resource requirements of the population (e.g., sugar for the Champagne yeast). As the population approaches its carrying capacity, resources per capita fall, reducing the fractional net birth rate and slowing population growth, forming the balancing "limits to growth" feedback, B1. In general, a population depends on many resources, each creating a balancing feedback that can limit growth. The constraints that are most binding determine which balancing loops will be most influential as the population grows: fermentation can be stopped by either depletion of the sugar the yeast consume or a rise in the concentration of the alcohol they produce as waste. The relationship between the ratio of population to carrying capacity and the net fractional birth rate must be downward sloping: when population is small relative to the carrying capacity, resources are abundant, each organism has all the resources it needs, and the population grows at its maximum fractional rate. As resources become scarce, however, fertility falls and mortality increases, lowering the net fractional birth rate, until, when the population equals the carrying capacity, resources are just scarce enough to halt growth. If resources became even more scarce—if, say, the carrying capacity dropped—then the population would fall, raising resources per capita until equilibrium is again reached. The relationship between net births and population is highly nonlinear. When population is small and resources abundant, the reinforcing population growth feedback dominates the dynamics of the system. But as the population grows, the balancing Limits to Growth loop becomes stronger, and eventually dominates the system dynamics. If the carrying capacity is fixed and there are no significant time delays in the balancing feedback then the population

[2]http://www2.dupont.com/Our_Company/en_US/glance/sus_growth/sus_growth.html, http://www.clintonfoundation.org/what-we-do/clinton-giustra-sustainable-growth-initiative, http://www.world.org/weo/growth.

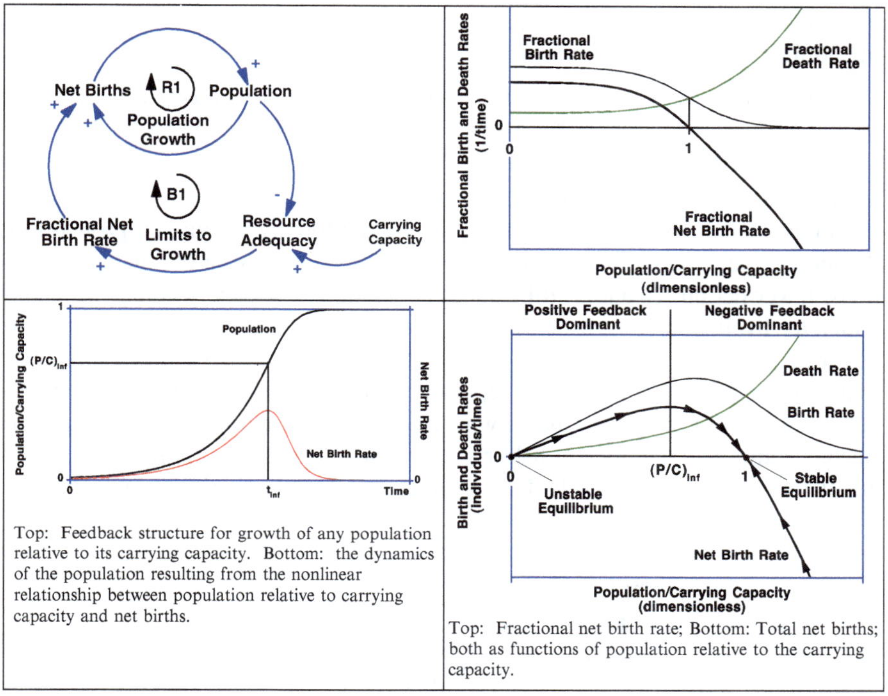

Top: Feedback structure for growth of any population relative to its carrying capacity. Bottom: the dynamics of the population resulting from the nonlinear relationship between population relative to carrying capacity and net births.

Top: Fractional net birth rate; Bottom: Total net births; both as functions of population relative to the carrying capacity.

Fig. 4 Nonlinear interaction of population and carrying capacity

will follow an S-shaped path, as seen in Fig. 4. However, for most species the carrying capacity is endogenous and there are long delays in the system; population growth can then degrade and overshoot the carrying capacity, leading to population decline or collapse.

Tipping Points

Nonlinear interactions of populations with their carrying capacity often create the possibility of sudden shifts in resource abundance. Consider the Atlantic cod, *Gadus morhua*. When Europeans began to fish in the rich waters off the east coast of North America, cod, a predator with few natural enemies, had reached the carrying capacity of their marine environment (Rosenberg et al. 2004). Stocks were immense: John Cabot, exploring Newfoundland in 1497, noted the cod were so thick they nearly blocked his ship (Kurlansky 1997). Fishers easily filled their holds, but because the total take was small, the balancing feedback dominated: by taking a few cod, fishers increased the abundance of food for those that remained, so stocks rapidly recovered. For decades, the more fishers caught, the more grew back, leading them to conclude that there were no limits. Many scientists agreed, including biologist T. H. Huxley (1883), who famously declared:

> I believe, then, that the cod fishery, the herring fishery, the pilchard fishery, the mackerel fishery, and probably all the great sea fisheries, are inexhaustible; that is to say, that nothing we do seriously affects the number of the fish.[3]

However, as the catch continued to grow, cod stocks eventually fell below the maximum sustainable yield. At that point, the reinforcing population growth loop dominated the dynamics, but now operating as a vicious cycle of smaller populations, fewer net births, and a still smaller population. The fishery collapsed, taking with it the livelihood of the fishers and the communities that depended on them, from St. Johns, Newfoundland to New Bedford, Massachusetts. Fig. 5 shows a simulation of the model in Fig. 4 configured to represent a fishery. The fleet grows exponentially at a modest rate of 2% y^{-1}. As the growing catch reduces the stock of cod, recruitment (the net addition to the stock of fish) rises, almost compensating for the catch, so the population declines only slowly. But by about year 225, the tipping point is reached:

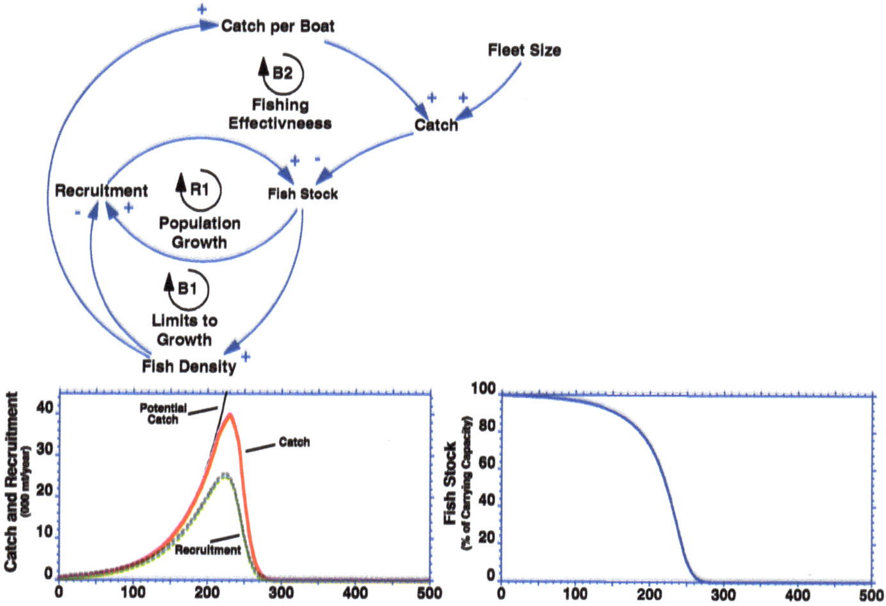

Fig. 5 Environmental tipping point: The fishing fleet (potential catch) grows at a constant rate of 2% y^{-1}. The fishery collapses when stocks fall enough to push the system into the regime in which recruitment falls because there are simply too few fish to replace the catch

---

[3] Huxley's analysis was, for the day, rather nuanced, and he did not believe that all fisheries were inexhaustible. He considered oyster beds and riverine salmon fisheries to be exhaustible, and recognized the tragedy of the commons, concluding that in such cases "Man is the chief enemy, and we can deal with him by force of law. If the stock of a river is to be kept up, it must be treated upon just the same principles as the stock of a sheep farm."

stocks have declined so much that recruitment begins to fall. Now the dynamics are dominated by the vicious cycle of fewer fish, lower recruitment, and a still smaller fish population. Although the precipitous decline in stocks soon forces catch per boat down (via the balancing Fishing Effectiveness loop, B2), the total catch still exceeds recruitment, so fish stocks continue to drop until the cod are extirpated.

Eroding Goals

Surely fishers would limit the catch before the tipping point is crossed. Because fish are common pool resources, stocks are vulnerable to the "Tragedy of the Commons" (Hardin 1968), in which overexploitation is the outcome of rational decision making because the benefits of taking more accrue to individual fishers, while the costs (lower future catch) are borne by all. Further, experimental studies (Moxnes 2000) show that even when the common pool resource problem is absent, people's poor understanding of resource dynamics leads to overexploitation and, often, collapse. More optimistically, Ostrom (2010) demonstrates how communities can establish sustainable harvesting for a variety of common pool resources. Communities that agree on and enforce limits on the harvest of their resources create an important balancing feedback, shown in Fig. 6 as the Catch Limit loop, B3.

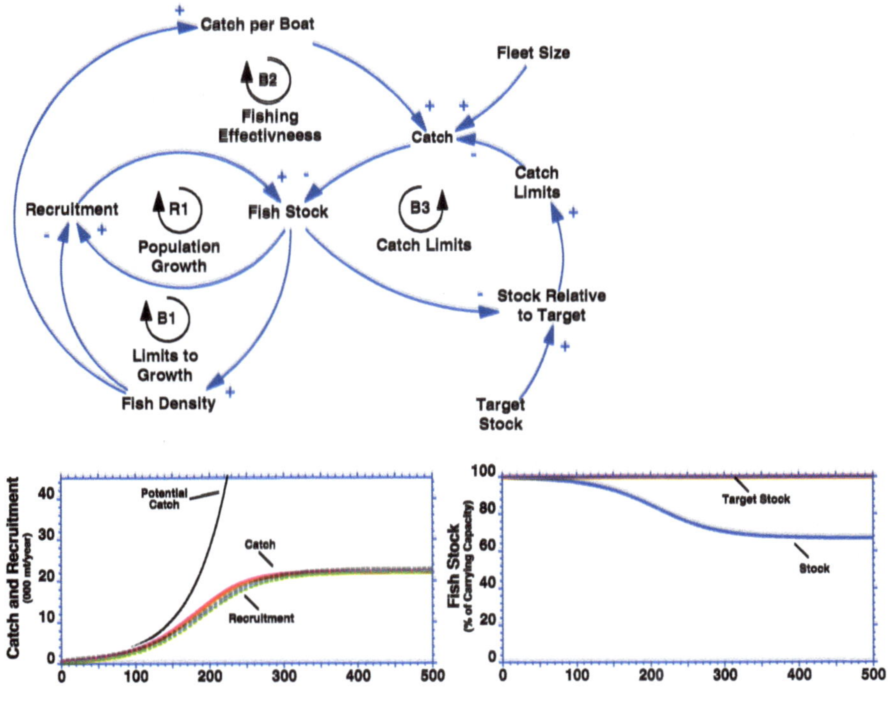

Fig. 6 Catch limits: fishery collapse can be avoided if the members of the community agree to lower the catch as stocks fall below a target (B3)—if the target is high enough

As the resource stock falls relative to a target, the community voluntarily limits its catch. If the target is high enough—and if enforcement or social norms prevent free riders from taking more than their share—the stock stabilizes above the tipping point and the resource can be sustained.

Few goals are absolute. How much should you weigh? What grades should a student earn? How much income is enough? When goals are not absolute, self-evident, and obvious, they co-evolve endogenously with the state of the system. To reduce cognitive dissonance (Festinger 1957), we adapt to our circumstances, redefining what we consider "normal" and "acceptable" to match our situation. Then, as our goals erode, so too do our efforts to improve system performance, a reinforcing feedback that can lead to path dependence and self-fulfilling prophecy (Forrester 1969; Sterman 2000). Thus, we buy larger clothes when we gain weight, literally easing the pressure that might motivate us to eat less and exercise more; teachers often adjust their beliefs about a student's ability toward that student's grades, providing more help to those they conclude are gifted and less to those they believe to lack potential, further boosting the achievement of the favored while the rest fall farther behind (Rosenthal and Jacobson 1968); as we consume more we become habituated to our higher standard of living, leading us to strive for still greater income (Kahneman et al. 1999; Layard 2005).

Eroding goals are particularly common in sustainability contexts due to our imperfect understanding of ecosystem dynamics. Due to limited information, natural variability, and limited knowledge of population dynamics, estimates of "normal" stocks and maximum sustainable yield are uncertain. Consequently, target stocks are vulnerable to political pressure from fishers, who often argue that current stocks are close to normal, so catch should not be limited. More important, people, including scientists, typically credit their own experience more than other information, while environmental changes are often slow relative to our lifespan. Fishery biologist Daniel Pauly (1995) describes the resulting "shifting baseline syndrome" in which "each generation of fisheries scientists accepts as a baseline the stock size and species composition that occurred at the beginning of their careers, and uses this to evaluate changes. When the next generation starts its career, the stocks have further declined, but it is the stocks at that time that serve as a new baseline. The result obviously is a gradual shift of the baseline, a gradual accommodation of the creeping disappearance of resource species, and inappropriate reference points for evaluating economic losses resulting from overfishing, or for identifying targets for rehabilitation measures" (Pauly 1995; see also http://www.shiftingbaselines.org). Fig. 7 shows the feedback structure and behavior of the shifting baseline syndrome. Now, as fish stocks fall, the beliefs of fishers and scientists about normal stock levels gradually drop. Pressure to limit the catch is reduced. Stocks fall further, and beliefs about the normal stock level fall still more in a vicious cycle—the Shifting Baseline feedback (R2). The simulation shows the result when the target stock adjusts to the actual stock over an average of 20 years. Goal erosion undermines the effectiveness of community efforts to limit the catch to a sustainable rate. The fishery still collapses.

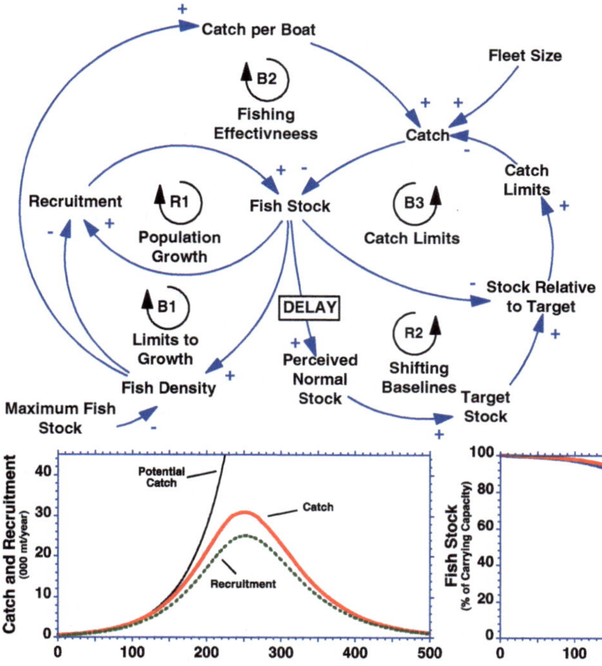

Fig. 7 Eroding goals: allowing target stocks to adjust to actual levels with a time constant of 20 years leads to collapse of the fishery through the reinforcing Shifting Baselines loop R2. All other conditions identical to Fig. 6

Eroding goals and shifting baselines are not limited to fisheries, but occur for a wide range of environmental issues, particularly those in which the dynamics are slow relative to human lifespan and in which the signal-to-noise ratio is low, for example, in climate change, where daily and seasonal fluctuations in local weather dominate our experience, while the slow rise in global average temperatures, loss of snow cover, changes in the behavior and range of species, and other impacts are hard to discern.

Time Delays

Time delays in complex systems are common and particularly troublesome. For example (Table 3), more than 6 decades have passed from the first undeniable evidence that air pollution from combustion of fossil fuels causes significant health problems, including death, yet the US EPA reports 145 million Americans live in so-called nonattainment areas—regions where air quality does not meet the standards of the Clean Air Act.

Table 3 Delays in societal response to air pollution in the USA

1800s:	widespread use of coal leads to growing air pollution in Europe and the USA
1948:	Smog in Donora, Pennsylvania, kills 20 people and sickens 6,000. Soon after, coal fumes kill nearly 800 in London
1955:	First US Federal Air Pollution Control Act laid primary responsibility for limiting air pollution upon states and cities, though it allocated $5 million for research
1963:	Federal Clean Air Act recognizes air pollution does not respect state boundaries; sets up regulations for interstate abatement; provides more assistance to state and local governments
1970:	Clean Air Act strengthened by defining "safe" standards for sulfur dioxide (SO_2), carbon monoxide (CO), particulates, volatile organic compounds (VOCs), nitrous oxides (NO_x), ozone (O_3), and lead (Pb). States are required to submit plans to meet standards by 1975
1977:	deadline postponed until 1982 as 78 cities were in violation of ozone standards
1988:	ninety urban areas with 150 million inhabitants exceed ozone standard; 40 violate CO standard
1990:	Amendments to Clean Air Act require all cities to meet ozone standard by 2007 (except Los Angeles, which had until 2010). Stricter regulations for auto emissions, gasoline, SO_2, and newly regulated pollutants
1990s:	medical evidence shows health problems and deaths from air pollution are caused by lower concentrations of key pollutants than previously thought
1997:	reflecting new science, and over strong opposition by industry, the National Ambient Air Quality Standards under the Clean Air Act tightened for six "criteria" air pollutants (O_3, CO, SO_2, NO_x, Pb, and PM_{10} and $PM_{2.5}$—particulates smaller than 10 and 2.5 μm). Industry groups immediately challenge the new regulations in court
2001:	US Supreme Court, in Whitman v. American Trucking Associations, Inc., unanimously upholds the bulk of the new regulations
2010:	145 million Americans live in "nonattainment" areas in violation of EPA standards for the six criteria pollutants. The National Research Council estimates that the pollutants released by use of fossil fuels, primarily O_3, NO_x, SO_2, and particulates, cause economic damage of at least $120 billion per year, including the cost of 20,000 premature deaths per year
Time from first clear signal of problem to initial Clean Air Act: 15 years	
Total delay from first clear signal to full compliance with law: 62 years and counting	

Source: paraphrased and condensed with permission from D. Meadows and A. AtKisson, *The Balaton Bulletin*, 1997; updated by J. Sterman from EPA Green Book (http://www.epa.gov/oar/oaqps/greenbk). Nonattainment population in 2010 given by the maximum population living in each nonattainment area across each criterion pollutant per EPA Green Book data as of December 17, 2010, http://www.epa.gov/oar/oaqps/greenbk/ancl3.html

The National Research Council (2010) estimated that the pollutants released by use of fossil fuels cost the US economy at least $120 billion per year, including the cost of 20,000 premature deaths per year. These estimates exclude the harms these pollutants cause to ecosystems and national security, the damage caused by other pollutants such as mercury, and the costs of climate change arising from anthropogenic GHGs.

Research shows people routinely ignore or underestimate time delays (Sterman 1989; Sterman 2000; Buehler et al. 2002; Faro et al. 2010). Underestimating time delays leads people to believe, wrongly, that it is prudent to "wait and see" whether a potential environmental risk will actually begin to cause harm, or until scientific research resolves uncertainty about whether something is harmful, then address it if it does. Consider climate change. Many people, including many who believe climate change is real and poses serious risks, nevertheless advocate a wait-and-see approach, reasoning that uncertainty about the causes and consequences of climate change means potentially costly actions to address the risks should be deferred. If climate change turns out to be more harmful than expected, policies to mitigate it can then be implemented, they argue.

Wait-and-see policies often work well in simple systems, specifically those with short lags between detection of a problem and the implementation and impact of corrective actions. In boiling water for tea, one can wait until the kettle boils before taking action because there is essentially no delay between the boiling of the water and the whistle of the kettle, nor between hearing the whistle and removing the kettle from the flame. Few complex public policy challenges can be addressed so quickly. To be a prudent response to the risks of climate change, wait-and-see policies require short delays in all the links in a long causal chain, stretching from the detection of adverse climate impacts to the decision to implement mitigation policies to emissions reductions to changes in atmospheric GHG concentrations to radiative forcing to surface warming and finally to climate impacts, including changes in ice cover, sea level, weather patterns, agricultural productivity, habitat loss and species distribution, extinction rates, and the incidence of diseases, among others. Contrary to the logic of "wait and see" there are long delays in every link of the chain (Fiddaman 2002; O'Neill and Oppenheimer 2002; Stachowicz et al. 2002; Alley et al. 2003; Thomas et al. 2004; Meehl et al. 2005; Wigley 2005; Solomon et al. 2009; Pereira et al. 2010). Similar delays exist for many environmental problems.

More problematic, the short- and long-run impacts of our policies often differ. Such "better-before-worse" behavior is common across many spatial and temporal scales. Credit card debt boosts your material standard of living today but forces a drop in consumption when the bills and interest must be paid. Smoking brings immediate pleasure but disease and death later. Forest fire suppression works in the short run, until the resulting fuel accumulation leads to more, hotter and more damaging fires decades later. DDT was a boon to agriculture and human health in the short run, while pest resistance and the harmful effects of chlorinated hydrocarbons on ecosystems and humans emerged only later. Similarly, what's best for the long term often imposes short-run costs, a "worse-before-better" pattern. Saving for retirement requires we sacrifice consumption in the short run. Restoring the cod fishery requires cutting the catch today. Because the long-term benefits and harms of current actions are uncertain, delayed, and diffuse, we are often biased toward actions that improve welfare in the short run at the expense of the future (better-before-worse). And the worse-before-better impact of policies required to improve long-run performance often causes them to fail (Repenning and Sterman 2001). The trade-off between the short- and long-term responses to policies is particularly problematic in the domain of sustainability because of the long time delays in ecological and economic processes.

Stocks and Flows

Stocks and the flows that alter them are fundamental in disciplines from accounting to zoology: Debt is increased by borrowing and reduced by repayment or default; the burden of mercury in a child's body is increased by ingestion and decreased by excretion; a population is increased by births and decreased by mortality. In physical and biological systems stocks are often tangible: the stock of fresh water in the Ogallala aquifer, the number of gasoline-powered vehicles, the amount of CO_2 in the atmosphere. The dynamics of our economic and social systems, however, are also determined by intangible resources such as technical knowledge, behavioral norms around littering and recycling, trust among extractors of common pool resources, and other forms of human, social, and political capital.

We should have good intuitive understanding of accumulation because stocks and the flows that alter them are pervasive in everyday experience: Our bathtubs accumulate the inflow of water through the faucet less the outflow through the drain, our bank accounts accumulate deposits less withdrawals, and we all struggle to control our weight by managing inflows and outflows of calories through diet and exercise. Yet research shows that people's intuitive understanding of stocks and flows is poor in two ways that perpetuate low leverage approaches to sustainability. First, despite their ubiquity, people have difficulty relating the flows into and out of a stock to the level of the stock, even in simple, familiar contexts such as bank accounts and bathtubs. Second, narrow mental model boundaries mean people are often unaware of the networks of stocks and flows that supply resources and absorb wastes.

Although the relationship between stocks and flows is a fundamental concept of calculus, knowledge of calculus is not necessary to understand their behavior. Any stock can be thought of as the amount of water in a tub. The water level accumulates the flow of water into the tub (the inflow) less the flow exiting through the drain (the outflow). The rate of change in the water level is the net flow, given by the difference between the inflow and outflow.[4] As everyone knows, the water level in the tub rises only when the inflow exceeds the outflow, falls only when the outflow exceeds the inflow, and remains the same only when the inflow equals the outflow. Yet even highly educated adults with strong background in STEM (Science, Technology, Engineering, and Mathematics) often do not understand these basic principles. Booth Sweeney and Sterman (2000) presented graduate students at MIT

---

[4] Consider any stock, S, with inflow I and outflow O. The stock at time T, S_T, is the integral of its net inflow from the initial time, t_0, plus the initial stock, S_{t_0}. Equivalently, the rate of change of the stock is the net inflow, $I–O$:

$$S_T = \int_{t_0}^{T} \text{Net Inflow}\, dt + S_{t_0} = \int_{t_0}^{T} (I - O) dt + S_{t_0};\ \frac{dS}{dt} = \text{Net Inflow} = I - O$$

Note the units of measure: stocks are measured in units, e.g., liters of water in a tub or tons of CO_2 in the atmosphere, while flows are measured in units/time period, e.g., liters/second or tons of CO_2/year.

with a picture of a bathtub and graphs showing the inflow and outflow of water and then asked them to sketch the trajectory of the stock of water in the tub. Although the patterns were simple, fewer than half responded correctly. The majority in these and subsequent experiments fail to apply the basic principles of accumulation. Rather, people often use the intuitively appealing "correlation heuristic" (Cronin et al. 2009), assuming that the output of a system should "look like"—be positively correlated with—its inputs. Although sometimes useful, correlational reasoning fails in systems with important accumulations. The US federal deficit and national debt have both risen dramatically since 1950. Correlational reasoning predicts that cutting the deficit would also cut the debt. However, because the national debt is a stock that accumulates the deficit, it keeps rising even if the deficit falls. The debt falls only if the government runs a surplus.

Poor understanding of accumulation leads to serious errors in reasoning about sustainability. Herman Daly (1991) articulated three fundamental, necessary conditions for sustainability in any finite environment, shown in Fig. 8 using standard stock and

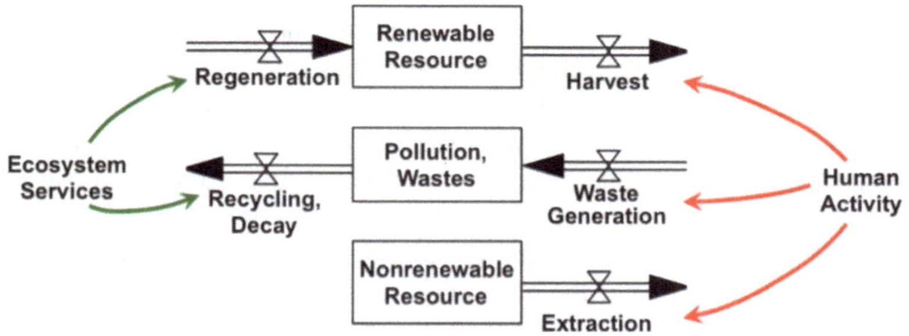

Fig. 8 Three necessary conditions for sustainability (Daly 1991) shown in stock and flow notation. Rectangles denote stocks; pipes and valves denote the flows. Here, the stock of renewable resources is depleted by harvest (e.g., logging) and filled by regeneration (e.g., forest regrowth). The harvest of renewables, generation of wastes, and extraction of nonrenewables are driven by human activity (the population and economy). Renewable resource regeneration and the processes that render wastes harmless (e.g., breakdown of sewage, removal of CO_2 from the atmosphere) are provided by ecosystem services. For simplicity the stocks that support activities are not shown but are themselves finite: there are no limitless sources and sinks on a finite planet

flow notation. In a sustainable society, (1) renewable resources cannot be used faster than they regenerate, (2) pollution and wastes cannot be generated faster than they decay and are rendered harmless; and (3), in the long run, nonrenewable resources cannot be used at all. These principles follow directly from the fundamental laws of accumulation: If a bathtub is drained at a higher rate than it fills, the level of water will fall; in exactly the same way, a sustainable society cannot harvest cod faster than they reproduce. Filling a tub faster than it drains raises the level of water, so a sustainable society cannot produce GHGs faster than they are removed from the atmosphere. And the level of water in a tub with an open drain but no inflow will fall until the tub is empty, so a sustainable society cannot rely on nonrenewables such as fossil fuels.

Our society is far from meeting any of these three fundamental requirements for sustainability (Wackernagel et al. 2002; Rockström et al. 2009). Any policy or program that purports to promote sustainability should be judged by whether it moves us closer to stabilizing stocks of resources and wastes. If it does not, then it does not advance the cause of sustainability. Experimental studies show that people do not understand these concepts. Sterman and Booth Sweeney (2007) gave 212 graduate students at MIT a description of the relationships among GHG emissions, atmospheric concentrations, and global mean temperature. The description, excerpted from the IPCC's "Summary for Policymakers," a document intended for nonspecialists, described the flows into and out of the stock of CO_2 in the atmosphere. Participants were then asked to sketch the emissions trajectory required to stabilize atmospheric CO_2. To highlight the stock–flow structure, participants were first directed to estimate future net removal of CO_2 from the atmosphere (net CO_2 taken up by the oceans and biomass), then draw the emissions path needed to stabilize atmospheric CO_2.

Knowledge of climatology or calculus is not needed to respond correctly. The dynamics are easily understood using the bathtub analogy. Like any stock, atmospheric CO_2 rises when the inflow to the tub (emissions) exceeds the outflow (net removal), is unchanging when inflow equals outflow, and falls when outflow exceeds inflow. Participants were informed that anthropogenic CO_2 emissions are now roughly double net removal, so the tub is filling. Yet, 84% drew patterns that violated the principles of accumulation. Nearly two-thirds asserted that atmospheric GHGs can stabilize even while emissions continuously exceed removal—analogous to arguing a bathtub continuously filled faster than it drains will never overflow. The false belief that stabilizing emissions would quickly stabilize the climate not only violates mass balance, one of the most basic laws of physics, but leads to complacency about the magnitude and urgency of the emissions reductions required to mitigate climate change risk (Sterman 2008).

Training in science does not prevent these errors. Three-fifths of the participants had degrees in science, technology, engineering, or mathematics (STEM); most others were trained in economics. Over 30% hold a prior graduate degree, 70% of these in STEM. These individuals are demographically similar to influential leaders in business, government, and the media, though with more STEM training than most. On a more hopeful note, it is possible for people to learn these principles: Sterman (2010) shows that a half-semester course on system dynamics modeling significantly improves people's ability to relate stocks and flows and reduces the prevalence of the correlation heuristic.

Understanding the principles of accumulation, while necessary, is not sufficient. People must also be able to identify the networks of stocks and flows through which resources and wastes move through the economy and ecosystem. The inability to recognize and map the network of stocks and flows in a system contributes to policy resistance and unsustainability by focusing people's attention on local conditions at the expense of the distal and delayed consequences of their actions. It allows us to externalize the environmental consequences of our actions by promoting the illusion that there are unlimited supplies of natural resources and limitless sinks into which wastes can be dumped. California's Air Resources Board promotes so-called zero emission vehicles (ZEVs; see http://www.arb.ca.gov/msprog/zevprog/zevprog.htm). True, ZEVs—likely to be electric vehicles—need no tailpipe. But the plants required to make the electricity to power them do generate pollution. California is actually promoting DEVs—*displaced* emission vehicles—whose wastes would blow downwind to other states or accumulate in nuclear waste dumps. Air pollution causes substantial harm, electric vehicles may prove to be an environmental boon compared to internal combustion, and eventually electricity must be produced from renewable sources such as wind and solar. But no technology is free of environmental impact, and no government can repeal the second law of thermodynamics.

Fig. 9 shows the stock and flow network mapping the movement of carbon from fossil fuels, through combustion, into the atmosphere, and from the atmosphere

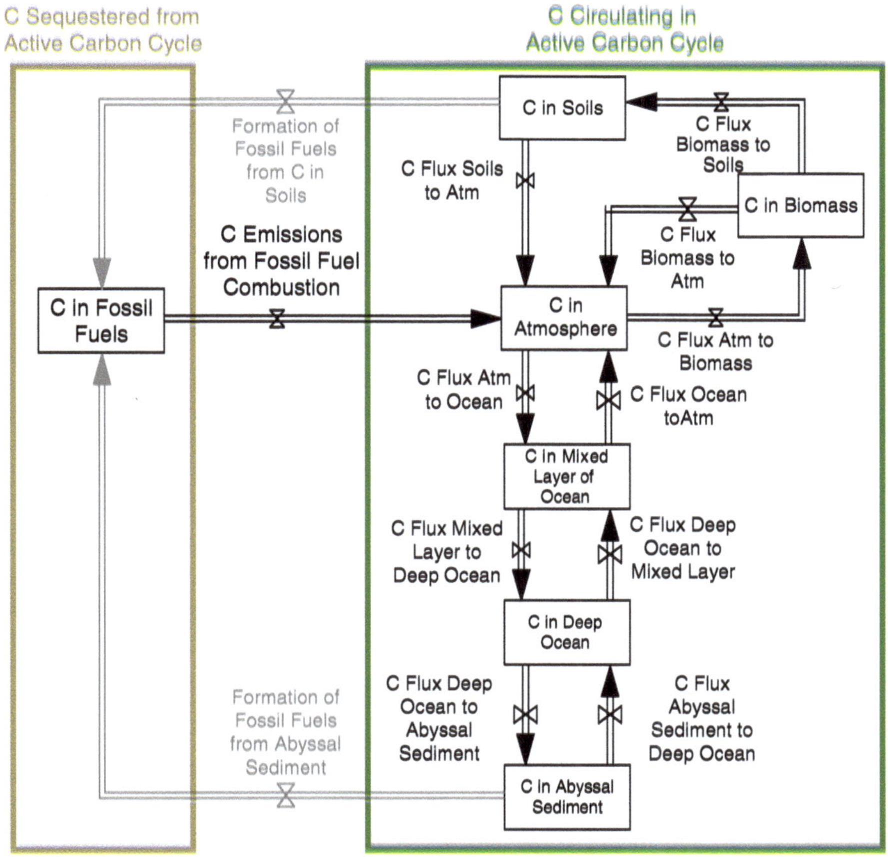

Fig. 9 Stock and flow structure of the carbon cycle. Combustion of fossil fuels injects carbon that has been sequestered for millions of years into the atmosphere, where it is taken up by biomass or dissolves in the ocean, but eventually cycles from these stocks back into the atmosphere. Flows showing the formation of fossil fuels from carbon in terrestrial soils and ocean sediments are shown in gray because these flows are, relative to human time scales, essentially zero. The diagram does not show flows of C associated with the formation and weathering of limestone and other rocks as these flows are unchanging over human time scales

into various stocks including carbon in biomass, soils, and the ocean. The fluxes of carbon among these reservoirs determine the concentration of CO_2 in the atmosphere, and thus anthropogenic global warming. Many argue that we can limit climate change by Reducing Emissions from Deforestation and land Degradation (REDD) (see e.g., http://www.un-redd.org). REDD policies often focus on carbon credits and offsets, through which polluters "offset" the CO_2 generated by the fossil fuels they burn by paying developing nations to preserve their forests or plant trees.

It's true that a growing forest removes carbon from the atmosphere. But what happens to that carbon? Within a few decades it returns to the atmosphere through several routes: First, as the forest grows, more and more leaves, pine needles, branches, and trees die and fall to the forest floor, where bacteria and fungi consume them, releasing CO_2 and methane back into the atmosphere. Second, if humans harvest the wood for fuel or clear that land for crops, the carbon stored in the trees returns to the atmosphere via fire or decay. Even if the forest is protected from legal and illegal logging, the larger the stock of carbon in the forest, the greater the chance of wildfire. Halting deforestation is essential in building a more sustainable world: it accounts for roughly 20% of total world carbon emissions and causes a multitude of other harms including erosion and mudslides; changes in regional albedo, cloud formation, and rainfall; and habitat loss that displaces indigenous peoples and accelerates species extinction. But allowing nations, firms, and individuals to "offset" their fossil fuel use by buying carbon credits to reduce deforestation is a fool's bargain. Burning fossil fuels injects carbon that has been sequestered for millions of years into the atmosphere. Such carbon remains in the active carbon cycle for eons, while afforestation removes carbon from the atmosphere only temporarily and does nothing to reduce the stock of carbon in the active carbon cycle.

Poor understanding of stock and flow networks is not limited to the public and policymakers but is all too common in academic research. In the early 1990s, William Nordhaus developed the DICE (Dynamic Integrated Climate Economy) model. DICE closes an important feedback: the economy generates GHGs, which alter the climate, which feeds back to reduce economic growth and emissions. But the DICE are loaded. The carbon cycle in the model represents only a single stock: carbon in the atmosphere (eq. 8 in Nordhaus 1992),

$$M(t) = bE(t) + (1 - d_M)M(t-1)$$

"where $M(t)$ is CO_2 concentrations relative to preindustrial times, b is the marginal atmospheric retention ratio, and d_M is the rate of transfer from the rapidly mixing reservoirs to the deep ocean" (p. 1316). The transfer rate d_M is constant, implying that the carbon sinks that accumulate the CO_2 removed from the atmosphere have infinite absorption capacity. As seen in Fig. 9, these sinks are finite; the carbon taken up by the land and oceans eventually makes its way back into the atmosphere. By omitting these stocks DICE ignores important nonlinear constraints on carbon uptake by biomass as primary production is constrained by other nutrients and as the partial pressure of CO_2 in the ocean rises. These feedbacks cause the fractional removal rate to fall as atmospheric CO_2 rises, as terrestrial and oceanic carbon sinks saturate, and as global mean temperature increases (e.g., IPCC 2007). Worse, the so-called marginal atmospheric retention ratio b is set to 0.64. A charitable interpretation is that 36% of total emissions is quickly absorbed out of the atmosphere (within a year), with the rest removed slowly, at the rate d_M. However, the emissions that leave the atmosphere quickly are absorbed by biomass or by the ocean. As these stocks fill, additional removal from the atmosphere is constrained. Since none of these carbon reservoirs are represented, however, Nordhaus has in fact assumed that 36% of total emissions disappear forever, without a trace. Expanding the model boundary to account for sink capacities and conserve carbon increases the warming generated by a given rate of CO_2 emissions, working against Nordhaus' conclusion that optimal carbon taxes are low (Fiddaman 2002; Solomon et al. 2009). Indeed, Solomon et al. (2009) show that even if GHG emissions fell to zero today, the carbon and heat already absorbed by the oceans would cause global mean surface temperature to remain roughly constant for at least 1,000 years: the impact of current GHG emissions on the climate are essentially irreversible.

Yet the narrow boundaries in resource models persist. For example, addressing the debate over future supply of minerals and energy, energy economist Morris Adelman (1993) declared:

"Minerals are inexhaustible and will never be depleted. A stream of investment creates additions to proved *reserves*, a very large in-ground inventory, constantly renewed as it is extracted....How much was in the ground at the start and how much will be left at the end are unknown and irrelevant" (p. xi). "The fixed stock does not exist" (p. xiii) "What exists, and can be observed and measured, is not a stock but a flow" (p. xiv).

Adelman's statements violate conservation of matter. Every ton of titanium and every barrel of oil added to the stock of proven reserves reduces the stock of titanium and oil remaining to be found in the future. Every ton and barrel extracted reduces the quantity remaining in the ground. *Ceteris paribus*, the smaller the stock of resources remaining to be discovered, the lower the productivity of exploration activity must be (on average), and the smaller the rate of addition to proven reserves will be for any investment rate. In the limit, if the stock of undiscovered resource fell to zero, the rate of additions to proven reserves would necessarily fall to zero.

Economists argue that a drop in proven reserves will raise prices, leading to substitution of other resources and inducing additional exploration activity and improvements in technology that increase exploration and recovery. But even if markets function well, additional exploration only drains the stock of undiscovered resource faster. Depletion must continue—the stock of resources in the ground must fall—as long as there is any extraction. Only if there is a so-called backstop technology that can fully substitute for all uses of the nonrenewable resource at a finite price, in finite time, will extraction fall to zero and halt depletion. The size of the resource base, the costs of substitutes, and whether new technologies can be developed before depletion reduces extraction and harms economic welfare are empirical questions, not matters of faith. The very possibility that depletion might matter cannot be assumed away, to be made untestable with models in which resources are infinite, the price system always functions perfectly, delays are short, and technology provides backstops at low cost.

Where Is the Leverage?

Integrating feedback, time delays, and stock–flow structures yields a simple conceptual framework to identify the key leverage points for the creation of a sustainable world. As shown earlier, the sustainability challenge arises from the collision of population and economic growth with the limits of our finite world. Fig. 10 shows

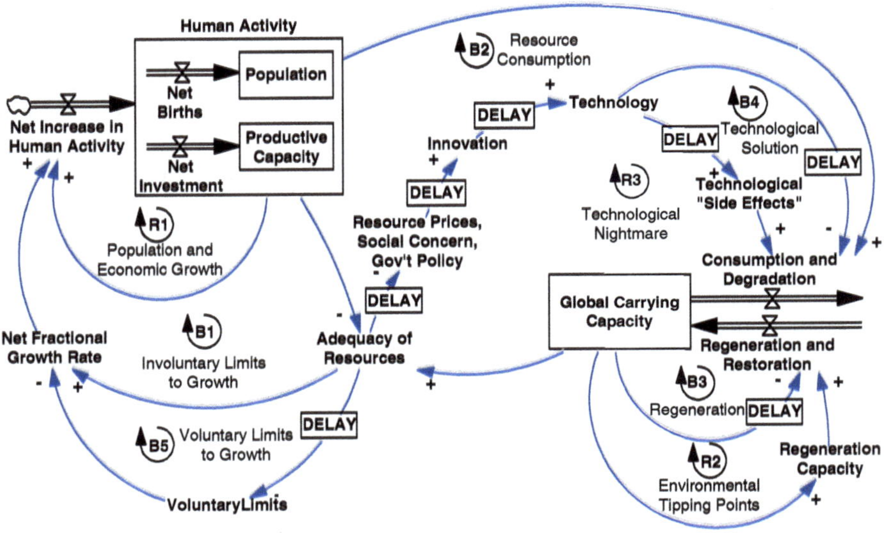

Fig. 10 Interactions of growth, carrying capacity, and technology

a simplified model integrating growth of human activity with the carrying capacity of the planet (see Meadows et al. 2004 for a more detailed model). On the left, human activity grows through the reinforcing feedbacks of population and economic growth described earlier (aggregated into reinforcing loop R1). If the environment were unlimited, growth could continue indefinitely. However, growth in human activity is constrained by the adequacy of resources (the ensemble of nonrenewable resources, renewable resources, and a healthy, clean environment shown in Fig. 8). As populations and economic activity grow relative to carrying capacity, the adequacy of those resources decline. Sufficient decline in resource adequacy lowers the net fractional growth rate in human activity, eventually causing growth to stop via "Involuntary Limits to Growth" (loop B1).

If the carrying capacity were constant growth would follow an S-shaped pattern (as in Fig. 4) in which resources per capita fall until they are just scarce enough to balance births with deaths: a subsistence equilibrium in which life would be nasty, brutish, and short. That naïve Malthusian model is simplistic because the carrying capacity of the earth is dynamic. On the one hand, the larger the population and the greater the economic impact of each person, the greater the consumption and degradation of the carrying capacity: a larger, richer population consumes more resources, generates more waste, uses more fossil fuels, emits more greenhouse gases, etc. On the other, the carrying capacity can regenerate: logging provides more light and nutrients for seeds and saplings; plants fix nitrogen and other nutrients in soils; chlorinated hydrocarbons such as DDT and dioxin eventually break down into harmless compounds. These processes are captured by the balancing Regeneration loop B3. Of course, there are delays in the regeneration process: acorns require decades to become mighty oaks; soils form at rates of a few millimeters per year; DDT degrades over decades. And some elements of the carrying capacity cannot be regenerated: fossil fuels and high-grade copper ores are nonrenewable; extinction is irreversible; stocks of plutonium and other nuclear wastes will remain with us far longer than any civilization on earth has yet endured.

Even for the renewable elements of the carrying capacity there are limits to regeneration and restoration, as illustrated by the fishery model earlier. Harvest a few cod and the population recovers, but take too many and the population collapses; take a few trees and the forest regenerates, but clear cutting can alter rainfall and surface albedo so that the land becomes savannah or desert. These processes are captured by the reinforcing Environmental Tipping Point feedback R2: degrade the earth's carrying capacity too much and its ability to regenerate withers, accelerating the collapse in a vicious cycle. Where these tipping points lie is usually uncertain—until they have been crossed, by which time it is too late.

Expanding the naïve Malthusian model, with its constant carrying capacity, to include the dynamics of the earth's resources and ecosystems changes the dynamics of the system. Now, as the population and economy grow, resources per capita fall in two ways: first, as before, there are more people relative to the resources of the earth. Second, the carrying capacity itself begins to fall as resource consumption and degradation exceed regeneration and restoration. If regeneration is rapid and regeneration capacity robust (loop B3 is strong and swift and the tipping point loop R2 is weak), then regeneration quickly rises to offset resource consumption and waste production and the decline in the carrying capacity is slight. Human population and activity still follow an s-shaped path, growing until resources are scarce enough to halt further increase. However, if regeneration is weak and slow, or the tipping points strong and close, then carrying capacity will fall. Unlike the prior case, the system does not reach equilibrium when the carrying capacity and human activity meet. Instead, the high level of human activity means consumption and degradation of the carrying capacity exceed regeneration, so the carrying capacity of the earth continues to fall. As it does, economic output and/or human population must fall. In the extreme, if the population remains dependent on nonrenewable resources or generates wastes that cannot be dissipated, the carrying capacity must continue to fall as long as there is any remaining activity, and the only equilibrium is zero population—extinction. Incorporating the dynamics of the carrying capacity changes the system dynamics from S-shaped growth to overshoot and collapse (Forrester 1971b; Meadows et al. 2004).

Many societies have degraded their carrying capacities and collapsed, for example, on Rapa Nui (Easter Island). Carbon dating puts the arrival of the first humans, intrepid Polynesian sailors, around the year 1200 (Wilmshurst et al. 2011). Pollen counts from soil cores and other records show that they found a lushly forested island with diverse fauna, particularly birds (Bahn and Flenley 1992; Steadman 1995). Population increased exponentially, and as it did, the islanders harvested trees to provide wood and fiber for fuel, boats, structures, ropes, and tools. The Polynesian rat, which arrived with the settlers, hastened the decline by killing birds and eating the seeds and nuts of the native palm. Deforestation was nearly complete by the year 1400, dramatically reducing the island's carrying capacity. The rain now washed the unprotected soil away. Wind speeds at ground level increased, carrying still more valuable soil into the sea. Erosion was so severe that sediment washed from higher elevations eventually covered many of the moai, so that European visitors thought these giant statues were just heads, when in fact they were complete torsos averaging 20 ft in height. Deforestation also increased evaporation from the soil and may have reduced rainfall, further reducing food production and fresh water supplies. Most of the bird species living on Easter Island became at least locally extinct. Only 1 of about 25 indigenous species still nests on the island today (Steadman 1995). Eventually, fishing, the other main source of food, fell as boats, lines, and hooks, all made from wood, could no longer be replaced. When the first Europeans arrived the islanders prized wood above all other items offered in trade.

As the carrying capacity declined, hardship slowed population growth. Population peaked between six and ten thousand around the year 1600. A precipitous decline set in by about 1680. Spear points and tools of war appeared for the first time and there is evidence of large battles. Some scholars believe there is evidence of cannibalism during this period. The first Europeans known to visit Easter Island arrived in 1722 and found a small, impoverished population. Similar overshoot and collapse dynamics befell other island societies (Kirch 1997) and much larger civilizations, such the Classic Maya (Webster 2002).

Technological optimists and many economists object that the model so far (loops R1, B1, B2, B3, and R2 only) ignores human ingenuity and the power of markets. The primitive people of Easter Island might have experienced overshoot and collapse because they lacked modern technology and could not leave the island, but, it is argued, today we have superior innovative powers and markets that can compensate for any resource shortages and environmental problems growth may create. Thus Julian Simon (1996) argued that

> Greater consumption due to increase in population and growth of income heightens scarcity and induces price run-ups. A higher price represents an opportunity that leads inventors and businesspeople to seek new ways to satisfy the shortages. Some fail, at cost to themselves. A few succeed, and the final result is that we end up better off than if the original shortage problems had never arisen.

Expanding the boundary of the model to capture prices and innovation creates a new balancing feedback, the Technological Solution loop B4. As a resource becomes scarce, its price rises, stimulating technical innovation that cuts demand and substituting more abundant resources for those that are scarce (e.g., drilling deep offshore oil wells in the Gulf of Mexico as shallower deposits on land are depleted; boosting the gas mileage of autos). Going further than free market advocates such as Simon, the technological solution feedback B4 also includes the possibility that scarcity may induce governments to increase research and development (e.g., R&D on alternative energy sponsored by the US Department of Energy), and correct market failures through regulation, stimulating innovation (e.g., CAFE standards and the cap and trade market in SO_2). Further, social norms may change in response to scarcity (e.g., recycling).

There are, however, important lags in these technological solution feedbacks, including delays in the detection of environmental problems, in recognizing the opportunity for profit when prices rise, and in the reallocation of capital and R&D resources. There are long delays before R&D yields new technologies and between laboratory demonstrations and commercialization. Once new technologies reach the market, there are even longer delays in adoption and the replacement of old infrastructure, and further delays before the carrying capacity responds.

In the technological optimist's model, innovation and technology are beneficial. But many technologies create unintended effects that intensify scarcity or environmental problems elsewhere. Taller smokestacks on Midwestern power plants reduced smog in Ohio and Pennsylvania, but caused acid rain in New York and New England; the Haber–Bosch process to fix nitrogen led to synthetic fertilizer, dramatically boosting crop yields (where farmers could afford it), but consumes huge amounts of fossil fuels while fertilizer runoff eutrophies rivers and lakes and creates dead zones in offshore waters. Dams generate electric power and reduce flooding, but the World Commission on Dams (2000) found that on balance, the impacts of large dams "are more negative than positive and, in many cases, have led to irreversible loss of species and ecosystems" because they halt the accumulation of fertile silt on downstream floodplains, cause deltas such as those of the Nile and Mississippi to subside, disrupt the lifecycle of riverine species such as salmon, and may contribute more to GHG emissions from the decay of inundated vegetation than they save through the generation of hydropower.

Expanding the boundary of the model to include the possibility of unintended harm from technological innovation creates the reinforcing "Technological Nightmare" feedback R3: as before, scarcity and environmental degradation caused by growth in human activity lead to higher prices for the affected resources, along with government and social responses. The resulting technological solutions have some benefits, but also lead, usually after delays, to harms that accelerate the erosion of the carrying capacity, leading to greater scarcity and new environmental problems, triggering still higher prices and still greater attempts to find a technological solution, in a vicious cycle.

How does the inclusion of the price system and technological innovation alter the dynamics of the system? Clearly, if markets are imperfect, if the delays in the social, economic, and technical response to scarcity and environmental degradation are long, or if the harmful "side effects" of technology dominate the benefits, then the result will still be overshoot and collapse: technological solutions will be "too little, too late" or will actually worsen the problem.

More interesting, what happens if markets work well, the delays in innovation are short and unintended harms absent? Will the result be, as Simon (1996) suggests, "that we end up better off than if the original shortage problems had never arisen?" No. Simon's conclusion would hold if the level of human activity that created the shortage were exogenous, unaffected by the availability of resources. But there is feedback: technological solutions increase the adequacy of resources and lower prices, enabling growth to proceed at a higher rate. The population and economy grow still further, reducing the adequacy of resources directly (loop B1) and indirectly, by increasing the rate of consumption and degradation of the carrying capacity (loop B2). The result: society is once again pushed up against one environmental limit or another. If markets and technology once again succeed in addressing those new limits, then human activity grows still further. To avoid involuntary limits to growth through technology, one must assume that technological solutions to all resource and environmental problems can be found, that the costs of these solutions are so low that they do not constrain economic growth; that the delays in the recognition of problems, in the innovation process, in adoption and diffusion of new technologies, and in the response of the carrying capacity are always short; that these solutions never generate significant unintended harms; and that technological solutions keep the carrying capacity from crossing important environmental tipping points. Most important, one must believe that, eventually, both population growth and people's desire for more income and wealth will end. If any of these conditions fail, then the carrying capacity will eventually drop, leading to overshoot and decline.

As is typical in complex systems, much of the debate between environmentalists and technological optimists focuses on the symptoms of the problem: resources and the resiliency of the environment. How much oil is there? How much nuclear and solar power can be produced? How much copper can be mined, and at what costs? And so on. That debate misses the point. It makes no difference how large the resource base is: to the extent technology and markets alleviate scarcity today, the result is more growth tomorrow, until the resource is again insufficient, some other resource becomes scarce, or some other environmental problem arises. Solve these, and growth continues until some other part of the carrying capacity is lost, some other limit reached. As long as growth is the driving force there can be no purely technological solution to the problem of scarcity. The high leverage points lie elsewhere, in the forces that cause population and economic growth. Even with significant potential for new technical solutions, a prosperous and sustainable future can only be built if growth of both population and material throughput cease voluntarily, before growth is stopped involuntarily by scarcity or environmental degradation (the balancing "Voluntary Limits" loop B5 in Fig. 10).

Fortunately, population growth may stabilize "voluntarily." Contrary to, or perhaps in response to, fears of a population explosion (as with China's one child policy), fertility and population growth have fallen in recent decades, a process known as the demographic transition. Prior to industrialization and economic development, both crude birth and death rates were high and variable. Infant mortality was high and life expectancy low, so women began to bear children early and bore many to ensure that a few would survive. Population growth was slow. The introduction of modern sanitation, public health systems, transportation, and medical care raised life expectancy. Death rates fell. Population growth accelerated dramatically. Eventually, however, birth rates also fell: more children survived to adulthood, so women did not need to bear as many to achieve their desired family size. Further, the economic benefits of children fall with industrialization because they enter the labor force later than in traditional agricultural societies and must be supported by their parents longer. Desired family size drops. Birth rates fall only slowly, however: family size, marriage age, and other determinants of fertility are strongly embedded in traditional culture, religious norms, and other social structures. Today, fertility and mortality are in rough balance in a number of developed economies. In still-developing nations, death rates have fallen, but birth rates continue to lag behind, so population growth remains high.

The UN's 2010 projections assume that the demographic transition will continue throughout the world, including the least developed nations, nearly stabilizing population by 2100 at more than ten billion. But even if population growth eventually stops, human impact on the environment will not: economic growth is projected to continue, and as the production of goods and services per capita rises, so too will the impact of each person. Endless economic growth is envisioned, indeed desired, by many technological optimists, for example, "The standard of living has risen along with the size of the world's population since the beginning of recorded time. There is no convincing economic reason why these trends toward a better life should not continue indefinitely" (Simon 1996). Endless economic growth is impossible, however. Resource use and environmental impact per person cannot fall to zero—people need a minimum amount of food, living space, energy, and waste disposal capacity, among other resources. The only way total impact can stabilize is for both population and economic output per person to stabilize. Yet no nation on earth seeks to end the growth of its economy.

These limits are not theoretical and distant. The best available science shows that humanity has already overshot the carrying capacity of the earth. The human family is projected to grow by more than two billion in less than 40 years, and most of these people will be born in the poor nations of the world. All of them, and the billions who today live in poverty, legitimately aspire to live the way those in developed nations do today, dramatically increasing the demand for resources and production of wastes. To take but one example, if the projected world population of 9.3 billion in 2050 lived like Americans did in 2008, there would be 7.8 billion vehicles on the world's roads, consuming 382 million barrels of oil per day, more than five times *total* world oil production in 2008, and spewing 60 billion tons of CO_2 per year into the atmosphere, nearly double *total* world CO_2 emissions from fossil fuels in that year.

Avoiding a decline in population or economic output will require all the technical and social innovation we can muster. We urgently require technologies to replace fossil fuels, cut GHG emissions, boost food production without use of toxic pesticides, create new antibiotics as pathogens evolve resistance, end deforestation, and protect biodiversity. We urgently need to create more effective markets to capture environmental and social externalities, providing businesses and consumers with the price signals that will drive innovation and stimulate efficient use of resources. We urgently require better science, environmental monitoring, and product testing so that the new technologies we develop and deploy do not create unintended consequences that worsen the very problems we seek to solve. But while necessary, technological innovation alone is not sufficient. We must also ask how much is enough? How much wealth, how much consumption do we each require?

With a few important exceptions (the work of Herman Daly and colleagues, e.g., Daly and Townsend 1993; see also Princen et al. 2002; Meadows et al. 2004; DeGraaf et al. 2005; Whybrow 2005; Victor 2008; Schor 2010), most of the research, teaching, and popular discourse on sustainability continues to focus on technological solutions—more energy, more resources, more efficient eco-friendly growth—while the actual leverage point—voluntarily limiting our consumption—remains largely undiscussable, particularly among our business and political leaders. That conversation is not an easy one. For many years I have asked my students "how much is enough" (Table 4).

Table 4 How much is enough?

1. **How much would you need to spend each year to be happy?** That is, how much consumption would be enough to satisfy you?

   Consumption spending here means expenditure to provide for the lifestyle you wish to have, including food, clothing, housing and furnishings, education, health care, travel, entertainment, and all other expenditures on goods and services.

   Consumption does not include charitable giving, but only what you spend on yourself and your immediate family (spouse and children).

   Consumption does not include saving or investment (e.g., to build future income for retirement or to leave an estate to your heirs).

   Consumption does not include payment of taxes, but only the cost of the goods and services you purchase.

   One way to think about this is to imagine that you are guaranteed an annuity for life, exempt from income and other taxes, and automatically adjusted for inflation. Under those conditions, what annuity would you require?

   Amount per year in US$:________________

   Select one of the following

   ☐ I need at least this much, but more is always better. ☐ This much would be enough.

2. **Imagine the following two worlds**

   World 1: last year you earned $150,000. This year you earned $200,000.

   World 2: last year you earned $200,000. This year you earned $150,000.

   The prices of all goods and services are the same in both worlds. The environmental impact of each world is the same, and, through use of green technologies, negligible.

   Which world do you prefer? ☐ World 1 ☐ World 2

3. **Imagine the following two worlds**

   World 1: you earn $150,000 per year. Everyone else earns $75,000 per year.

   World 2: you earn $250,000 per year. Everyone else earns $500,000 per year.

   The prices of all goods and services are the same in both worlds. The environmental impact of each world is the same, and, through use of green technologies, negligible.

   Which world do you prefer? ☐ World 1 ☐ World 2

Typical of results with diverse groups, the median response to Question 1 of 109 students at the MIT Sloan School of Management (primarily MBA students) in the fall term of 2010 was $200,000/year. The mean was over $2 million/year, skewed by 14% whose responses were $1 million/year or more. These responses are deeply disturbing. *Spending* $2 million (or even the median estimate of $200,000) per year dwarfs mean per capita *income* in the USA, with GDP per capita of $46,650, much less the GDP per capita of most African nations, which remains less than $1,000/year (2008$; see hdr.undp.org/en/statistics). The urge for *more* is strong: about half the students chose "more is always better." Among 156 similar students in my courses on sustainability in 2009 and 2010, an overwhelming 83% preferred to earn more next year than this year (Question 2). These students know they would be better off taking the extra $50,000 up front (the net present value of World 2 is higher: you could spend the same as in World 1, invest the extra $50,000 and have more than $200,000 the second year). When asked why they chose the less valuable option, many reported that it would be hard to reduce their standard of living if their income dropped, though there is nothing in the question that requires them to spend more in year 1 than year 2. Quite a few said they would feel they had somehow failed, would feel less worthy as a person, if their income dropped. Even more disturbing, 58% preferred to earn less each year—as long as they make *more* than everyone else (Question 3). People tend to judge how well off they are by social comparison, and are less happy when others have more than they do (Layard 2005). Of course, this is a zero sum game: everyone cannot be richer than everyone else. The struggle to keep up with the Jones' creates an obvious reinforcing feedback, an arms race of conspicuous consumption, egged on by advertising, enabled by borrowing and requiring us to work ever harder. As we do, we have less time for what matters most: exercising and staying healthy, spending time with family and friends, developing intellectually and spiritually, helping those in need.

Until we learn to end the quest for more—more income, more wealth, more consumption, more than last year, more than our neighbors—then a healthy, prosperous, and sustainable society cannot be created no matter how clever our technology. Innovation simply lets us grow until one or another limit to growth becomes binding. We are not accustomed to asking "how much is enough," uncomfortable connecting abstract debates about growth and scarcity with the way live, with our personal responsibility to one another and to future generations. Research, teaching, and action to promote sustainability must grapple with these issues if we are to fulfill Gandhi's vision of a world in which "there is enough for everyone's need but not for everyone's greed."

Teaching and Learning

Creating a sustainable science of sustainability requires changes not only in what we study but how we teach it and communicate it to policymakers, civic and business leaders, the media, and the public at large. How can people develop their systems thinking capabilities? How can people, including scholars, policymakers, business and civic leaders, educators, media, and members of the public, come to deeper understanding of complex systems and the high leverage points needed to build a sustainable society?

There is no learning without feedback, without knowledge of the results of our actions. Traditionally, scientists generate that feedback through controlled experimentation, an iterative process through which intuitions are challenged, hypotheses tested, insights generated, new experiments run. But experiments are impossible in many of the most important systems, including many critical for sustainability. When experiment is impossible, scientists rely on models and simulations, which enable controlled experimentation in virtual worlds. Simulation models have long been central in sustainability and environmental research, from models of the percolation of toxics through groundwater to ecosystem dynamics to climate change. But simulations are not only useful in knowledge creation. They must also become a main tool in knowledge communication. As scientists, we develop our understanding through an iterative, interactive learning process of experimentation in both the real world and the virtual world of simulations (Sterman 1994). But all too often we then turn around and tell the results to policymakers, our students and the public through reports, presentations, and lectures. We should hardly be surprised when these people, excluded from the process of discovery—unable to assess the evidence on their own, and presented with conclusions that conflict with deeply embedded mental models—become confused, ignore the results, and challenge our authority.

Interactive, transparent simulations for learning, rigorously grounded in and consistent with the best available science, are now available for a wide range of sustainability issues. To enable learning, these "management flight simulators" must give people control over assumptions and scenarios, encourage wide-ranging sensitivity analysis, and run nearly instantly online or on ordinary desktop and laptop computers, so that people receive immediate feedback. Examples range from simple models to help people develop their understanding of stocks and flows (e.g., http://bit.ly/atmco2), to Fishbanks, a simulation of the tragedy of the commons and "Eclipsing the Competition," a simulation of the solar photovoltaic industry (both available at http://bit.ly/mstir), to the C-ROADS climate policy simulation (http://climateinteractive.org) used not only for education but by policymakers and climate negotiators.

When experimentation is too slow, too costly, unethical, or just plain impossible, when the consequences of our decisions take years, decades, or centuries to manifest, that is, for most of the important issues we face in building a sustainable world, simulation becomes the main—perhaps the only—way we can discover *for ourselves* how complex systems work, where the high leverage points may lie. The alternative is rote learning based on the authority of a consultant, teacher, or textbook, a method that dulls creativity, stunts the very systems thinking and scientific reasoning skills we need, and thwarts implementation.

Conclusion

Our society is not sustainable. We harvest renewable resources faster than they regenerate, we create wastes and pollution faster than ecosystems can break them down into harmless substances, and we are, in the words of former US President George W. Bush, "addicted to oil" and other nonrenewable resources. Despite recycling, energy efficient light bulbs and other eco-friendly practices, these imbalances are getting worse, with world population projected to grow by more than two billion in just 40 years, and economic growth doubling the size of the real economy every 20 years.

Here I argued that the growing sustainability movement is neither effective nor itself sustainable. Most efforts by firms, individuals, and governments in the name of sustainability are directed at the symptoms of the problem rather than the cause. Many lead to improvement locally and in the short run at the expense of others and future generations. Such policy resistance is not unique to sustainability but common in complex systems at all scales, and arises from widespread failure of systems thinking. Our mental models have narrow boundaries and short time horizons. We commonly frame the sustainability challenge as a conflict, in which the economy, social justice, and the environment fight for primacy, when the economy and society are embedded in the ecosystems upon which all life depends. A healthy society and prosperous economy depend on a healthy environment, and the health of the environment depends on a healthy society and economy that fulfills people's needs. To move beyond slogans about interconnectedness and systems, however, we need to develop specific tools and methods to develop our systems thinking capabilities, methods that avoid both self-defeating pessimism and mindless optimism, while remaining true to scientific method and ecological realities.

Some may object that the call for systems thinking is futile, that most people, including our leaders, are incapable of understanding the complexities of the economy and environment. They caricature systems thinking as hoping that if we just understood how everything is connected to everything else we would all somehow stop living unsustainably, then criticize that cartoon as naïve. Cynics claim that humans are fundamentally selfish, greedy, and shortsighted. To the contrary, the problem is not the few who are truly uncaring. It is the failure of even those who sincerely care to understand the urgent need for action created by the long time delays, feedbacks, nonlinearities, and other characteristics of complex systems. It's the vast mass of us mindlessly going about our everyday business, oblivious to the consequences of our actions, our behavior shaped by the systems in which we are embedded, systems we created and that only we can change. It's the belief that we are helpless, that nothing we do makes a difference, that change is not possible—a belief that alienates and discourages us but that we also find comforting because it absolves us from the responsibility to act.

Overcoming policy resistance and building a sustainable world requires that we develop a meaningful systems thinking capability so that we can learn, collectively, how we can promote the common good. It requires an unswerving commitment to the rigorous application of the scientific method, and the inquiry skills we need to expose our hidden assumptions and biases. It requires us to face the ethical issues raised by growth and inequality, to speak, unafraid, of our deepest aspirations for a just, equitable and sustainable world. It requires that we listen with respect and empathy to others. It requires the humility we need to learn and the courage we need to lead, though all our maps are wrong. If we devote ourselves to that work we can move past denial and despair to create the future we truly desire—not just for us, but for our children. Not just for our children, but for all the children.

References

Adelman MA (1993) The economics of petroleum supply. The MIT Press, Cambridge, MA
Alley R, Marotzke J, Nordhaus W et al (2003) Abrupt climate change. Science 299:2005–2010
Axelrod R (1976) The structure of decision: the cognitive maps of political elites. Princeton University Press, Princeton, NJ
Bahn P, Flenley J (1992) Easter Island, Earth Island. Thames and Hudson, London, UK
Bäumer D, Vogel B (2007) An unexpected pattern of distinct weekly periodicities in climatological variables in Germany. Geophy Res Lett 34:L03819
Booth Sweeney L, Sterman JD (2000) Bathtub dynamics: initial results of a systems thinking inventory. System Dyn Rev 16:249–294
Booth Sweeney L, Sterman JD (2007) Thinking about systems: students' and their teachers' conceptions of natural and social systems. System Dyn Rev 23:285–312
Buehler R, Griffin D, Ross M (2002) Inside the planning fallacy: the causes and consequences of optimistic rime predictions. In: Gilovich T, Griffin D, Kahneman D (eds) Heuristics and biases. Cambridge University Press, Cambridge, UK
Cerveny R, Balling R (1998) Weekly cycles of air pollutants, precipitation, and tropical cyclones in the coastal NW Atlantic region. Nature 394:561–563
Intergovernmental Panel on Climate Change (2007) Climate change 2007: the physical science basis. Cambridge University Press, Cambridge, UK. http://www.ipcc.ch
Cronin M, Gonzalez C, Sterman J (2009) Why don't well-educated adults understand accumulation? A challenge to researchers, educators, and citizens. Organ Behav Hum Decis Proc 108:116–130
Daly H (1991) Steady-state economics, 2nd edn. Island Press, Washington, DC
Daly H, Townsend K (1993) Valuing the earth: economics, ecology, ethics. MIT Press, Cambridge, MA
DeGraaf J, Wann D, Naylor T (2005) Affluenza: the all-consuming epidemic, 2nd edn. Berrett-Kohler, San Francisco, CA
Diamond J (2005) Collapse: how societies choose to fail or succeed. Viking, New York, NY
Dörner D (1980) On the difficulties people have in dealing with complexity. Simulations Games 11:87–106
Dörner D (1996) The logic of failure. Metropolitan Books/Henry Holt, New York, NY
FAO (2008) The state of world fisheries and aquaculture 2008. UN Food Agric Organ, Rome
Faro D, McGill A, Hastie R (2010) Naïve theories of causal force and compression of elapsed time judgments. J Pers Soc Psychol 98:683–701
Festinger L (1957) A theory of cognitive dissonance. Row, Peterson, Evanston, IL
Fiddaman T (2002) Exploring policy options with a behavioral climate-economy model. System Dyn Rev 18:243–267

Fong I, Drlica K (2003) Reemergence of established pathogens in the 21st century. Kluwer/Plenum, New York, NY
Forrester JW (1969) Urban dynamics. Pegasus Communications, Waltham, MA
Forrester JW (1971a) Counterintuitive behavior of social systems. Technol Rev 73:52–68
Forrester JW (1971b) World dynamics. Pegasus Communications, Waltham, MA
Forster P, Solomon S (2003) Observations of a 'weekend effect' in diurnal temperature range. PNAS 100:11225–11230
Hardin G (1968) The tragedy of the commons. Science 162:1243–1248
Hoffman M, Hilton-Taylor C, Angulo A et al (2010) The impact of conservation on the status of the world's vertebrates. Science 300:1503–1509
Horn S, Sharkey P, Tracy D et al (1996) Intended and unintended consequences of HMO cost-containment strategies: results from the managed care outcomes project. Am J Manag Care 2:253–264
Huxley T (1883) Inaugural address. Fisheries exhibition, London http://aleph0.clarku.edu/huxley/SM5/fish.html
Kahneman D, Diener E, Schwarz N (1999) Well-being: the foundations of hedonic psychology. Russell Sage, New York, NY
Kirch P (1997) Microcosmic histories: island perspectives on 'global' change. Am Anthropol 99:30–42
Kurlansky M (1997) Cod: a biography of the fish that changed the world. Penguin, New York, NY
Laux P, Kunstmann H (2008) Detection of regional weekly weather cycles across Europe. Environ Res Lett 3:044005
Layard R (2005) Happiness: lessons from a new science. Penguin Press, New York, NY
McKibben B (2010) Earth. Times Books/Henry Holt, New York, NY
Meadows D, Randers J, Meadows D (2004) The limits to growth: the thirty year update. Chelsea Green, White River Junction, VT
Meehl GA, Washington WM, Collins WD et al (2005) How much more global warming and sea level rise? Science 307:1769–1772
Moxnes E (2000) Not only the tragedy of the commons: misperceptions of feedback and policies for sustainable development. System Dyn Rev 16:325–348
National Research Council (2010) Hidden costs of energy: unpriced consequences of energy production and use. National Academies Press, Washington, DC
Nordhaus W (1992) An optimal transition path for controlling greenhouse gases. Science 258:1315–1319
O'Neill B, Oppenheimer M (2002) Dangerous climate impacts and the Kyoto protocol. Science 296:1971–1972
Ostrom E (2010) Beyond markets and states: polycentric governance of complex economic systems. Am Econ Rev 100:641–672
Palumbi S (2001) Humans as the world's greatest evolutionary force. Science 293:1786–1790
Pauly D (1995) Anecdotes and the shifting baseline syndrome of fisheries. Trends Ecol Evol 10:430
Pereira H, Leadeley P, Proenca V et al (2010) Scenarios for global biodiversity in the 21st century. Science 330:1496–1500
Plous S (1993) The psychology of judgment and decision making. McGraw Hill, New York, NY
Princen T, Maniates M, Conca K (2002) Confronting consumption. MIT Press, Cambridge, MA
Repenning N, Sterman J (2001) Nobody ever gets credit for fixing problems that never happened: creating and sustaining process improvement. California Management Review 43(4):64–88.
Rockström J et al (2009) A safe operating space for humanity. Nature 461:472–475
Rosenberg A, Bolster W, Alexander K et al (2004) The history of ocean resources: modeling cod biomass using historical records. Frontiers Ecol Environ 3:84–90
Rosenthal R, Jacobson L (1968) Pygmalion in the classroom: teacher expectation and pupil's intellectual development, Expanded edn. Irvington Publishers/Ardent Media, Norwalk, CT
Schor J (2010) Plenitude: The new economics of true wealth. New York: Penguin.
Senge P, Smith B, Schley S, Laur J (2008) The necessary revolution. Doubleday, New York, NY

Simon J (1996) The ultimate resource 2. Princeton University Press, Princeton, NJ
Solomon S, Plattner G-K, Knutti R, Friedlingstein P (2009) Irreversible climate change due to carbon dioxide emissions. PNAS 106:1704–1709
Stachowicz J, Terwin J, Whitlatch R, Osman R (2002) Linking climate change and biological invasions: ocean warming facilitates nonindigenous species invasions. PNAS 99:15497–15500
Steadman D (1995) Prehistoric extinctions of Pacific island birds: biodiversity meets zooarchaeology. Science 267:1123–1131
Sterman JD (1989) Modeling managerial behavior: misperceptions of feedback in a dynamic decision making experiment. Manag Sci 35:321–339
Sterman JD (1994) Learning in and about complex systems. System Dyn Rev 10:291–330
Sterman JD (2000) Business dynamics: systems thinking and modeling for a complex world. Irwin/McGraw-Hill, New York, NY
Sterman JD (2002) All models are wrong: reflections on becoming a systems scientist. System Dyn Rev 18:501–531
Sterman JD (2008) Risk communication on climate: mental models and mass balance. Science 322:532–533
Sterman JD (2010) Does formal system dynamics training improve people's understanding of accumulation? System Dyn Rev 26:316–334
Sterman JD, Booth Sweeney L (2007) Understanding public complacency about climate change: adults' mental models of climate change violate conservation of matter. Clim Change 80:213–238
Suzuki D (2007) The sacred balance: rediscovering our place in nature, 3rd edn. Greystone Books, New York, NY
Tengs T, Ahma S, Savage J et al (2005) The AMA proposal to mandate nicotine reduction in cigarettes: a simulation of the population health impacts. Prevent Med 40:170–180
Thomas C, Cameron A, Green R et al (2004) Extinction risk from climate change. Nature 427:145–148
United Nations (2011) World population prospects: the 2010 revision. http://esa.un.org/unpd/wpp/index.html. Accessed 29 March 2012
US Forest Service (2003) Influence of forest structure on wildfire behavior and the severity of its effects. http://www.fs.fed.us/projects/hfi/science.shtml. Accessed 29 March 2012
Victor P (2008) Managing without growth. Edward Elgar, Cheltenham, UK
Wackernagel M, Schulz N, Deumling D et al (2002) Tracking the ecological overshoot of the human economy. PNAS 99:9266–9271
Wagenaar W (1978) Intuitive prediction of growth. In: Burkhardt D, Ittelson W (eds) Environmental assessment of socioeconomic systems. Plenum, New York, NY
Wagenaar W, Sagaria S (1975) Misperception of exponential growth. Percept Psychophys 18:416–422
Webster D (2002) The fall of the ancient Maya. Thames and Hudson, New York, NY
Whybrow P (2005) American mania: when more is not enough. WW Norton, New York, NY
Wigley T (2005) The climate change commitment. Science 307:1766–1769
Wilmshurst J, Hunt T, Lipo C, Anderson A (2011) High-precision radiocarbon dating shows recent and rapid initial human colonization of East Polynesia. Proceedings of the National Academy of Sciences 108(5):1815–1820
Wilde G (2002) Target risk 2: a new psychology of safety and health. PDE Publications, New York, NY
World Bank (2010) http://databank.worldbank.org. Accessed 29 March 2012
World Commission on Dams (2000) Dams and development. Earthscan, London, UK

A Landscape Approach for Sustainability Science

Jianguo (Jingle) Wu

Abstract The global life-support system for humans is in peril but no alternative to achieving sustainability is desirable. In response to this challenge, sustainability science has emerged in recent decades. In this chapter, I argue that to advance sustainability science a landscape approach is essential. Landscapes represent a pivotal "place" in the place-based research and practice of sustainability. Landscape ecology, as the science and art of studying and influencing the relationship between spatial pattern and ecological processes at different scales, can play a critically important role in the development of sustainability science. Global sustainability cannot be achieved without most, if not all, landscapes being sustainable. As landscapes are spatial units in which society and nature interact and co-evolve, it is more useful and practical to define landscape sustainability based on resilience rather than stability. Furthermore, the development of landscape sustainability measures can be facilitated by integrating landscape pattern metrics and sustainable development indicators.

Keywords Landscape sustainability • Sustainability science • Human–nature interactions • Sustainability metrics

Introduction

> This traditional dichotomy of humanity-vs.-nature is false and dangerous. On the one hand, it perpetuates our destructive mishandling of the biosphere. On the other hand, it scants the self-understanding that *Homo sapiens* needs to settle down on our home planet, hence as a

J. (Jingle) Wu (✉)
School of Life Sciences and Global Institute of Sustainability, Arizona State University, P.O. Box 874501, Tempe, AZ 85287-4501, USA
e-mail: Jingle.Wu@asu.edu

M.P. Weinstein and R.E. Turner (eds.), *Sustainability Science: The Emerging Paradigm and the Urban Environment*, DOI 10.1007/978-1-4614-3188-6_3,

> prerequisite to survival. Nature, to put the matter as succinctly as possible, is part of us, and we are part of nature
>
> E. O. Wilson (2007)

Human activities have transformed ecosystems and landscapes profoundly around the world, and the entire biosphere has been influenced in fundamental ways (Kareiva et al. 2007; Wu 2008). In search of solutions to a myriad of environmental and social problems, sustainability has become the defining theme of our time (Kidd 1992; Kates et al. 2005; Du Pisani 2006). Sustainability concerns our ability to maintain coupled human–environmental systems in a desirable state for multiple generations in the face of anthropogenic and environmental perturbations and uncertainties. To meet the needs and challenges of sustainability, a new kind of science has emerged in the past 2 decades—sustainability science—that focuses on the dynamic interactions between society and nature (Kates et al. 2001; Clark and Dickson 2003; Clark 2007; Weinstein 2010). The ultimate goal of sustainability science is not just to understand the human–environment relationship, but rather to improve it through producing knowledge and solutions for management, planning, and policy that are needed for a transition toward sustainability. Thus, sustainability science has to be integrative and pluralistic. As Reitan (2005) put it, sustainability science is "the cultivation, integration, and application of knowledge about Earth systems gained especially from the holistic and historical sciences (such as geology, ecology, climatology, oceanography) coordinated with knowledge about human interrelationships gained from the social sciences and humanities."

Three salient characteristics seem essential to sustainability science. First, sustainability science is multidimensional and transdisciplinary. This means that it deals with the nexus of environment, economy, and society, with integrative approaches cutting across natural and social sciences (Kates et al. 2001; Wu 2006). Second, sustainability is multiscaled and hierarchically linked in space and time. Sustainability can be defined at any scale from a local site (e.g., a household or a biological community) to the entire globe, although only the local, regional, and global scales have frequently appeared in the sustainability literature. Regardless of its specific definition, the sustainability of a system varies with scale in space and time and, as in other hierarchical systems, processes at different scales are linked in both bottom-up and top-down directions (O'Neill et al. 1986; Wu and Loucks 1995; Wu 1999). So, we not only need to ask the questions of what to sustain and what to develop, but also over what area and for how long. Third, sustainability science emphasizes use-inspired, placed-based research. Real-world problems occur in "places" and we must go "places" to understand and solve them. As Kates (2003) stated, "Sustainability science is regional and place based. ..., it is in specific regions, with distinctive social, cultural, and ecological attributes, that the critical threats to sustainability emerge and in which a successful transition needs to be based." This does not simply mean that sustainability science is an "applied" discipline; it is a transdisciplinary enterprise that bridges the traditional divide between basic and applied research by focusing on use-inspired and place-based problems (Clark 2007).

If the "place" in sustainability science is essential, then what is the "place?" Kates (2003) asked the same question: "What constitutes an appropriate classification of place? In part, the distinction is surely one of scale, and a grand query of sustainability will be these scale relationships." So, defining "place" in sustainability research is critically important to effectively dealing with the issues of scale and hierarchical linkages as well as integrating the environmental, economic, and social dimensions. In this chapter, therefore, I argue that, although "place" can be defined at any scale, "landscape" represents the most pivotal scale for sustainability research. I will first discuss what landscape is and then present a landscape perspective on sustainability, including conceptual and practical considerations.

Landscape as a Place for Sustainability

The term, "landscape," is a key concept in a number of fields, from social to geographical and ecological sciences. Because of the plurality of its origins and interpretations, landscape has acquired various connotations. The same word may refer to a natural landscape, a cultural landscape, a political landscape, an economic landscape, a mental landscape, an adaptive landscape, a landscape view, landscaping, or landscape painting (Fig. 1). "Landscape gives identity to place" and "landscape is where past and present meet" (Phillips 2007). Human geographers may think of landscape as "a work of human labor" or "an activity" of dynamic interactions between people and place (Mitchell 2000). As such, a landscape may also be

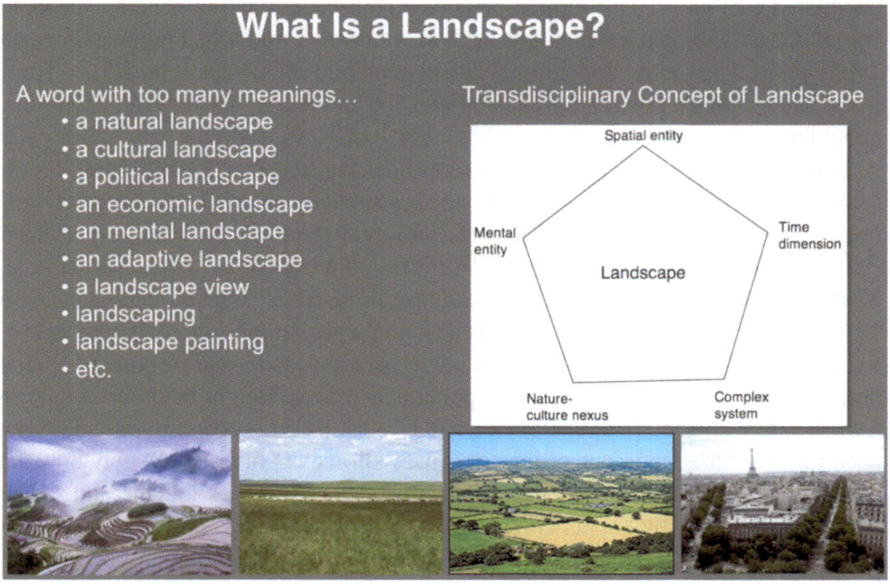

Fig. 1 A transdisciplinary concept of landscape based on discussion in Tress and Tress (2001)

considered as "a form of ideology" or "a way of carefully selecting and representing the world so as to give it a particular meaning," and thus it can be "an important ingredient in constructing consent and identity" (Mitchell 2000).

Geography has a long history of studying human–environment relationships, and a number of perspectives have been developed, with different research cores and methodologies that reflect a varying degree of affinity to either natural sciences or humanities (Turner 1997). The term, "cultural landscape," has been a fundamental concept in geography since its first use in Germany in the 1890s, referring to landscape modified by human activity as opposed to the primeval natural landscape. In his seminal publication, "The morphology of landscape," Sauer's (1925) defined cultural landscape as landscape "fashioned from a natural landscape by a cultural group." Since the 1960s, the concept of cultural landscape has been widely used in human geography (of which cultural geography is a part), anthropology, environmental management, and other related fields (Sauer 1925; Webb 1987). One of the major factors that contributed to the recent popularity of the term was the adoption of cultural landscapes in the International Convention for the Protection of the World's Cultural and Natural Heritage (or the World Heritage Convention) by the United Nations Educational, Scientific, and Cultural Organization (UNESCO) in 1992.

In the field of landscape ecology, the word "landscape" has different meanings. The main differences among various definitions reflect the different spatial scales at which a landscape is perceived and the different aspects of a landscape are emphasized (Wu and Hobbs 2007). For example, Forman and Godron (1986) defined landscapes as kilometers-wide geographic areas, which corresponds to the "human-perceived" landscape. This is the scale at which the field of landscape ecology was originally developed in Europe, and at which most landscape studies have been conducted ever since. This human-perceived landscape scale, in general, seems to coincide well with geographic units such as watersheds and urban regions (Forman 1995), as well as spatial domains of human perception (Gobster et al. 2007). Thus, it resonates with the public, the decision-makers, and researchers who are conscious about the environmental setting in which they live, work, and play.

However, most landscape ecologists consider landscape as a multiscale or hierarchical concept, meaning that a landscape is a spatially heterogeneous area of various sizes, depending on the subject of study and the research questions at hand (Urban et al. 1987; Wu and Levin 1994; Pickett and Cadenasso 1995; Turner et al. 2001). In this case, landscape is an "ecological criterion" (Pickett and Cadenasso 1995), and its essence does not lie in its absolute scale but in its internal heterogeneity. Different plant and animal species perceive, experience, and respond to spatial heterogeneity at different scales, and patterns and processes in landscapes tend to have different characteristic scales (Wu and Loucks 1995). Apparently, one does not need to consider a landscape of tens of square kilometers in order to study how the spatial patterning of grasses affects the movement of beetles (Wiens and Milne 1989) or is affected by gophers (Wu and Levin 1994).

Tress and Tress (2001) proposed a "transdisciplinary landscape concept" of landscape that encompasses five dimensions: (1) landscape as a spatial entity, (2) landscape as a mental entity, (3) landscape as a temporal dimension, (4) landscape as a nexus of nature and culture, and (5) landscape as a complex system (Fig. 1).

This is probably the most comprehensive of all landscape definitions. It is pertinent to cultural landscapes and implies a spatial scale that must be large enough to encompass key environmental, economic, and social processes that determine the sustainability of a place of interest. Following this notion, a landscape is more than just a geographic space as it has contents; a landscape is not merely a container as it shapes and is shaped by what it contains; a landscape is not just an environment modified by humans as it is a holistic system in which nature and culture co-evolve. Landscapes are endowed with and to foster the development of cultures, legacies, and stories. Today, most landscapes are "cultural landscapes" in which people interact or interfere with nature, whereas "natural landscapes" are found only as "islands" in an expanding sea of human land uses.

Scholars who study landscapes from either ecological or cultural perspectives seem to agree on the importance of landscape as an operational scale in sustainability research. For example, Forman (1990) argued that human-perceived landscapes, as a spatial scale for sustainable development, have significant advantages over broader scales such as the continent. Forman (1995) further pointed out that to deal with the "the paradox of management" (i.e., actions tend to be more effective at local scales whereas success often needs to be achieved at broader scales), "management and planning for sustainability at an intermediate scale, the landscape or region, appears optimum." The ordinary elements of human landscapes (e.g., forests, cropfields, urban land covers, residential areas, streams, and streets) also resonate well with human perception and thus facilitate decision-making (Nassauer 1997; Gobster et al. 2007). From a cultural geographer's perspective, Phillips (1998) commented that cultural landscapes are "places which can demonstrate that talk of sustainable development can be more than rhetoric."

In summary, the landscape represents a basic spatial unit of society–nature interactions and ought to be the primary "place" of study in sustainability science. It provides a multidimensional meeting ground for ecologists, geographers, social scientists, planners and designers, policy-makers who are all crucial to sustainability research. The landscape is large enough to incorporate key environmental, economic, and social processes and small enough to allow for in-depth and mechanistic studies that produce locally actionable solutions to sustainability problems.

Culture–Nature Relationship in Landscapes

As discussed in the previous section, landscapes, as commonly used in ecology and geography, represent a pivotal scale and place for sustainability. Beyond that, landscapes often shape, and are shaped by, the way we interact with nature. So, the structure and functionality of a particular landscape are reflective of the past and current relationships between humans and the environment in that region. As sustainability science is focused on the dynamic relationship between people and nature, landscapes have stories to tell, lessons to be learnt, and opportunities to offer.

Our perception and understanding of the relationship between people and nature in landscapes are often influenced by our philosophical roots and cultural traditions.

These traditions represent the historical antecedent to the modern technocratic approach to social and economic development. As Phillips (1998) stated: "The separation of culture and nature—of people from the environment which surrounds them—which has been a feature of western attitudes and education over the centuries, has blinded us to many of the interactive associations which exist between the world of nature and the world of culture." In contrast, one of the most influential Asian philosophies on the relationship between culture and nature—the "Unity of Man and Nature" ("天人合一")—advocates that people should be in harmony with the rhythms of nature (Chen and Wu 2009). Unity of Man and Nature was the quintessential theme shared by dominant ancient Asian cultures and has been described as the greatest contribution of Chinese culture to humanity (Ji 2007). While the contemporary roots of the concept of sustainability include the ideas of carrying capacity, biosphere conservation, and limits to growth (Kidd 1992), Unity of Man and Nature is one of its most relevant ancient philosophical roots (Fig. 2).

Fig. 2 Some key characteristics of sustainability science whose conceptual roots can be traced back to the ancient Chinese philosophy—the unity of man and nature. The focus of sustainability science is the dynamic relationship between nature and society, examined simultaneously from environmental, economic, and social dimensions at local, regional, and global scales. This transdisciplinary science is multiscale, multidimensional, and use-inspired and place-based. The unity of man and nature is its ultimate goal as well as its ancient philosophical root

The theme of Unity of Man and Nature is evident in some seminal works by western environmental scientists and landscape architects. For example, in his landmark book, "A Sand County Almanac," the conservation ecologist Aldo Leopold (1949) advocated for "a state of harmony between man and land," and a new land ethic that "changes the role of *Homo sapiens* from conqueror of the land-community to plain member and citizen of it." The landscape architect Ian McHarg (1969) developed the "design with nature" approach, based on the premise: "Let us then abandon the simplicity of separation and give unity its due. Let us abandon the self-mutilation which has been our way and give expression to the potential harmony of man-nature." Tress et al. (2001) argue that "The perceived division between nature and culture has dominated the academic world," and "In the case of landscapes, this divide is counter-productive and must be overcome since all landscapes are multidimensional and multifunctional."

To unite culture with nature in landscapes and to advance a landscape-based science of sustainability, four principles articulated by Nassauer (1995) should be borne in mind when we formulate our research questions: (1) human perception, cognition, and values of the landscape directly affect, and are affected by, the landscape; (2) cultural conventions have profound influences on both human-dominated and apparently natural landscapes; (3) cultural concepts of nature may differ from scientific concepts of ecological function; and (4) the appearance of landscapes communicates cultural values. In our attempt to integrating culture and nature in landscapes, we need to fully recognize the necessity and opportunities of taking pluralistic and ecumenical approaches, as no single perspective or approach is sufficient to understanding human–environment relationships (Turner 1997).

Defining Landscape Sustainability

If landscapes are pivotal, then how should sustainability be defined? Before defining the sustainability of landscapes, some discussion on the conceptualization of the structure and organization of landscapes should be helpful. Everything is related to everything else, but some are much more related to each than most others; and complexity often takes the form of hierarchical or modular structure (Simon 1962; Wu and Loucks 1995). From this hierarchical perspective, the world is a nested hierarchical system, in which smaller spatial units (e.g., individuals and local populations) form larger spatial units (e.g., ecosystems and landscapes) that in turn make up even larger spatial units (e.g., biomes and the entire biosphere). Many ecological, as well as socioeconomic, systems may be viewed as hierarchical patch dynamic systems whose behavior is determined by pattern–process interactions at different scales (Simon 1962; Wu and Loucks 1995; Wu 1999; Wu and David 2002). Wu and Loucks (1995) articulated five key elements of hierarchical patch dynamics: (1) ecological systems are spatially nested patch hierarchies, (2) dynamics of an ecological system can be studied as the composite dynamics of individual patches and their interactions, (3) pattern and process are scale dependent, (4) nonequilibrium and random processes are essential to ecosystem structure and function, and (5) ecological (meta)stability is often achieved through structural and functional redundancy and spatial and temporal incorporation of dynamic patches.

Landscapes are spatially nested hierarchical patch systems as each landscape is composed of different kinds of patches that in turn comprise smaller patches. As such, the sustainability of landscapes is not only influenced by the interactions among environmental, economic, and social components, but also by their spatial configurations and cross-scale linkages. In a similar way but on broader scales, human-perceived landscapes or cultural landscapes form a pivotal level in the hierarchy of study objects in sustainability science, which may include local communities/ecosystems, landscapes, nations/regions, and the entire world. In this context, the sustainability of a landscape is influenced both by upper levels (constraints) and lower levels (initiating processes and driving forces). From a hierarchical patch dynamics perspective, landscape sustainability is similar to landscape metastability—a shifting mosaic steady state in which macro-level structural and functional patterns are maintained through constant micro-level changes (patch dynamics).

Ecosystems and the biosphere are the prototypical examples of complex adaptive systems (Levin 1999), and so are landscapes. Interactions between spatial patterns and ecological and socioeconomic processes at differing scales are keys to the behavior of such systems. Key to the sustainability of any complex adaptive systems, including landscapes, is resilience. Holling (1973) defined resilience as the ability of a system to absorb change and disturbance without changing its basic structure and function or shifting into a qualitatively different state. This "ecological resilience" or "ecosystem resilience" stresses persistence, change, and unpredictability, and differs fundamentally from the equilibrium-based "engineering resilience" which focuses on efficiency, constancy, and predictability (Holling 1996).

More recent work has further refined Holling's (1973) definition by including the system's abilities to self-organize and adapt to changes, as well as expanding the concept to socioeconomic systems (Levin et al. 1998; Walker and Salt 2006). For example, social resilience is defined as the ability of a human community to withstand, and to recover from, external perturbations (Adger 2000). Resilience thinking frequently invokes the concepts of thresholds or tipping points, alternate stable states or regimes, regime shifts, complex adaptive systems, adaptive cycles, and transformability (Holling 2001; Walker and Salt 2006).

From a resilience perspective, landscape sustainability is not about maintaining the landscape at a steady state by reducing the variability in landscape dynamics or optimizing its performance, but rather focusing on the landscape's adaptive capabilities to cope with uncertainties. In the face of changing climatic conditions and intensifying land uses, the ability to self-organize and preserve system integrity is crucial to realizing landscape sustainability. Recent studies have suggested that high diversity of heterogeneous components, modular structures, and tight feedback loops often characterize resilient complex adaptive systems (Levin 1999; Levin and Lubchenco 2008). The hierarchical patch dynamics perspective corroborates this conclusion from complex adaptive systems theory and resilience research.

Based on the above discussion, it is tempting to define landscape sustainability as the capacity of a landscape to maintain its basic structure and to provide ecosystem services in a changing world of environmental, economic, and social conditions. To operationalize this rather general definition, different landscape types need to be distinguished because they each have different structural and functional characteristics. One common classification is the landscape modification gradient by Forman and Godron (1986): (1) natural landscape (without significant human impact), (2) managed landscape (where native species are managed and harvested), (3) cultivated landscape (with villages and patches of natural or managed ecosystems scattered), (4) suburban landscape (a town and country area with a heterogeneous patchy mixture of residential areas, commercial centers, cropland, managed vegetation, and natural areas), and (5) urban landscape (with remnant managed park areas scattered in a densely built-up matrix). Focusing more on characteristics related to system self-regenerative capacities, Naveh (1998) classified cultural landscapes into seminatural and managed multifunctional landscapes (e.g., protected areas, parks, recreation areas), traditional agricultural landscapes, rural and suburban landscapes, and urban landscapes. These landscapes are distinguished based on their energy inputs and self-organizing and regenerative capacities through the photosynthetic conversion of solar energy: (1) "solar-powered" seminatural and managed landscapes, ranging from protected areas, traditional agricultural landscapes, to contemporary organic farming systems, (2) "intensive agro-industrial" landscapes, including modern agricultural systems that are heavily subsidized by fossil energy, and (3) "technosphere" landscapes, including rural, suburban, and urban-industrial landscapes that are supported primarily by fossil energy, with all internal natural regenerative capacities lost.

Also, insight into landscape sustainability can be gained from examining traditional cultural landscapes, which are the products of long-term co-evolution between culture and nature. For example, based on a review of lessons from history, Forman (1995) observed that water problems, soil erosion, high population density, war, and a decline in exports are key attributes associated with decreased sustainability, whereas cultural cohesion, low population density, export–import trade, overall level and arrangement of the resource base, religious cohesion, varied linkages with adjacent areas, and a major irrigation or dike system are key attributes associated with increased sustainability. Selman (2007) suggested three propositions as a basis for assessing the sustainability of landscapes: (1) "cultural landscapes are sustainable if they are regenerative," (2) "landscape sustainability is characterized by ecological integrity and cultural legibility," and (3) "regenerative landscapes are distinguished by feedback loops leading to accumulation of cultural and ecological assets." Forman (1990) postulated that "for any landscape, or major portion of a landscape, there exists an optimal spatial configuration of ecosystems and land uses to maximize ecological integrity, achievement of human aspirations, or sustainability of an environment." More detailed studies need to be carried out to further test these observations, propositions, and hypotheses. This represents a promising future direction for operationalizing the science and practice of sustainability science.

Measuring Landscape Sustainability

For a landscape-based approach to sustainability to succeed in research and practice, measures must be developed to gauge sustainability at the landscape scale. A great number of sustainability indicators (or sustainable development indicators—SDIs) have been developed in the past several decades since the 1992 Earth Summit in Rio de Janeiro which proposed the fundamental principles and the program of action for achieving sustainable development. Especially after the World Summit on Sustainable Development (Earth Summit 2002) in Johannesburg in 2002, a number of international organizations, governmental agencies, NGOs, local communities and corporations, and academic scholars have devoted significant effort to the design and implementation of indicators that gauge the state and trajectory of environmental conditions and socioeconomic development. Today, hundreds of indicators and indices of sustainable development have been developed and used at the global, national, and local scales.

SDIs are indicators that provide information on the state, dynamics, and underlying drivers of human–environmental systems and represent arguably the most popular approach to gauging sustainable development. Landscape sustainability indicators should be developed based on the commonly recognized criteria, including: (1) an indicator set should cover the various dimensions of sustainability and their complex interactions; (2) indicators should be indicative of the state and changes of the targeted aspects of sustainability; (3) indicators should be informative based on available data; (4) indicators should be readily understandable and policy-relevant; and (5) the methods for weighting and aggregating variables should be transparent and unbiased (Wu and Wu 2011). A number of existing SDIs may be incorporated into landscape indicator systems (see examples in Table 1).

Table 1 A select group of sustainability indices commonly used in the assessment of sustainable development (Wu and Wu 2011)

Indicator	Description
Green GDP	Although GDP is the most popular measure of economic performance, it does not accurately reflect actual human or environmental well-being. Empirical data show that GDP is often negatively correlated with environmental quality, and its positive correlation with social well-being measures disappears after GDP reaches a certain level. Green GDP is a variant developed in the early 1990s in an attempt to factor in the effects of natural resource consumption and pollution on human welfare
Human development index (HDI)	HDI was created in the 1990s by the United Nations Development Program to assess the levels of human and social development. The index is composed of three primary aspects: life expectancy, education, and standard of living. HDI has become a standard and widely reported indicator in many official reports and academic publications. A major criticism of HDI is its abstraction from the environmental dimension of human welfare

(continued)

Table 1 (continued)

Indicator	Description
Inclusive wealth (IW)/genuine savings (GS)	Unlike the Green GDP, which is a "flow" measure, IW/GS are stock-based. The economic patterns of production and consumption are necessarily contingent upon the availability and configuration of the available resources, or "capital." Thus, inter-temporal transfers of economic opportunity are best represented by the value of capital stocks. The "inclusive" and "genuine" primarily refer to the inclusion of natural resources into economic accounting. According to this framework, a region or country is sustainable over a given period if its IW/GS per capita does not decline over that time
Genuine progress indicator (GPI) and index of sustainable economic welfare (ISEW)	GPI and ISEW are essentially equivalent metrics, although the former is more widely recognized than the latter. Like the Green GDP, they adjust the standard flow-based metric of economic performance to consider the role of environmental well-being. However, unlike Green GDP, GPI and ISEW divide economic transactions between those that make a positive contribution to human welfare and those that make a negative contribution (e.g., an oil spill). GPI and ISEW also include the imputed values of nonmarketed goods and services and adjust for income distribution effects
Material flows accounting (MFA)	MFA tracks the weight of a number of different material flows in the economy, including production inputs and outputs, matter moved in the environment to access resources, and residual material from the production process. By aggregating different material flows, MFA produces a single metric called the total material requirement (TMR), which gives a picture of the physical metabolism of the economic system. Although monetary accounting is still more widespread, the use of MFA is expanding
Ecological footprint (EF)	EF measures the land (and water) area that is required to support a defined human population indefinitely (Wackernagel and Rees 1996). The basic unit of measurement is the "global hectare," a normalized unit capturing the average biocapacity of all hectares of all biologically productive lands in the world. This comprehensive measure enables the comparison of human demands on the planet's ecosystems to the regenerative capacity of those ecosystems
Environmental sustainability index (ESI) and environmental performance index (EPI)	Published between 1999 and 2005, ESI was used as a measure of humanity's natural resource use. The computational methodology involved combining 76 variables into 21 metrics, which were then averaged to yield a single index. ESI was succeeded by EPI, which is developed by the same institutions and has been published in 2006, 2008, and 2010. EPI narrows its aims to two objectives: environmental health and ecosystem vitality. EPI is meant to provide a report of "more immediate value to policy-makers"

Indicator frameworks can help identify gaps in available data, indicator sets, and our overall understanding of the human–environmental relationship in landscapes (Wu and Wu 2011). Three indicator frameworks in the sustainability literature should be useful for developing landscape sustainability indicators: the Pressure--State-Response (PSR) framework, the theme- or issue-based frameworks, and the

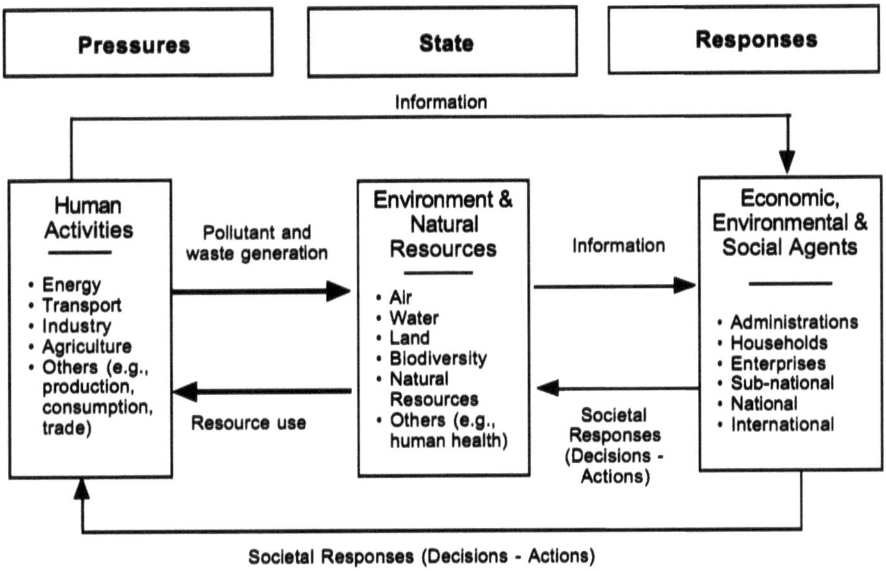

Fig. 3 Illustration of the pressure–state–response framework for the development of sustainability indicators

capital frameworks. With the PSR framework (Fig. 3), indicators of pressures represent forces that drive landscape changes; state indicators focus on current landscape conditions; and response indicators pertain to societal reactions to changes in the state of the landscape and underlying drivers. A theme-based framework organizes indicators around key issues, as illustrated in the 2001 indicator set by the United Nations Commission on Sustainable Development (CSD) (Fig. 4). The CSD theme-based framework has a hierarchical structure, with four dimensions of sustainable development (social, environmental, economic, and institutional), 15 themes, and 58 core indicators. The capital-based framework attempts to calculate the wealth of a region as a function of different kinds of capital: manufactured or built capital (all produced assets that form the human economy in a traditional sense), natural capital (the natural environment and resources), human capital (capacities of people to work, including knowledge, skills, and health), and social capital (stocks of social networks, trust, and institutional arrangements). Sustainability in this case depends heavily on whether a strong or weak sustainability perspective is pursued.

By modifying these frameworks to focus on the landscape scale, sustainability indicators can be developed for different kinds of landscapes. For example, the PSR framework may work better for natural and seminatural landscapes, whereas the theme- and capital-based frameworks seem more appropriate for human-dominated landscapes. Many existing landscape indices may find their places in these frameworks, but systematic efforts are needed to integrate SDIs and landscape pattern metrics. In addition, scalograms using landscape indicators may provide an effective approach to revealing hierarchical linkages and relating key elements of sustainability across multiple scales (Wu 2004).

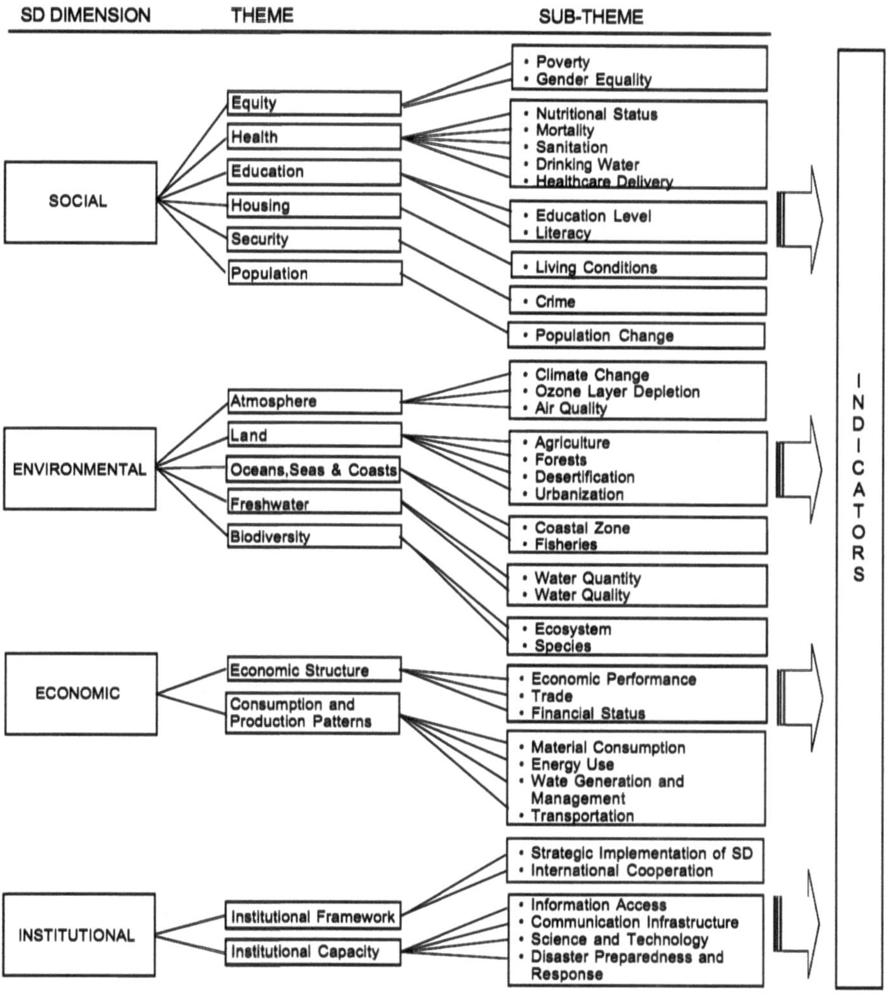

Fig. 4 The theme-based indicator framework developed by United Nations Commission on Sustainable Development (UNCSD 2001)

Landscape ecology has developed a large number of pattern metrics (or indices) to quantify the composition and configuration of landscapes (Li and Wu 2007). Many of these metrics have been successfully used to quantify how landscapes change over time and how different landscape compare and contrast. Landscape metrics can provide rich information on the diversity and relative abundance of different kinds of landscape components, as well as the shape complexity and spatial configuration of patch mosaics. Among the most commonly used ones are the number of patch types and their proportions, patch density, edge density, patch size, patch or landscape shape indices, connectivity indices, and fragmentation indices.

Some of these landscape metrics are conceivably useful in landscape sustainability assessment, although more research is needed to relate landscape metrics to sustainability variables and to develop sustainability-oriented landscape metrics (Wu and Hobbs 2002; Li and Wu 2004).

Landscape Ecology as a Cornerstone of Sustainability Science

If landscapes represent a pivotal scale of sustainability, then the ecology of landscapes ought to have something to offer to the science and practice of sustainability. Landscape ecologists have long considered the relevance of their science to sustainability (Naveh 1982; Forman 1990; Naveh 2007) and, more recently, to sustainability science (Potschin and Haines-Young 2006; Wu 2006; Musacchio 2009, 2011; Turner 2010). In this section, I briefly discuss some of the key ideas in landscape ecology and how this field can contribute to the development of sustainability science.

Although the term was coined in Europe in 1939, landscape ecology was not an established scientific field until the 1980s when remote sensing data and computers became widely accessible to scientists. The 1980s was also a time when ecological ideas of spatial heterogeneity and nonequilibrium dynamics flourished, and when landscape ecology took root in North America. Spatial heterogeneity is ubiquitous in all ecological systems, underlining the significance of pattern–process relations and scale. The main theme of contemporary landscape ecology, with an unmistakable focus on spatial heterogeneity, was articulated in Risser et al. (1984): "Landscape ecology focuses explicitly upon spatial pattern. Specifically, landscape ecology considers the development and dynamics of spatial heterogeneity, spatial and temporal interactions and exchanges across heterogeneous landscapes, influences of spatial heterogeneity on biotic and abiotic processes, and management of spatial heterogeneity." In addition, landscape ecology also fully recognizes the importance of the multidimensionality of landscapes and their cross-disciplinarity. Again, as Risser et al. (1984) put it: "A major forcing function of landscapes is the activity of mankind, especially associated cultural, economic, and political phenomena … Landscape ecology is not a distinct discipline or simply a branch of ecology, but rather is the synthetic intersection of many related disciplines that focus on the spatial-temporal pattern of the landscape."

Today, a general consensus seems to have emerged that landscape ecology is not simply an academic discipline, but rather a highly interdisciplinary field of study (Wu and Hobbs 2002, 2007). In an attempt to integrate the various connotations, Wu and Hobbs (2007) defined landscape ecology as the integration of the science and art of studying and influencing the relationship between spatial pattern and ecological processes on multiple scales. The "science" of landscape ecology focuses on the theoretical basis for understanding the formation, dynamics, and effects of spatial heterogeneity, whereas the "art" of landscape ecology reflects the humanistic and holistic perspectives necessary for integrating ecology, design and planning, socioeconomics, and management practices. Wu (2006) put forward a pluralistic and

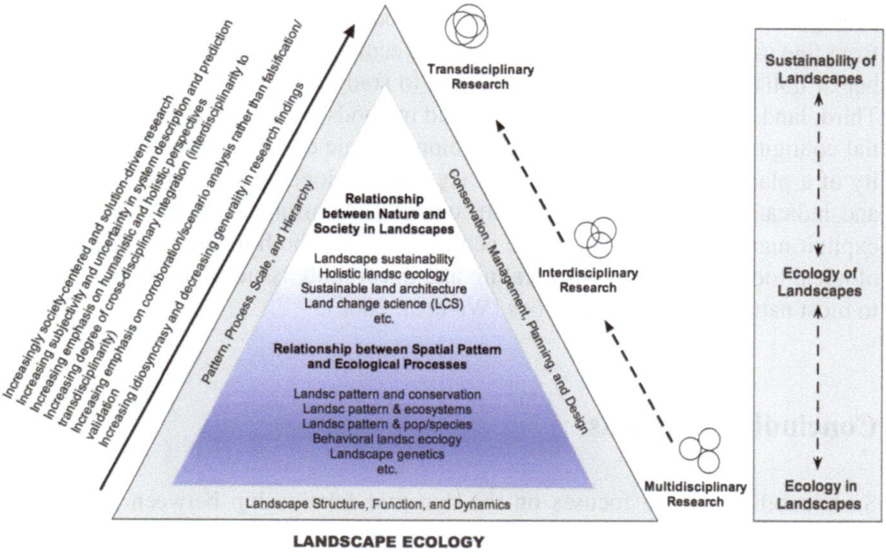

Fig. 5 The pyramid of landscape ecology as an interdisciplinary and transdisciplinary science

hierarchical framework that facilitates synergistic interactions between biophysical/pattern–process and holistic/humanistic perspectives in landscape ecology (Fig. 5). The "hierarchical" view here recognizes the varying scope and degree of cross-disciplinarity in landscape ecological studies, whereas the "pluralistic" view stresses the importance of different disciplines and perspectives. This pluralistic and hierarchical framework implies that all the five dimensions of landscape, as discussed in Tress and Tress (2001), are important in landscape studies.

Several key research areas in landscape ecology have been identified (Wu and Hobbs 2002, 2007). These include: quantifying landscape pattern and its ecological effects; the mechanisms of flows of organisms, energy, and materials in landscape mosaics; behavioral landscape ecology that focuses on how the behavior of organisms interacts with landscape structure; landscape genetics that aims to understand how landscape heterogeneity affects population genetics; causes and consequences of land use and land cover change; spatial scaling that deals with translation of information across heterogeneous landscapes; and optimization of landscape pattern for conservation or sustainability. Towards the transdisciplinary end of the spectrum landscape ecology is increasingly related to sustainability science in theory and practice (Fig. 5). The emerging "land-change science" focuses on observing and monitoring land use and land cover change, assessing its impacts on ecosystem processes and services, and understanding its causes and mechanisms (Rindfuss et al. 2005; Turner et al. 2007). Much of this has been part of key research topics and priorities (Wu and Hobbs 2002), and it is encouraging to see that ecologists and geographers converge on their views toward sustainability.

Overall, landscape ecology can contribute to sustainability science in several ways (Wu 2006). First, landscape ecology provides a hierarchical and integrative ecological basis for dealing with issues of biodiversity and ecosystem functioning from fine to broad scales. Second, landscape ecology has already developed a number of holistic and humanistic approaches to studying nature–society interactions. Third, landscape ecology offers theory and methods for studying the effects of spatial configuration of biophysical and socioeconomic component on the sustainability of a place. Fourth, landscape ecology has developed a suite of pattern metrics and indicators which can be used for quantifying sustainability in a geospatially explicit manner. Finally, landscape ecology provides both theoretical and methodological tools for dealing with scaling and uncertainty issues that are fundamental to most nature–society interactions (Wu et al. 2006).

Concluding Remarks

Sustainability science focuses on the dynamic relationship between society and nature, integrating environmental, economic, and social processes across scales of local communities, regions, and the entire global system. While it is difficult or implausible to pick a scale that is not relevant to sustainability, some scales may be more effective than others for studying and achieving sustainability. The importance of the global scale is given because global sustainability is the ultimate goal of the science and practice of sustainability. However, global-level studies usually have to be coarse-grained and lack details that are directly relevant to local actions. At the other end, studies at the scale of local communities, while extremely important, usually lack regional contexts and are difficult to scale up to the global scale.

To bridge this gap, landscapes represent an intermediate scale that is operational in research and actions and commensurate with human perception of the environment. Landscapes are not only the stage where environmental, economic, and social processes play out, but also the integrator of these processes. Landscapes are the products of interactions between human society and natural environment, representing a pivotal scale for linking local and global sustainability. Landscapes are arguably the most meaningful places in the place-based research in sustainability science. Also, landscapes provide a common ground for ecologists, geographers, planners and designers, and policy-makers to work together to shape and improve the society–nature relationship.

Sustainability science at the landscape scale will not only need to integrate the multiple dimensions of environment, economy, and society, but also should focus on elucidating the role of spatial heterogeneity in determining the sustainability of landscapes. Heterogeneity always makes scale matter. Thus, key research questions ought to address the issues on scale multiplicity, scaling relations, and hierarchical linkages. Consequently, landscape sustainability research will produce pattern–process–scale relations of places that are fundamental to sustainability science. To move forward with the landscape approach to sustainability, landscape ecology, as well as other related interdisciplinary fields, will continue to play an important role.

Acknowledgments I to thank Dr. Michael P. Weinstein for inviting me to give a presentation at the International Symposium on Sustainability Science at Montclair State University in October 2010, from which this paper has evolved. Also, I thank Tong Wu for a number of helpful discussions on sustainability, resilience, and environmental economics. My research in landscape ecology and sustainability has been supported by grants from National Science Foundation (DEB 9714833, DEB-0423704, BCS-0508002), US Environmental Protection Agency (R827676-01-0), and collaborative grants from National Natural Science Foundation of China and Chinese Academy of Sciences.

References

Adger WN (2000) Social and ecological resilience: are they related? Prog Hum Geogr 24:347–364

Chen X, Wu J (2009) Sustainable landscape architecture: implications of the Chinese philosophy of "unity of man with nature" and beyond. Landsc Ecol 24:1015–1026

Clark WC (2007) Sustainability science: a room of its own. PNAS 104:1737–1738

Clark WC, Dickson NM (2003) Sustainability science: the emerging research program. PNAS 100:8059–8061

Du Pisani JA (2006) Sustainable development—historical roots of the concept. Environ Sci 3:83–96

Forman RTT (1990) Ecologically sustainable landscapes: the role of spatial configuration. In: Zonneveld IS, Forman RTT (eds) Changing landscapes: an ecological perspective. Springer, New York

Forman RTT (1995) Land mosaics: the ecology of landscapes and regions. Cambridge University Press, Cambridge

Forman RTT, Godron M (1986) Landscape ecology. Wiley, New York

Gobster PH, Nassauer JI, Daniel TC et al (2007) The shared landscape: what does aesthetics have to do with ecology? Landsc Ecol 22:959–972

Holling CS (1973) Resilience and stability of ecosystems. Annu Rev Ecol Syst 4:1–23

Holling CS (1996) Engineering resilience versus ecological resilience. In: Schulze P (ed) Engineering within ecological constraints. National Academy Press, Washington

Holling CS (2001) Understanding the complexity of economic, ecological, and social systems. Ecosystems 4:390–405

Ji X (2007) Ji Xianlin on Chinese culture. China Books, Beijing

Kareiva P, Watts S, McDonald R et al (2007) Domesticated nature: shaping landscapes and ecosystems for human welfare. Science 316:1866–1869

Kates RW (2003) Sustainability science. In: IAP (Interacademy Panel on International Issues) (ed) Transition to sustainability in the 21st century: the contribution of science and technology. National Academies Press, Washington

Kates RW, Clark WC, Corell R et al (2001) Sustainability science. Science 292:641–642

Kates RW, Parris TM, Leiserowitz A (2005) What is sustainable development? Goals, indicators, values, and practice. Environment: Science and Policy for Sustainable Development 47(3):8–21

Kidd CV (1992) The evolution of sustainability. J Agric Environ Ethics 5:1–26

Leopold A (1949) A Sand County almanac. Oxford University Press, New York

Levin SA (1999) Fragile dominions: complexity and the commons. Perseus Books, Reading

Levin SA, Lubchenco L (2008) Resilience, robustness, and marine ecosystem-based management. Bioscience 58:27–32

Levin SA, Barrett S, Aniyar S et al (1998) Resilience in natural and socioeconomic systems. Environ Dev Econ 3:222–236

Li H, Wu JG (2004) Use and misuse of landscape indices. Landsc Ecol 19:389–399
Li H, Wu J (2007) Landscape pattern analysis: key issues and challenges. In: Wu J, Hobbs R (eds) Key topics in landscape ecology. Cambridge University Press, Cambridge
McHarg IL (1969) Design with Nature. Natural History Press, Garden City, New York
Mitchell D (2000) Cultural geography: a critical introduction. Blackwell, Oxford
Musacchio LR (2009) The scientific basis for the design of landscape sustainability: a conceptual framework for translational landscape research and practice of designed landscapes and the six Es of landscape sustainability. Landsc Ecol 24:993–1013
Musacchio LR (2011) The grand challenge to operationalize landscape sustainability and the design-in-science paradigm. Landsc Ecol 26:1–5
Nassauer JI (1995) Culture and changing landscape structure. Landsc Ecol 10:229–237
Nassauer JI (ed) (1997) Placing nature: culture and landscape ecology. Island Press, Washington
Naveh Z (1982) Landscape ecology as an emerging branch of human ecosystem science. Adv Ecol Res 12:189–237
Naveh Z (1998) Ecological and cultural landscape restoration and the cultural evolution towards a post-industrial symbiosis between human society and nature. Restor Ecol 6:135–143
Naveh Z (2007) Landscape ecology and sustainability. Landsc Ecol 22:1437–1440
O'Neill RV, DeAngelis DL, Waide JB et al (1986) A hierarchical concept of ecosystems. Princeton University Press, Princeton
Phillips A (1998) The nature of cultural landscapes—a nature conservation perspective. Landsc Res 23:21–38
Phillips A (2007) International policies and landscape protection. In: Benson JF, Roe M (eds) Landscape and sustainability. Routledge, New York
Pickett STA, Cadenasso ML (1995) Landscape ecology: spatial heterogeneity in ecological systems. Science 269:331–334
Potschin M, Haines-Young R (2006) "Rio+10", sustainability science and landscape ecology. Landsc Urban Plann 75:162–174
Reitan PH (2005) Sustainability science—and what's needed beyond science. Sustain Sci Pract Policy 1:77–80
Rindfuss RR, Walsh DJ, Turner BL II et al (2005) Developing a science of land change: challenges and methodological issues. PNAS 101:13976–13981
Risser PG, Karr JR, Forman RTT (1984) Landscape ecology: directions and approaches. Illini Natl Hist Surv Spec Publ. 2, Champaign, IL
Sauer CO (1925) The morphology of landscape. Publ Geog 2:19–53
Selman P (2007) Landscape and sustainability at the national and regional scales. In: Benson JF, Roe M (eds) Landscape and sustainability. Routledge, New York
Simon HA (1962) The architecture of complexity. Proc Am Philos Soc 106:467–482
Tress B, Tress G (2001) Capitalising on multiplicity: a transdisciplinary systems approach to landscape research. Landsc Urban Plann 57:143–157
Tress B, Tress G, De'camps H, d'Hauteserre A-M (2001) Bridging human and natural sciences in landscape research. Landscape and Urban Planning 57:137–141
Turner BL II (1997) Spirals, bridges and tunnels: engaging human-environment perspectives in geography. Ecumene 4:196–217
Turner MG (2010) A landscape perspective on sustainability science. In: Levin SA, Clark WC (eds) Toward a science of sustainability. Toward a Science of Sustainability Conference, Warrenton, Virginia
Turner MG, Gardner RH, O'Neill RV (2001) Landscape ecology in theory and practice: pattern and process. Springer, New York
Turner BL II, Lambin EF, Reenberg A (2007) The emergence of land change science for global environmental change and sustainability. PNAS 104:20666–20671
UNCSD (2001) Indicators of Sustainable Development: Guidelines and Methodologies. UN Commission on Sustainable Development, United Nations, New York
Urban DL, O'Neill RV, Shugart HH (1987) Landscape ecology: a hierarchical perspective can help scientists understand spatial patterns. Bioscience 37:119–127

Walker B, Salt D (2006) Resilience Thinking: Sustaining Ecosystems and People in a Changing World. Island Press, Washington

Wackernagel M, Rees WE (1996) Our Ecological Footprint: Reducing Human Impact on the Earth. New Society Publishers, British Columbia, Canada

Webb M (1987) Cultural landscapes in the National Park Service. Publ Histor 9:77–89

Weinstein MP (2010) Sustainability science: the emerging paradigm and the ecology of cities. Sustain Sci Pract Policy 6:1–5

Wiens JA, Milne BT (1989) Scaling of 'landscape' in landscape ecology, or, landscape ecology from a beetle's perspective. Landsc Ecol 3:87–96

Wilson EO (2007) Foreward. In: Penn D, Mysterud I (eds) Evolutionary perspectives on environmental problems. Aldine Transaction, Piscataway

Wu J (1999) Hierarchy and scaling: extrapolating information along a scaling ladder. Can J Remot Sens 25:367–380

Wu J (2004) Effects of changing scale on landscape pattern analysis: scaling relations. Landsc Ecol 19:125–138

Wu J (2006) Landscape ecology, cross-disciplinarity, and sustainability science. Landsc Ecol 21:1–4

Wu J (2008) Making the case for landscape ecology: an effective approach to urban sustainability. Landsc J 27:41–50

Wu J, David JL (2002) A spatially explicit hierarchical approach to modeling complex ecological systems: theory and applications. Ecol Modell 153:7–26

Wu J, Hobbs R (2002) Key issues and research priorities in landscape ecology: an idiosyncratic synthesis. Landsc Ecol 17:355–365

Wu J, Hobbs R (2007) Landscape ecology: the-state-of-the-science. In: Wu J, Hobbs R (eds) Key topics in landscape ecology. Cambridge University Press, Cambridge

Wu J, Levin SA (1994) A spatial patch dynamic modeling approach to pattern and process in an annual grassland. Ecol Monogr 64:447–464

Wu J, Loucks OL (1995) From balance-of-nature to hierarchical patch dynamics: a paradigm shift in ecology. Quart Rev Biol 70:439–466

Wu J, Wu T (2011) Sustainability indicators and indices. In: Madu CN and Kuei C (eds), Handbook of sustainable management. Imperial College Press, London

Wu J, Jones KB, Li H et al (eds) (2006) Scaling and uncertainty analysis in ecology: methods and applications. Springer, Dordrecht

It's OK to Talk About Sustainability

Alan D. Hecht

Abstract For much of the recent past, state and local governments and a number of businesses have led in making sustainability operational in the United States, but federal policies have lagged far behind. Today, however, environmental, economic, and social pressures are beginning to move governments and businesses to more urgently and effectively adopt sustainable management policies and practices. This shift in public policy and business strategy reflects a new reality that today's problems are more complex, involve new stressors and multiple environmental media, and thus require approaches that extend beyond traditional business practices or media-specific legislation. The transition to sustainability will not be easy. For the US Environmental Protection Agency (USEPA), this means going beyond the existing regulatory framework and advancing an environmental policy and research agenda that promotes sustainability science, innovation, and problem solving. For business and government alike, this means that innovation and sustainability science must be major drivers to advance economic growth while protecting the environment and human health. More than ever, it is "OK to talk about sustainability."

Keywords Sustainable business practices • US National Sustainability Strategy • Pressure–state–response model • USEPA

Views expressed in this article are those of the author and do not necessarily reflect the views or policies of the USEPA where he is employed. Mention of trade names or commercial products does not constitute Agency endorsement or recommendations for use.

A.D. Hecht (✉)
Office of Research and Development, US EPA,
Washington, DC 20460, USA
e-mail: Hecht.Alan@epamail.epa.gov

M.P. Weinstein and R.E. Turner (eds.), *Sustainability Science: The Emerging Paradigm and the Urban Environment*, DOI 10.1007/978-1-4614-3188-6_4,

Introduction

For much of the past few decades, state and local governments have led in making sustainability operational. Federal policies have lagged far behind. Since the early 1990s, the idea of sustainability has taken root in hundreds of state and local communities in the United States and around the world. These governments are natural laboratories for sustainability as they are responsible for development issues and are closer to the nexus of sustainability and quality of life. What happens at their level is an important measure of public support for sustainability.

On a global scale, the International Council for Local Environmental Initiatives (ICLEI) was established in 1990 by more than 200 local governments from 43 countries at the World Congress of Local Governments for a Sustainable Future. In 2003, the council expanded and revised its mission, charter, and name to reflect challenges faced by local governments. The new ICLEI–Local Governments for Sustainability includes more than 450 local governments, representing 300 million people in 63 countries. Today, its web site (http://www.iclei.org) provides invaluable information on best urban practices in topics such as planning for sustainable development, energy efficiency, transportation, green building, land use management, environmental management, and education.

Since 2005 the online resource SustainLane has also surveyed and ranked the 50 most populous US cities on the basis of sustainability (http://www.sustainlane.com/us-city-rankings). The city top-ranked by SustainLane in 2006 and 2008 was Portland, Oregon, which ranked above other cities in planning for clean technology, green building development, overall quality of life, and in planning and management for sustainability. The concept of sustainability has also been growing in small- and medium-size cities and a number of different regions in the US (Weiss 2009).

In parallel with growing support for sustainability at state and local levels, several major international firms are beginning to embrace sustainability principles and practices:

- In May 2005, the General Electric Company launched its program of "ecomagination," asserting that "things that are good for the environment are also good for business." GE was embarking on this initiative "not because it is trendy or moral, but because it will accelerate [economic] growth" (Sullivan and Schiafo 2005). In 2007, GE set a goal of generating $20 billion in revenue from ecomagination products by 2010.
- In similar fashion, the giant retail-goods manufacturer Proctor & Gamble set a goal in 2007 of gaining $20 billion in 5 years from innovative and sustainable products such as Tide Coldwater; 2 years later the firm raised its goal to $50 billion by 2012 (Winston 2009). In October 2010, P&G went further, announcing a new sustainability vision aiming to reach 5 billion consumers over the next 5 years, or roughly a billion more than the firm's present number of consumers. The vision is equally ambitious in aiming to use 100% renewable or recycled materials for all of its products and packaging and to achieve zero consumer

waste entering landfills. In order to meet those goals, P&G has established a set of benchmarks including renewable materials accounting for 25% of product/packaging materials, and for the firm's plants to rely on renewable power for 30% of its energy use.
- In the retail world, UK retailer Marks & Spencer (M&S) has set the goal of becoming the world's most sustainable major retailer by 2015. In 2007, M&S launched its "Plan A: Doing the Right Thing" ("because there is no Plan B"), a business-wide £200 million "eco-plan." M&S has made significant strides on its five goals of becoming carbon-neutral, sending no waste to landfill, extending sustainable sourcing, setting new standards in ethical trading, and helping customers and employees live a healthier lifestyle. Plan A began with 100 specific goals and in 2010 added 80 more (Marks & Spencer 2010a). M&S has reported that all the £50 million ($73 million) earned by Plan A activities in 2009 were invested back into the company and that by 2010 it had achieved 62 of its original 100 specific goals (Marks & Spencer 2010b).
- Among the many companies that have now adopted new management and technical strategies to advance sustainability, Walmart stands out for aggressively applying life cycle assessment (LCA) approaches to managing its global supply chain. In 2006, Walmart launched its Sustainability 360 program and established explicit goals to use 100% renewable energy sources, create zero waste, reduce greenhouse gas (GHG) emissions, and sell products that "sustain our resources and the environment."

These examples showcase businesses that have begun to see the goal of sustainability as a strategy for managing the rising cost of energy and material use and promoting competitiveness and innovation. A 2002 PricewaterhouseCoopers survey of industry stated that "companies that fail to become sustainable—that ignore the risks associated with ethics, governance and the triple bottom line of economic, environmental and social issues—are courting disaster" (PricewaterhouseCoopers 2002). Since then sustainability strategies have been incorporated into the management principles of dozens of Fortune 500 companies, with more of these firms based in Europe and Asia than in the United States.

The indispensable nature of sustainability was recognized by the European Union (EU) in 2000 when it made sustainability the goal of its Lisbon Strategy, aiming "to become the most competitive and dynamic knowledge-based economy in the world capable of sustainable economic growth with more and better jobs and greater social cohesion" (European Council 2000). The European Council declared that clear and stable objectives for sustainable development will present significant economic opportunities, which "has the potential to unleash a new wave of technology innovation, generating growth and employment" (Larsson et al. 2002). Today this view of sustainability as an economic strategy is becoming more widespread in industry. After examining sustainability initiatives in energy and manufacturing at 30 large corporations, a 2009 study published in the *Harvard Business Review* concluded that "sustainability is a mother lode of organizational and technological innovations that yield both bottom-line and top-line returns" and that "there is no alternative to sustainable development" (Nidumolu et al. 2009).

Given progress toward sustainability in Europe and at state and local government levels and in the business community in the United States, why has the concept of sustainability been so difficult to advance in the US federal government and what drivers are operating today to change this? Can the goal of sustainability become a more integral part of US national policy? How can science and innovation advance sustainable solutions? And what role can the US Environmental Protection Agency (USEPA) play in moving the United States to the next level of environmental protection with a focus on sustainability? Use of a conceptual pressure–state–response model described in the following sections sheds light on these questions.

Pressure–State–Response Model Leads to Sustainability

Making sustainability operational in the federal government has been limited by a number of factors largely related to fears of adding economic burdens to business and society. The current debate on regulating GHGs and climate change is a good example of divergent views on the economic impacts of climate change (Hecht 2009). However, notwithstanding the debate on climate change, sustainability is now being recognized as a possible source of innovation in business and government. This recognition has been driven by a number of social, economic, and environmental factors.

Beginning in 2007, I started to sketch out factors exerting pressures on business strategies and government policies to move toward sustainability (Hecht 2007). I used a conceptual systems model (Fig. 1) often called a "pressure–state–response

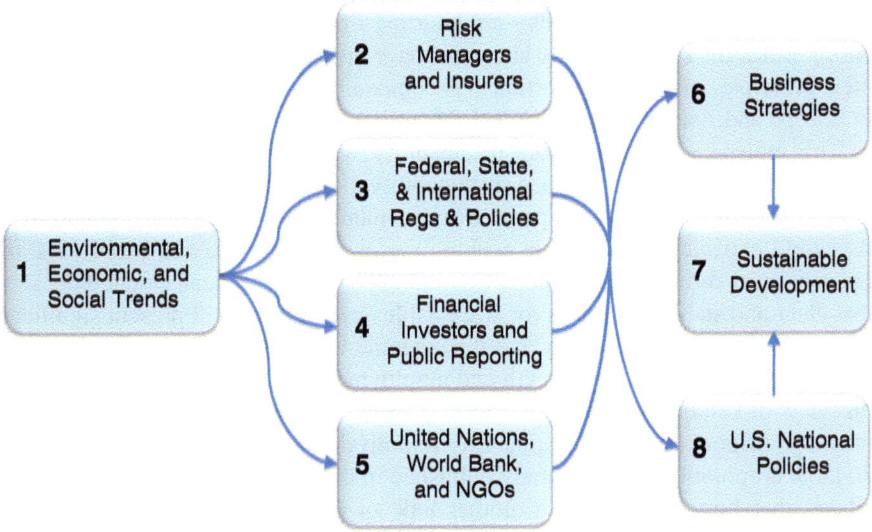

Fig. 1 A conceptual pressure–state–response model

model," which identifies how social, economic, and environmental stressors (Box 1) were shaping business strategies and government policies by their effects on four stakeholder groups (Boxes 2, 3, 4, and 5): Risk Managers and Insurers, Policy Makers and Regulators, Financial Investors, and the UN and Global Society. The result was to move business strategies and government policies toward sustainability (Boxes 6, 7, and 8).

Expanding this diagram to comprehensively model a real system Fig. 1 would show many positive and negative impacts affecting each rectangle. Conflicts between state and federal policies regulating GHG emissions would be seen as prominent positive feedback that has moved federal policies toward sustainability. Such conflicts led to the 2007 US Supreme Court case disputing whether the Clean Air Act gives the USEPA the authority to regulate carbon dioxide gas as a pollutant. The several states that had initiated the lawsuit argued that The USEPA has such authority, while the USEPA under the George W. Bush administration opposed such an interpretation. Following the Supreme Court's ruling in favor of the states, the USEPA under the Obama administration announced it will apply the Clean Air Act to regulate GHG emissions.

This figure would become very complicated if it depicted all the relevant positive and negative feedbacks. For example, rising GHG emissions—a result of current business strategies and government policies—are affecting insurance practices, corporate strategies, and government policies. Investors and financial managers reacting to climate risks are encouraging companies to reduce their carbon footprint. Government feedback is both positive (as it sets targets for emissions reduction or GHG intensity) and negative (as it resists certain business and international pressures).

Global environmental, economic, and social pressures are among the major drivers of this model (Box 1). The world's nations now use resources equivalent to 1.5 planets like earth to support their populations and economies. Drawing on UN scenarios, the Global Footprint Network suggests that if current trends in population and consumption continue, by the 2030s we will need the equivalent of two earths to support the world's population. The ecosystem will be under even more pressure by 2050 when global population will reach about nine billion, some 30% higher than in 2000. Shortages and deterioration of natural resources and the impacts of climate change will increasingly limit our ability to attain or maintain sustainable economic growth (Global Footprint Network 2010).

These pressures on the planet date back far before the recognitions of environmental problems that led to the creation of the USEPA and the calling of the World Environmental summits of 1972, 1992, and 2002. In the early 1970s, environmental protection was largely focused on addressing issues related to industrial emissions and occupational safety: environmental challenges were highly visible and easy to understand. Congress addressed the obvious problems of air and land pollution and water contamination through media-specific environmental laws; consequently our air and water are cleaner, many damaging industrial waste sites have been restored, and we are producing less hazardous waste.

New pressures are now threatening the well-being and resilience of both human society and the natural environment. These pressures include growth in population and the economy, resulting in increased use of energy and materials and significant changes in land use. These not only drive climate change but also threaten biodiversity and integrity of vital natural resources such as clean air and water, soil, forests, and wetlands.

Today, government and business leaders cannot easily ignore economic and social statistics such as the 2.9 billion people living on less than $2 a day, and the 2.5 billion people having no access to proper sanitation. Many social stressors affect those at the "bottom of the economic pyramid"—the four billion people in developing countries with annual incomes under $3,000. While the individual income of these four billion is very low, together they have purchasing power of $5 trillion (Hammond et al. 2007). The distress of the world's less fortunate people affects not only the stability of nations but also business operations and opportunities, with the result that international firms operating in many developing nations must find ways to address issues involving social and economic well-being in order to maintain what they call their "license to operate." Unilever's global operations appear to demonstrate such concerns: Business Week has suggested that the whole world constitutes "Unilever's laboratory" because in Brazil the global conglomerate operates a free community laundry, provides financing for drip irrigation, and recycles seventeen tons of waste; in Bangladesh it funds a floating hospital; in Ghana it teaches sustainable practices to deprived communities; in India it helps women start micro-enterprises; and globally it discloses how much carbon dioxide and hazardous waste its factories produce (Business Week 2007).

The health of the environment and its ecosystems both affect and are affected by the behavior of business and government. While the full costs of the loss and degradation of ecosystem services—including access to clean water and sanitation—are difficult to measure, the available evidence demonstrates that the costs are substantial and growing: the 2005 United Nations *Millennium Ecosystem Assessment* determined that fifteen of the twenty-four significant ecosystem services are being degraded or used unsustainably (Millennium Ecosystem Assessment Program 2005). Many of the losses in ecosystem services are a consequence of actions taken to increase the supply of other societal services, especially food production. These trade-offs often shift the costs of ecosystem degradation from one group of people to another; the greatest costs will likely be borne by future generations.

For the past decade, external pressures on the environment and awareness of its impacts have been growing. The economic recession since 2008 has created additional stress on business and has also been giving some firms (and Congress) an excuse to oppose any new regulation on climate change as an untimely extra burden on the economy. Such a position is very shortsighted, for the long-term picture without energetic intervention is more threatening: the combined impact on society of continuing population growth, urban development, and increased use of materials and energy is dramatic. Now more than ever, new business strategies and government policies promoting sustainability are needed. The many stressors included in Box 1 impact activities in Boxes 2, 3, 4, and 5.

Risk Mangers and Insurers (Box 2)

Risk managers are paid for avoiding costly problems and the insurance industry has quickly come to understand that unsustainable development is costly. Floods, droughts, earthquakes, hurricanes, and tornados are the expected sources of most major insurance losses. Because changes in the frequency of such events are critical in anticipating risk, effective techniques to understand and evaluate future risk are essential to the viability of insurance firms.

A number of insurers have been leaders in the study of natural catastrophes. Aiming to describe the new risk landscape, insurers such as Swiss Re operate extensive research programs on the early detection and assessment of environmental and health risks, while Munich Re publishes an annual review of disasters and catastrophes and has set up a foundation to support sharing knowledge bearing on risk.

Munich Re, Swiss Re, and other major insurance and reinsurance firms are bringing new attention to issues of environmental sustainability. In reacting to expected pressures from climate change, these firms have adjusted their rate structures and called for government action. In addition, insurance firms now commonly offer businesses the option of reducing their insurance premiums by adopting innovative "green" programs based on improving their risk profile and commitment to sustainability. But the insurance industry cannot address the challenges of climate change on its own. Government regulations are clearly needed.

Regulations and Policy (Box 3)

A combination of environment regulations and business standards is impacting how businesses think about sustainability. EU countries began a GHG cap-and-trade program in 2003. The impact of this approach is getting mixed reviews: from 2007 to 2009 emissions significantly dropped, but the decline may be due largely to the concurrent economic recession. In areas other than climate change, the EU has set new standards on addressing the growing amounts of wastes and toxic chemicals in the environment. The EU has also enacted several directives with important global environmental implications, including directives for the Restriction of Hazardous Substances (RoHS), Waste Electrical and Electronic Equipment (WEEE), and Registration, Evaluation and Authorization of Chemicals (REACH). RoHS and WEEE, designed to tackle the rapidly increasing waste stream of electrical and electronic equipment, complement EU action on landfills and incineration of waste. The REACH regulation gives industries greater responsibility to manage risks from chemicals and to provide related safety information; it requires manufacturers and importers to gather information on the properties of substances, which will help them to manage the substances safely and to register the information in a central database.

From a sustainability perspective, the EU directives regulate product inputs rather than outputs, manage materials rather than waste, promote use of LCA and cradle-to-grave management, apply green engineering and green chemistry principles, shift the burden of proof to industry, and measure and manage future financial risk and liabilities. Combined with pressures from insurers and risk managers, these directives advance the movement toward sustainability.

The EU is also now moving ahead on regulating the use of nanomaterials. Its NanoSustain project aims to develop innovative solutions for the sustainable design, use, recycling, and final treatment of nanotechnology-based products. These goals will be achieved by comprehensively gathering data, generating relevant missing data, and evaluating and validating data for specific nanoproducts or product groups based on their potential threat to human and environmental health hazards and to impacts that may occur after their production. NanoSustain will set the stage for the development of new sustainable products and industrial applications and hence for making the European nanotechnology industry more competitive. A new EU rule—part of a 397-page cosmetics regulation approved in November 2009 by the EU Council—will require cosmetics manufacturers to list any nanoparticles contained in products marketed within the EU.

In the United States, the USEPA is confronting a host of issues related to climate change, economic growth, demographics and aging, urban development and redevelopment, energy and material use, non-point sources of pollution, ecosystem destruction, and new chemical and biological risks. In response to 2008 Congressional appropriations (H.R. 2764; Public Law 110-161), the USEPA has issued the Mandatory Reporting of Greenhouse Gases Rule. Intended to collect accurate and timely emissions data to inform future policy decisions, the rule requires large US sources and suppliers to report GHG emissions. Under the rule, suppliers of fossil fuels or industrial GHGs, manufacturers of vehicles and engines, and facilities that emit 25,000 metric tons or more per year of GHG emissions are required to submit annual reports to the USEPA.

While Congress debates the content and scope of national legislation on GHGs, this USEPA reporting system will provide a better understanding of the sources of GHGs and will guide development of the best possible policies and programs to reduce emissions. In addition to new efforts to regulate emissions, the USEPA has come to recognize that, while regulating dangerous pollution and toxics certainly remains a necessary and vital task, eliminating the use of toxic materials altogether is a better, more sustainable approach. It is therefore not surprising that as pressures grow and new risks are identified, USEPA programs have been inching toward life cycle analysis, green chemistry, green design, green engineering, smart growth, and industrial ecology.

Environmental regulators are not alone in responding to growing social pressures on business. Disclosure requirements for industry have been strengthened significantly in recent years. Significant pressure for transparency in business operations has come from the Financial Accounts Standards Board (FASB) of the American Institute of Certified Public Accountants (AICPA). In December 2006, FASB issued FIN 47, an interpretation of its Accounting for Asset Retirement Obligations Standards, which has prodded firms that had been slow to record obligations for the anticipated expenses of retiring physical assets in an environmentally safe and sound manner. The FASB accounting procedures require firms to identify assets such as building sites, mines, chemical plants, and nuclear power facilities that may cause long-term environmental damage and which the firms may be legally required to restore to their original conditions. Firms are now clearly required to recognize those future obligations as they purchase, construct, and use their physical assets. The FASB accounting procedures also require that firms estimate the potential risk and liability of operating facilities that produce environmentally dangerous products. Such procedures reinforce the movement toward more sustainable management practices by reducing long-term risk.

Finance and Investors (Box 4)

Environmental and social pressures are also pushing bankers, pension fund managers, and individual investors toward more sustainable and socially responsible investing. As the availability and quality of natural resources are under mounting threat, commodities and public goods such as food, clean air, and water are becoming increasing scarce. The financial sector has the potential and the instruments to play a vital catalytic role for the conservation and value enhancement of natural resources (Alms and Schanz 2008). For example, banks that adhere to the Equator Principles must assess the social and environmental impacts of projects that they finance. Influenced by actions and pressures from groups like the activist Rainforest Action Network, Citigroup has gone beyond the Equator Principles by committing to refuse funding for projects that could result in illegal logging, other environmental damage, or harm to indigenous people. Such actions demonstrate the potential power of social pressures on business. Reacting to rising business and public pressures on climate change, the Security and Exchange Commission (SEC) in January 2010 voted to require firms to provide information to investors about risks to their businesses associated with climate change.

Changing perspectives on Wall Street and among pension fund managers and millions of institutional and individual investors are also evident in the growth of socially responsible mutual funds and from the evolving definition of fiduciary responsibility to include environmental risk and performance. The 2007 *Fiduciary Guide to Toxic Chemical Risk*, one of an increasing number of reports about chemicals, foods, and other products, noted the "growing concern about the impact on human health of relatively small amounts of chemicals in everyday products." The *Guide* notes that some of the largest law firms in the world have "definitely concluded that considering environmental, social, and governance issues is at the core of Fiduciary Duty of Prudence and that fiduciaries have an affirmative duty to consider toxic chemical issues that impact corporate risk, returns and shareholder values (Ambachtsheer et al. 2007)." If Fig. 1 were designed to capture all the positive and negative feedbacks in the system, then this new interpretation of fiduciary responsibility would be seen as a positive feedback of the changing risk landscape.

UN, World Bank, NGOs, and Global Society (Box 5)

The lower rectangle in the center of Fig. 1 includes pressures coming from the international community that are impacting the convergence of business and government toward sustainability. Since 1972 the United Nations has been at the center of championing environmental and social issues by collecting data, encouraging national reporting, organizing world conferences and summits, and fostering international agreements. The UN-rooted activities have focused global attention on a suite of social and environmental issues that are increasingly affecting business strategies and government policies. While UN conferences may not lead to concrete and binding actions, they have elevated public debate on strategic issues and exerted significant pressure for member governments to take action.

The World Bank similarly provides a variety of lending and advisory services to support the energy, transport, water, and information, and communication technology sectors in client countries. The World Bank actions are outlined by the Sustainable Infrastructure Action Plan (SIAP) and the Infrastructure Recovery and Assets Platform (INFRA). In 2009, support for infrastructure represented 38% of all World Bank commitments (World Bank Group 2008).

Concurrent with the growth of UN activities has been the increase in non-government organizations focusing on environmental and social issues. Today these organizations are key partners with government and business in efforts to bring clean water, sanitation, clean energy, and medical care to billions of people around the world. Non-government organizations are also exerting considerable pressure on business by using modern satellite and Internet technology. For example, GeoEye has become one of the major global providers of real-time satellite data for business sectors seeking information on illegal logging and mining (http://www.geoeye.com/CorpSite).

Convergence of Business Strategies and Government Policies (Boxes 6, 7, and 8)

The conceptual systems model shown and described above is in many ways a variant of a pressure–state–response model. In this model, human activities exert pressures on the environment (such as pollution, land use change, or increased demand for livestock products). These result in changes in the environment (e.g., changes in pollutant levels, habitat diversity, and livestock production) which in turn impact economic, social, and environmental conditions. How society responds to these changes is reflected in business strategies and practices and government policies that are slowly converging on sustainability. This convergence reflects a new understanding that innovation and sustainable practices can boost the economy.

Analyzing business stressors and responses to the recession, Winston (2009) highlights four key factors that can accelerate movement toward sustainable practices. He argues that businesses must (1) get lean and generate immediate bottom-line savings by reducing energy use and waste; (2) get smart by using value-chain data to cut costs, reduce risks, and focus innovation efforts; (3) get creative by posing heretical questions that force companies to find solutions to tomorrow's challenges today; and (4) get engaged by giving employees ownership of environmental goals and the tools to act on them. Winston argues that green initiatives can ratchet up a company's resource efficiency, creativity, and employee motivation. He concludes that sustainability is at the very *core* of recovery: no company or society, he insists, can last unless it cares for all of its human, financial, and environmental resources and capital (Winston 2009).

Dozens of companies (like P&G, Walmart, GE, and others discussed in the previous section) are now evolving their business models by setting sustainability goals and metrics and reporting annually on progress toward those goals. Some companies are taking regulatory actions that go beyond existing federal guidelines. For example, Walmart has set a standard for lead in toys that is 85% lower than required by US regulations. Winston (2009) notes that Toys "R" Us, Target, and Sears have phased out products containing certain chemicals (such as BPA or phthalates) that studies indicate are dangerous to human health.

All of the above suggest a new economic model that Lubin and Esty (2010) call the "Sustainability Imperative." The key point here is that sustainable management is not a threat to the economy but a necessary force for innovation and competiveness. If in fact politics is all about money, then sustainability should drive economic development and accelerate this business–government convergence. This will still take time since long-standing conflicts between business and government over the economic impacts of regulations and policies continue. The price to be paid by industries releasing GHGs remains today the major test case for this convergence.

The pressures of the recession and projections of future energy and resource needs clearly strengthen the argument for a different way of managing our economy. Along with social and economic factors, these pressures in turn influence federal policy, which today is putting greater attention on ways to achieve a "green economy." This means more efficient operations in government management as well as in advancing science, technology, and innovation.

In sum, because of domestic and international environmental, economic, and social pressures, federal policy is now overcoming past resistance to the concept of sustainability. Can the goal of sustainability now become a more integral part of US national policy? And how can science and innovation advance sustainable solutions? This challenge for the USEPA and other agencies is discussed in the next section.

Sustainability at the USEPA: Promoting Sustainability Science and Innovation

The enactment of the National Environmental Policy Act in 1970 formally established as a national goal the creation and maintenance of "conditions under which [humans] and nature can exist in productive harmony, and fulfill the social, economic and other requirements of *present and future generations of Americans*" [emphasis added]. This language is remarkably similar to the UN-sponsored Brundtland Report's definition of "sustainable development" 17 years later (UN General Assembly 1987). Implementing this goal and policy begs a number of practical questions such as "What kind of regulations, policies, strategies, and practices are needed to advance sustainability?" and "How will such changes impact economic development?" These questions have often led to conflict between business and government over regulatory policies and their implementation, including policies aimed at reducing pollutants and GHG emissions (Hecht 2009).

For much of its history, the USEPA has wrestled with how to define the optimum regulatory framework for implementing sustainability policies. Many USEPA administrators have inched the agency forward, adapting to changing environmental issues and slowly moving to make sustainability a key element of environmental policy (Grossarth and Hecht 2007).

One historic effort by the USEPA to promote sustainability came in its 1993 report to Congress, "Sustainable Development and the Environmental Protection Agency." Prompted by international events such as the 1992 Rio Earth Summit, Congress was "interested in USEPA's effort to explore the concept of sustainable development." In particular, the committee was interested in how environmental concerns can be best incorporated in national, State, and local development and economic planning and decision-making processes (USEPA 1993).

Acknowledging that the USEPA "has not employed the concept of sustainability explicitly in an overall policy framework or programmatic objective," the 1993 report saw the problem as a consequence of a number of concerns including the "minor role that sustainability plays in USEPA's statutory authority"—a factor that remains highly relevant today. The Report to Congress also noted that "the full scope of planning and implementation of sustainable development policies extended well beyond the purview of USEPA."

While this is obviously true, the role that the USEPA can play in organizing and integrating its own programs is a more practical challenge. Today almost every federal agency is wrestling with how to make sustainability operational. Consequently, the White House Office of Science and Technology Policy has expanded the goal of the Committee on Environment and Natural Resources to include Sustainability (CENRS). Major thrusts of the broadened CENRS are to coordinate across federal agencies and to promote the use of sustainability science in advancing a greener or more sustainable economic growth (http://www.whitehouse.gov/administration/eop/ostp/nstc/committees/cenrs).

In its 1993 report to Congress, the USEPA concluded that the concept of sustainable development "provides a useful framework for discussion of the Nation's long-term environment and economic priorities, although these concepts have not been developed yet to the extent that they provide a basis for EPA's operational planning" (USEPA 1993). The 1993 report failed to recognize the important role that USEPA science and technology could play in achieving sustainability.

Ten years later, the Office of Research and Development (ORD) under the leadership of Paul Gilman launched a renewed effort on sustainability. The independent-minded Gilman knew that the task of advancing sustainability was not going to be easy. His political advisor and communications director confirmed this, stating in 2003 that the concept of sustainability had "no political traction." Fortunately, this atmosphere slowly changed and by the end of 2007, the same political appointee assured me that it was "OK to talk about sustainability." Gilman's vision was to move ORD research beyond its decades-long focus on supporting regulatory development through risk assessment and management, which had gained currency during the 1990s, as the USEPA faced and had to prioritize a large set of responsibilities and as advanced technology allowed for improved detection of potentially toxic chemicals.

Paul Anastas, USEPA's Assistant Administrator for Research and Development from 2009 to early 2012 and a widely respected researcher and author on green chemistry, vigorously promoted ORD's research programs to address sustainability declaring that USEPA science and research must inform, enable, and empower sustainable solutions to the challenges posed to human health and the environment. He emphasized that understanding problems is important and essential, but the only reason to understand a problem deeply is to empower its solution. A diagnosis alone is not a cure, Anastas insisted: we must facilitate solutions to the environmental problems we face.

ORD has moved through five phases, each aimed at advancing sustainability science and innovation. In the mid-1990s, it promoted its "Pollution Prevention Research Strategy" aimed "at implementing a program for systematic research and development activities to carry pollution prevention into the twenty-first century and toward the realization of sustainable development." A key objective of this research program was to improve and develop genetic tools and methodologies such as LCA, which today is a major decision support tool in industry and government. Recognizing the importance of consumer and public support for sustainability, the strategy pioneered new efforts to "develop economic, social, and behavioral tools to improve environmental policies and programs." (http://www.epa.gov/ord/htm/documents/p2.pdf).

This pioneering work was later transferred into the ORD "Sustainability Research Strategy" which attempted to make sustainability *an integrating concept* across its programs. It used the concept of *living laboratories* (regional and state projects) to transfer sustainability concepts to users. It began such transfer through its Collaborative Science and Technology Network for Sustainability (CNS) program and by funding scores of CNS projects that connected diverse sets of partners including universities, federal agencies, and local governments. It also began research to focus on metrics, decision support tools, and technology development.

In a third phase from 2005 to 2007, ORD continued its move toward a more systems-based approach as it developed a sustainability research strategy that focused on systems management. Toward this end, ORD transitioned its Pollution Prevention and New Technologies research program into the Science and Technology for Sustainability (STS) research program. In a fourth phase from 2007 to 2010, ORD responding to guidance from its Science Advisory Board (SAB) and the Board of Scientific Counselors (BOSC) began to apply sustainability research to particular areas of national significance, selecting the goal of sustainable biofuels as an initial area of emphasis.

Today ORD is aiming to make sustainability its "true north." Toward this goal, it is developing research linkages and themes around transdisciplinary research and systems analysis (Fiksel et al. 2009). Making sustainability operational will require realignment of USEPA science into a more systems-oriented approach and acknowledgement of the need for developing models that advance the concept of *resilience*—the capacity for an enterprise to survive, adapt, and grow in the face of turbulent change (Fiksel 2006). In a complex, connected, and uncertain world, resilience will enable human systems to cope successfully with continual waves of change.

In the USEPA and across ORD, making sustainability operational will require an integrated organizational management strategy so that science and management can reinforce each other and lead to a more innovative regulatory and policy framework. ORD has already taken initial steps toward this type of transformation and has asked the National Academies of Sciences to consider how to incorporate the theme of sustainability into all of USEPA's activities. An ad hoc committee under the NRC's STS program has prepared a consensus report addressing several central questions:

- What should be the operational framework for sustainability for USEPA?
- How can the USEPA decision-making process rooted for more than two decades in the risk assessment/risk management paradigm be integrated into this new sustainability framework?
- What scientific and analytical tools are needed to support the framework?
- What set of strategic metrics and indicators should the USEPA build to determine if sustainable approaches are or are not being employed successfully?
- Which assessment techniques and accounting protocols should the USEPA adopt to inform ongoing efforts to improve its sustainability practices and procedures?

The NRC study aims to help the USEPA overcome its stove-piped and fragmented organization. NRC panel member Terry Davies describes the challenges USEPA faces, noting that the laws and the agency focus on pollution control, whereas the emphasis needs to be on prevention. Summing up his preliminary remarks to the NAS committee, Davies said what the agency needs is a global perspective, a fast response time, a focus on products rather than waste, a foundation of science rather than law, a sympathetic approach to economic growth, an anticipatory rather than reactionary stance, a system for self-evaluation, and a renewed emphasis on data.

One of the major challenges for moving sustainability forward and making USEPA an agency committed to sustainability has been the question of how this regulatory agency created to address pollution control could evolve over 40 years and undertake activities to address new problems resulting from population increases, urbanization, and global economic growth. The USEPA has made substantial progress over the decades in addressing obvious and highly visible pollution issues. But in many cases, the USEPA has been reactive to issues rather than getting out in front of them. The sustainability challenge is in effect anticipating future problems, seeing them in an integrated manner, and using all available tools to address them.

Developing sustainable solutions to existing and future environmental and human health problems raises complex scientific and technological issues that cannot be addressed using traditional approaches. If the USEPA is to solve these challenging problems, it must employ integrated systems thinking that complements traditional single-discipline approaches. In all aspects of our work, from problem identification and definition to research design and implementation, ORD must involve the widest span of disciplines to bring different perspectives to the table.

Need for a National Sustainability Policy

While many federal agency reports deal with a range of sustainability issues, there is no government-wide management strategy that focuses on key national issues related to sustainability. "Measuring the Green Economy," a new Department of Commerce baseline report published with contributions from many agencies, begins to advance a collective strategy to accelerate the green economy (US Department of Commerce, Economics and Statistical Administration 2010). Data in this report reveal that green products and services comprised only 1–2% of the total business economy in 2007 and the economy has between 1.8 and 2.4 million green jobs—indicating that we have a long way to go to achieve a green economy. The modest numbers in the Commerce report are a starting point for the use of the well-defined metrics necessary for measuring future growth in green sectors of the economy. National policy is essential for promoting renewable energies, regulating GHGs, and adapting to climate change, which are evolving and cross-cutting dimensions that affect virtually all federal agencies.

Two executive orders—one issued by President George W. Bush and the other by President Barack Obama—have attempted to make sustainability operational in managing government buildings and other facilities. In January 2007, President Bush signed Executive Order 13423, "Strengthening Federal Environmental, Energy, and Transportation Management," which sets goals in the areas of energy efficiency, acquisitions, renewable energy, toxics reductions, recycling, sustainable buildings, electronics stewardship, vehicle fleets, and water conservation. This Order explicitly directs heads of federal agencies to implement sustainable practices in these areas, and specifies that "sustainable" means "creat[ing] and maintain[ing] conditions, under which humans and nature can exist in productive harmony, that permit fulfilling the social, economic, and other requirements of present and future generations of Americans" (Bush 2007).

In 2009, President Obama issued Executive Order 13514, "Federal Leadership in Environmental, Energy, and Economic Performance," which directs each federal agency to appoint a sustainability czar to oversee efforts to reduce GHGs and enhance energy efficiency (Obama 2009). Managing federal facilities is a narrower and thus easier task than creating sustainability polices that many agencies would manage under the constraints of federal statutes.

The collective impact of federal policies is only now leading to recognition that sustainability is an integrating concept, tool, and objective that calls for coordinating policies affecting land, water, and air policies must be coordinated. For example, a successful national energy policy is not based on technology alone, but also on effective management of policies affecting land, water, and air. Policies and regulations must be linked to create sustainable national strategies. And we need a national sustainability policy taking into account both national and international issues.

To serve in much the same way that the annual National Security Strategy guides federal policies in the security area, we need a *National Sustainability Strategy (NSS)*. The 2010 National Security Strategy lays out a strategic approach for advancing American interests, including the security of the American people, a growing US economy, support for our values, and an international order that can address twenty-first century challenges. In an analogous fashion, the NSS would serve as a strategic outline for achieving a greener economy through a convergence of business practices and federal policies and regulations. The NSS should define long-term goals and define a set of indicators or metrics to measure results. It should parallel the National Intelligence Council's role in anticipating and preparing for future challenges. For example, the NIC's "Global Trends 2025: A Transformed World" looks at how key global trends might impact world events in the coming 15 years (http://www.dni.gov/nic/NIC_2025_project.html).

Most federal agencies are reassessing their roles in advancing sustainability. The USEPA will obviously play a critical role in achieving these goals but the issues extend beyond the USEPA with its environmental regulatory focus to nearly all federal agencies—hence the need for a coordinated national strategy. For example, the USEPA's work in the water area is heavily dependent upon activities of the US Geological Survey and state agencies. The success of endangered species protection programs relies on collaboration among the Department of the Interior's Fish and Wildlife Service, the USEPA, the states, and many non-government organizations. The effectiveness of climate change programs depends on USEPA collaboration with the National Atmospheric and Oceanic Administration in the Department of Commerce and with the states. The USEPA should be one of several federal agencies helping us move toward a new environmental management approach that is better suited to the complex and urgent environmental problems of today and the future (Fiksel et al. 2009). At its core the agency must embrace and institutionalize sustainability.

Conclusions

This paper has reviewed why has the concept of sustainability been so difficult to advance in the US federal government and how new environmental, economic, and social drivers are operating to better define sustainability and to make it operational in business and in government. Much of the business world now sees sustainability as a means to reduce long-term risk, reduce costs, and enhance competitiveness.

Many government leaders now see sustainability as essential to domestic well-being, economic growth, and international security as it relates to poverty and social unrest, food security, energy use, and availability of material resources. Advances in science and technology are essential to promote innovation and sustainable solutions. Public understanding and support is also crucial if sustainability is to become operational. Only with effective federal coordination of these vital and interacting elements—green business strategies, regulations and policies, science and technology, and public support—can we achieve sustainability.

Moving toward sustainability will require overcoming bureaucratic stovepipes and fostering coordination within and across agencies and between government and business. For the USEPA this means using science not only to fulfill its mandate to develop and enforce regulations to protect human health and the environment but also to move beyond the current regulatory framework in order to develop and implement a more integrated, systems-based, and cross-media approach to address environmental management.

A promising approach to making sustainability operational in the United States would be the creation of an annual NSS that would define long-term goals and better inform the public about the emerging global sustainability issues and how to effectively deal with them. Like the National Security Strategy, the NSS would outline a coordinated national strategy to achieve crucial short- and long-term national goals—in this case, that of a more sustainable economy. The strategy should define a set of indicators or metrics to measure results and parallel what the National Intelligence Council does in anticipating and preparing for future challenges. More than ever, it is "OK to talk about sustainability."

Acknowledgment The author is grateful to Edward Fallon for his valuable suggestions and editing. Thanks to Joseph Fiksel for drafting Fig. 1.

References

Alms H, Schanz K-U (2008) Capitalising on natural resources: new dynamics in financial markets. Swiss Re Centre for Global Dialogue, The Sustainability Forum, Zürich

Ambachtsheer J, Kron J, Liroff RA, Little T, Massey R (2007) Fiduciary guide to toxic chemical risk. Investor Environmental Health Network and the Rose Foundation for Communities and the Environment, Arlington, VA

Bush GW (2007) Executive order 13423: strengthening federal environmental, energy, and transportation management. Federal register 72(17) (Jan 24), 3919–3923

Business Week (2007) Beyond the green corporation (Jan 29). http://www.businessweek.com/magazine/content/07_05/b4019001.htm. Accessed 26 March 2012

European Council (2000) Lisbon European Council 23 and 24 March 2000 presidency conclusions. Lisbon. http://www.europarl.europa.eu/summits/lis1_en.htm. Accessed 26 March 2012

Fiksel J (2006) Sustainability and resilience: toward a systems approach. Sustain: Sci, Pract, and Policy 2:14–21

Fiksel J, Graedel T, Hecht AD et al (2009) USEPA at 40: bringing environmental protection into the 21st century. Environ Sci Technol 43:8716–8720

Global Footprint Network (2010). www.footprintnetwork.org. Accessed 26 March 2012

Grossarth SK, Hecht AD (2007) Sustainability at the US Environmental Protection Agency, 1970-2020. Ecol Eng 30:1–8

Hammond A, Kramer W, Tran J et al (2007) The next 4 billion: market size and business strategy at the base of the pyramid. World Resources Institute and International Finance Corporation, Washington, DC

Hecht AD (2007) The next level of environmental protection: business strategies and government policies converging on sustainability. Sustainable Dev Law Pol 7(Fall):19–25

Hecht AD (2009) Resolving the climate wars. Sustain Dev Law Pol 4:4–5

Larsson A, Azar C, Sterner T et al (2002) Technology and policy for sustainable development. Centre for Environment and Sustainability at Chalmers University of Technology and the Göteborg University, Goteborg

Lubin DA, Esty DC (2010) The sustainability imperative: lessons for leaders from previous game-changing megatrends. Harv Bus Rev (May):42–50

Millennium Ecosystem Assessment Program (2005) Millennium Ecosystem Assessment. Island Press, Washington, DC

Nidumolu RA, Prahalad CK, Rangaswami MR (2009) Why sustainability is now the key driver of innovation. Harv Bus Rev (Sept):87:56–64

Obama B (2009) Executive Order 13514, Federal leadership in environmental, energy, and economic performance. Federal register74/194 (Oct 5), 52117–52127

PricewaterhouseCoopers (2002) Sustainability survey report. PricewaterhouseCoopers, New York

Marks & Spencer (2010a) 100 ways to make a difference: the plan. http://plana.marksandspencer.com/about/the-plan. Accessed 26 March 2012

Marks & Spencer (2010b) Your M&S: how we do business report. Marks & Spencer, London. http://corporate.marksandspencer.com/documents/publications/2010/How_We_Do_Business_Report_2010. Accessed 18 Feb 2011

Sullivan N, Schiafo R (2005) Talking green, acting dirty. The NY Times, June 12

UN General Assembly (1987) Report of the world commission on environment and development: our common future. United Nations General Assembly, New York

US Department of Commerce, Economics and Statistical Administration (2010) Measuring the green economy. National Institute of Standards and Technology, Washington

USEPA (1993) Sustainable development and the Environmental Protection Agency. USEPA report to congress. USEPA 230 R-93-005, Washington, DC

Weiss J (2009) Local governance and sustainability: major progress, significant challenges. In: Dernbach J (ed) Agenda for a sustainable America. Environmental Law Institute, Washington, DC

Winston A (2009) Green recovery: get lean, get smart, and emerge from the downturn on top. Harvard Business Press, Boston

World Bank Group (2008) Sustainable infrastructure action plan FY 2009-2011. International Bank for Reconstruction and Development/World Bank Group, Washington, DC

Part II
Balancing Ecology and Economy: Natural Capital and Quality of Life

In addressing the issue of sustainability, John Peterson Meyers posed the fundamental question: "how much of the earth's ecological integrity can we disrupt before we pass a threshold in the loss of life-support services?" In this section, we consider ecosystem services and natural capital as the conditions and processes through which undisturbed ecosystems, and the species that comprise them, sustain and fulfill human life. Not only do these services maintain biodiversity, but also they produce ecosystem goods that are of direct value to the world's economies. Relative to sustainability, new and truly integrated assessments and models of the quality, quantity, and spatial and temporal dynamics of ecosystem services and the various aspects of their connection to human well-being in the long run are needed.

The Value of Natural and Social Capital in Our Current Full World and in a Sustainable and Desirable Future

Robert Costanza

Abstract Ecosystem services are the benefits people obtain from ecosystems. These include provisioning services, such as food and water; regulating services, such as regulation of floods, drought, and disease; supporting services, such as soil formation and nutrient cycling; and cultural services, such as recreational, spiritual, and other nonmaterial benefits. These benefits may or may not be fully perceived by people. Most are outside the market exchange system and are best thought of and managed as public goods (the commons). Ecosystems are experiencing serious degradation in regard to their capability of providing services. At the same time, the demand for ecosystem services is rapidly increasing as populations and standards of living increase.

Keywords Quality of life • Happiness • Sustainable well-being • Full world • Lisbon principles • Genuine Progress Indicator

Introduction

As the world has moved from one relatively empty of humans and their artifacts to one increasingly full of humans and their artifacts, the value of our natural and social capital assets (the commons) has become significantly more important to sustaining human happiness and well-being than marketed goods and services (as measured by GDP). In this world we must better assess, model, and value our natural and social capital assets. A sustainable and desirable future is one that respects biophysical boundaries, distributes resources and responsibilities fairly, and adequately values and balances built, human, social and natural capital assets.

R. Costanza (✉)
Institute for Sustainable Solutions, Portland State University,
Portland, OR 97201, USA
e-mail: Robert.Costanza@pdx.edu

M.P. Weinstein and R.E. Turner (eds.), *Sustainability Science: The Emerging Paradigm and the Urban Environment*, DOI 10.1007/978-1-4614-3188-6_5,

From an Empty to a Full World

The mainstream model of development (also known as the "Washington consensus") is based on a number of assumptions about the way the world works, what the economy is, and what the economy is for (Table 1). These assumptions were created during a period when the world was still relatively empty of humans and their built infrastructure. In this "empty world" context, built capital was the limiting factor, while natural capital and social capital were abundant. It made sense, in that context, not to worry too much about environmental and social "externalities" since they could be assumed to be relatively small and ultimately solvable. It made sense to focus on the growth of the market economy, as measured by GDP, as a primary

Table 1 Basic characteristics of the current development model and the emerging sustainable and desirable "ecological economics" development model

	Current development model: the "Washington Consensus"	*Sustainable and desirable development model*: an emerging "Green Consensus"
Primary policy goal	*More*: economic growth in the conventional sense, as measured by GDP. The assumption is that growth will ultimately allow the solution of all other problems. More is always better	*Better*: focus must shift from merely growth to "development" in the real sense of improvement in quality of life, recognizing that growth has negative by-products and more is not always better
Primary measure of progress	GDP	GPI (or similar)
Scale/carrying capacity	Not an issue since markets are assumed to be able to overcome any resource limits via new technology and substitutes for resources are always available	A primary concern as a determinant of ecological sustainability. Natural capital and ecosystem services are not infinitely substitutable and real limits exist
Distribution/poverty	Lip service, but relegated to "politics" and a "trickle down" policy: a rising tide lifts all boats	A primary concern since it directly affects quality of life and social capital and in some very real senses is often exacerbated by growth: a too rapidly rising tide only lifts yachts, while swamping small boats
Economic efficiency/ allocation	The primary concern, but generally including only marketed goods and services (GDP) and institutions	A primary concern, but including both market and nonmarket goods and services and effects. Emphasizes the need to incorporate the value of natural and social capital to achieve true allocative efficiency

(continued)

Table 1 (continued)

	Current development model: the "Washington Consensus"	*Sustainable and desirable development model*: an emerging "Green Consensus"
Property rights	Emphasis on private property and conventional markets	Emphasis on a balance of property rights regimes appropriate to the nature and scale of the system, and a linking of rights with responsibilities. A larger role for common property institutions in addition to private and state property
Role of government	To be minimized and replaced with private and market institutions	A central role, including new functions as referee, facilitator and broker in a new suite of common asset institutions
Principles of governance	*Laissez faire* market capitalism	Lisbon principles of sustainable governance

means to improve human welfare. It made sense, in that context, to think of the economy as only marketed goods and services and to think of the goal as increasing the amount of these goods and services produced and consumed.

But the world has changed dramatically. We now live in a world relatively full of humans and their built capital infrastructure. In this new context, we have to reconceptualize what the economy is and what it is for. We have to first remember that the goal of the economy is to sustainably improve human well-being and quality of life. We have to remember that material consumption and GDP are merely means to that end, not ends in themselves. We have to recognize, as both ancient wisdom and new psychological research tell us, that material consumption beyond real need can actually reduce our well-being. We have to better understand what really does contribute to sustainable human well-being, and recognize the substantial contributions of natural and social capital, which are now the limiting factors to sustainable human well-being in many countries. We have to be able to distinguish between real poverty in terms of low quality of life and merely low monetary income. Ultimately, we have to create a new vision of what the economy is and what it is for and a new model of development that acknowledges this new full world context and vision (Table 1).

Quality of Life, Happiness, and the Real Economy

There is a substantial body of new research on what actually contributes to human well-being and quality of life (Costanza et al. 2008). This new "science of happiness" clearly demonstrates the limits of conventional economic income and consumption in contributing to well-being. Kasser (2003) points out, for instance, that people who focus on material consumption as a path to happiness are actually less happy and even suffer higher rates of both physical and mental illnesses than those who do not. Material consumption beyond real need is a form of psychological "junk food" that only satisfies for the moment and ultimately leads to depression.

Easterlin (2003) has shown that well-being tends to correlate well with health, level of education, and marital status, and not very well with income beyond a certain fairly low threshold. He concludes that:

> People make decisions assuming that more income, comfort, and positional goods will make them happier, failing to recognize that hedonic adaptation and social comparison will come into play, raise their aspirations to about the same extent as their actual gains, and leave them feeling no happier than before. As a result, most individuals spend a disproportionate amount of their lives working to make money, and sacrifice family life and health, domains in which aspirations remain fairly constant as actual circumstances change, and where the attainment of one's goals has a more lasting impact on happiness. Hence, a reallocation of time in favor of family life and health would, on average, increase individual happiness.

Layard (2005) synthesizes many of these ideas and concludes that current economic policies are not improving happiness and that "happiness should become the goal of policy, and the progress of national happiness should be measured and analyzed as closely as the growth of GNP."

Frank (2000) also concludes that some nations would be better off—overall national well-being would be higher, that is—if we actually consumed less and spent more time with family and friends, working for our communities, maintaining our physical and mental health, and enjoying nature.

On this last point, there is substantial and growing evidence that natural systems contribute heavily to human well-being. Costanza et al. (1997a) estimated the annual, nonmarket value of the earth's ecosystem services at $33 trillion/year, substantially larger than global GDP at the time and yet an almost certainly a conservative underestimate. The UN Millennium Ecosystem Assessment (MEA) (2005) is a global compendium addressing the status and trends of ecosystem services and their contributions to human well-being.

So, if we want to assess the "real" economy—all the things which contribute to real, sustainable, human well-being—as opposed to only the "market" economy, we have to measure and include the nonmarketed contributions to human well-being from nature; from family, friends, and other social relationships at many scales; and from health and education. One convenient way to summarize these contributions is to group them into four basic types of capital that are necessary to support the real, human-well-being-producing economy: built capital, human capital, social capital, and natural capital.

The market economy covers mainly built capital (factories, offices, and other built infrastructure and their products) and part of human capital (spending on labor, health, and education), with some limited spillover into the other two. Human capital includes the health, knowledge, and all the other attributes of individual humans that allow them to function in a complex society. Social capital includes all the formal and informal networks among people: family, friends, and neighbors, as well as social institutions at all levels, like churches, social clubs, local, state, and national governments, NGOs, and international organizations. Natural capital includes the world's ecosystems and all the services they provide. Ecosystem services occur at many scales, from climate regulation at the global scale, to flood protection, soil formation, nutrient cycling, recreation, and aesthetic services at the local and regional scales.

Natural Capital and Ecosystem Services

"Ecosystem services" (ES) are the ecological characteristics, functions, or processes that directly or indirectly contribute to human well-being—the benefits people derive from functioning ecosystems (Costanza et al. 1997a; MEA 2005). Ecosystem processes and functions may contribute to ecosystem services but they are not synonymous. Ecosystem processes and functions describe biophysical relationships and exist regardless of whether or not humans benefit (Boyd and Banzhaf 2007; Granek et al. 2010). Ecosystem services, on the other hand, only exist if they contribute to human well-being and cannot be defined independently.

The ecosystems that provide the services are sometimes referred to as "natural capital," using the general definition of capital as a stock that yields a flow of services over time (Costanza and Daly 1992). In order for these benefits to be realized, natural capital (which does not require human activity to build or maintain) must be combined with other forms of capital that *do* require human agency to build and maintain. These include (1) built or manufactured capital, (2) human capital, and (3) social or cultural capital (Costanza et al. 1997b).

These four general types of capital are all required in complex combinations to produce any and all human benefits. *Ecosystem services thus refer to the relative contribution of natural capital to the production of various human benefits, in combination with the three other forms of capital.* These benefits can involve the use, nonuse, option to use, or mere appreciation of the existence of natural capital.

The following categorization of ecosystem services has been used by the MEA (2005):

(a) *Provisioning services*—ecosystem services that combine with built, human, and social capital to produce food, timber, fiber, or other "provisioning" benefits. For example, fish delivered to people as food require fishing boats (built capital), fisherfolk (human capital), and fishing communities (social capital) to produce.
(b) *Regulating services*—services that regulate different aspects of the integrated system. These are services that combine with the other three capitals to produce flood control, storm protection, water regulation, human disease regulation, water purification, air quality maintenance, pollination, pest control, and climate control. For example, storm protection by coastal wetlands requires built infrastructure, people, and communities to be protected. These services are generally not marketed but have clear value to society.
(c) *Cultural services*—ecosystem services that combine with built, human, and social capital to produce recreation, aesthetic, scientific, cultural identity, sense of place,

or other "cultural" benefits. For example, to produce a recreational benefit requires a beautiful natural asset (a lake), in combination with built infrastructure (a road, trail, dock, etc.), human capital (people able to appreciate the lake experience), and social capital (family, friends, and institutions that make the lake accessible and safe). Even "existence" and other "nonuse" values' require people (human capital) and their cultures (social and built capital) to appreciate.

(d) *Supporting "services"*—services that maintain basic ecosystem processes and functions such as soil formation, primary productivity, biogeochemistry, and provisioning of habitat. These services affect human well-being *indirectly* by maintaining processes necessary for provisioning, regulating, and cultural services. They also refer to the ecosystem services that have not yet or may never be intentionally combined with built, human, and social capital to produce human benefits but that support or underlie these benefits and may sometimes be used as proxies for benefits when the benefits cannot be easily measured directly. For example, net primary production (NPP) is an ecosystem function that supports carbon sequestration and removal from the atmosphere, which combines with built, human, and social capital to provide the benefit of climate regulation. Some would argue that these "supporting" services should rightly be defined as ecosystem "functions," since they may not yet have interacted with the other three forms of capital to create benefits. I agree with this in principle, but recognize that supporting services/functions may sometimes be used as proxies for services in the other categories.

This categorization suggests a very broad definition of services, limited only by the requirement of a contribution to human well-being. Even without any subsequent valuation, explicitly listing the services derived from an ecosystem can help ensure appropriate recognition of the full range of potential impacts of a given policy option. This can help make the analysis of ecological systems more transparent and can help inform decision makers of the relative merits of different options before them.

Are We Really Making Progress?

Given this definition of the real economy, are we really making progress? Is the mainstream development model really working, even in the "developed" countries? One way to tell is through surveys of people's life satisfaction, which have been relatively flat in the USA and many other developed countries since about 1975. A second approach is an aggregate measure of the real economy that has been developed as an alternative to GDP called the Index of Sustainable Economic Welfare (ISEW—Daly and Cobb 1989) and more recently renamed the Genuine Progress Indicator (GPI—Cobb et al. 1995).

Let's first take a quick look at the problems with GDP as a measure of true human well-being. GDP is not only limited—measuring only marketed economic activity or gross income—it also counts all of this activity as positive. It does not separate desirable, well-being-enhancing activity from undesirable well-being-reducing activity. For example, an oil spill increases GDP because someone has to clean it up, but it obviously detracts from society's well-being. From the perspective of GDP, more crime, more sickness, more war, more pollution, more fires, storms, and pestilence are all potentially good things, because they can increase marketed activity in the economy.

GDP also leaves out many things that *do* enhance well-being but are outside the market. For example, the unpaid work of parents caring for their own children at home does not show up, but if these same parents decide to work outside the home to pay for child care, GDP suddenly increases. The nonmarketed work of natural capital in providing clean air and water, food, natural resources, and other ecosystem services does not adequately show up in GDP, either, but if those services are damaged and we have to pay to fix or replace them, then GDP suddenly increases. Finally, GDP takes no account of the distribution of income among individuals. But it is well known that an additional $1 worth of income produces more well-being if one is poor rather than rich. It is also clear that a highly skewed income distribution has negative effects on a society's social capital.

The GPI addresses these problems by separating the positive from the negative components of marketed economic activity, adding in estimates of the value of nonmarketed goods and services provided by natural, human, and social capital, and adjusting for income-distribution effects. While it is by no means a perfect representation of the real well-being of nations, GPI is a much better approximation than GDP. As Amartya Sen and others have noted, it is much better to be approximately right in these measures than precisely wrong.

Comparing GDP and GPI for the USA shows that, while GDP has steadily increased since 1950, with the occasional dip or recession, GPI peaked in about 1975 and has been flat or gradually decreasing ever since. From the perspective of the real economy, as opposed to just the market economy, the USA has been in recession since 1975. As already mentioned, this picture is also consistent with survey-based research on people's stated life satisfaction. The USA and several other developed countries are now in a period of what Herman Daly has called "un-economic growth," where further growth in marketed economic activity (GDP) is actually reducing well-being on balance rather than enhancing it. In terms of the four capitals, while built capital has grown, human, social, and natural capital have declined or remained constant and more than canceled out the gains in built capital.

A New Sustainable, Ecological Model of Development

A new model of development consistent with our new full world context (Table 1) would be based clearly on the goal of sustainable human well-being. It would use measures of progress that clearly acknowledge this goal (i.e., GPI instead of GDP). It would acknowledge the importance of ecological sustainability, social fairness, and real economic efficiency.

Ecological sustainability implies recognizing that natural and social capital are not infinitely substitutable for built and human capital, and that real biophysical limits exist to the expansion of the market economy. Climate change is perhaps the most obvious and compelling of these limits.

Social fairness implies recognizing that the distribution of wealth is an important determinant of social capital and quality of life. The conventional development model, while explicitly aimed at reducing poverty, has bought into the assumption that the best way to do this is through growth in GDP. This has not proved to be the case and explicit attention to distribution issues is sorely needed. As Frank (2007) has argued, economic growth beyond a certain point sets up a "positional arms race" that changes the consumption context and forces everyone to consume too much of easily seen positional goods (like houses and cars) at the expense of nonmarketed, nonpositional goods and services from natural and social capital. Increasing inequality of income actually reduces overall societal well-being, not just for the poor, but across the income spectrum (Wilkenson and Pickett 2009).

Real economic efficiency implies including all resources that affect sustainable human well-being in the allocation system, not just marketed goods and services. Our current market allocation system excludes most nonmarketed natural and social capital assets and services that are huge contributors to human well-being. The current development model ignores this and therefore does not achieve real economic efficiency. A new, sustainable ecological development model would measure and include the contributions of natural and social capital and could better approximate real economic efficiency.

The new development model would also acknowledge that a complex range of property rights regimes are necessary to adequately manage the full range of resources that contribute to human well-being. For example, most natural and social capital assets are public goods. Making them private property does not work well. On the other hand, leaving them as open access resources (with no property rights) does not work well either. What is needed is a third way to *propertize* these resources without privatizing them. Several new (and old) common property rights systems have been proposed to achieve this goal, including various forms of common property trusts.

The role of government also needs to be reinvented. In addition to government's role in regulating and policing the private market economy, it has a significant role to play in expanding the "commons sector" that can propertize and manage nonmarketed natural and social capital assets. It also has a major role to play as facilitator of societal development of a shared vision of what a sustainable and desirable future would look like. Strong democracy based on developing a shared vision is an essential prerequisite to building a sustainable and desirable future (Prugh et al. 2000). This new vision implies a core set of principles for sustainable governance.

Principles of Sustainable Governance

The key to achieving sustainable governance in the new full world context is an integrated (across disciplines, stakeholder groups, and generations) approach based on the paradigm of "adaptive management," whereby policy-making is an iterative experiment acknowledging uncertainty, rather than a static "answer." Within this paradigm, six core principles (the Lisbon principles) that embody the essential criteria for sustainable governance have been proposed (Costanza et al. 1998). Some of them are already well accepted in the international community (e.g., Principle 3); others are variations on well-known themes (e.g., Principle 2 is an extension of the subsidiary principle); while others are relatively new in international policy, although they have been well developed elsewhere (e.g., Principle 4). The six Principles together form an indivisible collection of basic guidelines governing the use of common natural and social capital assets.

- *Principle 1: Responsibility.* Access to common asset resources carries attendant responsibilities to use them in an ecologically sustainable, economically efficient, and socially fair manner. Individual and corporate responsibilities and incentives should be aligned with each other and with broad social and ecological goals.
- *Principle 2: Scale-matching.* Problems of managing natural and social capital assets are rarely confined to a single scale. Decision-making should (1) be assigned to institutional levels that maximize input, (2) ensure the flow of information between institutional levels, (3) take ownership and actors into account, and (4) internalize costs and benefits. Appropriate scales of governance will be those that have the most relevant information, can respond quickly and efficiently, and are able to integrate across scale boundaries.
- *Principle 3: Precaution.* In the face of uncertainty about potentially irreversible impacts to natural and social capital assets, decisions concerning their use should err on the side of caution. The burden of proof should shift to those whose activities potentially damage natural and social capital.
- *Principle 4: Adaptive management.* Given that some level of uncertainty always exists in common asset management, decision-makers should continuously gather and integrate appropriate ecological, social, and economic information with the goal of adaptive improvement.
- *Principle 5: Full cost allocation.* All of the internal and external costs and benefits, including social and ecological, of alternative decisions concerning the use of natural and social capital should be identified and allocated. When appropriate, markets should be adjusted to reflect full costs.
- *Principle 6: Participation.* All stakeholders should be engaged in the formulation and implementation of decisions concerning natural and social capital assets. Full stakeholder awareness and participation contributes to credible, accepted rules that identify and assign the corresponding responsibilities appropriately.

Some Policies to Achieve Real, Sustainable Development

The conventional development model is not working, for either the developed or the developing world. It is not sustainable and it is also not desirable. It is based on a now obsolete empty world vision and it is leading us to disaster. We need to accept that we now live in a full world context where natural and social capital are the limiting factors. We could achieve a much higher quality of life, and one that would be ecologically sustainable, socially fair, and economically efficient, if we shift to a new sustainable development paradigm that incorporates these principles.

The problem is that our entire modern global civilization is, as even former President Bush has acknowledged, "addicted to oil" and addicted to consumption and the conventional development model in general. An addictive substance is something one has developed a dependence on, which is either not necessary or harmful to one's longer term well-being. Fossil fuels (and excessive material consumption in general) fit the bill. We can power our economies with renewable energy, and we can be happier with lower levels of consumption, but we must first break our addiction to fossil fuels, consumption, and the conventional development model, and as any addict can tell you: "that ain't easy." But in order to break an addiction of any kind, one must first clearly see the benefits of breaking it and the costs of remaining addicted, facts that accumulating studies like the IPCC reports, the Stern Review (2007), the MEA (2005), and many others are making more apparent every day.

What else can we do to help break this addiction? Here are a few suggestions.

- Create and share a vision of a future with zero fossil fuel use and a quality of life higher than today. That will involve understanding that GDP is a means to an end, not the end itself, and that in some countries today more GDP actually results in less human well-being (while in others the reverse is still true). It will require a focus on sustainable scale and just distribution. It will require an entirely new and broader vision of what the economy is, what it's for, and how it functions.
- Convene a "new Bretton Woods" conference to establish the new measures and institutions needed to replace GDP, the World Bank, the IMF, and the WTO.

  These new institutions would promote:

  - Shifting primary national policy goals from increasing marketed economic activity (GDP) to maximizing national well-being (GPI or something similar). This would allow us to see the interconnections between built, human, social, and natural capital, and build real well-being in a balanced and sustainable way.
  - Reforming tax systems to send the right incentives by taxing negatives (pollution, depletion of natural capital, overconsumption) rather than positives (labor, savings, investment).
  - Expanding the commons sector by developing new institutions that can *propertize* the commons without privatizing them. Examples include various forms of common asset trusts, like the atmospheric (or sky) trust (Barnes et al. 2008) payments for depletion of natural and social capital and rewards for protection of these assets.
  - Reforming international trade to promote well-being over mere GDP growth. This implies protecting natural capital, labor rights, and democratic

self-determination first and *then* allowing trade, rather than promoting the current trade rules that ride roughshod over all other societal values and ignore nonmarket contributions to well-being.

We can break our addiction to fossil fuels, overconsumption, and the current development model and create a more sustainable and desirable future. It will not be easy, and it will require a new vision, new measures, and new institutions. It will require a directed evolution of our entire society (Beddoe et al. 2009). But it is not a sacrifice of quality of life to break this addiction. Quite the contrary, it is a sacrifice not to.

Acknowledgments Versions of parts of this chapter have appeared in previously published works with a range of co-authors. I thank the co-authors of those works and Michael P. Weinstein for helpful comments on earlier drafts.

References

Barnes PR, Costanza P, Hawken D et al (2008) Creating an earth atmospheric trust. Science 319:724
Beddoe R, Costanza R, Farley J et al (2009) Overcoming systemic roadblocks to sustainability: the evolutionary redesign of worldviews, institutions and technologies. Proc Natl Acad Sci USA 106:2483–2489
Boyd J, Banzhaf S (2007) What are ecosystem services? Ecol Econ 63:616–626
Cobb C, Halstead T, Rowe J (1995) The genuine progress indicator: summary of data and methodology. Redefining Progress, San Francisco
Costanza R, Daly HE (1992) Natural capital and sustainable development. Conserv Biol 6:37–46
Costanza R, D'Arge R, de Groot R et al (1997a) The value of the world's ecosystem services and natural capital. Nature 387:253–260
Costanza R, Cumberland JC, Daly HE et al (1997b) An introduction to ecological economics. St. Lucie Press, Boca Raton
Costanza R, Andrade F, Antunes P et al (1998) Principles for sustainable governance of the oceans. Science 281:198–199
Costanza R, Fisher B, Ali S et al (2008) An integrative approach to quality of life measurement, research, and policy. Surv Perspect Integr Environ Soc 1:1–5
Daly HE, Cobb J (1989) For the common good. Beacon Press, Boston
Easterlin RA (2003) Explaining happiness. Proc Natl Acad Sci USA 100:11176–11183
Frank R (2000) Luxury fever. Princeton University Press, Princeton
Frank R (2007) Falling behind: how rising inequality harms the middle class. University of California Press, Berkeley
Granek EF, Polasky S, Kappel CV et al (2010) Ecosystem services as a common language for coastal ecosystem-based management. Conserv Biol 24:7–216
Kasser T (2003) The high price of materalism. MIT Press, Cambridge
Layard R (2005) Happiness: lessons from a new science. Penguin, New York
Millennium Ecosystem Assessment (MEA) (2005) Ecosystems and human well-being: synthesis. Island Press, Washington, DC
Prugh T, Costanza R, Daly HE (2000) The local politics of global sustainability. Island Press, Washington, DC
Stern N (2007) The economics of climate change: the Stern review. Cambridge University Press, Cambridge
Wilkenson R, Pickett K (2009) The spirit level: why greater equality makes societies stronger. Bloomsbury Press, New York

Steps Towards Sustainability and Tools for Restoring Natural Capital: Etang de Berre (Southern France) Case Study

James Aronson, Florian Claeys, Vanja Westerberg, Philippe Picon, Guillaume Bernard, Jean-Michel Bocognano, and Rudolf de Groot

Abstract Communities, nations, not-for-profit groups, and some mining, infrastructure, and energy corporations are catching on to the fact that the ecological restoration of degraded ecosystems is vital to their search for sustainability and ecological accountability. The science of restoration ecology can provide the tools and major building blocks necessary to develop a transdisciplinary sustainability science and is a problem-solving toolkit used on the road to global, regional, national, and local sustainability. We discuss a landscape-scale restoration program for the large

J. Aronson (✉)
Centre d'Ecologie Fonctionnelle et Evolutive (CNRS-UMR 5175),
1919, Route de Mende, Montpellier 34293, France
e-mail: james.aronson@cefe.cnrs.fr

F. Claeys
École Normale Supérieure, Département de Biologie, 45 rue d'Ulm, 75005 Paris, France
e-mail: florian.cleays@gmail.com

V. Westerberg
Ecole nationale supérieure agronomique de Montpellier (SupAgro),
Laboratoire Montpelliérain d'Économie Théorique et Appliquée (LAMETA),
2 place Pierre Viala, 34060 Montpellier, Cedex 1, France
e-mail: vanja.westerberg@supagro.inra.fr

P. Picon • G. Bernard
Groupement d'Intérêt Publique pour la Réhabilitation de l'Etang de Berre,
GIPREB, Cours Mirabeau, Berre-l'Etang 13130, France
e-mail: philippe.picon@gipreb.fr; guillaume.bernard@gipreb.fr

J.-M. Bocognano
Grand Port Maritime de Marseille, 23 Place de la Joliette, BP 81965, Marseille 13226, France
e-mail: Jean-Michel.Bocognano@marseille-port.fr

R. de Groot
Environmental Systems Analysis Group, University of Wageningen, PO Box 47,
Wageningen 6700 AA, The Netherlands
e-mail: dolf.degroot@wur.nl

M.P. Weinstein and R.E. Turner (eds.), *Sustainability Science: The Emerging Paradigm and the Urban Environment*, DOI 10.1007/978-1-4614-3188-6_6,

(155 km^2) and heavily polluted *Étang de Berre* (Berre Lagoon) to illustrate these ideas. This lagoon is situated between Marseille, Salon-de-Provence, and Aix-en-Provence in southern France. We illustrate the use of (a) sequential references, which is a technique from the field of restoration ecology that helps clarify goals and develop consensus among stakeholders and scientists of differing backgrounds; and (b) HMCA (historical multicriteria analysis), which is a variation of MCA that is often used in ecological and environmental economics. We show how to use a HMCA to synthesize ecological, social, and economic criteria across different historical time periods and be applied to a large scale, multifaceted project of this sort when a sequential reference exercise has been performed. Lastly, we note that ecological restoration is the key means for restoring natural capital (RNC) and to simultaneously recover and revitalize social capital. In the ecologically and economically beleaguered and vulnerable area as the one considered here, and indeed many others around the world, the road to sustainability passes through a portal of what we call "RNC thinking."

Keywords Sustainability • Restoring natural capital • Mediterranean lagoon • Historical MCA • Sequential reference method • Berre lagoon

Introduction

Many communities, nations, not-for-profit organizations, and some large corporations have understood that the ecological restoration of degraded ecosystems is important—perhaps even essential—to achieve sustainability. Ecological restoration programs can create jobs and livelihood opportunities and boost the flow of ecosystem services to society while also contributing to the conservation of biodiversity and functional ecosystems (Bullock et al. 2011). Additionally, and across a very wide range of biomes or ecosystem types, ecological restoration can be cost-effective provided that the full range of benefits from restored ecosystem are fully accounted for, along with the upfront costs (Neßhöver et al. 2011; de Groot et al. in review; Elmqvist et al. in review). For these reasons and others, ecological restoration actions are increasingly being deployed, developed, and implemented worldwide, where they are supported by global policy commitments such as the Convention on Biological Diversity, the United Nations Framework Convention on Climate Change's negotiations to update the Clean Development Mechanism Kyoto Protocol with something called "Reducing Emissions from Deforestation and Forest Degradation (REDD)" as an international fund- or credit-based mechanism for reducing CO_2 and other greenhouse gas emissions and protecting forest ecosystems (Rey Benayas et al. 2009; Bullock et al. 2011; Alexander et al. 2011). Legislation in some countries is getting underway that requires true restoration (Aronson et al. 2011). In many other countries, the long-standing regulations and environmental legislation are getting tougher, which is a good first step. Many mineral extraction and transformation companies, energy utilities and other corporate giants, as well as insurance companies and large lending banks are taking note of this trend, even though hard data from the corporations are hard to find.

Measures aimed at ecological restoration, however, are still generally seen as net-cost projects, despite evidence showing that investing in the restoration of renewable natural capital* (=ecosystems and biodiversity) (note: definitions of terms in the text followed by "*" the first time they appear are provided in Box 1), and cultivated natural capital (production-oriented systems), makes economic sense once the full range of benefits provided by biodiversity and ecosystem services are taken into account. Ecological restoration*, ecological rehabilitation* and, more broadly, the restoration of natural capital* at the landscape* scale, can help reconcile the objectives of nature conservation, on the one hand, and sustainable economic development goals on the other, and thereby help society move towards a more sustainable, more just, and more desirable future. But, the road is not smooth, and there will almost always be conflicts of interest.

Box 1 Definitions of Terms as Used in This Chapter

Ecosystem degradation is the loss of biodiversity and the simplification or disruption of the structure, function, and composition caused by disturbances that are too frequent or severe to allow the natural regeneration or recovery to occur. Degradation results from various factors that are often interlinked, such as human activities, climate perturbations, and extreme events (e.g., drought, fire, and storms), which reduce the quality and flow of ecosystem goods and services.

Ecological restoration is "The process of assisting the recovery of an ecosystem that has been degraded, damaged, or destroyed" (SER 2004). It is an intentional activity that initiates or accelerates those ecosystem processes that return complexity and resilience, reinstate structure and function, and reestablish a trajectory of self-sustaining maturation or, in a special case, reestablishes arrested succession. The term is often used, in a very broad and rather vague way, to mean to return a site or system to "predisturbance conditions."

Ecological rehabilitation, in the broad sense, is the improvement of ecosystem functions without necessarily achieving a return to "predisturbance" conditions. Emphasis is generally on restoring *ecosystem processes and functions* so as to increase the flow of services and benefits to people (SER 2004; Clewell and Aronson 2007). Care must be paid not to so heavily favor one process or function with the result of rendering the ecosystem more fragile or vulnerable than it was before. However, when returning an ecosystem to a former state or ideal condition is not possible, rehabilitation is often the best option.

Historical multicriteria analysis (HMCA) is an invention of the authors intended to facilitate the collective choice of an ecological reference (sensu SER 2004; Clewell and Aronson 2007) and is based on a sequential reference study of the historical stages involved in the degradation and transformation of a given ecosystem or landscape.

(continued)

Box 1 (continued)

Landscape is an assemblage of ecosystems that are arranged in recognizable patterns and that exchange organisms and materials such as water (note: there are many other acceptable definitions of this concept) (Forman and Gordon 1986). Notably, most landscapes today are mosaics of interacting ecosystems that may be natural, near-natural, or production systems, in which spaces or landscape units managed for social and economic use without any specific systems thinking.

Natural capital is an economic metaphor for the limited stocks of physical and biological natural elements found on Earth. Some of these stocks are of direct use to society (resources), and some are not. According to Rees (1995) and MA (2005), there are four, partially overlapping types: (1) renewable (living species and ecosystems), (2) nonrenewable (subsoil assets, e.g., petroleum, coal, diamonds), (3) replenishable (e.g., the atmosphere, potable water, fertile soils), and (4) cultivated (e.g., crops and forest plantations). Natural capital "provides the basis for all life," and it is a highly useful metaphor for the "stocks," "assets," or reserves of physical and biological elements found on earth (MA 2005), some of which are used by people, and then are called "resources." If natural capital is a stock or an asset, then the "dividend" is the flow in ecosystem goods and services derived from the assets; e.g., the dividend from renewable natural capital is ecosystems and biodiversity. We note that many species or attributes of natural ecosystems are not directly useful to people and are not normally considered as resources. The concept of renewable natural capital emphatically includes those attributes, which exist within the ecosystems as a whole, and evolve. Natural capital is a not just the obviously marketable bits and pieces.

Restoration of natural capital (*RNC*) is an investment in natural capital stocks to improve the sustainability of both natural and human-managed ecosystems, while contributing to the socioeconomic well-being of people (Aronson et al. 2007, 2010). Renewable, replenishable, and cultivated natural capital delivers ecosystems goods and services. RNC is required when delivery of ecosystem services is interrupted or impeded. The RNC includes the ecological restoration or rehabilitation of ecosystems, ecologically sound improvements to production systems, ecologically sound improvements in the utilization of biological resources, or nonrenewable natural capital, and efforts to increase public awareness and appreciation for the importance of natural capital and the wisdom of investing in its restoration, which is sometimes also referred to as "restoring ecological infrastructure" (TEEB 2010; Neßhöver et al. 2011).

RNC thinking is needed to complement "Resilience Thinking" (Salt and Walker 2006). Restoration of Natural Capital (RNC) is panoply of investments, interventions, improvements in current practices, and new ideas and social programs to promote and accelerate national and international transitions towards sustainability. RNC is a broader concept than ecological restoration.

Social capital is the institutions, relationships, social networks, and shared cultural beliefs and traditions that promote hope and mutual trust.

The preceding chapters of this book have many approaches to address the question about how scientists can help society move forward on the path to sustainability through and within the emerging paradigm of sustainability science. Sustainability science comprises the different fields of scientific endeavors that bring together scientists and professionals to study the dynamic interactions between "nature" and human society. From the perspective of restoration ecology, this endeavor is undertaken with a view to helping society maintain and restore both natural capital and social capital*. Sustainability science, therefore, should contribute to the goals set by the Brundtland Commission of the UN in 1987: "Sustainable development is development that meets the needs of the present without compromising the ability of future generations to meet their own needs" (WCED 1987).

We argue that restoration ecology, when coupled with the practice of ecological restoration, can provide major building blocks for the development of a transdisciplinary science of sustainability and to support society's search for problem-solving processes carried on the rocky road to sustainability. This contribution holds at all spatial and temporal scales. The concepts of the restoration of natural capital* (RNC) and RNC thinking* will be required as well.

To illustrate these ideas, we discuss a landscape-scale restoration program located in and around the large (155 km²) and heavily polluted *Étang de Berre* (Berre Lagoon) that is situated between three of the major cities in southern France: Marseille, Salon-de-Provence, and Aix-en-Provence. But, first let us clarify an interesting historical twist that bears on our tale.

The closest equivalent in English to the French word *étang* is "lake," but it is often translated as "lagoon." Indeed, the ambiguity is well founded in the present case. The body of water known as the *étang de Berre* includes the *étang de Bolmon* (see below) which was formed 8,000 years ago when sea rise slowed. Human remodeling only began about 2,000 years ago when people with an interest in fishing deepened a channel connecting the western edge of the lake to the Mediterranean Sea, thereby turning the *étang* into a coastal lagoon characterized by regular influxes of seawater. From a restoration or sustainability perspective, this is one of many complexities concerning this large Mediterranean wetland that needs to be taken into account, because most people today consider it a lagoon and do not want it to go back to being a lake. What most of the very large local population does want is to see this very large "lagoon" rehabilitated, restored, and revitalized. But, there is little agreement on how to do it, who should do it, who should pay for it, or even what should be done.

Our goal here is to present some concepts and tools that are being employed in the Berre landscape restoration program in order to help in planning, budgeting, and, above all, in consensus-building. These tools include a sequential reference system to guide ecological restoration efforts that is based on a semiquantitative scenario of ecological degradation* and transformation of the area over the past two millennia. We also briefly discuss the use of what we term a HMCA* that is used to synthesize ecological, social, and economic criteria across different historical time periods applicable to a large scale, multifaceted project of this sort. In this chapter, we summarize the main features of the project. A more detailed presentation of the results will be provided in a subsequent journal article. Finally, we advance the basic concept of RNC thinking, i.e., that restoration "pays," especially if we consider that ecological restoration is a means to restore natural capital, while also restoring social capital. In such an ecologically and economically beleaguered and vulnerable area as the one considered here, the "high" road to sustainability passes through the portal of RNC thinking.

The Target Area: The Berre Lagoon and the Berre-Bolmon-Rove Landscape

Water is an "organ of the Earth" (the physiological metaphor of Bachelard 1942). And coastal lagoons? They are filters and regulators at the interface of continents and oceans. They are the kidneys of the planetary plumbing system, concentrating a large portion of ecosystem outputs and energy flows cascading down from the watersheds for which they serve as the natural sinks and receivers. As a result, they are endowed with exceptional biological and ecological richness, productivity, and attractiveness to people.

In spite of these attributes, coastal lagoons were all too often used in conflicting ways. Not only were they cherished for fishing, hunting, navigation, and recreation, but also as a free-of-charge and free-for-all dumping area. Sadly, this is true even in the *Mare Nostrum* ("Our Sea") as the Mediterranean was known by the Romans, and the Berre Lagoon is a case in point (Fig. 1). Coastal lagoons are resilient, yet, like all ecosystems, they are vulnerable to degradation and more or less irreversible

Fig. 1 Aerial view of the Berre Lagoon (southern France). The three superimposed images are of the Saint Chamas hydroelectric plant (*top*), the Rove tunnel (*right*), and the Caronte channel (*lower left*). See text for explication (*source*: Géoportail 2011)

transformation. One of the most serious pollution and degradation problems they face is eutrophication, which occurs when a body of water acquires a high concentration of nutrients, especially phosphates and nitrates. In the Berre Lagoon, nutrient enrichment is far advanced, thanks to unabated urban, industrial, and agricultural waste disposal over many decades.

The Berre Lagoon is approximately 15,500 ha, is the third largest inland body of water in Europe, and is the largest of the 25 or 30 lagoons nestled along the Mediterranean Basin coastline (Papayannis 2008). It is located just a few kilometers northwest of Marseille, which is France's oldest city, and quite near Salon-de-Provence, Aix-en Provence, and many other smaller urban areas. Unlike many coastal wetlands, the Berre Lagoon has never been drained and removed for farmland. However, it has been used by people in many ways over many millennia.

The Berre Lagoon was cherished since its Chalcolithic Age origins for the abundant biological resources it supplied and its attraction as a dwelling place. Based on digs at on the western end of the lagoon, archeologists date the oldest permanent human settlements back to 800 BCE. The human presence in the Berre region increased steadily after the Phoenician port of Massalia (now Marseille) was founded in ca. 600 BCE (Bellet 1979). At some point during their stay in the region, the Romans deepened a 6.5 km long canal to connect the western end of the lagoon to the Mediterranean Sea, probably to increase the fishery productivity (Caronte Channel; Fig. 1). Since the seventeenth to nineteenth centuries, the intensification of resource extraction, then industrialization, followed by urbanization of the shorelines in the twentieth century, has rendered the ecological state of the lagoon somewhat poor, to say the least.

A major event took place in 1925, when the Rove tunnel and canal were opened after 15 years of government-funded construction (Fig. 1). The goal was to connect the port of Marseille to the Rhône River in order to enhance river traffic via an inland route that was safer for merchandise-laden barges than was motoring along the Mediterranean coast to the mouth of the river. The canal was created along the southern shore of both the Berre and Bolmon lagoons. The Bolmon, situated in the southeast part of the wetland complex, is partially isolated by a natural sandbar that has been sparsely inhabited for decades, if not centuries. This canal + tunnel passageway functioned for 38 years until the tunnel collapsed in 1963.

French public policy for the region was strongly biased in favor of industrialization during the entire twentieth century, when three oil refineries and many other factories were implanted with heavy public subsidies. This development increased job opportunities, and the population size rose steadily in the ten municipalities bordering the lagoon. It doubled from 1950 to 1990 and is now 350,000 in the larger landscape we call BBR (for Berre and Bolmon Lagoons and the Rove tunnel and canal). During this 40-year period, everything from effluents from sewage treatment plants, agricultural residues, to industrial and urban wastes was dumped into the lagoon.

Another major event took place in 1966, when a hydroelectric power plant opened at Saint Chamas on the northern shore of the lagoon (Fig. 1). The Saint-Chamas plant is at the tail end of no less than 20 interconnected power plants stretching all the way up the Durance River, whose Verdon tributary begins high in the Alps. The course of the Durance River was artificially diverted to empty into the northern portion of the Berre Lagoon as part of this industrial effort, which is managed by the state-controlled utility company, EDF. This company had a monopoly on energy distribution in France until 2010 and was responsible for producing almost all energy not derived directly from nuclear power. The chain of hydroelectric plants culminating at Saint Chamas can provide a significant input in the electrical power grid for the region in a matter of just minutes to help cope with surges in consumer demand (Clébert and Rouyer 1991; Collomp 2002). In this way, brown-outs and black-outs can be avoided during peak demand periods throughout the heavily populated SE portion of France. Additionally, in a country where 80% of electrical power comes from nuclear energy, and where the future of that industry is increasingly called into question because of the problems related to wastes and occasional leaks, all alternative sources of power—including hydroelectric—are of great public and strategic interest.

Historically, the Berre lagoon was more or less salty based on how exchanges with the sea were managed. The deepening of the channel and the opening of the Rove tunnel, for commercial and industrial development in the early 20th century, have confirmed the marine dimension of the lagoon, leading to an important revival of traditional activities (e.g., fishing, shellfish gathering, algae harvests, salt production and soap industries), and also contributed to the building the cultural dimensions of nearby residents. Consequently, most people in the area today agree that it should stay that way. One of the major concerns regarding the Saint Chamas plant is that its effluents not only bring 500 tons of silt per annum from the Durance river (up to 1.6 million tons in 1977), but also huge quantities of nonsaline water that are dumped each year into the lagoon. This resulted in the decline of surface water salinity from 24–36 to 1–22 ppt. The annual average of freshwater inputs from the Saint Chamas plant is now limited to 1.2 billion m^3. Indeed, more than 85% of the fresh water coming into the lagoon derives from the EDF plant, with the remainder coming from the Arc, Cadière, and Touloubre rivers. In 1977, 6.6 billion m^3 of fresh water were dumped into the lagoon, equal to seven times its volume. The annual average discharge into the lagoon is now 1 billion m^3, but with large variations both seasonally and interannually.

This enormous annual input of fresh water and silt into the lagoon has led to the stratification of both salinity and oxygen levels in the bottom of the lagoon, maintaining large areas of anoxia. The huge inputs of nitrates from the Durance amplified the eutrophication problem and led to virtual demise of the benthic biota, and to the collapse of ecological functionality. To top it off, algal decomposition in the lagoon is the source of malodorous odors every summer, which creates a severe nuisance for the inhabitants of the surrounding communities and for the day visitors visiting the shores of the lagoon to sunbathe, sail, or swim. In retrospect, over the course of the last century, there has been a veritable cascade of ecological degradation and fragmentation, that spread to the entire socioecological system encompassing the lagoon and BBR landscape, leading to, among other things, the end of traditional hunting, fishing, and recreational activities on and near the lagoon, and the tarnishing of Berre Lagoon's public image at regional and national levels.

Several local associations were created in the 1980s to call for the closure of the Saint Chamas plant, or at least a redirection of the waters spewed forth from the facility to the nearby Rhône River, rather than continuing to dump them into the Berre Lagoon. At the very least, these associations—with the support of various town councils—argued for a significant cleanup effort on the part of the state and regional government, not to mention the energy utility, oil refineries, and chemical industries, in order to ameliorate environmental conditions in and around the lagoon.

GIPREB Mission and Trajectory to Date

A Public Interest Group for the Rehabilitation of the Etang de Berre (GIPREB) was created (http://www.etangdeberre.org/) in 1990, in response to a decade of public outcry about the polluted and stinky lagoon. Starting in 2000, the goal of GIPREB focussed on stimulating and coordinating actions aiming at the restoration or rehabilitation of the aquatic components of the Berre and Bolmon lagoons, and the clogged-up and malodorous Rove canal.

The results in some areas were quick and dramatic. A 1993 grassroots-inspired referendum led to legislation (the Barnier plan) that imposed the first quotas on the EDF concerning the volume of fresh water that could be dumped into the Berre Lagoon: 1.2 billion m^3 $year^{-1}$, instead of the three or more billion m^3 $year^{-1}$ in preceding years. A new series of even stricter quotas were imposed in 2005, following European litigation in which France was condemned to pay penalties for the degradation to the Berre Lagoon (Maljean-Dubois and Truilhe-Marengo 2005). The technical and political debates were strident, however, as described below.

Today, the GIPREB's goal has expanded to include the ecological restoration of the Berre and Bolmon portions of the lagoon, as well as the aquatic habitat created by the Rove canal and tunnel. The socioeconomic reintegration and revitalization of the watershed and coastal territories is implicitly sought as part of this restoration.

The most controversial and expensive operation concerns the proposed diversion of the waters and silt from Saint Chamas to the Rhône River. The feasibility studies initiated by the GIPREB yielded very detailed reports and calculations. The cost would be close to two billion euros according to an unpublished study commissioned by the GIPREB, and negotiations are likely to be stiff. In the meantime, a new and important component of the project is the experimental reopening of the Rove tunnel in 2013. The purpose of this project is to artificially renew seawater circulation in the lagoon using pumps installed at shoreline where the tunnel begins in the harbour of Marseille in the Estaque cove, in the Mediterranean Sea (Fig. 2). This influx of highly saline water will offset—at least partially—the ongoing influx of freshwater coming into the Berre Lagoon at Saint Chamas. But the main objective of the reopening of the Rove tunnel and canal is to improve the ecological quality of the Bolmon lagoon, which is presently highly confined and eutrophic. It is hoped to "breath" life back into this portion of the lagoon by increasing the water renewal rate.

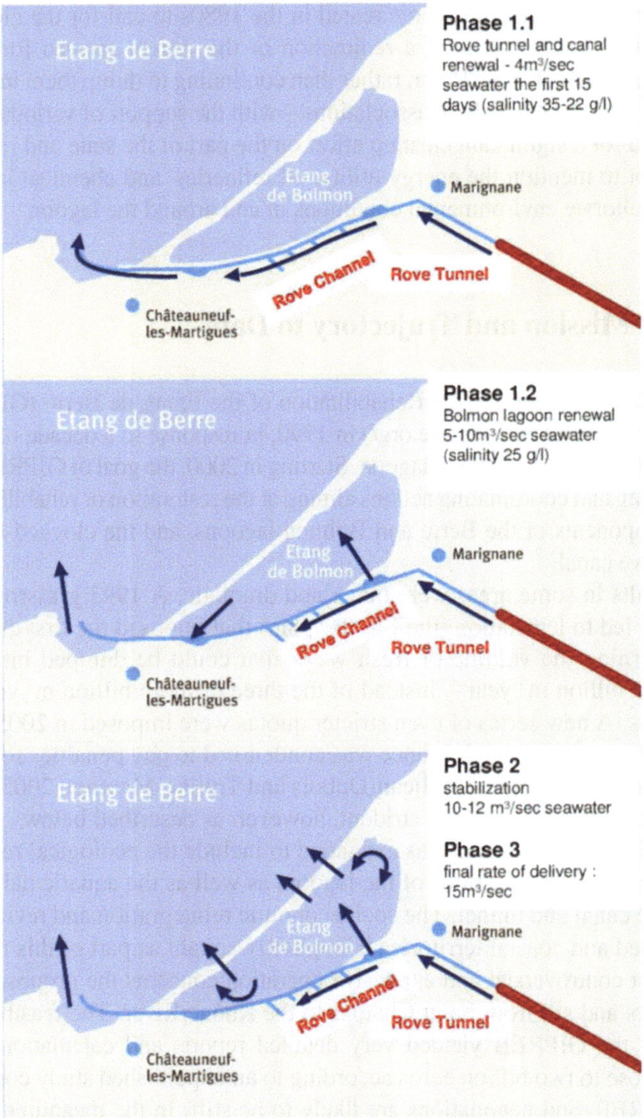

Fig. 2 The experimental reopening of the Rove tunnel and channel is depicted in four phases, 1.1, 1.2, 2, and 3, with the anticipated rates of seawater inflow and salinity for each phase. Note that "Marignane" is the site of the international airport of Marseille

The results of preliminary studies (the "Ramade studies") showed that the project was viable and timely, provided that certain critical technical issues were respected. Yet debate broke out on which level of water flow to choose. The results of these first studies revealed that a 20 $m^3 s^{-1}$ flow would be required to meet the stated restoration objectives. Despite the clear fact that failure to respect this level might induce undesirable effects, some stakeholders—including the French government—argued for a more modest flow rate that would be cheaper to install and easier to maintain. This debate led to a delay of several months culminating in a decision taken by the GIPREB General Assembly in favor of the flow rate initially recommended. Several more months were spent to decide who should manage the project and what specific objectives should be sought.

Soon after the project began slowly taking shape, the owner and manager of the Bolmon lagoon (the public coastal conservation entity, *Le Conservatoire du Littoral*; see http://www.conservatoire-du-littoral.fr/front/process/Home.asp) questioned the validity of the project. This major stakeholder argued that the introduction of seawater to the Bolmon lagoon would modify the "original" fresh water ecosystem, but provided no scientifically grounded basis for this assertion. The positive impact of this assertion was to stimulate debate, discussions, and consultations with experts within and among all the various stakeholders of the BBR, for which GIPREB played the key role of negotiator.

Two important points we wish to emphasize are the great public demand for restorative action, and that the State has called on the GIPREB and the Port of Marseille to intervene and proceed to clean-up, rehabilitate, and restore the lagoon in undefined ways. These actors also understand that their mission is to strive to improve the socioecological sustainability of the lagoon. In 2009, after having noticed significant confusion about basic concepts and the conflict among stakeholders and actors, the directors of GIPREB and the Port de Marseille concluded that new conceptual and analytical tools were required. They engaged a restoration ecologist who brought along an economist, a landscape ecologist, and a student of RNC. In the next section, we present the three major tools under development and application by this transdisciplinary team: (1) sequential references, (2) HMCA, and (3) RNC thinking.

Three Tools for Restoring Natural Capital and Improving Quality of Life

The first of the new tools now being applied to the BBR program comes from restoration ecology; the second one comes from ecological economics; and the third one is an outcome of the early interactions of ecological economists with restoration ecologists addressing the practical challenge of implementing ecological restoration sustainably, and at a large scale.

All three tools can help elucidate the complexity and conflicts in situations where sustainability will not come easily. They can, in fact, help advance the outreach and consensus-building process that is essential to restore natural capital and social capital, and to make a transition towards sustainability at any spatial scale.

Tool No. 1: Sequential References

Some of the major conceptual issues in any ecological restoration project or program are: "Why invest in this, and, how much will it cost?" Additionally, one should ask "How does one plan, organize and monitor the progress of a restoration project within a complex biophysical and socioeconomic setting?" Although not all restoration scientists and practitioners of ecological restoration agree with this approach, one school of thought follows the SER Primer of Ecological Restoration (SER 2004), which includes an argument for the use of an ecosystem of reference model to help for planning and project integration (cf. Aronson et al. 1995; White and Walker 1997; Egan and Howell 2001). From this perspective, the agreed-upon reference system is the core element of a restoration project, because it serves to define the goals, the methodology, and the "vision" of the stakeholders and operations, as well as inform the choice of diagnostic and monitoring protocols. In principle, it helps to synthesize all information obtained at each step and adjust the management of the project and project accordingly.

A series of refinements of the reference model idea have gradually emerged in the last few years that lead to a sequential reference model (see Aronson and Vallejo 2006; Clewell and Aronson 2007). The basic ideas are simple: if an ecological restoration or a RNC project is likely to be long and complicated, then a sequential series of references may be a better planning and consensus-building tool than a single reference. A series of stages can be identified to define a degradation and transformation scenario occurring in the past that brought the ecosystem (or landscape) to its current undesirable state, and which now drives the impetus to invest in restoration. Finally, conceptual and strategic links can be made between each successive stage in the projected restoration process and one or more of the different historical stages identified in the degradation and modification scenario. We present this scenario-building process in Fig. 3 in a very schematic fashion. It should be noted that socioeconomic variables are included with the nested ecological variables.

The target ecosystem shown in Fig. 3, at any given stage of development, is represented as a circle within two matrices. One is a biophysical matrix (i.e., the landscape), and the other is a socioeconomic matrix (comprising the way that people use and manage the ecosystem). Ecosystem goods and services are represented as triangles situated around the outer circle of each "star" on the model. As a triangle get longer or shorter, this change represents what people have done or are doing or extracting from the ecosystem in terms of goods and services. Fig. 3, therefore, provides a multidimensional chart of a society's move away from sustainability, or back towards it. In other words, ecosystem goods and services wax and wane in response to human use, management, and investment in the restoration of natural capital.

On the basis of this schematic model, an expert committee was set up to identify the historical periods and reference states of our study area. The data serving as

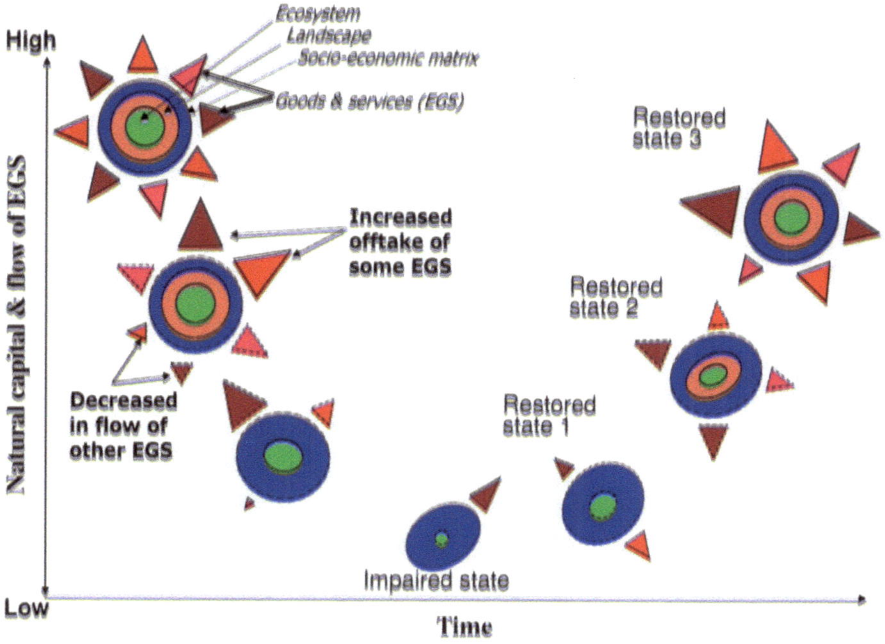

Fig. 3 Sequential references in ecological restoration. As described in the text, concentric *circles* in each "star" represent an Ecosystem within a Landscape and a Socio-economic matrix; *triangular appendages* represent Ecosystem Goods and Services (EGS) that wax and wane in response to human behaviors. The figure is modified from Clewell and Aronson (2007). Copyright © 2007 Andre F. Clewell and James Aronson. Reproduced by permission of Island Press, Washington, DC

input to Fig. 4 were compiled, the descriptive, qualitative figure was elaborated, and the general schematic model was applied to the BBR landscape.

Fig. 3 is a schematic diagram that is intentionally general. Fig. 4, in contrast, provides a more detailed and site-specific application of the general model of sequential references. Fig. 4 illustrates the complexity of the problem facing managers and would-be restorationists working in the Berre Lagoon and in the larger BBR landscape. As noted earlier, the Berre Lagoon was a lake before it became a lagoon. No one expert could affirm that there is one single historical period that perfectly corresponds to an ideal reference state for the RNC program of the BBR landscape. Arbitrary and subjective choices must, therefore, be made on the basis of negotiation and consensus-building. Along the axis of natural capital and the flow of EGS, for example, we see that the historical periods dating from 5000 BCE to 1949 are associated with higher values relative to after 1949. In that same time period, however, human well-being and the degree of socioeconomic development was well below 2010 levels. More light can be shed on this issue by use of the second tool in our toolkit, as described in the next section.

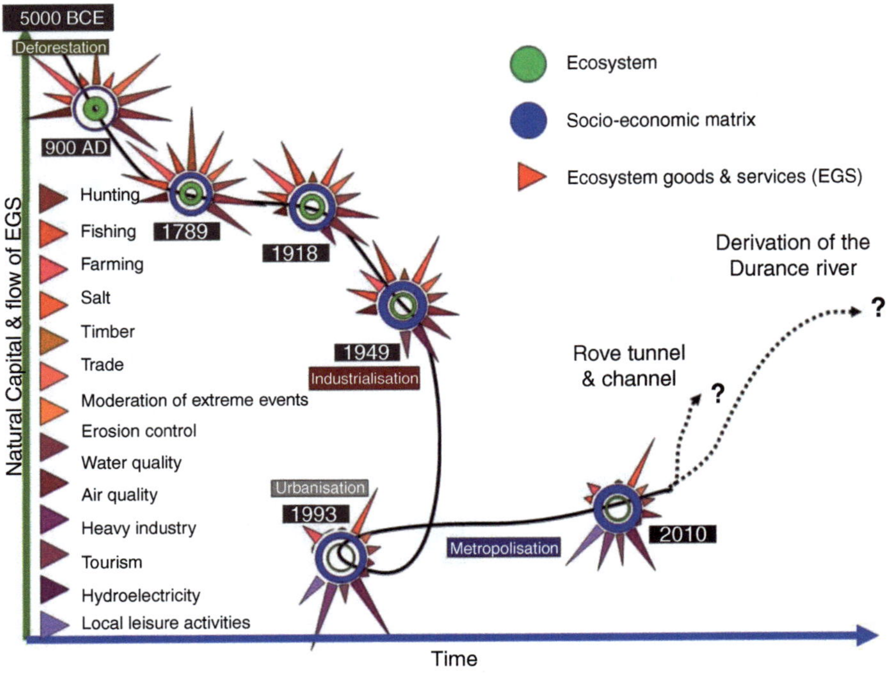

Fig. 4 Sequential References model applied to the BBR Landscape (Berre Lagoon, Bolmon Lagoon, and Rove tunnel and channel). Six major stages are shown in the historical transformation of the landscape since deforestation began five millennia BCE. The two major interventions under discussion are also indicated in the *lower right*. See the text and legend of Fig. 3 for an explanation

Tool No. 2: Historical Multicriteria Analysis

We undertook a multicriteria analysis (MCA) that has a strong historical dimension (HMCA) to aid in the decision-making process when faced with the range of alternative reference states or phases revealed in the complex process depicted in Fig. 4. The aim of our HMCA—an exercise never before attempted, to the best of our knowledge—was to identify an historical period that could be judged as optimal by a majority of stakeholders, in terms of both its socio-cultural-economic and ecological aspects or dimensions. As for any reference system (sensu Clewell and Aronson 2007), however, the definition of an optimal historical state does not imply that we should unequivocally strive for fulfilling those same socio-cultural-economic and ecological dimensions today. This is because there are certain sociocultural and economic parameters that we are not able to alter today. In each reference state, however, there are other parameters revealed by the HMCA that are fully at the disposition of communities and planners to work towards and implement. The HMCA presented below thus aims to bring to light—in a methodical, analytical, and transdisciplinary fashion—the historical period of the BBR landscape that best balanced the ecological, sociocultural and economic factors of interest. This HMCA, like all MCAs, was elaborated on the basis of the experts' opinions. But, the tool is now being applied to enrich a "bottom-up" debate about the most desirable future of the Berre Lagoon and the larger BBR landscape, which is a debate aimed at consensus-building among stakeholders.

Tool No. 3: RNC Thinking

The RNC is a panoply of investments, interventions, improvements in current practices, and new ideas meant to promote and accelerate national and international transitions towards sustainability. The RNC is a broader concept than ecological restoration. The RNC aligns ecological restoration with all other efforts to reduce waste and pollution within production and resource exploitation systems, as well as in and around cities and their support systems, and along modern transport networks. It also refers to educational programs that teach and reinforce the importance of natural capital to human economies and well-being. Having presented the three tools, let us now consider how this will be accomplished.

HMCA to Inform the Choice of a Reference Model Among Varying Alternatives

As should be clear by now, the history and current status of the BBR landscape is the result of a complex net of interactions between successive human societies and a panarchy of ecological processes. As a prerequisite for a holistic landscape-level approach to restoration and RNC, a thorough and transdisciplinary "brainstorming" effort about the historical references of the BBR project was undertaken (described by the three tools). The results of this reflection was applied and described below.

Methodology

A MCA provide techniques to compare and rank different outcomes through the use of a variety of indicators or criteria. It is increasingly used in areas such as natural resources, management, climate change and adaptation, and water management (Bell et al. 2003; Paneque Salgado et al. 2009; Sheppard and Meitner 2005; Martin-Ortega and Berbel 2010). In this project, we used the DEFINITE 3.0 software developed by the Environmental Studies Institute of the Free Amsterdam University (Janssen et al. 2003) to conduct the MCA (see Box 2 for a note of caution about MCA). Furthermore, because we wanted to select a historical period as an ecological reference, we decided to call our tool a Historical MCA (HMCA). The software

Box 2 A Word of Caution About MCA

A MCA involves selecting a set of "criteria" to be achieved by a project or policy alternative, and assigning a score to each criterion on the basis of the predicted impact or effect. These criteria are usually measured in a range of different units. The scores are then adjusted by multiplying them by weights representing the analyst's or expert's assessment of the relative importance of each variable's impact. The scores are then standardized and summed to provide an indication of net benefit. The critics of an MCA argue the results can be biased in favor of an alternative when those who are consulted about the determination of weights or the attribution of scores are stakeholders or subject-matter experts (Dobes and Bennett 2009). In the same way, the solution provided might be considered acceptable only by the stakeholders who noted high values (or weights) for certain variables, while other stakeholders/experts may well indicate something else. As an alternative, a cost–benefit analysis (CBA) is often advocated because of its "whole-of-society perspective" whereby all members of the concerned society are included by aggregating the utilities (as measured by willingness-to-pay, for example) of a representative sample of individual citizens. While these criticisms have some grounding, it is the authors' opinion that they do not justify the rejection of a MCA (we note that a CBA has many critics as well). As is the case with the application of any analytical method, criticism of a MCA serves to remind us to be prudent when interpreting and using results. In particular, a MCA is one of several tools to help stakeholders organize available information, think about consequences, explore their own wishes, and advance a collective decision-making process (Belton and Stewart 2002). Therefore, we stress that this MCA project was undertaken as a first attempt to evaluate the desirability of different possible historical reference periods, and as a tool to enrich an ongoing debate regarding the future of the Berre Lagoon and the BBR landscape; it was not to be used to dictate policy in a "top-down" fashion on the basis of would-be "scientific results." The situation is far too complex for such an arrogant approach.

presents the advantage of assembling a diversified panel of multicriteria methods that have a high modularity on standardization and sensitivity analysis.

The HMCA was conducted in two steps. First, we sought to define and delineate the outstanding historical periods or phases that could include the historical reference period that maximized the provision of ecosystem goods and services, but without sacrificing sustainability. This was done on the basis of the nine ecological attributes of a restored ecosystem, which are proposed as useful criteria to orient any ecological restoration project (SER 2004; Clewell and Aronson 2007). In the second step, we used a more global set of ecological, socioeconomic, and cultural criteria to identify the historically based ecological reference state that could maximize human well-being and ecosystem "health" simultaneously. We describe our methods next.

Identification of the Alternatives

The first step was to identify a set of alternative historical reference states and a set of criteria to evaluate these reference states. In this analysis, the alternatives refer to different time periods characterized according to the successive transformations and permutations of the BBR landscape. The groundwork undertaken in the sequential reference model (conducted by the expert committee who engaged in a 3-day structured conversation process) was used to identify the historical reference-period alternatives of the BBR landscape. Despite the multimillennial perspective offered by Fig. 4, we decided to refer only to the mid-nineteenth century (see also ENSAML 2001), under the assumption that, for the great majority of actors and stakeholders, this period was more than sufficient for our purposes. We briefly outline the main characteristics of the pertinent reference periods in Box 3.

Box 3 Details Used in the Historical MCA Analysis

Before 1863: The Berre Lagoon is a brackish water ecosystem with a rich and diversified flora and fauna. The abundance of fish and molluscs supports numerous traditional activities of fishing and shellfish breeding, and underpin a strong feeling of local identity. However, a first step of industrialization, namely heavy chemistry, has already appeared in the western part of the lagoon.

1863–1924: The Caronte channel is reworked and deepened increasing salinity in the lagoon and the marine fauna and flora is reestablished. A large investment in the construction of harbors and waterways begins that gradually imposes an industrial vocation on the lagoon, while the rural world in France is hard hit by various economic, social, and cultural crises, not least of which was WWI.

1925–1965: The fauna and flora of the lagoon and surrounding ecosystems are not yet profoundly affected by industrial activities, but the overall health of the lagoon slowly begins to deteriorate. The increasing chemical pollution, as well as the beginning of eutrophication, leads to the contamination of fish and mollusks that then is followed by a prohibition of commercial fishing in 1957. Industrialization continues with the support of the aeronautic industry in the 1920s and the petrochemical industries in the 1930s. The opening of the Rove tunnel and canal, on the SE side of the lagoon, together with the Caronte channel created a major waterway connecting Marseille's port to the Rhône valley and, thence, with Lyon, Paris and the rest of France.

(continued)

Box 3 (continued)

1966–1992: The collapse of the Rove tunnel in 1963, and near-simultaneous opening of the hydroelectric plant at Saint Chamas in 1966, led to the destruction of aquatic communities and ecosystems in the lagoon, all of which were already heavily affected by various sources of pollution. Salinity in both the Berre and the Bolmon Lagoons declined, leading to the local extinction of numerous marine species and to the regression of all benthic faunal assemblages. There were socioeconomic consequences as well. The traditional interactions between the lagoon and people were replaced by industrialization and urbanization, which transformed the character of the region and eroded its overall attractiveness for the local populations and visitors.

1993–2005: The mobilization of local grassroots organization led to the adoption in 1993 of the Barnier Plan. This plan imposed quotas on freshwater dumping from the Saint Chamas plant. Water quality then improved, but the faunal and floral habitats remained deeply impaired. A restructuring of heavy industry was carried out through a growing externalization or outsourcing of various processes, while a cultural transformation accompanied the growing awareness of the negative environmental impacts of the industrial activities.

2006–2010: Ecosystems in the lagoon were heavily affected by eutrophication. Industry lost ground, but the local economy was relieved by the well-developed tertiary sector. The public and NGO management of the remaining intact or seminatural habitats and scenic sites around the Lagoon led to the development of a dense fabric of NGOs. When the local consequences of the economic crisis began in 2008, the precariousness and poverty indicators put forward a strong deterioration of the social situation in the whole region and the specialization of the Berre Lagoon economy increased the effects of the crisis (DROS 2010).

Identification of Criteria Used to Judge the "Performance" of Each Reference State

The experts' points of view were considered adequate to establish comparisons. There were 21 criteria selected for the final HMCA (Table 1). These criteria were inspired, on the one hand, by the general model shown in Fig. 3 in which an ecological system is included in a socioeconomic matrix and with ecosystem services at the interface, and, on the other hand, by the nine identified attributes which are a necessary prerequisite for a restored ecosystem (SER 2004). Of these nine attributes, "auto-maintenance" and "indigenous composition" were not used because of a lack of data. We readily acknowledge that the pool of experts consulted was small; further interviewing will be necessary to complement their input.

Table 1 Description of criteria and subcriteria used in the HMCA

Criteria categories	Criteria	Subcriteria in the socioecological performance matrix	Unit	Source
Ecological complex	Characteristic composition	Biotic composition, specific richness of benthic macrofauna, seagrass area, average abundance indexes of macrophytes	–	GIPREB archives
	Landscape integration	Artificialized coast length, total freshwater inflow	–	GIPREB archives
	Physical environment	Average salinity, average depth of euphotic zone, dioxygen saturation	–	GIPREB archives
	Functioning	Various concentrations, linked to eutrophication	–	GIPREB archives
	Potential threats	Solid and liquid inputs of Saint Chamas plant	–	GIPREB archives
	Resilience	Flood resilience, summer stress resilience	0–1	Expert opinion
	Functional composition	Pico-phytoplankton, micro-phytoplankton, zooplankton, benthic macrofauna, macrophytes, seagrasses, fishes	0–1	Expert opinion
Socioeconomic matrix	Economic development	Public infrastructure and purchasing power	0–1	Expert opinion
	Social cohesion	Unemployment, precariousness, and health	0–1	Expert opinion
	Demographic dynamism	Population density	0–1	Expert opinion
	Institutional environment	Environmental protection and social policy	0–1	Expert opinion
	Cultural strength	Cultural vitality and diversity	0–1	Expert opinion
Ecosystem services provision and exploitation	Provisioning	Relative abundance of fishes, shellfish, wrack (seaweed), and salt	0–1	Expert opinion
	Regulating	Water quality and waste treatment	0–1	Expert opinion
		Erosion and flood control	0–1	
	Habitat	Waterways, nursery	0–1	Expert opinion
	* Cultural services	Tourism	0–1	Expert opinion
		Local leisure activities and water sports	0–1	

* There are 4 provisioning services, and 4 regulating ones included in the 21 criteria used in this analysis

For ease of interpretation, we have grouped the criteria and the respective subcriteria in distinguishable components under the main headings: the ecological complex, socioeconomic matrix, ecosystem services provision and exploitation (Table 1). The chosen set of criteria was intended to cover the revelation of the most pertinent ecological, economic, social and cultural parameters, and ecosystem services. No one criterion reflected the same concept as any other. In this way, we sought to avoid double counting and overvaluation of a single aspect.

Assessing the Value Associated with the Alternative Reference States

The frame of reference in any MCA with finite alternatives is the performance matrix in which the performance of each alternative is described in comparison to all other criterion. We constructed a performance matrix for the ecological complex in this survey by drawing on ecological time series data from the GIPREB archives, and another derived from the socioecological and ecosystem service matrix (Table 2). In the "Ecological performance matrix" the multiple

Table 2 Raw ranking of historical time periods

Alternative reference periods	Impact score ecological matrix	Impact score socioecological
Before 1863	0.93	0.48
1863–1924	0.93	0.54
1925–1965	0.86	0.64
1966–1992	0.53	0.40
1993–2005	0.43	0.36
2006–2009	0.35	0.39

impacts of each historical alternative were standardized to a common metric by using the linear standardization technique suggested by Howard (1991). As for the comprehensive socioecological matrix, standardization was done (for those criteria not held within the GIPREB archives) by asking stakeholder experts to score the above criteria impacts in Table 1 along a unitary scale that ranged from 0 = no impact to 1 = full impact. Finally, the criteria were weighted to denote the influence that each criterion has in building up the total preference relation. As a working hypothesis for this preliminary study, equal weighting was applied to avoid imposing value judgments on the relative importance of one criterion vs. any other. This seemed to be the best course given that the number of interviewed experts remains limited (see Wang and Yang 1998; Janssen 2001; Hämäläinen and Alaja 2008 for a discussion of the merits of various equal weighting vs. various weighting methods).

Aggregation of Scores, Sensitivity Analysis, and Examination of Results

The second and final step before conducting a sensitivity analysis consisted of aggregating and analyzing the total performance scores. Upon combining the weights and the standardized scores for each alternative, the HMCA yielded a clear ranking for the ecological and the socioecological matrix (Table 2).

A sensitivity analysis was conducted to confront uncertainty stemming from hesitations during the problem structuring process (e.g., which alternatives should be used? How important are the criteria? Should weight be employed? etc.). The sensitivity analysis confirmed the results in Table 2. If ecological criteria were the only criteria, then the best historical reference was "Before 1863." However, with regards to the entire socioecological complex, the HMCA indicated that the 1925–1965 period was preferred, followed by the 1865–1924 period (Box 3).

Discussion

The results of the HMCA suggested that the optimal time period, if judged purely from ecological criteria, was before 1836. Prior to this time, the lagoon was not subject to eutrophication, its coastline was only slightly modified, and the native fauna and flora were largely intact. The results of the HMCA are not surprising but, nonetheless, are of great value because they show one way to quantitatively integrate socioeconomic and cultural variables in an analysis undertaken as part of a decision-making process related to sustainability. Although these are preliminary results and based on a small number of expert opinions, the HMCA ranking in Table 2 points to a reference state for the BBR landscape restoration program in the 1925–1965 time period. At that time, the lagoon suffered from increasing eutrophication, but the fauna and flora were, nonetheless, typical of a marine habitat, and the surrounding ecosystems were not yet profoundly affected by industrial activities. This period was also marked by an improvement of socioeconomic attributes compared to previous periods. To wit, the Caronte channel and the Rove tunnel were both completed by then, and yet the traditional activities exploiting the lagoon and the associated cultural identity were maintained (at least until 1957, when fishing was banned because of water pollution). Lastly, it is during this period that annual paid leave was instituted in France, which brought about a rapid rise in tourism, local leisure activities, and the enjoyment of the various cultural services delivered by the BBR landscape (Fig. 5).

Fig. 5 Panorama of ecosystem services furnished by the Berre Lagoon. (**a**) Wind surfing © Annick Amabile. (**b**) Sailing © M. Torres GIPREB. (**c**) Wildlife watching © M. Torres GIPREB. (**d**) Fishing © M. Torres GIPREB. (**e**) Rowing © M. Torres GIPREB. (**f**) Duck hunting © M. Torres GIPREB

Conclusions and Perspectives

There is an urgent need in our rapidly changing and increasingly crowded world, including in the Mediterranean region, to find ways to reconcile nature conservation with sustainable economic development objectives. Although the outcome of the HMCA was the agreement that there was an optimal reference state between 1925 and 1965, it does not dictate that we should seek—or think that we ever could go "back in time." The way of life of people in the region has changed, as it has everywhere; cultural norms and expectations have evolved, and so going back is not an option. What remains as an option, however, is to strive for sustainability by employing those features and resources of modern life and what social capital is available to us, to move towards a state highlighted by the HMCA when, among other things, local marine species and benthic fauna started to repopulate the lagoon. Arriving at this state would permit, or at least facilitate, recovery of the Berre Lagoon's formidable capacity to provide regulating, provisioning, supporting, and cultural ecosystem services. This objective is even more pertinent today than half a century ago. Southern France's economy today is largely service-dominated, and its human population density is increasing rapidly. There is a potential to increase tourism revenues in the BBR landscape—especially through the development of ecotourism associated with the provisioning services of a restored Berre Lagoon. Such a development would offer a new source of wealth and pride for the surrounding communities of the Berre and Bolmon portions of the Lagoon, and potentially reduce the pressure on the BBR from industrial activities that are vulnerable to upheavals in the globalized economy.

The HMCA analysis supports the hypothesis that improved ecological conditions could be achieved by increasing the salinity of the lagoon. The proposed reopening of the Rove tunnel and canal would accomplish this objective simply by artificially renewing seawater circulation into the lagoon. Ideally, a way will soon be found to divert the huge fresh water and silt inputs from the hydroelectric plant at Saint Chamas into the Rhône River (Fig. 4), thereby permitting a continued supply of renewable electricity to more than 200,000 households in the region while, at the same time, contributing to the restoration of natural capital and improvement of quality of life to the large local populations next to Europe's largest lagoon.

We suggest, therefore, that a cost–benefit analysis (CBA) based on the total economic value is now required to compare the financial costs associated with the divergence of fresh water and silt at Saint Chamas with the benefits of: (1) the abolition of restrictive quotas on freshwater dumping from the Saint Chamas plants, (2) increased ecosystem provisioning and cultural services, in particular in terms of new recreational capacity, (3) the removal of the foul-smelling odors, and (4) improved tourism revenues. The benefit of such a CBA would be the opportunity to engage the BBR population in a valuation exercise—to develop an optimal RNC alternative that could be debated and ultimately embraced by policy makers and citizens. The missing link, however, may be what we have called RNC thinking and the awareness of the value of natural capital and quality of life. A summary of the best data available to support this assertion is in Box 4. Readers interested in these details should also consult TEEB Foundations (2010).

One valuable and overarching step toward a sustainability transition would be to recognize that the RNC is a good investment for future generations (Box 4, cf. de Groot et al. in review; Elmquist et al. in review). A true transition to "RNC thinking" at societal and international levels will require revising unregulated markets and market failures that lead to ecosystem destruction and biodiversity loss. This development will vastly improve collaboration among scientists and nonscientists, including community leaders, business leaders, and policy-makers.

Box 4 The Value of Natural Capital: Wetlands

Human well-being depends upon the "free" services provided by Natural Capital. For example, wetlands purify water, forests clean the air and regulate climate, bees pollinate agricultural crops, mangroves support fisheries and protect coasts, etc. An overview of the monetary value of the main services provided by wetlands is in Fig. 6, which is based on a large number of case studies.

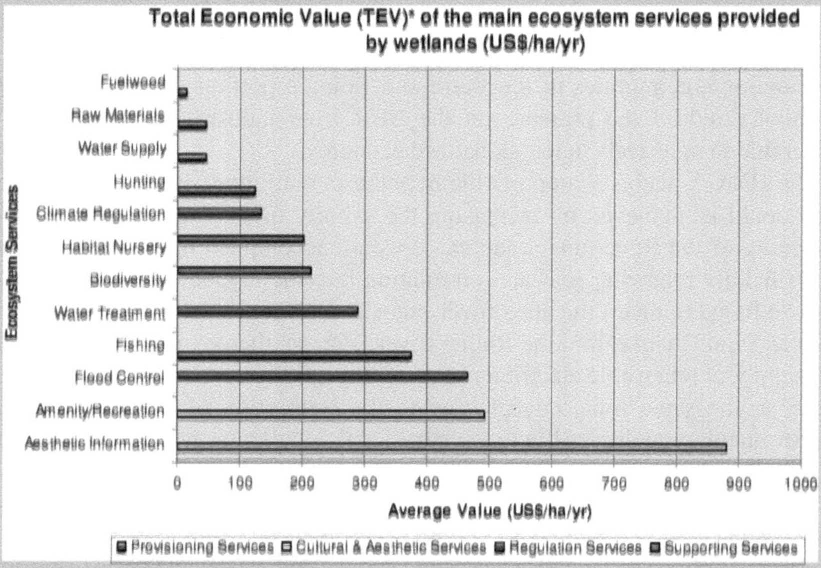

Fig. 6 The total economic value (TEV) of the main ecosystem services provided by wetlands (US$/ha/year). All figures are average global values based on sustainable use levels and are derived from Schuijt and Brander (2004) (calibrated for 2000), and Costanza et al. (1997) (calibrated for 1994). The two studies reviewed over 200 case studies

Unfortunately, most of these are public services that have no proper market and, thus, no market price. As a result, their benefits are ignored or only partially taken into account in our current economic system. Scientific studies, however, are increasingly demonstrating that, in addition to the dependence of the livelihoods and health of millions of people, the contribution of natural capital to the economy is immense (TEEB 2010). Evidence is also mounting that, when all services of our Natural Capital are properly accounted for, the benefits of money spent on conservation and restoration by far outweigh the costs (Fig. 7).

(continued)

Box 4 (continued)

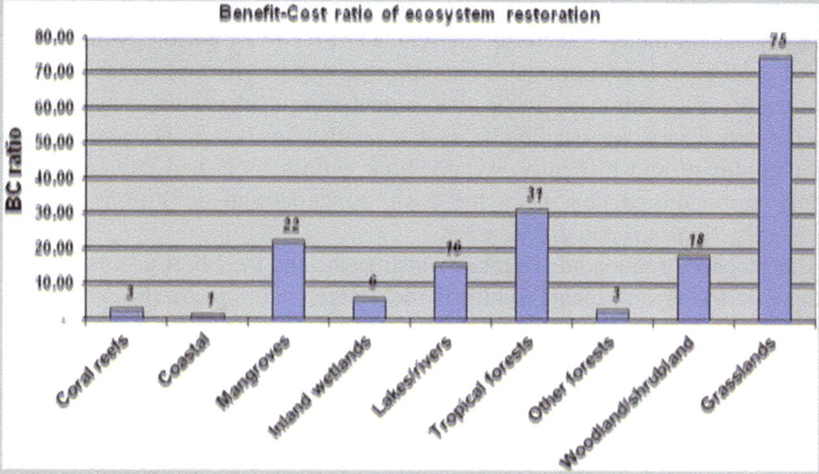

Fig. 7 A benefit–cost ratio of ecosystem restoration. The ratio is based on data in Neßhöver et al. (2011). Depending on the type of ecosystem and socioeconomic conditions the benefit–cost ratio of ecosystem restoration ranges between a factor of 3–75. Similar benefit–cost ratios apply to "green and blue" space in urban ecosystems

Biodiversity and natural ecosystems are a crucial source of nonmaterial well-being through their influence on mental health and the historical, national, ethical, religious, and spiritual values. While the conceptual and methodological developments in economic valuation have aimed at covering a broad range of values, including intangible ones (see the concept of Total Economic Value below), it can be argued that sociocultural values cannot be fully captured by economic valuation techniques and have to be complemented by other approaches in order to inform decision-making. This is notably the case where some ecosystem services are considered essential to a people's very identity and existence. To obtain at least a minimum (baseline) measure of importance of sociocultural benefits and values, several metrics have been developed such as the Human Well-being Index.

It also will build on the intuitive understanding that there is a direct connection between ecosystem health and public health, and social capital. Restoring natural capital implies, and, in fact requires, restoring social capital, which includes large amounts of hope and trust.

Ways Towards Sustainability

Meadows (1999) said: "Complex systems are, well, complex. It's dangerous to generalize about them." Likewise, the science of restoration ecology and the practice of ecological restoration are complex—no matter where you are. There is no one-size-fits-all solution to fix or restore degrading ecosystems, especially when they are very clearly socioecological systems with many historical layers. In spite of this complexity and difficulty, we do think certain tools and strategies can be widely applicable if and when the basic concepts and definitions have been thoroughly discussed and agreed upon.

The proposed sequential reference construction and the MCA process constitute an analytic framework meant to inform the design, and improve the coherence and acceptability, of a RNC project. This process starts with the definition of the potential reference conditions, and then characterizes the socioecological system to restore, and then adopts the restoration strategy and tactics, and project management, and then initiates the monitoring evaluation, communication and aftercare of the restoration rehabilitation and reintegration work.

Moreover, because the set of proposed criteria can be filled with a large spectrum of data, the proposed HMCA approach can be applied to any kind of socioecological system, and can constitute a support for comparative studies or meta-analyses. These tools—plus what we call RNC thinking—clearly can help constitute a participatory, consensus-building process, whether in the developing countries of the geopolitical South, or in the affluent, industrialized countries of the North (Kates et al. 2000). This should help advance what Kates et al. (2000) memorably called "sustainability science" and is the heart of transdisciplinarity, which is when scientists, engineers, and professional practitioners, along with nonprofessional and nonscientist stakeholders, come together to work towards the protection and restoration of the natural capital for the sake of biodiversity maintenance, sustainable and adaptable ecosystem functioning, and the quality of life for human society.

Acknowledgments We warmly thank R. Eugene Turner and Michael P. Weinstein for their helpful comments, and Bérengère Merlot and Elisabeth Le Corre for help with the manuscript and the figures.

References

Alexander S, Nelson C, Aronson J et al (2011) Opportunities and challenges for ecological restoration within REDD+. Rest Ecol 19(6):683–689

Aronson J, Vallejo R (2006) Challenges for the practice of ecological restoration. In: Van Andel J, Aronson J (eds) Restoration ccology: the new frontier. Wiley-Blackwell, Oxford, UK

Aronson J, Dhillion S, Le Floc'h E (1995) On the need to select an ecosystem of reference, however imperfect: a reply to Pickett & Parker. Rest Ecol 3:1–3

Aronson J, Milton SJ, Blignaut JN (eds) (2007) Restoring natural capital: science, business and practice. Island Press, Washington, DC
Aronson J, Blignaut JN, de Groot RS et al (2010) The road to sustainability must bridge three great divides. Annals New York Acad Sci 1185:225–236
Aronson J, Brancalion PHS, Durigan G et al (2011) What role should government regulation play in ecological restoration: ongoing debate in São Paulo State, Brazil. Rest Ecol 19(6):690–695
Bachelard G (1942) L'eau et les rêves. Editions José Corti, Paris, France, p 12
Bell ML, Hobbs BF, Ellis H (2003) The use of multi-criteria decision-making methods in the integrated assessment of climate change: implications for IA practitioners. Socio-Econ Plann Sci 37:289–316
Bellet M-E (1979) Guide archéologique des rives de l'étang de Berre. Édisud, Aix-en Provence, France
Belton V, Stewart TJ (2002) Multi-criteria decision analysis: an integrated approach. Kluwer Academic Publishers, Dordrecht, The Netherlands
Bullock J, Aronson J, Rey Benayas JM et al (2011) Restoration of ecosystem services and biodiversity. Trends Ecol Evol 26:541–549
Clébert J-P, Rouyer J-P (1991) La Durance. Privat, Toulouse, France
Clewell AF, Aronson J (2007) Ecological restoration: principles, values, and structure of an emerging profession. Island Press, Washington, DC
Collomp A (2002) La découverte des gorges du Verdon: Histoire du tourisme et des travaux hydrauliques. Édisud, Aix-en-Provence, France
Costanza R, d'Arge R, de Groot R et al (1997) The value of the world's ecosystem services and natural capital. Nature 387:253–260
de Groot R, Blignaut J, van der Ploeg S et al (in prep.) Restoration pays: evidence from the Field
Dobes L, Bennett J (2009) Multi-criteria analysis: "good enough" for government work? Agenda 16:7–30
DROS—Dispositif Régional d'Observation Sociale Provence-Alpes-Côte d'Azur (2010) Baromètre social 2009. Les Cah DROS 9:1–16
Egan D, Howell EA (2001) The historical ecology handbook: a restorationist's guide to reference ecosystems. Island Press, Washington, DC
Elmqvist T, Setälä H, Handel S et al (in prep.) Benefits of restoring ecosystem services in cities
ENSAML—École Nationale Supérieure d'Architecture de Marseille-Luminy (2001) L'étang de Berre: De la mer au lac. Éditions générales du CAUE des Bouches-du-Rhône, Marseille, France
Forman RTT, Gordon M (1986) Landscape ecology. Wiley, New York
Géoportail (2011) Institut Géographique National (IGN), France. http://ww.geoportail.fr. Accessed 10 September, 2011
Hämäläinen RP, Alaja S (2008) The threat of weighting biases in environmental decision analysis. Ecol Econ 68:556–569
Howard AF (1991) A critical look at multiple criteria decision making techniques with reference to forestry applications, Can J For Res 21:1649–1659
Janssen R (2001) On the use of multi-criteria analysis in environmental impact assessment in The Netherlands. J Mult-Crit Decis Anal 10:101–109
Janssen R, van Herwijnen M, Beinat E (2003) DEFINITE case studies and user manual. Institute for Environmental Studies, Vrije University, Amsterdam, Netherlands
Kates RW, Clark WC, Corelle R et al (2000) Sustainability science. Science 292:641–642
MA (Millennium Ecosystem Assessment) (2005) Ecosystems and human well-being: synthesis. Millennium ecosystem assessment series. Island Press and World Resources Institute, Washington. http://www.millenniumassessment.org/. Accessed 10 September, 2011
Maljean-Dubois S, Truilhe-Marengo E (2005) Le conflit entre les pêcheurs de l'étang de Berre et EDF: Quelques remarques à propos des arrêts CJC des 15 juillet et 7 octobre 2004. Droit Environ 131:186–190

Martin-Ortega J, Berbel A (2010) Using multi-criteria analysis to explore non-market monetary values of water quality changes in the context of the Water Framework Directive. Sci Total Environ 408:3990–3997

Meadows DH (1999) Leverage points: places to intervene in a system. Sustainability Institute. http://challenge.bfi.org/reference/PlacesToIntervene.pdf. Accessed 10 September, 2011

Neßhöver C, Aronson J, Blignaut JN et al (2011) Investing in ecological infrastructure. In: Ten Brink P (ed) The economics of ecosystems and biodiversity in national and international policy making. Earthscan, London

Paneque Salgado P, Corral Quintana S, Guimaraes Pereira A et al (2009) Participative multi-criteria analysis for the evaluation of water governance alternatives. A case in the Costa del Sol (Malaga). Ecol Econ 68:990–1005

Papayannis T (2008) Action for culture in Mediterranean wetlands. Med-INA, Athens, Greece. afoutri@med-ina.org. Accessed 10 September, 2011

Rees WE (1995) Cumulative environmental assessment and global change. Environ Impact Assess Rev 15:295–309

Rey Benayas JM, Newton AC, Diaz A, Bullock JM (2009) Enhancement of biodiversity and ecosystem services by ecological restoration: a meta-analysis. Science 325:1121–1124

Salt D, Walker B (2006) Resilience thinking: sustaining ecosystems and people in a changing world. Island Press, Washington, DC

Schuijt K, Brander L (2004) The economic value of the world's wetlands. Conserving the source of life. WWF Living Waters, Gland, Switzerland

SER (2004) The SER primer on ecological restoration. Society for Ecological Restoration International, Science and Policy Working Group. http://www.ser.org

Sheppard SRJ, Meitner M (2005) Using multi-criteria analysis and visualisation for sustainable forest management planning with stakeholder groups. For Ecol Manage 207:171–187

TEEB Foundations (2010) The economics of ecosystems and biodiversity: ecological and economic foundations. In: Kumar P (ed) An output of TEEB. Earthscan, London

Wang M, Yang J (1998) A multi-criterion experimental comparison of three multi-attribute weight measurement methods. J Mult-Crit Decis Anal 7:340–350

WCED (World Commission on Environment and Development) (1987) Our common future: the Brundtland report. Oxford University Press, New York

White PS, Walker JL (1997) Approximating nature's variation: selecting and using reference information in restoration ecology. Restor Ecol 5:338–349

Sustainability of Biodiversity Under Global Changes, with Particular Reference to Biological Invasions

Daniel Simberloff

Abstract Five major interacting global forces heavily influence changed distribution and abundance of biodiversity: (1) biological invasions, (2) overharvest, (3) changes in climate, (4) biogeochemical cycles, and (5) habitat. Modified land use, overharvest, and climate change have already affected distributions over large areas, regionally eliminating or substantially reducing natural resources such as fishing stocks, forestry trees, and traditional food plants. Further, habitat and climate change indirectly affect biodiversity patterns by their effects on invasions, biogeochemical cycles, and overharvest. Species-level biodiversity—the number of species on the planet—is unsustainable unless these impacts are ameliorated. Insufficient attention has been paid to how invasions and changed biogeochemical cycles directly affect distribution and abundance of biodiversity, as opposed to indirectly affecting biodiversity by modifying land use. Several direct, large-scale impacts of invasions and altered biogeochemical cycles on distributions and abundances of important species have been documented. These suggest that these global phenomena warrant much more research. Forestalling these global changes, or simply retarding them, has proven difficult, perhaps because of the immense scale of the enterprises causing them. However, great improvements in preventing or minimizing impacts of invasions are attainable, though we have been slow to develop effective policies and management strategies. Risk assessment procedures for planned introductions and invasion pathways are improving and can be modified to account for predicted climate and biogeochemical changes. Early warning systems, an underused tool in invasive species management, can be expanded and tied to effective rapid response procedures. Many approaches to managing established nonnative populations have produced successful control, but these successes result from projects highly tailored to particular species rather than from "silver bullets" that target multiple invaders

D. Simberloff (✉)
Department of Ecology and Evolutionary Biology, University of Tennessee,
Knoxville, TN 37996, USA
e-mail: dsimberloff@utk.edu

M.P. Weinstein and R.E. Turner (eds.), *Sustainability Science: The Emerging Paradigm and the Urban Environment*, DOI 10.1007/978-1-4614-3188-6_7,

simultaneously. Such successes will be possible even as other global changes proceed, so long as we remain committed to the effort.

Keywords Biogeochemical cycles • Climate change • Habitat destruction • Invasions • Overharvest

Introduction

What exactly does it mean to say "sustainability of biodiversity?" Biologists typically define "biodiversity" as the variety of life at three levels: the number of species (species richness), the number of distinct genotypes within species (genetic diversity), and the number of distinct communities of species (community-level diversity) (Office of Technology Assessment 1987). Although data are rapidly accumulating on genetic diversity of certain species, the relative dearth of knowledge about both genetic diversity and community-level diversity is still too great even to draw patterns of current status, much less to ask whether such biodiversity is sustainable under current human practices. Nevertheless, the numbers of species of some groups of organisms in many areas is quite well known, and much effort has gone into producing credible estimates of species richness even for relatively poorly known groups (e.g., May 1997). Further, at the species level of biodiversity, telling data exist or can be cogently deduced that permit us to depict past patterns and current trajectories. In this chapter, the term "biodiversity" will refer to the species level.

Biologists and paleontologists distinguish between background extinction and mass extinction. Species have always been going extinct during the 3.5 billion years that life has existed on earth; between 98 and 99% of all species that ever lived are extinct, and they lasted an average of only 4–5 million years (May 1997). However, species have not always gone at extinct at the same rate. Throughout the history of life, there has always been a slow process of extinction, a relatively few species going extinct each century, century after century. This is background extinction, and its rate has gradually been falling. This type of extinction has been more than balanced by the origination of new species, leading to a generally upward trend in species-level biodiversity (Raup and Sepkoski 1982). However, at five separate times there have been cataclysmic mass extinctions of large fractions of biodiversity in relatively short periods of time, over and above background extinction rates (Sepkoski 1984; cf. Alroy 2008), such as the mass extinction precipitated by an asteroid at the end of the Cretaceous era, about 65 million years ago, that killed off most dinosaurs over a period of just a few decades (Schulte et al. 2010). There have also been several less dramatic but still marked excursions in extinction rate above background levels.

We are now in the midst of what is often termed the "sixth mass extinction" (Leakey and Lewin 1995; Glavin 2007). Because so many groups of species in so many parts of the world are so poorly known, it is difficult to get precise estimates of the extinction rate, but if we stick to relatively well-known groups of species, a dark picture emerges. For instance, birds are probably the best known vertebrate group, because an army of birdwatchers supplements a substantial number of ornithologists. A conservative reading of fossil data suggests that we could expect, as a background rate, about one bird species to go extinct every 100–1,000 years (American Museum of Natural History 1998; Pimm 1998, 2007; BirdLife International 2000). But 128 bird species are known to have gone extinct in the last 500 years, and 103 of these extinctions have occurred since 1800 (BirdLife International 2000). This is 50–500 times the background rate. It is believed that 1,186 bird species are currently at risk of extinction within the next few decades (BirdLife International 2000), more than 12% of the total 9,700 species of birds. These threatened birds are not uniformly distributed over the surface of the earth. They are disproportionately located in certain regions and especially on islands (Groombridge 1992; Birdlife International 2000). One could repeat this exercise with other well-studied groups such as freshwater fish and Australian mammals and derive similarly disturbing findings (Pimm 1998).

In the United States, a nation in which species status is particularly well studied, about a third of all species are believed to be in some danger of extinction, and about 20% of those are in great danger of extinction within the next few decades (Master et al. 2000). For some groups, especially those inhabiting freshwater, percentages are even higher. For freshwater mussels and crayfishes, for example, over 50% of species face some threat to existence. For birds, the figure is 14%—close to the global percentage. Plants have by far the greatest number of species known to be imperiled in the United States (Master et al. 2000), partly because there are so many more known species of plants (18,100) than of, say, freshwater mussels (292). In fact, the number of imperiled insect species is undoubtedly much greater than the number of imperiled plant species, but so little is known about the biology and status of most groups of insects, even in a nation such as the United States with a relatively well-studied entomofauna, that one cannot even estimate the number of species that are imperiled, except for a few groups such as dragonflies (Odonata: Anisoptera) and tiger beetles (Coleoptera: Cicindelinae ; Stein et al. 2000). Since the termination of a moratorium imposed by Congress in 1995 on federally listing species under the Endangered Species Act, the list has gradually grown by a few percent each year. However, except for mammals, reptiles, and birds, the cumbersome nature of the listing process and lack of adequate biological knowledge suggest that far more species are actually imperiled than are listed—for example, four times as many amphibians and flowering plants and over ten times as many for most other taxa (Master et al. 2000; Wilcove and Master 2005). In sum, if humankind continues to behave as it has for the past two centuries, current levels of species-level biodiversity are definitely not sustainable.

Causes of Biodiversity Loss

In order to know what actions might redress this accelerating loss of biodiversity, it is necessary to know why all these species are going extinct. Five ongoing major global changes are drastically modifying the distribution and abundance of biodiversity: (1) biological invasions, (2) overharvest, (3) climate change, (4) modified biogeochemical cycles, and (5) habitat change (U.S. National Research

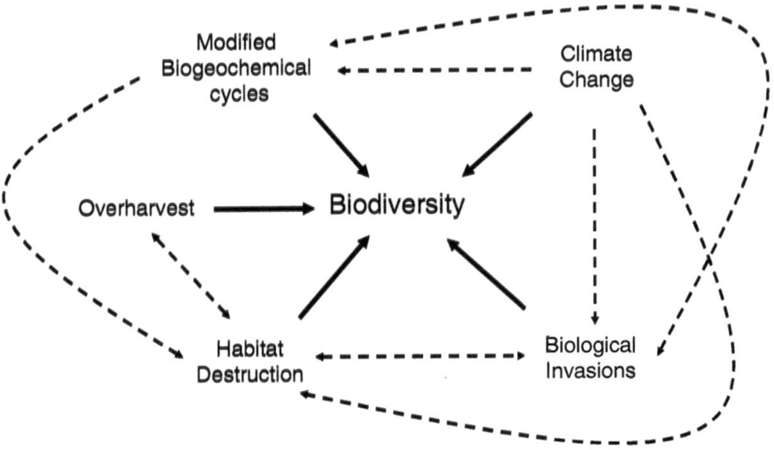

Fig. 1 The five major causes of species-level biodiversity loss and the major interactions between them (*dashed line*)

Council 2000). Furthermore, these global changes often interact in ways that exacerbate the impacts on biodiversity (Fig. 1). Biodiversity is not sustainable at anywhere near present levels without ameliorating the effects of these changes.

Most attention has been on the conservation consequences of habitat change—habitat destruction, conversion, and fragmentation—until the last 15 years, when public concern about the implications of global climate change rapidly increased. However, the impact of global climate change on regional and especially global biodiversity over the past two centuries has actually been far less than that of habitat change, biological invasions, overharvest, and probably changed biogeochemical cycles, and that is likely to be the case at least for the next two decades. Only one species-level extinction has been convincingly attributed to climate change recently (Pounds et al. 1999; Thomas et al. 2004), while for 680 animal species for which a cause for recent extinction could be assigned, invasive species were a contributing factor in 91 cases, habitat destruction in 82 cases, and overharvest in 77 cases (Clavero and García-Berthou 2005). It is quite possible that, by the year 2050, far more species will be "committed to extinction" by projected climate change (Thomas et al. 2004), perhaps as many as will be committed to extinction by then because of habitat change. However, how long before the commitment is realized is an unknown factor; it will vary among species, but it will surely take decades for most of these species to disappear. The other factors, however, have already caused many extinctions and are continuing to do so (see, e.g., Clavero and García-Berthou 2005).

Habitat Destruction

A staggering amount of both terrestrial and aquatic habitat has been transformed in the last two centuries, and the rate of transformation has accelerated rapidly over the past few decades (Millennium Ecosystem Assessment 2005). In large parts of the United States, more than half of the original vegetation type has been destroyed, converted to agriculture, housing, industrial use, and plantation forestry. Losses of wetlands are equally dramatic. This pattern is not unique to the United States; it is a global trend. Of 14 biomes studied worldwide by the Millennium Ecosystem Assessment project (2005), more than half had between 20 and 50% of their area converted to human use. For instance, three-fourths of Mediterranean forests and temperate grasslands worldwide have been placed in cultivation, while a third of all mangroves have been destroyed just in the last two decades (Millennium Ecosystem Assessment 2005). It is believed that half of all wetlands that existed in 1900 have since been lost, but data are inadequate to confirm or refute that figure. A fundamental ecological principle is that each species can tolerate only a limited range of habitat variables, so the massive amount of habitat destruction has led to an enormous toll of extinction and endangerment. Many species are specialized to use only a narrow range of habitats, and such habitat specialists have been particularly ravaged by habitat destruction. A forest species cannot survive if all forest is felled, nor can an aquatic species survive when a wetland is drained.

For one particularly well-studied group, out of 116 bird species and subspecies whose cause of recent extinction is known, 68 (59%) were driven to extinction by habitat change (Simberloff 1986). For birds worldwide, for 1,008 of the 1,186 imperiled species (85%), habitat loss/degradation is a major threat (BirdLife International 2000). Of 98 bird species of the United States whose causes of imperilment are believed to be known, habitat degradation or loss was wholly or partly responsible for the plight of 88 species (Wilcove et al. 2000). For 1,880 imperiled species of all types in the United States for which causes of imperilment are established, 85% are wholly or partly threatened by habitat loss or degradation. It is possible that birds are particularly sensitive to habitat alteration (see, e.g., Simberloff 1986), but many other taxa include large numbers of habitat specialists. Of imperiled species in the United States, more than 80% of plants, mammals, reptiles, amphibians, fishes, freshwater mussels, tiger beetles, and butterflies are threatened wholly or partly by changed habitat (Wilcove et al. 2000).

Not only is habitat being lost, but for many species the remaining natural area is increasingly fragmented (Primack 2010). Patches of forest and prairie are increasingly smaller and more isolated in many areas. One reason is land cleared for agriculture and pasture. Similarly, remaining bogs are increasingly separated as more of them are drained. For some species, even if enough habitat exists, the fragmented state can threaten their persistence—for example, they can be preyed upon as they move from patch to patch, or they can experience great difficulty finding food or mates. Highways and railroads can serve as formidable impediments to movement for some species (Forman and Alexander 1998; Spellerberg 2002). Further, the patches themselves, if they are small enough, might be heavily influenced by edge effects, so that large fractions of the patch differ subtly from the habitat in the center (Ries et al. 2004). Because fragmentation of a habitat is inevitably accompanied by loss of area of the habitat (e.g., the area of a road corridor bisecting a formerly continuous forest), the relative impacts of habitat loss and habitat fragmentation have been difficult to tease apart (Fahrig 2003; Ewers and Didham 2006). However, there is no doubt that, in certain cases, fragmentation per se constitutes a substantial threat to a species existence.

Invasions

Biological invasions are an enormous, multifaceted environmental and conservation threat (Mack et al. 2000; Baskin 2002) with substantial but sometimes subtle impacts on biodiversity. For instance, invasive species—especially predators such as rats, cats, and the small Indian mongoose—have entirely or partly caused most bird extinctions over the past two centuries, and invasive species trail only habitat destruction and direct exploitation (especially for human food) as a threat to currently imperiled bird species, affecting ca. 350 of the 1,186 threatened species (Birdlife International 2000). Not only introduced predators, but also introduced competitors, herbivores, and plants threaten various bird species. Invasions are a key factor in biodiversity decline for other taxa as well. On islands, invasions are the leading cause of extinction over the past 20 years, and in freshwater habitats, invasions are the second leading cause (Millennium Ecosystem Assessment 2005). In the United States, invasions are the second leading threat to imperiled species (after habitat destruction), affecting almost half of all such species (Wilcove et al. 2000).

Single invasive species can have a huge effect on local and even regional biodiversity. A chilling example is the impact of the predatory rosy wolf snail, *Euglandina rosea*, introduced deliberately from Florida and Central America to many islands in the Pacific and elsewhere as a biological control agent for the previously introduced giant African snail, *Lissachatina fulica*. In fact, the rosy wolf snail has not controlled the giant African snail, which grows too big for it to be consumed by the predator. Rather, the rosy wolf snail is responsible for the global extinction of at least 50 species of land snails on Pacific islands, such as the Hawaiian Islands and Tahiti, and remains a threat to many others (Civeyrel and Simberloff 1996; Cowie 2002).

The small Indian mongoose, *Herpestes auropunctata*, was introduced to Jamaica in 1872 to control rats in cane fields, then deliberately spread from Jamaica to many islands around the world (e.g., Hawaiian islands, many of the West Indies) for the same purpose or in an attempt to control snakes (Simberloff et al. 2000). Other introductions of this mongoose (e.g., to Fiji, Okinawa, Amami, Mauritius, Croatia) make this species a global plague. Although it is difficult to separate its total impact from that of rats, cats, and mustelids introduced to the same islands, the mongoose has surely contributed to the extinction of several mammal and bird species and subspecies, and it continues to threaten others, as well as amphibians (Hays and Conant 2007).

Many other invasive animals have eliminated native species or threaten to do so (Simberloff and Rejmánek 2011). However, some invasive plants transform entire landscapes or water bodies and generally have the greater impacts (Simberloff 2002; Ehrenfeld 2010) and may completely eliminate species locally. Each such event brings them closer to global extinction. Often introduced plants simply overgrow and thereby replace native vegetation. For instance, Australian paperback (*Melaleuca quinquenervia*) and Brazilian pepper (*Schinus terebinthifolius*) have overgrown almost half a million hectares of forest and prairie in south Florida, and *Hydrilla verticillata* has blanketed 40,000 ha of open water throughout the state (Schmitz et al. 1997). South American water hyacinth (*Eichhornia crassipes*), though controlled in Florida, covers much of Lake Victoria in Africa (Matthews and Brand 2004a) and many water bodies in the southeastern United States (Schardt 1997) and Asia and Australia (Matthews and Brand 2004b), often smothering native vegetation. Water hyacinth forms a thick mat that blocks sunlight and kills plants and animals. This impact, in turn, causes hypoxia, which leads to a cascade of destruction through the food web. Introduced plants can have many other impacts, some of them quite subtle (e.g., changed nutrient cycles [Vitousek 1986]) and others very obvious (e.g., intensified fire regime [D'Antonio and Vitousek 1992]), that can affect entire ecosystems to the great detriment of many of the native animal and plant species (Ehrenfeld 2010).

Introduced species can interact with one another to exacerbate the total impact on native species (Simberloff 2006). For instance, on Christmas Island, populations of the long-present introduced yellow crazy ant (*Anoplolepis gracilipes*) increased dramatically when a scale insect was subsequently introduced (O'Dowd et al. 2003; Abbott 2004). The ants feed on the honeydew the scales produce and protect the scales from predators and parasites, increasing their populations. The ants devastate populations of the native red land crab (*Gecarcoidea natalis*), which in turn leads to massive growth of ground cover plants, seeds and seedlings of which had been removed by crabs. The honeydew spurs growth of a sooty mold that causes canopy dieback of large trees.

Biogeochemical Cycles

Because the nitrogen, phosphorus, and carbon cycles are crucial to life on earth, it is not surprising that changes in them can drastically affect biodiversity. Toxic pollutants, such as heavy metals and DDT, were the key focus of conservation concern about chemicals during the 1960s through the 1980s. Today, greatly increased nutrient loading, particularly by nitrogen and phosphorus compounds, has come to be recognized as a critical and pervasive global change (United States National Research Council 2000; Millennium Ecosystem Assessment 2005), and the climate impacts of the changed carbon cycle now dominate our attention. Humans now produce more reactive nitrogen than all natural processes combined, and as much of half of all nitrogen fertilizer may end up in the environment, primarily through runoff (Millennium Ecosystem Assessment 2005). Similarly, phosphorus application has increased threefold since 1960 (Millennium Ecosystem Assessment 2005). Accumulation of phosphorus in the soil changes plant communities, while runoff and leaching into water bodies lead to eutrophication.

Enhancement of the nitrogen cycle has received the most attention. Some impacts of this enhancement affect biodiversity indirectly. For instance, emissions of nitric oxides (from many human activities, including fertilizers, land clearing, and various manufacturing processes) have increased greatly; these are important precursors of acid rain, which in turn negatively affects aquatic organisms in many water bodies (Vitousek et al. 1997). The direct effects of enhanced nitrogen on biodiversity probably weigh most heavily on the many native plant species that have evolved to function best at lower levels; these are often replaced by invasive species adapted to the new nutrient regime (Vitousek et al. 1997). Nitrogen enrichment has produced severe eutrophication of many estuaries and coastal areas, which in turn has led to local or even regional loss of species of fish, shellfish, and planktonic and benthic organisms (Vitousek et al. 1997; Howarth et al. 2000).

Although phosphorus also contributes substantially to eutrophication of many coastal ecosystems, phosphorus enrichment is the key threat in freshwater, because producers in freshwater systems are generally phosphorus-limited. Some progress has been made in limiting phosphorus wastewater effluent from industrial and municipal sources, but excess fertilization of agricultural crops plus manure production by livestock lead to phosphorus accumulation in the soil, some of which is transported in runoff into aquatic ecosystems (Carpenter et al. 1998). The most visible effect of lake eutrophication is proliferation of algae, which can lead to loss of plant beds that are critical habitats for certain aquatic animals. The algal growth in turn leads to proliferation of bacterial decomposers that break down dead plant matter. The resulting oxygen reduction (hypoxia) can cause fish kills. Thus, local biodiversity is an early casualty of lake eutrophication.

The increasing amount of carbon dioxide in the atmosphere has many direct impacts on plants, some of which likely affect animals that feed or live on them (Bezemer and Jones 1998). For instance, elevated atmospheric carbon dioxide causes leaf nitrogen content to decrease in most plant species, and leaf-chewing insects respond by increasing their food consumption, often by substantial amounts. Another impact of increased atmospheric carbon dioxide is increased ocean acidification. This change, in turn, threatens species that use calcium carbonate to form structures such as shells or coral skeletons (Orr et al. 2005). Among these are cold-water corals, which provide essential fish habitat, and shelled pteropods (a group of marine gastropods), which are a major food source for marine predators. These direct effects on the carbon cycle have not yet threatened species' existence. However, an indirect effect of the buildup of carbon dioxide and other greenhouse gases is global climate change, which is beginning to threaten biodiversity directly.

Climate Change

The list of species in the United States threatened or endangered under the Endangered Species Act cites many as affected by land use change, many by introduced species, but only one clearly affected by global warming—the polar bear, listed as threatened in 2008. A current lawsuit by the Center for Biological Diversity aims to force the U.S. Fish and Wildlife Service to list it as endangered. The polar bear story is quite well known, and it has become a poster child for global warming; with melting ice, the bears must swim further and live on beaches, where they are far from their main food—seals. Several polar bear populations have already suffered substantial declines. Two coral species—staghorn coral (*Acropora cervicornis*) and elkhorn coral (*Acropora palmata*)—were also listed as threatened (2006), but ocean warming was cited as only one of several threats (and not the foremost).

Listing of several other species as threatened by global warming has become a politically contentious issue in the United States in the face of climate change skeptics. For example, avian ecologists are convinced that declines in several penguin species are partially due to global warming and that warming constitutes a grave and imminent threat. In response to a petition and a lawsuit by the Center for Biological Diversity and the Turtle Island Restoration Network, the U.S. Fish and Wildlife Service in 2010 listed five penguin species (Humboldt penguin, yellow-eyed penguin, white flippered penguin, Fiordland penguin, and erect-crested penguin) as threatened but declined to cite climate change as a part of the threat, even though the petition and lawsuit argued that it is the primary threat.

Other species on the U.S. Endangered Species list will surely be affected by global warming, but that is not one of the threats yet listed for them. For instance, species adapted to higher elevations of mountains will have to move constantly higher as the climate warms, unless they can reach colder areas—for instance, climbing down the mountain and going north. But most cannot do that, because they cannot move quickly enough and because the intervening habitat is unsuitable—one can consider how plants would accomplish this feat in a landscape heavily occupied by humans. And eventually, as climate continues to warm, even the mountain peaks will not be suitable habitat—such species will go extinct. A prominent example is the American pika (*Octotona princeps*). This mammal occupies high elevations of mountains in the American West and is believed to be in danger of extinction from global warming (Guralnick 2007), leading to a petition to the Fish and Wildlife Service in 2007 to list it under the Endangered Species Act. The petition was denied in 2010, based on agency scientists' assessment that pikas will either adjust to warmer temperatures or migrate higher upslope (if there is an upslope left!). Another high-elevation species is the endangered Uncompahgre fritillary butterfly (*Boloria acrocnema*), restricted to a few peaks in the San Juan Mountains of southwestern Colorado. Britten et al. (1994) contend that a warming trend well within the range of those predicted by general circulation models of climate could easily eliminate its entire habitat.

Many changes in both geographic and elevational range, as well as timing of life history events such as migration or flowering, have already been recorded over the past few decades in response to warming (Parmesan 2006). The substantial predicted temperature changes over the next century (Metz 2007) imply a further increase in the number of species threatened directly by climate change.

Overharvest

Overexploitation, particularly hunting and fishing, ranks fourth in the United States (after habitat change, biological invasions, and chemical stressors) as a threat to imperiled species (Wilcove et al. 2000). It plays a similar role globally; for instance, it is the second leading threat, after habitat change, to bird species (BirdLife International 2000), with 233 imperiled species hunted for food and 111 imperiled species trapped for the pet trade. Overharvest is a particular threat to mammal and reptile species, hunted for sport, food, or the pet trade, but many other types of species are threatened by overharvest—for example, butterflies, which are prized by collectors (Wilcove and Masters 2005). Even plants are threatened by overharvest. For instance, in the United States *Echinacea tennesseensis*, a medicinal plant listed under the U.S. Endangered Species Act, is partly threatened by collecting for medicinal purposes, and several cycads and cacti are imperiled by harvest for use as ornamentals. Even edible fungi are threatened by collecting. In China, three fourths of all imperiled vertebrate species are threatened by overexploitation (Li and Wilcove 2005). Among marine fishes, the situation is not more promising. For instance, five heavily fished species of large groupers in the genus *Epinephelus* are listed as endangered in the IUCN Red List (World Conservation Union 2010); the U.S. Fish and Wildlife Service simply lists them as "Species of Special Concern" because they consider the available data insufficient to judge the degree of threat. Of a total of 133 global, regional, and local extinctions of marine fish populations, overexploitation was the main cause, contributing to 55% of these cases (Dulvy et al. 2003).

Interactions Among the Key Extinction Forces

Often two or more factors combine to threaten a species' existence. For instance, in addition to species threatened directly by global warming, many animal species will be threatened as rising temperatures plus the predicted increasingly severe and frequent droughts in certain regions (Intergovernmental Panel on Climate Change 2007) change vegetation that is the required habitat. Climate change also interacts with biological invasions. For example, three major invasive plants in the United States—kudzu (*Pueraria lobata*), privet (*Ligustrum sinense*, *Ligustrum vulgare*), and cogongrass (*Imperata cylindrica*)—are predicted to spread widely as climate warms (Bradley et al. 2010). Of course, some invasive species will have their ranges contract with predicted climate change, but many more, such as the red imported fire ant (Fitzpatrick et al. 2007), are predicted to be able to spread further.

Species introductions often interact with habitat change, as conversion of a natural landscape to one dominated by agriculture or some other human use makes the habitat less suitable for native species and more suitable for one or more introduced ones (Lockwood et al. 2007). For instance, the use of what had been prairie throughout much of the American West for livestock grazing has led to the destruction of native bunchgrasses, which has opened the way for Eurasian cheatgrass (*Bromus tectorum*) to spread and transform tens of millions of hectares into virtual cheatgrass monocultures. This landscape transformation was a disaster for native plant diversity, and cheatgrass is of poor nutritional quality for cattle as well (Mack 2011).

Biological invaders are also often favored by nitrogen deposition from air pollution. For example, in Minnesota grasslands, nitrogen-fertilized plots were quickly overgrown by invasive European quackgrass (*Elymus repens*), and most of the diverse native grass community disappeared (Vitousek et al. 1997). Nitrogen enhancement and invasion can also interact in a different manner to facilitate further invasions. The Atlantic shrub *Morella faya* was introduced to the Hawaiian Islands in the nineteenth century as an ornamental, then widely spread for forestry purposes through the 1920s. This nitrogen-fixing plant has invaded geologically young, nitrogen-poor areas on the volcanic island of Hawaii (Vitousek 1986; Vitousek and Walker 1989; Vitousek et al. 1987). Hawaii lacks native nitrogen fixers, so native plants have evolved adaptations to nitrogen-poor soil. Native plants have gradually been replaced, and nitrogen and water content of the canopy have doubled in the wake of invasion by *M. faya* (Asner and Vitousek 2005). The previous absence of nutrients had impeded the spread of a number of introduced ornamentals and weeds not adapted to the nutrient-poor conditions. However, now that *M. faya* has invaded, the habitat has become suitable for other introduced plants. A similar phenomenon has occurred in areas of Australia where low soil phosphorus has limited invasion by many introduced species. However, the introduction of three plant species that are able to concentrate phosphorus in the soil underneath them has facilitated the invasion of previously restricted nonnatives (Fisher et al. 2006; Turner et al. 2008).

Overharvest can similarly interact with other forces to bring a species to the threshold of extinction. For instance, habitat destruction, by reducing cover, exacerbates the threat posed by overhunting to many bird species used for food. It also reduces nest site availability, worsening the threat to imperiled cage bird species (BirdLife International 2000). Similarly, damage inflicted on marine substrates, including coral reefs, during the process of overfishing greatly compounds the impact on both the target species and many others (Dayton et al. 1995). The catalog of ways in which the various forces interact to exacerbate the impact on biodiversity is far longer than the examples adduced here, but the important point is clear: the impacts of these forces are more than simply additive.

What to Do About Declining Biodiversity

It appears that there is nearly universal concern with the decline of biodiversity and eagerness to try to stop it. The Convention of Biological Diversity, opened for signature at the Earth Summit in Rio de Janeiro in 1992, has as its first two goals the conservation of biodiversity and the sustainable use of its components. It has been ratified by 193 nations; only the United States among major nations has failed to do so. The United Nations declared 2010 to be the International Year of Biodiversity. Yet, as demonstrated earlier, the wave of extinctions spreads and even accelerates, seemingly inexorably.

If biodiversity is not sustainable given current practices, what can be done to make it sustainable? The huge scale of global climate change makes it seem almost futile to address, and a number of "climate skeptics" are in fact not skeptical about the existence of anthropogenic climate change but rather doubt that any feasible human actions can be mounted to retard it. They are incorrect; many partial solutions that have been suggested would lessen the rate of climate change substantially (Noble et al. 2005). Similarly to climate change, alterations in biogeochemical cycles are diffused and not geographically restricted to sites engendering the change. And the scale of changes in biogeochemical cycles like the nitrogen cycle is enormous. This diffusion and the enormous scale do not mean that such forces cannot be stemmed, but effective actions would entail international agreements and global cooperation at a level not yet achieved. Again, a number of feasible ways to ameliorate this problem greatly have been proposed (Howarth et al. 2005). Although many cases of habitat change involve two or more neighboring nations, for the most part habitat destruction is a national problem and lessening it rests on how governments encourage or constrain citizens in their use of lands and waters. This is a subject well beyond the scope of this essay; suffice it to say that various successful approaches (see, e.g., Koh and Gardner 2010; Tallis et al. 2010) could be replicated or adapted to many other regions and, in sum, could greatly alleviate this greatest of threats to biodiversity. Overharvest is also largely a national issue, although certain cases (e.g., marine fisheries) are international. The solutions again largely reside in the interactions between governments and their citizens, although international instruments are required where a common resource (e.g., whales and marine fishes) is exploited by different nations (Peres 2010).

Impacts of biological invasions, like those of overharvest, are primarily experienced at the national or even local level, because each invasion involves a specific species in a specific new place, and there are many ways to prevent or slow these down and much recent progress (Simberloff 2010). Because each invasive species arriving in a new site has come from some other site, and because an invasive species, once established, does not respect national boundaries, international agreements and joint efforts are also often important.

Reducing the Impact of Biological Introductions

The best way to deal with introduced species is to prevent their introduction in the first place. If an invader defeats these efforts and establish populations, we can try to eradicate them, especially if we detect them quickly. If eradication proves impossible, we can attempt to manage these populations at low enough levels that they are not problematic.

Introductions can be either planned or unplanned, and prevention differs somewhat between these categories. For planned introductions, such as new game species or ornamental plants, an appropriate law would entail a risk analysis and consist of either a black list (list of forbidden species), a white list (list of approved species), or a combination of the two. For these lists to work, the risk analyses must be sufficiently accurate that most species that would become invasive are recognized and put on black lists or kept off white lists. Such analyses for invasive species are in their infancy, but much active research (e.g., Kolar and Lodge 2002) is focused on improving them. For plants, various versions of the Australian Weed Risk Assessment (Pheloung et al. 1999) are in use worldwide and are believed to be effective.

Preventing unplanned introductions entails identifying pathways by which they occur (Ruiz and Carlton 2003): ballast water for marine organisms, untreated wooden packing for some insects, hitchhiking insects and pathogens on ornamental plants, and the like. These pathways must then be constricted. For instance, ballast water can be exchanged in the open ocean and ornamental plants can be fumigated or refrigerated. None of these methods will eliminate every potential invader, but many of them greatly reduce their number, and technologies are constantly being improved. Of course, inspections at ports of entry are critical for both planned and unplanned introductions, and various technologies, including both trained detection dogs and increasingly accurate X-ray equipment, are in use in some nations (Baskin 2002). However, expense, including labor costs, has hindered their deployment in many nations.

An ongoing monitoring system combined with an early warning/rapid-response system is key to eradicating introduced species inexpensively, but labor and training expense hinder the efforts of most nations to construct such a system. Web-savvy, engaged citizens can certainly reduce the monitoring cost greatly. Early warning by alert citizens allowed the eradication of the Asian longhorned beetle (*Anoplophora glabripennis*) in the Chicago area and the "killer alga" (*Caulerpa taxifolia*) in California (Simberloff 2010). In both instances, authorities mobilized a response almost immediately after discovery and maintained the effort for several years to ensure complete eradication. Maintaining the response effort even when the initial population of the invader has been largely suppressed is a key step whose failure has sometimes stymied eradication projects. Another key requirement is clear lines of authority so that an agency can compel cooperation and prevent individuals from subverting an eradication effort if they object to the eradication per se or to the methods chosen to pursue it. The biology of the target organism must be well understood, and the probability of quick reinvasion must be low.

Where these criteria have been met, many eradication projects have succeeded. Frequently these are on islands, where reinvasion is unlikely. Rats have been eradicated from islands as large as 113 km^2 (Pascal et al. 2010), and feral goats and pigs have been eradicated from Santiago Island (585 km^2) in the Galapagos (Cruz et al. 2005). Even larger eradications are possible. For instance, the African mosquito (*Anopheles gambiae*) vector of malaria was eradicated from ca. 50,000 km^2 in northeastern Brazil (Davis and Garcia 1989), and several fly species have been eradicated from large regions, especially in the tropics (Klassen 2005). Plant eradications have proven more challenging, and most successful projects are small scale (e.g., Gardener et al. 2010). However, the weed *Kochia scoparia* was eradicated from a large area of Western Australia (Randall 2001), and the 50-year-long campaign to eradicate witchweed (*Striga asiatica*) over thousands of hectares in the southeastern United States is nearing success (Eplee 2001). Plant eradications that failed usually did so because of lack of support or cooperation rather than technical infeasibility (Gardener et al. 2010). Successful eradication projects have used methods running the gamut from hand-pulling of weeds and hunting and trapping of mammals to use of sterilized males for insects and fishes and chemical pesticides and herbicides for many species.

For established invasive populations, three main technologies alone or in combination have often provided adequate control: mechanical or physical control, chemical control, and biological control. With sufficient labor, even massive invasions can sometimes be curtailed by physical means alone. In South Africa, the invasive Australian tree *Acacia cyclops* has been effectively controlled by mechanical means alone—cutting and pulling roots (Matthews and Brand 2004a). Herbicides have often maintained invasive plants at low levels. In Florida, water hyacinth (*E. crassipes*) was drastically reduced and eventually controlled by use of the herbicide 2,4-D along with some mechanical removal (Schardt 1997). In South Africa, a large public works program, Working for Water, has used physical, mechanical, and chemical methods to clear thousands of hectares of land of introduced plants that use prodigious amounts of water (Matthews and Brand 2004a).

Long-term use of chemicals may lead to the evolution of resistance, and chemicals are expensive. They may also have nontarget impacts, although many later-generation herbicides and pesticides have fewer of these. These issues have led to great interest in biological control—deliberate introduction of a natural enemy (predator, parasite, or disease) of an introduced pest. Most such projects have not succeeded, and some have had substantial nontarget impacts; both the aforementioned rosy wolf snail and the small Indian mongoose were introduced as biological control agents. However, successes can be dramatic. For instance, on the island of St. Helena a tropical American scale insect (*Orthezia insignis*) had threatened the existence of the endemic gumwood tree (*Commidendrum robustum*), but a predatory South American lady beetle (*Hyperaspis pantherina*) now limits the scale insect population (Booth et al. 2001). Many other projects have been successful, primarily in control of agricultural pests, but the technology is increasingly aimed at invaders of natural systems (Van Driesche et al. 2008).

Conclusion

Human activities have rendered current levels of species-level biodiversity unsustainable, primarily because of (1) biological invasions, (2) overharvest, plus anthropogenic global changes: (3) climate change, (4) modified biogeochemical cycles, and (5) habitat change. Further, these five factors interact with one another in complex and often idiosyncratic ways to exacerbate their impacts on biodiversity. The impacts on biodiversity of biological invasions and modified biogeochemical cycles are less well publicized than are overharvest and changed climate and habitat. The scope and, for climate change and biogeochemical cycles, diffuse nature of these factors can easily induce pessimism about humankind's ability to ameliorate them sufficiently to stem biodiversity loss. However, technologies and policies to alleviate all of these factors do exist. For biological invasions, although international agreements on the movement of goods and people are needed to restrict pathways for hitchhiking organisms and to limit the deliberate introduction of species presenting a high risk of invasion, many national, regional, and local actions can contribute greatly to solving this problem. Probably most useful would be an effective early warning/rapid-response system, plus sufficiently centralized and enabled government agencies to act quickly and to coordinate both policies and on-the-ground actions. Although economic costs will be associated with the development of effective machinery to tackle biological invasions, such costs are not astronomic and drastic lifestyle shifts are unnecessary. Certainly the costs of failing to bring invasions under control will be greater.

Acknowledgment I am grateful to Billy Goodman, Michael P. Weinstein, and an anonymous referee for many useful comments on an early draft of this manuscript.

References

Abbott KL (2004) Alien ant invasion on Christmas Island, Indian Ocean: the role of ant-scale mutualisms in the dynamics of supercolonies of the yellow crazy ant, *Anoplolepis gracilipes*. Ph.D. Thesis. Biological Sciences, Monash University, Melbourne

Alroy J (2008) Dynamics of origination and extinction in the marine fossil record. PNAS 105(suppl 1): 1536–1542

American Museum of Natural History (1998) http://www.amnh.org/science/biodiversity/extinction/Intro/OngoingProces.html. Accessed 15 October 2010

Asner GP, Vitousek PM (2005) Remote analysis of biological invasion and biogeochemical change. PNAS 102:4383–4386

Baskin Y (2002) A plague of rats and rubbervines. Island Press, Washington

Bezemer TM, Jones TH (1998) Plant-insect herbivore interactions in elevated CO_2: quantitative analyses and guild effects. Oikos 82:212–222

BirdLife International (2000) Threatened birds of the world. Lynx Edicions and BirdLife, Barcelona and Cambridge

Booth RG, Cross AE, Fowler SV et al (2001) Case study 5.24. Biological control of an insect to save an endemic tree on St. Helena. In: Wittenberg R, Cock MJW (eds) Invasive alien species: a toolkit of best prevention and management practices. CAB International, Wallingford

Bradley BA, Wilcove DS, Oppenheimer M (2010) Climate change increases risk of plant invasion in the Eastern United States. Biol Invasions 12:1855–1872

Britten HB, Brussard PF, Murphy DD (1994) The pending extinction of the Uncompaghre fritillary butterfly. Conserv Biol 8:86–94

Carpenter S, Caraco NF, Correll DL et al (1998) Nonpoint pollution of surface waters with phosphorus and nitrogen. Ecol Appl 8:559–568

Civeyrel L, Simberloff D (1996) A tale of two snails: is the cure worse than the disease? Biodiv Conserv 5:1231–1252

Clavero M, García-Berthou E (2005) Invasive species are a leading cause of animal extinctions. Trends Ecol Evol 20:110

Intergovernmental Panel on Climate Change (2007) Climate Change 2007: synthesis report. Contribution of working groups I, II and III to the fourth assessment report of the Intergovernmental Panel on Climate Change (Core Writing Team, Pachauri RK, Reisinger A, eds) IPCC, Geneva

Cowie RH (2002) Invertebrate invasions on Pacific islands and the replacement of unique native faunas: a synthesis of the land and freshwater snails. Biol Invasions 3:119–136

Cruz F, Donlan CJ, Campbell K et al (2005) Conservation action in the Galápagos: feral pig (*Sus scrofa*) eradication from Santiago Island. Biol Conserv 121:473–478

D'Antonio CM, Vitousek PM (1992) Biological invasions by exotic grasses, the grass/fire cycle, and global change. Annu Rev Ecol Syst 23:63–87

Davis JR, Garcia R (1989) Malaria mosquito in Brazil. In: Dahlsten DL, Garcia R (eds) Eradication of exotic pests. Yale University Press, New Haven

Dayton PK, Thrush SF, Agardy MT et al (1995) Environmental effects of marine fishing. Aquat Conserv Marine Freshw Ecosyst 5:205–232

Dulvy NK, Sadovy Y, Reynolds JD (2003) Extinction vulnerability in marine populations. Fish Fish 4:25–64

Ehrenfeld JG (2010) Ecosystem consequences of biological invasions. Annu Rev Ecol Evol Syst 41:59–80

Eplee RE (2001) Case study 2.10. Co-ordination of witchweed eradication in the U.S.A. In: Wittenberg R, Cock MJW (eds) Invasive alien species: a toolkit of best prevention and management practices. CAB International, Wallingford

Ewers RM, Didham RK (2006) Confounding factors in the detection of species responses to habitat fragmentation. Biol Rev 81:117–142

Fahrig L (2003) Effects of habitat fragmentation on biodiversity. Annu Rev Ecol Evol Syst 34:487–515

Fisher JL, Veneklaas EJ, Lambers H et al (2006) Enhanced soil and leaf nutrient status of a Western Australian *Banksia* woodland community invaded by *Ehrharta calycina* and *Pelargonium capitatum*. Plant Soil 284:253–264

Fitzpatrick MC, Weltzin J, Sanders NJ et al (2007) The biogeography of prediction error: why does the introduced range of the fire ant over-predict its native range? Glob Ecol Biogeogr 16:24–33

Forman RTT, Alexander LE (1998) Roads and their major ecological effects. Annu Rev Ecol Syst 29:207–231

Gardener MR, Atkinson R, Rentería JL (2010) Eradications and people: lessons from the plant eradication program in Galapagos. Restor Ecol 18:20–29

Glavin T (2007) The sixth extinction: journeys among the lost and left behind. Thomas Dunne Books, New York

Groombridge B (1992) Global biodiversity. Status of the earth's living organisms. Chapman & Hall, London

Guralnick R (2007) Differential effects of past climate warming on mountain and flatland species distributions: a multispecies North American mammal assessment. Glob Ecol Biogeogr 16:14–23

Hays WST, Conant S (2007) Biology and impacts of Pacific island invasive species. 1. A worldwide review of effects of the small Indian mongoose, *Herpestes javanicus* (Carnivora: Herpestidae). Pac Sci 61:3–16

Howarth R, Anderson D, Cloern J et al (2000) Nutrient pollution of coastal rivers, bays, and seas. Issues Ecol 7:1–15

Howarth R et al (2005) Nutrient management. In: Chopra K et al (eds) Ecosystems and human well-being: policy responses: findings of the Responses Working Group of the Millennium Ecosystem Assessment. Island Press, Washington

Klassen W (2005) Area-wide integrated pest management and the sterile insect technique. In: Dyck VA, Hendrichs J, Robinson AS (eds) Sterile insect technique. Principles and practice in area-wide integrated pest management. Springer, Dordrecht
Koh LP, Gardner TA (2010) Conservation in human-modified landscapes. In: Sodhi NS, Ehrlich PR (eds) Conservation biology for all. Oxford University Press, New York
Kolar CS, Lodge DM (2002) Ecological predictions and risk assessment for alien fishes in North America. Science 298:1233–1236
Leakey RE, Lewin R (1995) The sixth extinction: patterns of life and the future of humankind. Doubleday, New York
Li Y, Wilcove DS (2005) Threats to vertebrate species in China and the United States. Bioscience 55:147–153
Lockwood JL, Hoopes MF, Marchetti MP (2007) Invasion ecology. Blackwell, Malden
Mack R (2011) Cheatgrass. In: Simberloff D, Rejmánek M (eds) Encyclopedia of biological invasions. University of California Press, Berkeley
Mack R, Simberloff D, Lonsdale M et al (2000) Biotic invasions: causes, epidemiology, global consequences, and control. Ecol Appl 10:689–710
Master LL, Stein BA, Kutner LS et al (2000) Vanishing assets. Conservation status of U.S. species. In: Stein BA, Kutner LS, Adams JS (eds) Precious heritage. The status of biodiversity in the United States. Oxford University Press, New York
Matthews S, Brand K (2004a) Africa invaded. Global Invasive Species Programme, Cape Town
Matthews S, Brand K (2004b) Tropical Asia invaded. Global Invasive Species Programme, Cape Town
May RM (1997) The dimensions of life on earth. In: Raven PH, Williams T (eds) Nature and human society. The quest for a sustainable world. National Academy Press, Washington
Metz B (2007) Climate change 2007: mitigation of climate change. Contribution of working group III to the fourth assessment report of the Intergovernmental Panel on Climate Change. Cambridge University Press, New York
Millennium Ecosystem Assessment (2005) Ecosystems and human well-being: biodiversity synthesis. World Resources Institute, Washington
Noble I et al (2005) Climate change. In: Chopra K et al (eds) Ecosystems and human well-being: policy responses: findings of the responses working group of the Millennium Ecosystem Assessment. Island Press, Washington
O'Dowd DJ, Green PT, Lake PS (2003) Invasional "meltdown" on an oceanic island. Ecol Lett 6:812–817
Office of Technology Assessment (US Congress) (1987) Technologies to maintain biological diversity, OTA-F-330. US Government Printing Office, Washington
Orr JC, Fabry VJ, Aumont O et al (2005) Anthropogenic ocean acidification over the twenty-first century and its impact on calcifying organisms. Nature 437:681–686
Parmesan C (2006) Ecological and evolutionary responses to recent climate change. Annu Rev Ecol Evol Syst 37:637–669
Pascal M, Le Guyader H, Simberloff D (2010) Biological invasions and the conservation of biodiversity. World Org Anim Health (OIE) Sci Tech Rev 29:387–403
Peres CA (2010) Overexploitation. In: Sodhi NS, Ehrlich PR (eds) Conservation biology for all. Oxford University Press, New York
Pheloung PC, Williams PA, Halloy SR (1999) A weed risk assessment model for use as a biosecurity tool evaluating plant introductions. J Environ Manage 57:239–251
Pimm SL (1998) Extinction. In: Sutherland WJ (ed) Conservation science and action. Blackwell, Oxford
Pimm SL (2007) The extinction puzzle. http://www.projectsyndicate.org/commentary/pimm1/English. Accessed 15 October 2010
Pounds JA, Fogden MLP, Campbell JH (1999) Biological response to climate change on a tropical mountain. Nature 398:611–615
Primack RB (2010) Essentials of conservation biology, 5th edn. Sinauer, Sunderland

Randall R (2001) Case study 5.5. Eradication of a deliberately introduced plant found to be invasive. In: Wittenberg R, Cock MJW (eds) Invasive alien species: a toolkit of best prevention and management practices. CAB International, Wallingford

Raup DM, Sepkoski JJ (1982) Mass extinctions in the marine fossil record. Science 215:1501–1503

Ries L, Fletcher RJ Jr, Battin J et al (2004) Ecological responses to habitat edges: mechanisms, models, and variability explained. Annu Rev Ecol Evol Syst 35:491–522

Ruiz GM, Carlton JT (eds) (2003) Invasive species: vectors and management strategies. Island Press, Washington

Schardt JD (1997) Maintenance control. In: Simberloff D, Schmitz DC, Brown TC (eds) Strangers in paradise: impact and management of nonindigenous species in Florida. Island Press, Washington

Schmitz DC, Simberloff D, Hofstetter RH et al (1997) The ecological impacts of nonindigenous plants. In: Simberloff D, Schmitz DC, Brown TC (eds) Strangers in paradise: impact and management of nonindigenous species in Florida. Island Press, Washington

Schulte P, Alegret L, Arenillas I et al (2010) The Chicxulub asteroid impact and mass extinction at the Cretaceous-Paleogene boundary. Science 327:1214–1218

Sepkoski JJ Jr (1984) A kinetic model of Phanerozoic taxonomic diversity. III. Post-Paleozoic families and mass extinctions. Paleobiology 10:247–267

Simberloff D (1986) The proximate causes of extinction. In: Raup D, Jablonski D (eds) Patterns and processes in the history of life (Dahlem Conference L536). Springer, Berlin

Simberloff D (2002) Ecological and economic impacts of alien species. In: Claudi R, Nantel RP, Muckle-Jeffs E (eds) Alien invaders in Canada's waters, wetlands, and forests. Natural Resources Canada, Canadian Forest Service, Ottawa

Simberloff D (2006) Invasional meltdown six years later—important phenomenon, unfortunate metaphor, or both? Ecol Lett 9:912–919

Simberloff D (2010) Invasive species. In: Sodhi NS, Ehrlich PR (eds) Conservation biology for all. Oxford University Press, New York

Simberloff D, Rejmánek M (eds) (2011) Encyclopedia of biological invasions. University of California Press, Berkeley

Simberloff D, Dayan T, Jones C et al (2000) Character displacement and release in the small Indian mongoose, *Herpestes javanicus*. Ecology 81:2086–2099

Spellerberg IF (2002) Ecological effects of roads. Science Publishers, Enfield

Stein BA, Kutner LS, Adams JS (eds) (2000) Precious heritage: the status of biodiversity in the United States. Oxford University Press, New York

Tallis H, Goldstein JH, Daily GC (2010) The natural capital project. In: Sodhi NS, Ehrlich PR (eds) Conservation biology for all. Oxford University Press, New York

Thomas CD, Cameron A, Green RE et al (2004) Extinction risk from climate change. Nature 427:145–148

Turner PJ, Scott JK, Spafford H (2008) The ecological barriers to the recovery of bridal creeper (*Asparagus asparagoides* (L.) Druce) infested sites: impacts on vegetation and the potential increase in other exotic species. Austral Ecol 33:713–722

United States National Research Council (2000) Global change: ecosystems research. National Academy Press, Washington

Van Driesche R, Hoddle M, Center T (2008) Control of pests and weeds by natural enemies: an introduction to biological control. Blackwell, Malden

Vitousek PM (1986) Biological invasions and ecosystem properties: can species make a difference? In: Mooney HA, Drake JA (eds) Ecology of biological invasions of North America and Hawaii. Springer, New York

Vitousek PM, Walker LR (1989) Biological invasion by *Myrica faya* in Hawai'i: plant demography, nitrogen fixation, ecosystem effects. Ecol Monogr 59:247–265

Vitousek PM, Walker LR, Whittaker LD et al (1987) Biological invasion by *Myrica faya* alters ecosystem development in Hawaii. Science 238:802–804

Vitousek PM, Aber J, Howarth RW et al (1997) Human alteration of the global nitrogen cycle: causes and consequences. Issues Ecol 1:1–16

Wilcove DS, Master LL (2005) How many endangered species are there in the United States? Front Ecol Environ 3:414–420

Wilcove DS, Rothstein D, Dubow J et al (2000) Leading threats to biodiversity: what's imperiling U.S. species. In: Stein BA, Kutner LS, Adams JS (eds) Precious heritage. The status of biodiversity in the United States. Oxford University Press, New York

World Conservation Union (2010) http://www.iucn.org/about/work/programmes/species/red_list/search_iucn_red_list/. Accessed 15 October 2010

Part III
From Science to Policy: Managing the Commons, Social Learning, and Social Responsibility

Population growth and related land-use changes continue to exert local and regional pressures on the sustainability of ecosystems. To more effectively manage human life-support systems, governing agencies need better capacity to plan for and guide growth. Because policies cannot manage one activity, or part of a system, without considering its connections with all the other parts, future management scenarios should focus on multiple activities within specific areas defined by ecosystem rather than political boundaries. It also places humans in the landscape, within the broader context of the biological and physical environment, and ultimately combines ecology and human dimensions into "society-integrated" ecosystem management (CBD 1993; Naveh 2002, 2005). Scientific understanding of both ecosystem processes and of the underlying role of variability in maintaining ecosystem resiliency (that might otherwise descend irreversibly into degraded states; Holling 2000) has improved in the past several decades. As a result, emerging management approaches can begin to conform more closely to ecological and societal values rather than being driven by purely political constraints.

There is urgent need to better understand the consequences of our actions. This is stated quite succinctly by John Sterman from MIT who comments: "As the world changes, decision makers and the scientific community increasingly recognize that we are not only failing to resolve the persistent sustainability problems we face, but are in fact causing them" (Sterman 2002). Our well-intentioned efforts with global initiatives, environmental "summits," and dialogue to confront environmental issues not only fail to resolve sustainability issues, but create unanticipated side effects. Today's decisions often provoke reactions (feedbacks) that we could not foresee, and they often become tomorrow's problems. Sterman calls this policy resistance (see his contribution herein) the tendency for interventions to be defeated by the response of the system to the intervention itself. As we invoke "green technology" as the panacea for the world's sustainability ills, these same words become powerful sponsors of advice and caution.

The National Research Council (2002) has called for informed dialogue on goals for the sustainability transition, a dialogue that is necessary if societies are to adopt a measure of responsibility for their choices, i.e., where it should be headed rather than passively navigate the currents of demographic, economic, and environmental change. Social learning will be a centerpiece of this transition.

References

CBD (Convention on Biological Diversity) (1993) Handbook of the convention on biological diversity, 2nd edn. Transcontinental Printing, Montreal

Holling CS (2000) Theories for sustainable futures. Conserv Ecol 4:7–13

National Research Council (NRC) (2002) Our common journey: a transition toward sustainability. National Academy Press, Washington

Naveh Z (2002) A transdisciplinary education program for regional sustainable development. Int J Ecol Environ Sci 28:167–191

Naveh Z (2005) Epilogue: towards a transdisciplinary science of ecological and cultural landscape restoration. Restor Ecol 13:228–234

Sterman JD (2002) All models are wrong: reflections on becoming a systems scientist. System Dyn Rev 18:501–531

"Post-sustainability": The Emergence of the Social Sciences as the Hand-Maidens of Policy

Michael R. Redclift

Abstract The background to the current financial crisis and the problems surrounding policies to combat climate change through transitions out of dependence on carbon are examined. This review provides an example of how the quest for sustainability has invoked new policy tools, and the limitations of these tools in accounting for human behavior and agency. After providing a critique of current sustainable development policy, I suggest that there are fundamental flaws in the way policy has addressed both agency and structure in relation to climate change. I argue for a need to draw away from the path dependence that has served to define mainstream policy initiatives focused on individual consumer behavior, and argue for a stronger recognition of structural inequalities at the international and national level, as the cornerstone of an alternative, more sustainable, political stance. If we have now arrived at a "tipping point" on climate change, then we need to address the problem of decarbonization through an approach that goes well beyond market "mechanisms," and requires both social and political mobilization.

Keywords Sustainability • Markets • Carbon • Environmental policy • Path dependency • Normative science

Introduction

The last 2 decades have witnessed a series of crises over the economy and the environment. In this paper, I argue that they are linked and, although this is sometimes acknowledged in policy circles, and by scientists, that the policy response to the "twin crisis" has been inadequate. During this period, the apparent incompatibilities between

M.R. Redclift (✉)
Department of Geography, Kings College London,
The Strand, London, WC2R 2LS, UK
e-mail: michael.r.redclift@kcl.ac.uk

M.P. Weinstein and R.E. Turner (eds.), *Sustainability Science: The Emerging Paradigm and the Urban Environment*, DOI 10.1007/978-1-4614-3188-6_8,

economic growth and environmental sustainability came to mean two things. First, the limits of resource capacity were in danger of being exceeded. Resource shortages were a constraint on further economic growth and development and there was a need to curtail economic growth which was costly in natural resource and energy use. We should conserve resources to facilitate growth. This was essentially the "Limits to Growth" position articulated in the early 1970s (Meadows et al. 1972).

From this starting point in the 1970s, there was another, distinct position developed which became associated with the term "sustainable development" (Redclift 1987, 2008). This position took the argument one stage further. Clearly, the existing levels of economic growth represented a threat to the environment and resources but, in addition, we needed to identify and (increasingly) measure the trade-offs between economic growth and environmental conservation. The "weak" view of sustainable development suggested that when these two elements were in conflict, it was often possible to substitute human-made capital, such as physical infrastructure and capital goods, for the "natural" capital and resources which were being lost. A "stronger" view of sustainable development would imply much lower levels of substitution between human-made and natural capital, because it was often impossible to substitute for natural capital, notably what Pearce and associates referred to as "critical natural capital" (such as primary tropical forests, mangroves, and coral reefs) whose loss could not be recovered (Pearce 1991).

The first position on the economy/environment imbalances lost support partly because it was a product of high energy prices (the oil hikes of the 1970s), which did not continue, and for which substitutes could be found. As hydrocarbons became relatively cheaper, and the effects of the Green Revolution in expanding food staples to meet population growth began to be acknowledged, it was no longer clear that the Malthusian position held—that population exceeded the resources necessary to feed this growth. And the drive for economic development in the South which had been urged by the Brandt Commission was overtaken by events (Brandt 1980). At first, it was put in jeopardy by the Debt Crises of the 1980s, the Structural Adjustment Programmes, and, "post-recovery" of the early 1980s, which included the deregulation of markets, and the "retreat of the state" from some economic activities, in favor of the market. Even more importantly, perhaps, was the rise of the new (and increasingly deregulated) economies, of the so-called BRICS (Brazil, Russia, India, and China) and, especially, China and India which began to enjoy historically high rates of economic growth. The problem for the rapidly growing economies of Asia was not the absence of economic growth, but its transformation along more sustainable lines using fewer natural resources and less energy to produce the same amount of economic growth.

The genius of the second position ("sustainable development" in both weak and strong forms) was that almost everybody could sign up to it. There were very few dissenting voices (Redclift 1987; Norgaard 1988; Adams 1990). The mechanisms which were unleashed via deregulation and the neoliberal ascendancy became the favored instruments of policy in seeking to achieve "sustainable development" (Stiglitz and Serra 2008). Their economic policy component was often talked about in terms of the "Washington Consensus," although it was not clear that the consensus included all of the "beneficiaries" of interventions from the International Monetary Fund and the World Bank.

The environmental measures which paralleled economic deregulation and the development of new markets were as follows and took several forms: First, attempts were made to internalize what economists identified as environmental "externalities" in products and services, that is the usually unintended consequences of economic activities that bore heavily on the environment. The code name for this process, although one used much more in Europe than in the United States, is "Ecological Modernization." This was viewed as a competitive strategy by the European Union, in seeking to give Europe a competitive advantage over the United States and any newly developing rivals in Asia. This approach relied on counting the embodied carbon in products, seeking to reduce energy and material throughput, and consequently making a "win/win" gain, by reducing energy costs (hydrocarbon prices were rising again by then) and environmental damage. It was envisioned that future trade arrangements would also take account of "embodied carbon," and that the first nations to acknowledge this would prove the trade "winners." Some of the more imaginative policies of the European Union facilitated this outcome in the 1990s.

Second, the development of carbon markets, both within industries and, more importantly, between countries, was an important new development. These new markets represented a challenge for entrepreneurship, new market opportunities for investors, and required very little government action. Carbon markets were thus popular among devotees of free-market economics and environmentalism, unlike other interventions such as carbon taxes (Simms 2005). It is worth adding, perhaps, that a decade ago few paused to consider what might happen when markets fall and the price of carbon, like that of other traded commodities would drop significantly. Both these developments are examined later in this paper.

The conversion of governments to a more or less uncritical view of markets was even more evident in the international efforts to "protect" biodiversity. The biodiversity regime was expressed in the Convention on Biological Diversity (1992) and the Cartagena Protocol on Biosafety (2000). This development demonstrated a shift away from a focus on the loss of *species* diversity, and thus the loss of complex ecosystems, and toward a focus on the preservation of *genetic* diversity, where the principal gains were in the pharmaceutical industries and agriculture (Paterson 2008). Again the almost imperceptible shift was from *nature conservation* to *nature as commodity*. The main opposition to the latter was from groups—principally Non-Governmental Organizations—which argued that marginalized people had *rights* in nature which governments and the pharmaceutical industry ignored. However, the industry lobby won much of the political and ideological struggle, insisting that *ex situ* conservation in gene banks should be treated as equivalent to *in situ* conservation in ecosystems. In effect, the pharmaceutical companies improved their access to biologically diverse plants under new international regimes of trade and intellectual property.

The third element in the redesign of "sustainable development" policy was the creation of the "consumer-citizen," the idea that the individual could best express their preferences for goods and services through their own (and their household's) personal consumption. Parallel with the development of both cleaner technology and carbon markets came the concern with sustainable consumption. Partially as a result of their insufficient understanding of the link between social structures and consumer habits, and the awkward politics of wealth redistribution, governments came to favor consumer encouragement to live more sustainably and to reduce household "footprints." This implied the design of new "lighter" consumer goods, evocations to act in more environmentally responsible ways, and an accent on "life-style" and the consumer, at the expense of livelihoods and citizenship.

From the perspective of those most critical of market-based environmental valuation, the conjunction of newly "liberated" markets and environmental concern was a necessary contradiction of capitalism seeking a resolution, and could with hindsight be seen as a "managed senescence," if we continue with the biological metaphors of "development" (Smith 2007; Bellamy-Foster 2010; Woodgate 2010). A more mainstream view, however, would be that they addressed system failures, and could even lead to a rejuvenated, if scarcely recognizable, type of materials "light" capitalism.

A Novel Crisis: Financial Markets and the Environment

The hopes that markets and technology, together with more informed personal choices, would solve the environmental problems associated with accelerated economic growth and the enormous rise in global consumption were about to be challenged by a number of events. First, it was extremely difficult to square increased consumption and easy credit with longer term sustainability. The financial crisis which began in 2007/2008 was a "crisis" fed by the personal greed of many bankers and financial managers, and fueled by the virtually unregulated production of credit—not because interest rates were low but because the price attached to housing equity (especially in the United States, the United Kingdom, Spain, and Ireland) was unrealistically high. The rise in "sub-prime" lending and borrowing, took place under systems of ineffective governance that emphasized everybody's right to owner-occupied property regardless of collateral and debt levels. Politically, these policies were "sold" to consumers as everybody's right to credit rather than their right to debt.

The financial crisis also revealed that the model was completely unsustainable in the long term. The policy response from the developed world paid lip service to the rapidly disappearing Green policy agenda, but did not support this rhetoric with effective interventions. Only China and South Korea allocated more than one third of their fiscal stimulus to Green investment (HSBC 2009). The fiscal stimulus for most countries merely deepened the environmental crisis, as production was "out-sourced" to less clean suppliers and a brake was put on the conversion to more sustainable production if it came at an initial economic cost.

There was also now considerable evidence of the effects of the financial down-turn on migration, as well as poverty, and on a continental scale, especially in China.

China supported the United States' debt through buying in to its financial packages and through the expansion of its consumer goods exports raised consumption in the West and assisted the cost of living to most consumers. Despite the global economic recession, internal migration between the interior and the coast of China continued to accelerate, while China maintained an impressive annual growth rate of almost 10%. The large and newly developing economies, especially in Asia, were increasingly looked to for the stimulus that would lead the economies of Europe and North America out of their recession.

Another process that gathered speed is that of transnational sourcing of food, minerals, and other resources. The internationalization of capital movements and the need to secure resources led to increased transnational acquisition of land and minerals principally in Africa, on the part of China and some of the Gulf States. Rather than depend exclusively upon trade relations to meet their domestic resource deficiencies—because trade contracts during an economic recession—the advantages of acquisition of land, water sources, and food (via "virtual water") became evident, especially for their geopolitical reach (Allan 2003). Land displacement for crops like soya had already changed international food/land imbalances, by providing cheap fodder for animals in Europe and North America, but the leasing and ownership of land in transactions between countries of the Global South was new.

Each of these processes appeared to have undermined the international economic stability that had been taken for granted since the 1980s, albeit with a few wobbles. They also suggested connections between the financial and environmental crises that had only been dimly perceived in the past. Other questions remained relevant: could one generalize a "successful" model of economic development to the global scale without doing irreparable harm to the environment? And, were individual country economies placed in jeopardy when pursuing international policies to limit the damage of climate change (Copenhagen 2009)?

The Financial Crisis, Climate Change, and Consumption

The changes in the way that materials, food, and energy are sourced globally have usually been discussed without much reference to sustainable consumption in the countries of the Global North. Today the expansion of credit in much of the developed world and the associated levels of personal and corporate debt are necessarily linked with the banking crisis that has affected most financial institutions since September 2008, leading to an economic downturn and period of recession. An understanding of the "limits" imposed by shifts in demand needs to be complemented by an analysis of the rising levels of personal consumption and debt, not only in the developed world but also in many middle-income and fast-growing developing economies (Durning 1992; Redclift 1996; Princen et al. 2002). As we shall see, these issues are closely linked to the global problems surrounding climate change and are manifested at different spatial scales.

The "toxicity" of many financial institutions was triggered by excessive lending, and low interest rates prompted by the costs of increased international conflict after "9/11," especially in Iraq and Afghanistan. Financial institutions linked to housing equity in a number of countries helped create an unrealistic credit profile, especially in the United States, the United Kingdom, Spain, and Ireland. The rise of the credit economy, as well as the popularization of new instruments like home equity loans with which to draw on expected capital gains, contributed to dramatic changes in commercial credit markets. There are clear indications from the literature over a decade ago that savings behavior shifted most sharply in countries with more liberal access to personal credit, notably in the United States and the United Kingdom (Calder 1999; Parker 1999; Manning 2001; Guidolin and La Jeunesse 2007). As we have seen, this brought about a loss of confidence in the ability of the lending institutions to recoup their assets, and national governments acted to guarantee the private banking sector against a feared "run on the banks." These developments occurred within a context of relatively high personal (and institutional) indebtedness since financial deregulation was initiated in the 1980s.

At the same time, another shift has been occurring in consumer policy. This shift was prompted by the much wider acknowledgement of global climate change, especially after the Stern Report was published in 2007 (Stern 2007). The need to pursue "low-carbon" solutions to economic growth rapidly altered the policy discourses surrounding consumption and the environment, and it has become an article of faith for public policy that economic growth is only tolerable if it does not exacerbate existing concentrations of carbon in the atmosphere. In 2008, the United Kingdom's Climate Change Bill was introduced, establishing a very ambitious target for carbon reductions of 80% by 2050 (Appendix Two, therein). This policy activity has been accompanied by sustained lobbying on the part of NGOs and others, including Rising Tide, Friends of the Earth, the Campaign Against Climate Change, and the series of Climate Camps that have repeatedly mobilized sections of the public. It is worth adding that "low-carbon" solutions take different forms and meet different levels of support and opposition from the business and citizen communities. Examples exist of "no-regrets" public policies in the United States, such as energy conservation measures, that can serve to reduce some greenhouse gas emissions, and which have gained wide support, often from groups opposed to new environmental regulation and even incentives.

This new policy perspective is seen clearly in the document which, more than any other, represents the high-water mark of free-market environmentalism. The Stern Review (*Stern Review: The Economics of Climate Change*) noted that:

> The transition to a low-carbon economy will bring challenges for competitiveness but also opportunities for growth... Reducing the expected adverse impacts of climate change is therefore both highly desirable and feasible.

This quotation illustrates the way in which what had previously been viewed as a "threat" could quickly become an "opportunity," although the quotation fails to say for whom the opportunities exist. Unsurprisingly, the immediate responses to Stern (and the IPCC 4th Assessment of 2007) were optimistic in tone. One commentator on business and the environment wrote that (Welford 2006):

> People would pay a little more for carbon-intensive goods, but our economies could continue to grow strongly... The shift to a low-carbon economy will also bring huge opportunities... Climate change is the greatest market failure the world has seen.

The characterization of climate change as a "market failure" immediately offered economists, businesses, and Government a lifeline. Rather than necessitating expensive and comprehensive restructuring in new systems of provision, or even reduced volumes of production and consumption, Stern's neoclassical view was that sustainability could be delivered through *increased* consumption of particular kinds of products, simultaneously. Feeding the economy has come to typify the mainstream environment and consumption discourse.

These developments in the economy and in public policy raise some awkward questions for our understanding of the policy discourses that have characterized the field. There is still considerable confusion over the most effective way of reducing consumption and the accompanying carbon emissions, and several of the assumptions about consumer behavior—such as the role of an "information deficit" surrounding the environmental costs of products and services are, at best, questionable (Redclift and Hinton 2008). Remarkably, the assumptions about personal behavior being triggered by available information are also largely untested. Whilst policy-makers and pundits alike tend to measure progress toward sustainable consumption in terms of the numbers of purchases of particular "green" or "ethical" commodities, where success is framed in terms of market share, an alternative discourse suggests that the most effective forms of sustainable consumption rest on other facets of behavior, such as frugality, thrift and what has been termed "voluntary simplicity" (Soper et al. 2008). If this is indeed the case, then a focus on lowering economic growth may still be preferable to pursuing "sustainable growth" strategies.

As the quote from the Stern Report above suggests, climate change is now regarded as a "given," and markets are now considered more relevant to policy solutions than ever before. The reduced dependency on hydrocarbons is widely regarded as deliverable through changing consumer policy. In effect, the language of "Green Consumerism" has reduced the politics of climate change to the size of a Green consumer product. The evidence of an impending economic recession in the developed world might only serve to intensify this process, creating policy tensions where once there were only "policy opportunities."

I began this paper by arguing that the "contradictions" of thinking about sustainability and development have merged into distinct policy discourses, around the idea of "natural limits," resource capacity and (un)sustainable consumption. Each of these discourses can be usefully informed by recent work in the social sciences which explores the changing role of science policy. One route is to point up the limitations of a policy agenda apparently driven by faith in the verities of science itself. A realistic, science-driven policy agenda, such as the work detailed in this and other volumes from natural scientists, has been paralleled by a "science-sceptical" post-modern academic discourse in the social sciences (Yearley 1996; Demeritt 1998). This latter position begins by confronting the objectivity claims of science, suggesting instead that biases enter into the way normal science is undertaken, and that scientists themselves are often poorly equipped to understand the limitations and assumptions of their method. "Science-skeptical" positions part company with the realism that defines most of the natural science projects, and prefers to regard "environmental" problems as socially constructed, problems for society rather than problems of nature.

Neither position—"science-driven" and "science skeptical"—represents a threat to the other since they inhabit quite different epistemological terrain, and address different audiences. Most natural scientists remain blissfully ignorant of the meanderings of social constructivism and hold to a firm account of evidence-based, experimental procedure. In the process, however, we have seen an enlarged academic debate, and one that closely examines the way environmental language is deployed, while at the same time recognizing that public policy discourses themselves carry weight. The policy discourse surrounding "sustainability" has real consequences in itself, by becoming a point of reference for those who agree and those who differ as to its importance. However, issues around the social authority of science, and the way it is employed politically, are linked to structural shifts both in the "formal" economy and in the "informal" and "virtual" economies that have emerged recently. I suggest that these shifts enable us to speak of a "post-sustainability" political and policy discourse; one that recognizes "sustainability" not as a scientifically verifiable fact, but as a social and political construct within a specific historical conjuncture. These issues and connections are examined later in this paper.

I have argued that the policy debate has proceeded through assumptions about "choice" and "alternatives," that they have been largely devoid of any critical, structural analysis, and frequently narrow the sociological field of opportunity, by assuming that people act primarily as consumers, rather than citizens (Redclift 2010). There is clearly room for more rigorous analysis of what is a very broad social terrain beginning with the assumptions of the model implicit in most environmental and consumer policy.

Assumptions of the Policy Model

The neoliberal model characterizing the 1980s and 1990s was viewed by many as a liberating model. It removed "government" as the engine of economic momentum, and opened up activities to the market, or introduced "shadow" markets which encouraged individuals to behave as if markets operated, in the process not merely shifting economic activities to the private sector, but implementing a new logic for the public sector (a sector that continued to grow in most developed countries). The new policies also deregulated financial flows, facilitating the movement of capital particularly finance capital, and reduced the burden on capital, through reducing the presumed penalties on growth-like corporate taxes. The model also removed many of the politically negotiated rights that organized labor had gained in the developed world, and reconfigured the frontiers of the "welfare state." Among the existing capitalist economies, only those of the European Union sought to combine this market-based model with measures in favor of labor, consumer, and environmental protection, producing a hybridization of neoliberal thinking and traditional welfare support.

Rethinking the role of the state and the consumer in economic growth held importance for the environment, too. The new policy emphasis, especially within the European Union, was on moving away from the management of capitalist growth on more environmentally sustainable lines and toward enabling private actors to pursue their interests while *simultaneously* promoting sustainability. Policy increasingly sought to structure incentives for actors, believing that the "agency" of the individual, if it existed at all, consisted of a kind of "consumer-agency," rather than the battery of roles that constituted "citizenship." This wider view of the multifarious roles performed by the "citizen" had been pioneered by social democratic (and some Christian Democrat) governments. However, as I have argued, the new model envisaged the individual as reducible to their "consumer self," and this applied as much to the way environmental externalities were treated, as to the loosening up of credit, and (in the case of some economies) the burden of equity-based housing.

These changes came at a cost, of course. The movement of neoclassical economics into more mainstream environmental policy left several concerns at the margin of policy and politics. The challenges of reducing material throughput, and reducing carbon emissions, converted environmental policy into a technical question, while the agency of social movements and their pursuit of alternative social and cultural objectives was effectively sidelined. Unlike the position in the first half of the twentieth century, for the discursive politics of the decades after 1980, the term "utopia" was treated pejoratively, as irrelevant, and out of phase with the realities of the "enabling market." The apparent need to reassure the public that the impending environmental dystopias were not inevitable seems to have led policymakers to emphasize individual contributions (e.g., Tesco's advertising line that "every little helps") above collective political action. Unlike other policy domains, notably education, the environment was not considered "aspirational" by many leading politicians during the years of New Labour Government in the United Kingdom (1997–2010).

The underlying assumptions of the dominant model transposed the supposed "barriers" to market freedom and choice in the formal economy, to the new terrain of environmental and sustainability policy. Policy interventions assumed that similar barriers, this time "social" rather than economic, existed to people acting more sustainably in everyday life (Redclift and Hinton 2008). It was suggested that these social barriers were constituted by habit, poor education, a lack of information and state bureaucracy, and could be rectified by policy. The solution was to introduce more choice of products and services, new "Greener" technologies, and market opportunities which could maximize utility while placing more responsibility on the individual. The individual consumer could regard herself as "Greener" through encouragement, or even, in the current argot "nudging," that is, being leant upon by government to behave better. This solution rendered the individual as a consumer, rather than a fully reflexive citizen and their environment solely in terms of products and services, rather than social processes or structures.

At the same time, science was viewed as part of the solution, rather than the "problem" confronting societies threatened by climate change. The decisions were only obliquely political, and technical solutions held the promise of removing politics from environmental policy entirely. As demonstrated in the Stern Report, we were embarking on what has been termed a "post-political" future (Swyngedouw 2009), which is one in which consensus science came to exercise normative authority, and political judgments about the way resources and rights to them were distributed could be left to (supposed) independent rational discussion.

The market research approach to modeling consumer behavior regards consumer attitudes, obtained through surveys and focus groups, as proxy for social and economic structures. It matters little whether a consumer is a poor single parent living in a high-rise housing complex or a wealthy household living in a rural area, using two cars to do the shopping and ferry the children to school. What matters is that the attitudes displayed influence the household's level and type of market engagement. The task then is to tailor policy for different consumer profiles.

Capitalism "Lite"

At a more "macro" level, the development of carbon markets, both within industries and, more importantly, between countries, represent a mature version of the "market solution" model. On the one hand, the development of carbon markets was welcomed by many sectors of industry; indeed, they were heralded as a "challenge for entrepreneurship," providing new "market opportunities" (Lovins et al. 2000). At the same time, as we have seen, they required very little government action and were consistent with the largely deregulatory model being widely pursued.

Carbon markets were thus popular among devotees of free-market economics and those who recognized the urgency of environmental action, but who bemoaned the shifts in behavior that this might imply. As one "progressive" think tank in the UK put it, "they (provide) the political opportunity to highlight, secure and celebrate wealth creation. The benefits from the low-carbon transition are waiting to be grasped" (Policy Network 2010). Notwithstanding the endorsement of carbon markets by large sections of political opinion, they also raised other questions which were an anathema to the more radical Green opinion, raising the possibility, following Oscar Wilde's famous dictum, of "knowing the price of everything and the value of nothing."

The existence of carbon markets contributed to the new middle-ground consensus that has come to characterize business-friendly environmental policy during the first decade of the twenty-first century. Organizations like the Carbon Trust advertised heavily in publications like "The Economist," where individual entrepreneurs were singled out for compliments and communicated their endorsement of carbon trading. "What was I thinking when I cut our carbon and joined the standard?" asks the Chris Pilling of HSBC. The answer is a conclusive "win/win" piece of advocacy: "I saved money, gained a competitive edge, improved efficiency and shared the tangible benefits of accreditation."

The clear benefits of encouraging industry to enter the new carbon markets, however, only represented one part of the equation. The downsides of carbon trading were perhaps less "tangible" but equally compelling. Once the financial recession became apparent, the benefits of carbon *markets* began to recede.[1] The "cap-and-trade" model was beginning to lose ground by late 2009 in precisely those economic systems which had earlier favored it. Under President Obama in the United States, electricity utilities looked likely to use "cap and trade" but transport emissions were more likely to be taxed and industrial emissions regulated. The "new tools" of the market was less in evidence in 2010 than 10 years earlier. By the same token, the appeal of the "old" policy instruments of taxation and regulation was more apparent during "bad times" when governments, especially in the United States and Europe, needed to raise income, particularly for much needed new investments in energy (including renewables). The Economist put it this way: "climate action may come to lean more heavily on the command-and-control techniques than the market-based approaches they were intended to replace" (The Economist March 20, 2010).

What is the significance of carbon markets for individual consumers, whose attention has increasingly been drawn toward ways of reducing carbon "footprints"—which is the mechanism favored by many mainstream commentators? Carbon footprints appear to provide a ready-made and measurable way of enabling individuals to make choices about travel, in particular, leading some of them to "offset" some choices against others and improve their sustainability "profile." This has led some commentators to advocate individual carbon budgets as the logical consequence of carbon measurement.

However, there are a number of problems associated with carbon foot-printing that are not always discussed. First, although it is a technique that allows comparisons between indicators, carbon footprints cannot be converted into monetary or social values, and so they are of only limited use to policy (OECD 2004; Schmidtt 2009). In addition, measuring an individual's carbon footprint does not help us to understand what an acceptable rate of carbon is for an individual, or how their personal contribution might contribute to the wider society. It provides no interpretative framework through which policy can be guided. Finally, carbon foot-printing uses no standard placement for the boundaries of the system in which it is deployed. Most calculations use "cradle-to-gate," or "cradle-to-site/plate" as the system boundary, while the least used framework, and probably the most inclusive, is from "cradle-to-grave."

[1] There have been several reports suggesting that the European Union's Emissions Trading Scheme (ETS) will do little to encourage investment to reduce emissions during the economic recession. On the present course, emissions trading is likely to produce only a 3% reduction in emissions within the EU by 2020. Two effects will be observed. First, the cap on emissions will exceed projected EU emissions providing no economic incentive to move to clean technology and infrastructure before 2012. Second, because the EU allows unused permits and offsets under phase three (2013–2020), any claimed economic incentive during this later period will also be reduced. ('Recession plus ETS ≡ fewer carbon emissions in the EU; National Audit Office 2009).

Another consumer-led policy initiative to close the "carbon loop," and one triggered by the intergovernmental agreements at the first Earth Summit in 1992, which heralded the Clean Development Mechanism (CDM), is the development of voluntary carbon offsets. Carbon offsetting was seen as an approach with considerable appeal to environmentally conscious consumers, which might help assuage the guilt of people who traveled frequently by plane, but were painfully aware of the carbon cost of doing so. Offsetting flights is widely promoted as a solution to emissions reduction. It involves travelers paying a fee on top of their airfare to "offset" the carbon emitted by the journey. There is, however, considerable confusion surrounding carbon offsets: the way that emissions are measured, the fees charged for managing offsets, and the methods employed in calculating them, are all contentious and complex calculations (Gossling 2000). In addition, the main target of voluntary offsets has been tourists rather than the more significant business traveler, for whom there is evidently less appeal in "guilt-free flying" (Francis 2009). The operator Responsible Travel, which has pioneered ethical tourism in Europe, has recently dropped its offsetting choices, on the grounds that some tourists might travel *more* because they believe the effect of their flights has been neutralized. Critics of offsetting argue that it has a negligible effect on carbon sinks in the global South and, indeed, removes the responsibility for preventing deforestation in the developing countries themselves (Draper et al. 2009; Dawson et al. 2010).

Finally, in all the discussion of carbon accounting, trading, and offsetting, there is a beggar at the feast. What might happen when markets fall and the price of carbon drops significantly? This possibility eventuality had not received much attention in the optimistic decade that proceeded the economic recession. In addition, some commentators argued that those in the European Union are now faced with a "sub-prime" market in carbon as the price drops, and that investors will lose the benefits of government support. This is a situation not entirely dissimilar to that in the housing market a decade earlier.

The shift toward more conventional policy tools, especially regulation, might also have political consequences, because the environmental movement in all its complexity assumes the lobbying role that has been the specialism of business and the environment since the ascent of Ecological Modernization. If President Obama's Bill in the United States does not pass through Congress in 2010, then this role will arguably become even more critical, strengthening the hand of more radical grassroots organizations in the political system rather than more business-friendly Green lobbyists.

Underlying Structural Issues

On closer examination, the sociological consequences of these developments are profound. The pricing of individual household-level technology, like wind turbines and solar panels, makes some Green innovations look more like "positional goods" than public goods. This occurs when the gains accrue mainly to people who can put up the initial capital, utilize government subsidies, and, ultimately, reap the income benefits (Hirsch 1977). The contribution of these consumer "fixes" to national energy production and the emerging "energy gap" in the second decade of the century is likely to be very modest, indeed. The observation that has been made about other forms of Green or Ethical consumerism, such as "responsible travel" and "fair trade," also applies to these new forms of Green consumerism too: they require a higher outlay and yield most benefits for the middle-class consumer. Meanwhile, the big decisions on energy generation and conservation are stalled, and the nuclear power "option" is resurrected.

The emergence of Green technology as a positional good, rather than what economists call a "public good," is only one of several indications that climate policy is insufficiently grounded in our knowledge of social structures. The existence of embodied carbon, and its acknowledgement in the discussions (but not the policies) surrounding global trade agreements is another (Kejun and Murphy 2008). Climate policy, and the piecemeal attempt to provide incentives for individuals to reduce their own carbon dependency, is rarely linked to a wider global experience outside the OECD countries. In what ways does it contribute to the transfer of much needed cleaner technology to the global South? What are the international and distributive consequences *within* the global South of our attempts at limited decarbonization in the North?

We might, indeed, dig deeper still. What other forms of human agency, other than those of the "informed" consumer, have been left out of the equation? What are the wider social and cultural implications of placing so much emphasis on trading in a "bad" (pollution) rather than a "good" (such as cleaner technology)? What forms of human agency, innovation, and collective action lie outside the compass of "entrepreneurship," but help distill community support and engage environmental citizens (Dobson 2003)? Climate scientists are seen as the "guardians of the dogma" on climate change, but there is evidence of low levels of public trust in science, including climate science. What is required, then, to mobilize areas in which there are high levels of public mistrust, such as climate change, while other institutions and practices do command widespread public support, such as community-based credit unions and some of the financial mutuals? New forms of Web communication and networking suggest widespread support for organizations which are embedded locally in communities and which acknowledge, rather than ignore, social and economic inequalities. As in previous historical periods, addressing international and national structural inequalities might become the engine of new transitions, creating new social solidarities, and means of liberation from the path dependency associated with our heavy reliance on hydrocarbons (Redclift 2008).

Is There a "Bright Narrative?"

I have argued here that a meaningful transition to a low-carbon economy is impossible as long as we rely on models of market choice and normative science policy that leave little room for collective and group behavior and ignore the underlying social commitments that govern our everyday lives. The dual crises of global financial debt and climate change are reaching a "tipping point" beyond which it will be difficult to move.

Already there is evidence that some behavioral responses to the environmental and financial crisis are taking forms that are not easily accommodated to the prevailing approaches to environmental policy favored by most governments. They lay in challenges to conventional food systems, alternative recycling and re-use activities, and small-scale attempts to provide sustainable renewable energy at the level of communities, as well as individual households, and the brave efforts of enthusiasts to hold back ecological damage. Much of this activity is "informal" in a new sense, too; it is often funded within the "formal" market economy, but depends heavily on Web-based organization and group and community loyalties without formal institutional ties. These partial, but evolving, challenges to conventional thinking and behavior are often only weakly connected to each other, because they cover a number of apparently isolated social fields. What they do reveal are fissures in the fabric of governance and the management of nature, and a need felt by some third-sector organizations to transcend anxiety over the environment. They reveal ways in which conventional path dependency is shifting, allowing new kinds of social organization and governance to emerge, often in unexpected places, to build new forms of social and ecological resilience. One can observe in more resilient human and natural systems the ability to "self-organize" among the components or constituent parts of the system (Adger et al. 2005). This resilience has also been observed in relation to apparently "marginalized" groups in the Global South, notably urban "squatter" populations, but similar trends have occurred during periods of economic recession or war in the industrialized world, when families had no alternative to finding their own solutions. This was particularly the case during and immediately after the Second World War in the United Kingdom, when housing was not available and food was rationed (Redclift and Hinton 2008, 2009). Can alliances today be built from these small innovative "alternatives?" Can a "brighter narrative" be developed for the future?

In the recent past, extreme traumas such as World War II have transformed many of the taken-for-granted assumptions that characterize modern industrial societies. Major shifts in behavior, such as rationing, women's employment and dramatic changes in resource and energy use, have come about as path dependency has been transformed by events on the world stage. Societies and economies have been mobilized for different purposes. Although historically conjunctural, such experiences might help inform us today. The challenges of the "new" conflicts associated with climate change today are much greater, of course, and carry fewer political imperatives than the wartime policy on the Home Front. The "tipping point" is no longer the prospect of military occupation by an enemy, but a "retreat" in the face of a self-induced problem: anthropogenic climate change. In exploring the possibilities of transition to a post-carbon future, we might begin by examining the "pieces"—fragmented, often virtual and inevitably local—with which such a narrative might be constructed. They need to be constructed from peoples' lives and the resilience of their households and communities, rather than simply from their performance in consumer markets that are often transitory and unstable. But this is unlikely to be enough to redirect economic development in ways that are genuinely sustainable. The "Bright Narrative" may elude this generation, but without re-examining solidarities and social commitments in the next generation it will be too late to even contemplate.

References

Adams WM (1990) Green development: environment and sustainability in the third world. Routledge, London

Adger WN, Arnell NW, Tompkins EL (2005) Successful adaptation to climate change across scales. Glob Environ Change 15:77–86

Allan JA (2003) Virtual water—the water, food and trade nexus: useful concept or misleading metaphor? Water Int 28:4–11

Bellamy-Foster J (2010) Marx's ecology and its historical significance. In: Redclift MR, Woodgate G (eds) The international handbook of environmental sociology, 2nd edn. Edward Elgar, Cheltenham

Brandt Commission (1980) North-South: a programme for survival. Pan Books, London

Calder L (1999) Financing the American dream: a cultural history of consumer credit. Princeton University Press, Princeton

IPCC (Intergovernmental Panel on Climate Change) (2007) Fourth assessment. http://www.ipcc.ch/publications_and_data/publications_ipcc_fourth_assessment_report_synthesis_report.html. Accessed 7 June 2010

Dawson SEJ, Lemelin H, Scott D (2010) The carbon cost of polar bear viewing tourism in Churchill, Canada. J Sustain Tour 18:319–336

Demeritt D (1998) Science, social constructivism and nature. In: Braun B, Castree N (eds) Remaking reality: nature at the millennium. Routledge, New York

Dobson A (2003) Citizenship and the environment. Oxford University Press, Oxford

Draper S, Goodman J, Hardyment R et al (2009) Tourism 2023: four scenarios, a vision and a strategy for UK outbound travel and tourism. Forum for the Future, October. http://www.scribd.com/doc/21001489/UK-Tourism-2023-Full-Report-Web-Version. Accessed 31 July 2010

Durning A (1992) How much is enough? The consumer society and the future of the earth. Norton, New York

Francis J (2009) Responsible travel has ditched offsetting flights and holidays for environmental reasons. Guardian (London), 16 October

Gossling S (2000) Sustainable tourism development in developing countries: some aspects of energy use. J Sustain Tour 8:410–425

Guidolin M, La Jeunesse E (2007) The decline in the US personal savings rate: is it real and is it a puzzle? Federal Reserve Bank of St. Louis Review, November/December 2007

Hirsch F (1977) The social limits to growth. Routledge, London

Kejun CA, Murphy D (2008) Embodied carbon in traded goods. Trade and Climate Change Seminar, June 18–20. IISD, Copenhagen

Lovins A, Hawken P, Lovins LH (2000) Natural capitalism. Earthscan, London

Manning R (2001) Credit card nation: the consequences of America's addiction to credit. Basic Books, New York

Meadows DH, Randers J, Meadows DL (1972) The limits to growth. Pan Books, London

National Audit Office (2009) European Union emissions trading scheme. National Audit Office, London

Norgaard R (1988) Sustainable development: a co-evolutionary view. Futures 20:606–620

OECD (Organisation for Economic Cooperation and Development) (2004) Measuring sustainable development: integrated economic, environmental and social frameworks. OECD, Paris
Parker J (1999) Spendthrift in America? On two decades of decline in the US savings rate. NBER Macroecon Ann 14:305–370
Paterson M (2008) Global governance for sustainable capitalism? The political economy of global environmental governance. In: Adger N, Jordan A (eds) Governing sustainability. Cambridge University Press, Cambridge
Pearce D (1991) Blueprint two. Earthscan, London
Princen T, Maniates M, Conca K (eds) (2002) Confronting consumption. MIT Press, Cambridge
Redclift MR (1987) Sustainable development: exploring the contradictions. Routledge, London
Redclift MR (1996) Wasted: counting the costs of global consumption. Earthscan, London
Redclift MR (2008) The environment and carbon dependence: landscapes of sustainability and materiality. Curr Sociol 57:369–387
Redclift MR (2010) The transition out of carbon dependence: the crises of environment and markets. In: Redclift MR, Woodgate G (eds) The international handbook of environmental sociology, 2nd edn. Edward Elgar, Cheltenham
Redclift MR, Hinton E (2008) Progressive governance conference: living sustainably. Policy Network, London
Redclift MR, Hinton E (2009) Austerity and sustainability: the road from wartime austerity to the consumer citizen. Working Paper no 17. Department of Geography King's College, London
Schmidtt H (2009) Carbon foot-printing, labelling and life cycle assessment. Int J Life Cycle Asses suppl 1:S6–S9
Simms A (2005) Ecological debt: the health of the planet and the wealth of nations. Pluto Press, London
Smith N (2007) Nature as accumulation strategy. In: Panitch L, Leys C (eds) Coming to terms with nature. Socialist Register 43:16–36
Soper K, Thomas L, Ryle M (eds) (2008) Counter-Consumerism and its pleasures. Palgrave, Basingstoke
Stern N (2007) The economics of climate change: the Stern Review. Cambridge University Press, Cambridge
Stiglitz J, Serra N (2008) The Washington consensus reconsidered. Oxford University Press, Oxford
Swyngedouw E (2009) The antinomies of the post political city: in search of a democratic politics of environmental production. Int J Urban Reg Res 33:601–620
Welford R (2006) The economics of climate change: an overview of the Stern review. Intl J Innovation Sustainable Development 1:260–262
Woodgate G (2010) Introduction. In: Redclift MR, Woodgate G (eds) The international handbook of environmental sociology, 2nd edn. Edward Elgar, Cheltenham
Yearley S (1996) Rethinking the global. In: Yearley S (ed) Sociology, environmentalism, globalisation: rethinking the globe. Sage Publications, London

The Purpose and Politics of Ecosystem-Based Management

Judith A. Layzer

Abstract Ecosystem-based management (EBM) emerged in the late 1980s as an alternative to the piecemeal, jurisdiction-by-jurisdiction approach to natural resource management that dominated the twentieth century. EBM features three central attributes: (1) planning at a landscape scale, (2) collaboration with stakeholders, and (3) adaptive and flexible implementation. According to its proponents, EBM can generate management that is not only ecologically sensitive and responsive to new scientific information but also widely accepted. Application of EBM has yielded some important environmental benefits, including improvements in scientists' understanding of large-scale ecosystems. Those advances in knowledge, however, have not necessarily translated into the kinds of political and policy changes that the proponents of EBM had hoped for. Nor have they yielded more resilient ecosystems. Instead, in prominent cases of EBM, powerful interests have dominated the collaborative planning process, and flexible implementation has allowed those who are not committed to evade responsibility for implementing environmental sustainability measures. Simply enhancing scientific models to better assess complex risks will not ensure that EBM yields genuine ecological restoration. Also important are a credible and stringent regulatory framework and political leaders who place a premium on ecological integrity.

Keywords Ecosystem-based management • Landscape-scale planning • Adaptive management • Collaborative planning • Evaluation

J.A. Layzer (✉)
Department of Urban Studies and Planning, MIT, 77 Massachusetts Avenue, 9-326, Cambridge, MA 02139, USA
e-mail: jlayzer@mit.edu

M.P. Weinstein and R.E. Turner (eds.), *Sustainability Science: The Emerging Paradigm and the Urban Environment*, DOI 10.1007/978-1-4614-3188-6_9,

Introduction

Ecosystem-based management (EBM) emerged in the late 1980s as an alternative to the piecemeal, jurisdiction-by-jurisdiction approach to natural resource management that predominated during the twentieth century.[1] EBM was originally the brainchild of scientists and natural resource managers concerned with the deterioration of natural systems. They pointed out that the "conventional" approach to natural resource management—a product of environmental laws and procedures layered on top of conservation-era mandates to produce commodities—was not only generating conflict but also failing to stem the decline of natural systems. In fact, despite the efforts of environmentalists, environmentally oriented scientists, and land-use managers, species extinctions were accelerating, as high-quality habitat disappeared or was fragmented and degraded. Evidence from a variety of sources supported the conclusion that the cumulative impact of human activity was reducing the carrying capacity of terrestrial and aquatic ecosystems, and destroying the ecological services on which humans and other species depend (MEA 2006).

Critics suggested that there were three main reasons conventional management was failing: (1) its top-down, one-size-fits-all, expert-driven approach; (2) an emphasis on either exploiting or protecting individual resources; and, (3) a lack of coordination among entities with authority over land and natural resources. As a remedy, they proposed an approach widely known as "ecosystem management," which involves planning across large landscapes; collaborating with stakeholders; and flexible, adaptive management. EBM eventually became the preferred term for this set of practices because it conveys the idea of managing the activities of humans within ecosystems, and not the ecosystems themselves. Proponents anticipated that EBM would enhance the integration of science and management and therefore lead to more resilient landscapes. In practice, EBM has yielded some important environmental benefits, including improvements in scientists' understanding of large-scale ecosystems. But those advances in knowledge have not necessarily translated into the kinds of political and policy changes that proponents of EBM had hoped for. Nor have they yielded more resilient ecosystems. Instead, in prominent cases of EBM, powerful interests have dominated collaborative planning processes, and flexible implementation has allowed those not fully committed to EBM principles to evade responsibility for implementing environmental restoration measures.

[1] EBM is just one of the new approaches that emerged at this time. Smaller, more local initiatives constitute what Weber (2003) calls grass-roots ecosystem management (GREM). According to Weber, GREM arises in rural, natural-resource-dependent communities, primarily in the West, in an effort to overcome gridlock in public lands management. The lines dividing different types of initiatives are blurry, and some authors cast a wide net (Gordon and Coppock 1997; Meffe et al. 2002), while others draw a distinction between initiatives that are large scale and often led by government and more ad hoc, small-scale efforts driven primarily by local activists (Cestero 1999). This chapter focuses on large-scale, government-led initiatives former. It also focuses on EBM in the USA, although the concept has taken hold around the world.

Origins of EBM

The scientific community was particularly influential in developing the concept of EBM. Ecologists and conservation biologists had long complained that the conventional regulatory framework for managing natural resources treated complex, diffuse phenomena as though they were separable into clearly defined problems that were linear with respect to cause and effect. This critique was rooted in a series of conceptual and value shifts within the ecological sciences. During the 1970s and 1980s, ecologists had moved away from the classical, equilibrium-based paradigm, in which ecological systems were viewed as closed, self-regulating systems operating within a single stable state. In its place, they adopted a "flux-of-nature" view that treated ecosystems not as self-regulating but as limited by external sources. Furthermore, ecologists had recognized that disturbances—such as fire, floods, droughts, and storms—played a central role in shaping ecosystem dynamics, and that humans were an inextricable part of natural systems (Perry and Amaranthus 1997; Pickett and Ostfeld 1995). From this perspective, ecosystems were depicted as open, unpredictable, and unique, rather than as closed and comprehensible. In addition, process, dynamics, and context were more important than endpoint stability (Meffe and Carroll 1994). Parallel to these developments in ecology was the emergence of conservation biology, whose practitioners focused on conserving biological diversity and ecological integrity, rather than commodity production (Noss and Cooperrider 1994).[2]

Ecologists and conservation biologists charged that the conventional management paradigm, with its emphasis on controlling natural variation, was producing brittle ecosystems that were unable to deal with external shocks (Holling and Meffe 1996). "The command-and-control approach," wrote Holling and Meffe (1996) "implicitly assumes that the problem is well-bounded, clearly defined, relatively simple, and generally linear with respect to cause and effect." But in a complex, nonlinear, and poorly understood world, such an approach could have disastrous ecological, social, and economic consequences. Attempting to manage fisheries using the concept of "maximum sustainable yield," for example, caused the widespread collapse of commercial fish stocks and, consequently, declines in marine ecosystems. Benign neglect, an alternative approach also rooted in the equilibrium-based paradigm, was untenable as well. Often practiced in nature preserves, benign neglect led to increasing populations of wildlife that, when confined within park boundaries, decimated vegetation and then declined precipitously (Botkin 1990).

Environmentalists and environmentally oriented policy scholars were also dismayed by the approach to natural resource management that prevailed in the

---

[2] According to Haeuber (1996), the EBM concept has historical roots that predate the flux-of-nature paradigm in ecology. He notes that in the early 1930s the Ecological Society of America's Committee for the Study of Plant and Animal Communities recommended protecting ecosystems rather than just species, incorporating natural disturbance regimes into management, and using a core reserve/buffer design approach for natural area protection.

1970s and 1980s. Drawing on the philosophies articulated by George Perkins Marsh and Aldo Leopold, they insisted that planning should conform to ecological boundaries rather than political jurisdictions, and that policymaking should address environmental problems in a comprehensive and holistic fashion. They had hoped the environmental laws passed in the 1960s and 1970s would transform resource management agencies from commodity maximizers into protectors of forest, range, and coastal lands. In practice, however, those agencies' institutionalized emphasis on commodity extraction and commercial development was resistant to change. Rather than transforming their practices, public officials simply tried to accommodate environmental demands within the existing paradigm of multiple-use resource management. The typical result was conflict over rules and decisions, with development interests defending their historical prerogatives and environmentalists seeking to enforce new statutory mandates through administrative challenges and litigation.

Further complicating matters was the fact that, in any given landscape, multiple entities had jurisdiction over land and natural resources, creating an incoherent patchwork of rules and practices. The Forest Service and Bureau of Land Management (BLM) were trying to balance demands for commodity production, recreation, and wilderness preservation on their properties. Adjacent lands managed by the Fish and Wildlife Service or the National Park Service aimed to conserve habitat for wildlife and protect scenic vistas. Other federal agencies—including the Bureau of Reclamation and Army Corps of Engineers—had authority over federal irrigation and flood-control projects, both of which affected surrounding lands. Although often operating within a single landscape, these various agencies were making little effort to coordinate their activities. Exacerbating the management challenge, state and privately owned tracts—often comprising the most ecologically valuable land—were scattered among the federally owned land, but were subject to their own regulations, including local zoning ordinances, property tax laws, and state forestry and wildlife laws.

Critics from a variety of disciplines charged that the cumulative result of perpetual conflict, complex landownership patterns, and an overwhelming focus on individual resources or species was the steady decline of wildlife and vital ecological services.[3] To remedy these deficiencies, they prescribed an ecosystem-based approach emphasizing cumulative impacts, as well as connections among ecosystem elements and between ecological and social systems. Their aim was to conserve or restore the long-term ability of a particular ecosystem to deliver ecological goods and services and recover from broad-scale perturbations. Some of the earliest U.S. examples of EBM actually preceded efforts to define the approach formally. For example, although not explicitly identified as EBM, the multistate efforts to restore the Great Lakes and the Chesapeake Bay that were initiated in the late 1970s and early 1980s adopted an ecosystem-based approach. Two events in the late 1980s,

[3] Ecological services are the benefits supplied by natural systems; they range from clean air and water to the cycling of nutrients that are essential to life.

however, underscored the need for, and catalyzed high-level interest in, EBM: the controversy surrounding efforts to conserve grizzly bear habitat in the Greater Yellowstone region, and the fight to save the northern spotted owl and its old-growth forest habitat in the Pacific Northwest.

The Greater Yellowstone Ecosystem (GYE) spans more than 18 million acres across three states and is the largest nearly intact natural ecosystem in the temperate zone. It provides critical habitat for the world's largest elk herds, as well as for free-roaming bison, grizzly bears, whooping cranes, bald eagles, peregrine falcons, and trumpeter swans (Keiter 1991). The GYE is also a "complex patchwork of management and ownership" (Goldstein 1991) that includes 28 different political units. For decades, a lack of shared conservation goals among the region's many stakeholders led to habitat fragmentation, disruption of ecological processes, and human–wildlife confrontations (Glick and Clark 1998). In the mid-1980s, after legislators castigated the National Park Service and Forest Service for failing to coordinate their planning and management activities, the two agencies reinvigorated the Greater Yellowstone Coordinating Committee (GYCC), an interagency group created in the 1960s. The GYCC eventually issued a draft vision statement that called for ecosystem management that would conserve the region's "naturalness" (Fitzsimmons 1999). After an intensely negative reaction from local politicians and economic interests, however, the committee dramatically revised the document, ultimately producing a bland and ineffectual final version.

The Northwest Forest Plan, crafted in response to the spotted owl crisis in the Pacific Northwest, marked the start of a more successful EBM initiative. A series of disputes in the late 1980s over preserving spotted owl habitat culminated in litigation that virtually halted logging in the region's federally owned old-growth forests. In the spring of 1993, newly elected President Clinton charged a team of experts with crafting a forest management plan that would end the stalemate. The resulting plan embodied important EBM attributes. First, its boundaries were defined not by political jurisdictions but by the range of the spotted owl and other endemic species. Second, it was based on a comprehensive, state-of-the-art scientific assessment and sought to establish a network of preserves and corridors to facilitate the dispersal of native species and to re-establish natural disturbance regimes. Third, it proposed creating ten adaptive management areas where land managers could experiment with novel interventions. Because the plan reduced logging to well below historic levels, it also included measures to help dislocated timber workers and their communities adjust to the new regime.

Theoretical Benefits of EBM

By the mid-1990s, EBM had become the preferred strategy of many professional societies and the dominant paradigm—at least rhetorically—among the nation's land management agencies (Beattie 1996; Christensen et al. 1996; Dombeck 1996;

Interagency Ecosystem Management Task Force 1995; NAPA 1995; PCSD 1996; Society of American Foresters 1993; Thomas 1996; USEPA 1994; Western Governors' Association 1998). Scholars and practitioners offered a variety of definitions that, although not identical, consistently emphasized three elements: (1) landscape-scale planning with an emphasis on restoring ecosystem processes and functions; (2) collaboration with stakeholders; and (3) flexible, adaptive implementation (Browman et al. 2004; Christensen et al. 1996; Clark 1999; GAO 1994; Gordon and Coppock 1997; Franklin 1997; Grumbine 1994; Keiter 1998, 2003; Lamont 2006; Lee 1993; Meffe et al. 2002; Wallace et al. 1996; Yaffee 1999, 2002). Proponents hoped that, in combination, these three elements would lead to more comprehensive management at larger spatial scales on a longer time frame than conventional management and would therefore lead to a more sustainable and resilient landscape (Table 1).[4]

Table 1 Ecosystem-based management vs. traditional management

Attribute	Traditional natural Resource management	Ecosystem-based Management
Underlying view of nature	A collection of resources to be controlled	Complex, dynamic, inter-related, and inherently unpredictable systems
Relevant science	Equilibrium perspective: succession leads to stable climax communities; reductionist methods; goal is predictability	Flux-of-nature perspective: disturbance is normal; holism; embrace of uncertainty and surprise
Goal(s) of management	Maximum sustainable yield of commodities	Sustainable ecosystems, ecological integrity OR Balance between commodity production, amenities, and ecological integrity
Decision making	Centralized, top-down, expert-driven	Decentralized, participatory, collaborative
Implementation/ solutions	Prescriptive, uniform, piecemeal, technology-based; emphasis on control and remediation of damage	Incentive-based or voluntary, locally tailored, and performance-based; emphasis on prevention
	Management that is rigid and aims for control	Management that is experimental, adaptive

[4] Haeuber and Franklin (1996) argue that sustainability is at the core of EBM, its essential element and precondition. Franklin (1997) defines sustainability as "the maintenance of the potential of our terrestrial and aquatic ecosystems to produce the same quantity and quality of goods and services in perpetuity."

Landscape-Scale Planning

A central feature of EBM is planning at a landscape scale. Forest ecologists Perry and Amaranthus (1997) explain: "The critical role of landscapes and regions in buffering the spread of disturbances, providing pathways of movement for organisms, altering climate, and mediating key processes such as the hydrologic cycle means that the fate of any one piece of ground is intimately linked to its larger spatial context." Ecologists acknowledge that there is no single, widely accepted scientific definition of an ecosystem (Szaro et al. 1998), but there are distinctive landscapes, such as watersheds, that are widely recognized as meaningful because of their distinctive biogeographical features. The point is that planning, and therefore management, should be organized around the problem(s) to be solved, not around political units or property lines, and should focus on the relationships among landscape elements, rather than on individual elements in isolation.

Theoretically, landscape-scale planning should yield significant environmental benefits for two main reasons. First, it requires scientists to develop integrative assessments illuminating the relationships within coupled human-ecological systems, as well as the ecological structures and functions that are critical to a system's long-term resilience. It also requires scientists to incorporate multiple scales and the dynamic character of ecosystems. Such assessments, in turn, raise policymakers' and stakeholders' awareness of critical ecological processes. Armed with this knowledge, they should be more likely to design solutions that are holistic and comprehensive—and, therefore, more effective at conserving biological diversity than are uniform, single-media-oriented, national-level policies (Christensen et al. 1996; Meffe and Carroll 1994; Murphy 1999). Second, landscape-scale planning requires coordination among the numerous entities with jurisdiction over the landscape. Such coordination, in turn, should alleviate the problems that arise when federal and state agencies operating within a single ecosystem pursue inconsistent policies. It should also avert the "death by a thousand cuts" that occurs when localities make decisions that disregard spillovers across jurisdictional boundaries and facilitate urban sprawl (Beatley and Manning 1997).

Stakeholder Collaboration

Another critical element of EBM is collaboration among stakeholders—an attribute that gained prominence as proponents contemplated the challenges of implementation (Duane 1997). Specifically, stakeholders should select the desired states of the ecosystem and formulate the means to achieve those states. In most collaborative planning processes, participants deliberate with the aim of reaching consensus, generally defined as willingness by all to accept the decision of the group. When properly structured, consensus-based problem solving identifies solutions that promise gains for all of the participants, even though no one group is likely to get everything it wants.

In theory, engaging stakeholders in a collaborative process of defining the goals, objectives, and outputs of EBM will produce several environmental benefits. Repeated interactions among stakeholders are likely to increase their knowledge and understanding of one another's interests and eventually foster trust among participants (Axelrod 1984; Dryzek 1990; Innes and Booher 1999; Susskind and Cruikshank 1987). Trust, in turn, generates more creative interactions, which can yield innovative solutions (Dryzek 1990; Innes 1996; Wondollek and Yaffee 2000). Brick and Weber (2001) explain: "Instead of a system premised on hierarchy, collaboratives devolve significant authority to citizens, with an emphasis on voluntary participation and compliance, unleashing untapped potential for innovation latent in any regulated environment."

Proponents of collaboration also expect that it will yield solutions that are more effective at solving environmental problems than top-down approaches, because the process incorporates more and better information, and does so more thoroughly. Collaboration is likely to engage scientists more productively than adversarial processes because, in a deliberative forum, reasoning, rather than tactics, is paramount (Andrews 2002; Ozawa 1991). Moreover, unlike decision-making by narrowly trained experts, collaboration incorporates local knowledge, which is based on extended, close observation of how an ecosystem behaves (Berkes 1999; Brunner et al. 2005; Fischer 2000). In the process, it filters out the biases and broadens the perspectives of experts, while simultaneously enhancing the technical expertise of citizens (Brick and Weber 2001; Susskind and Cruikshank 1987).

Involving all interested parties can also ease implementation because everyone who might obstruct the implementation of a collaborative decision will have participated in formulating the solution (Blumenthal and Jannink 2000; Meffe et al. 2002). Gordon and Coppock (1997) point out that "The inclusiveness of the process broadens the base of support, making it harder for die-hard opponents to overturn agreements as soon as they see a political advantage." By contrast, local stakeholders tend to perceive mandates issued by federal officials as unfair and illegitimate and therefore to resent and resist them (Susskind and Cruikshank 1987). This is why Leslie and McLeod (2007) argue that "meaningful engagement with stakeholders is needed to create management initiatives that are credible, enforceable, and realistic."

Flexible, Adaptive Implementation

The third critical element of EBM is implementation that is flexible and adaptive. A flexible implementation strategy is one that employs information, incentives, performance standards, and voluntarism, rather than prescriptive rules and deterrence (Fiorino 2004). Such flexibility is important because next-generation environmental problems are fundamentally different from those tackled in the

1970s. Centralized rules may have been appropriate for problems caused by large factories, for example, but they are inadequate for dealing with suburban sprawl, agricultural runoff, and other problems caused by myriad individual decisions (Esty and Chertow 1997; Graham 1999). Adaptive management entails designing interventions to illuminate ecosystem responses, in light of which management can be continually refined, which is essential because both ecological and social systems are complex, dynamic, and inherently unpredictable (Cortner and Moote 1999; Karkkainen 2002). Ideally, adaptive management begins with the establishment of baseline conditions and the identification of gaps in knowledge about a system. Next, scientists devise management interventions as experiments that test hypotheses about the behavior of the system and monitor the results of those interventions. Finally, managers modify their practices in response to information gleaned from monitoring (Holling 1978).

Flexible, adaptive implementation promises at least two major environmental benefits. In theory, flexibility fosters a sense of stewardship among regulated entities, increasing the likelihood they will take protective measures that exceed minimum legal requirements (Fiorino 2004). By contrast, according to critics of the status quo, traditional regulatory approaches appear unreasonably burdensome and arbitrary, so provoke resistance or efforts to circumvent the rules. Those who do comply are likely to engage in the minimum legally required behavior change (Fiorino 2004; Freeman 1997). Adaptive management promotes continuous learning, which is essential given our limited ability to comprehend the dynamic and unpredictable natural world and the impacts of our interventions (Holling 1995, 1996; Lee 1993). Adjustments in our responses to new information should result in management practices that are more reflective of the best scientific understanding of ecosystem process and function.

Critiques of EBM

Despite great enthusiasm about the potential benefits of EBM, its critics worried that the concept was too ambiguous to bring about genuine environmental protection, and that absent a shift in values EBM would yield more of the same while breeding complacency (Lackey 1998; Ludwg et al. 1993; Stanley 1995). Some critics suggested that existing statutory frameworks that give precedence to commodity production or species-level obligations might impede ecosystem-based approaches (Keiter 1998; Tarlock 2003). Others worried that institutional factors, particularly longstanding agency missions and standard operating procedures, would obstruct EBM initiatives (Cortner and Moote 1999; Keiter 1998). Still others complained that flexible implementation would allow evasion of protective measures by recalcitrant managers and stakeholders (Lowi 1999; Steinzor 2000), while adaptive management, although desirable in theory, would encounter resistance in practice (Johnson 1999; Stankey et al. 2003; Walters 1997).

The gravest fears about EBM, however, focused on stakeholder collaboration.[5] Critics argued that collaboration aimed at consensus would yield lowest-common-denominator solutions rather than environmentally protective ones. According to this logic, collaboration undermines efforts to depart dramatically from the status quo because, in an effort to attain consensus, planners exclude or marginalize those with "extreme" views, skirt contentious issues, focus on the attributes of the ecosystem that are easiest to control, and avoid considering solutions that impose costs on participating stakeholders (Beierle and Cayford 2002; Coglianese 2001; Eckersley 2002; Peterson et al. 2005; Stanley 1995). Some skeptics charged that collaboration actually exacerbates the power imbalance between environmental and development interests and, therefore, generates *worse* outcomes than the traditional regulatory approach (Amy 1990; Coggins 2001; McCloskey 1996; Stahl 2001; Steinzor 2000).

Ambivalence about stakeholder collaboration is reflected in the extent to which proponents are explicit about the preeminence of restoring ecological integrity and biological diversity. Some proponents of EBM, including many scientists, emphasize sustaining ecosystems and moderating human behavior to accommodate natural constraints (Callicott 2000; Christensen et al. 1996; Grumbine 1994; Lamont 2006; Wood 1994). Wood (1994), for example, argues that "To embrace the ecosystem management concept is to accept that ecological factors such as maintaining biological diversity, ecological integrity, and resource productivity dictate strict limits on social and economic uses of the land." Similarly, Grumbine (1994) contends that "ecosystem management integrates scientific knowledge of ecological relationships within a complex sociopolitical and values framework toward the general goal of protecting native ecosystem integrity over the long term." And Salwasser (1998, p 90) argues that "The aim of ecosystem management on national forests should be to sustain healthy land first, then to provide people with the variety of benefits and options they need and want, consistent with basic land stewardship." This perspective acknowledges the strong possibility of trade-offs between environmental protection and development goals, particularly in the short run (Rosenberg and Sandifer 2009; Levin and Clark 2010).

Many others, however, have proffered a view of sustainability in which social, economic, and ecological benefits are pursued simultaneously and, apparently, harmoniously. For example, in 1996 the Keystone Center defined ecosystem management as "a collaborative process that strives to reconcile the promotion of economic opportunities and livable communities with the conservation of ecological integrity and biological diversity" (Fitzsimmons 1999). Szaro et al. (1998) argue that "the mandate [of EBM] should be to protect environmental quality while also producing the resources that people need. Therefore, ecosystem management cannot simply be a matter of choosing one over the other." The EBM Tools Network, a

[5] While critics on the left suggested that EBM would yield watered-down, and therefore insufficiently protective, solutions, critics on the right argued that EBM was a vehicle of nature-worshipping environmentalists to elevate protection of ecosystems above all else (Fitzsimmons 1999).

web-based alliance of EBM researchers and practitioners seeking to promote EBM for coastal and marine environments, says that EBM "is concerned with the ecological integrity of coastal-marine systems and the sustainability of both human and ecological elements." Such formulations elide the possibility of short-term economic or ecological losses that may result from the pursuit of multiple goals simultaneously.

Cortner and Moote (1999) acknowledge the ambiguity that pervades definitions of EBM when they say that "while ecosystem management explicitly recognizes that social goals and objectives play a central role in framing management direction, it also presumes that humans will decide to make protection of ecological processes their overriding social objective." Similarly, the U.S. Government Accountability Office (GAO 1994) observes: "Proponents of ecosystem management believe that coordinating human activities across large geographic areas to maintain or restore healthy ecosystems…would, among other things, better address declining ecological conditions and ensure the sustainable long-term use of natural resources, including the production of natural resource commodities." The GAO also recognizes, however, that "In the absence of a clear statement of federal priorities for sustaining and restoring ecosystems and the minimum level of ecosystem health needed to do so, ecosystem management has come to mean different things to different people."

EBM in Practice

Despite its ambiguities, EBM has become the strategy of choice among natural resource managers worldwide. By 1999, federal officials in the United States had stopped referring to ecosystem management, retreating in the face of vitriolic reactions from commodity interests and western "wise use" advocates. But the concepts that underpin EBM persisted and initiatives continued under different names, such as "collaborative conservation." In 2006, for example, the Cooperative Sagebrush Initiative began engaging stakeholders across eleven western states in an ostensibly comprehensive effort to reconcile resource use with conservation of the sage grouse, whose numbers were dwindling as a result of habitat fragmentation.

Even as the term fell out of favor among U.S. land managers, enthusiasm was growing for applying EBM principles to marine ecosystems (Browman et al. 2004; McLeod et al. 2005; Rosenberg and McLeod 2005; UNEP and GPA 2006; Ruckelshaus et al. 2008). California's 1999 Marine Life Management Act required EBM for managing all marine wildlife in the state's waters. Two prestigious scientific panels (Pew Oceans Commission 2003; US Commission on Ocean Policy 2004) recommended taking an ecosystem-based approach to managing all marine systems. In fact, by the early 2000s EBM had become the dominant paradigm for managing natural resources around the world, and several international organizations, including the World Conservation Union (Pirot et al. 2000), United Nations (UNCB 2003; UNDP et al. 2003), and the Millennium Ecosystem Assessment

(MEA 2006), developed case studies and principles for successful implementation of EBM.[6]

A Systematic Assessment of EBM

Although there is widespread enthusiasm about EBM, systematic assessments of its efficacy have been few and far between. This is in part because, until the 2000s, few initiatives had existed long enough for evaluators to assess them. The complexity and heterogeneity of those initiatives that were under way made evaluation particularly challenging. Beginning in the mid-2000s, however, I sought to conduct a systematic assessment of the results of EBM and the mechanisms by which those results were produced. I chose four prominent cases of EBM, two terrestrial and two aquatic-system initiatives, in rapidly urbanizing regions of the United States. It became clear upon further investigation that, although they differed in their particulars, all four cases were generating results that were only minimally environmentally beneficial. In hopes of clarifying which attributes of EBM were responsible for these disappointing outcomes, I identified three additional cases that were similar in terms of locations and the problems being addressed but seemed to be producing more substantial environmental benefits (Table 2).[7]

Table 2 Case selection

	Minimal environmental benefit	Substantial environmental benefit
Terrestrial ecosystem	Austin (Texas) Balcones Canyonlands Conservation Plan (BCCP)	Pima County (Ariz.) Sonoran Desert Conservation Plan (SDCP)
	San Diego (Calif.) Multiple Species Conservation Plan (MSCP)	
Aquatic ecosystem	Comprehensive Everglades Restoration (Fla)	Kissimmee River Restoration (Fla)
	California Bay-Delta Program (Calif.)	Mono Basin Restoration (Calif.)

The results of my investigation were surprising. All seven of the initiatives I examined yielded concrete policies and practices that appeared likely, over time, to produce some environmental benefits. Each had prompted the creation of a deeper and more holistic understanding of how specific ecosystems work which, in turn, had fostered a more widespread recognition among policymakers and stakeholders of the relationships among the landscape's ecological elements and functions. Without exception, the programs had furnished participants with a rationale for raising large sums of money that were used to acquire ecologically valuable land or to

[6] Elsewhere, EBM went by other names, including Integrated Coastal Management, Integrated Water Resources Management, and Integrated River Basin Management.

[7] For analysis, see Layzer (2008).

undertake activities aimed at restoring ecological functions. Each program had empowered environmentally oriented personnel within agencies and jurisdictions, some of whom had tried to institutionalize more environmentally beneficial practices. Only those projects that did not rely on collaborative planning, however, had yielded policies and practices that appeared likely to conserve and restore biological diversity and, therefore, ecological resilience.

A comparison among the seven cases suggested that a landscape-scale focus was an important catalyst for the adoption of more environmentally protective policies and practices. In each of the cases, scientists described a defining "moment" when they realized that what happens in one part of a system affects the other parts; for the first time, they saw the system as a whole, not just a set of parts. In addition, scientists identified key drivers of ecological damage and documented the mechanisms by which that damage was occurring. They recommended measures to conserve key species and the ecological processes they depend on. Furthermore, in every case, trying to address problems at a landscape scale prompted planners to adopt more comprehensive approaches to environmental problem solving and led to new forms of coordination among disparate agencies and jurisdictions.

The beneficial effects of collaborating with stakeholders were more elusive, however. In the four cases where policymakers deferred to stakeholders to set goals, the policies and practices that emerged appeared unlikely to conserve or restore ecological health because, to gain consensus, planners skirted trade-offs and opted instead for solutions that promised something for everyone. The resulting plans typically featured management-intensive approaches with little buffering. As a consequence, they imposed the risk of failure on the natural system. There are several explanations for this result. First, although collaboration did enhance trust, there is little evidence that stakeholders' interests were genuinely transformed or that the collaborative process generated innovative solutions. Instead, consensus-oriented groups tended to marginalize advocates who espoused "extreme" views. Even with a carefully selected stakeholder group, negotiations often resembled bargaining more than deliberation, particularly as plans became more specific. Stakeholder groups tended to avoid the most difficult issues or to mask differences by using vague language—decisions that ultimately haunted implementation.

Stakeholder collaboration also did not ensure that the best information would prevail. The four cases that involved collaboration featured scientific enterprises that were difficult to penetrate, so that local knowledge was often ignored. Nor did collaboration put an end to bickering among stakeholders over scientific issues. (It is noteworthy that the plans that did not rely on collaboration were actually more recognizably grounded in science than those that did.) The evidence also failed to support the notion that collaboration ensures durable implementation. Instead, implementation exposed many of the differences papered over during the collaborative planning processes, as stakeholders sought to prevent or modify projects that threatened their interests.

A commitment to flexible, adaptive implementation did not compensate for the failings of these four environmentally risky plans and, in fact, sometimes exacerbated them. Adaptive management did not yield a willingness to alter policies in the face of new information, partly because minimalist plans actually provided little room for adjustment, but also because management and monitoring were insufficiently funded, and learning by scientists did not translate automatically into management changes. Flexible implementation allowed managers with missions that were incompatible with ecological restoration to resume user-friendly practices when political conditions shifted.

By contrast, when policymakers—elected officials, administrators, or judges—endorsed an environmentally protective goal and used regulatory leverage to prevent development interests from undermining that objective, as they did in the three comparison cases, the resulting policies and practices were more likely to conserve or restore ecological integrity. In these cases, a willingness among political leaders to make ecological health the preeminent aim changed the balance of power and altered perceptions of what was politically feasible. When restoring ecological health was the paramount goal, planners were more likely to approve, and managers to implement, approaches that relied less on energy-intensive manipulation and more on enhancing the ability of natural processes to sustain themselves, even if doing so imposed costs on some stakeholders.

Generalizing the Results

The insights generated by this comparative analysis should be taken as cautionary, rather than definitive. In particular, it is important to note that the four cases of genuine EBM were more complex, both geographically and organizationally, than the comparison cases. That said, the findings are consistent with the results of other research on EBM or its components. Several detailed examinations of the Chesapeake Bay Program, for example, have yielded comparable results. The Chesapeake Bay Program was established in 1983 to restore the resilience of the nation's largest and most productive estuary. But, on January 5, 2010, the 26-year-old program missed its second major cleanup deadline. Careful analysis of monitoring data suggested that efforts to reduce pollution of the bay had fallen more than 40% short of their goals, despite spending of nearly $6 billion, because the impacts of relentless growth were overwhelming pollution control efforts.

Most observers attribute the Chesapeake Bay Program's failures to its reliance on collaborative planning and flexible implementation (Ernst 2003; Horton 2003; Layzer 2012). Historically, the program has operated as a multistate cooperative, with the Environmental Protection Agency in a supporting role and the Bay states following different paths depending on their political culture and proximity to the Bay. While the state of Maryland has typically enacted more stringent measures, Virginia and Pennsylvania have taken advantage of the program's flexibility to adopt minimally protective policies and practices. Furthermore, the states' heavy reliance on nonregulatory approaches—including educational programs, incentives, and voluntary stewardship initiatives—has yielded few measurable results. In fact, the single most effective measure taken in the watershed to date has been the ban on phosphate detergent adopted by the Bay states in the 1980s and 1990s.

Similarly, in the Great Lakes a multibillion-dollar international effort to restore ecological integrity has been under way for decades. Yet ecological recovery has been limited, and progress is likely to be negated by increasing population, urban sprawl, and ongoing problems with invasive species, all of which will be exacerbated by global warming (Shear 2006). There are signs of progress in the Great Lakes: phosphorus inputs have declined; there has been a dramatic recovery of the walleye as a result of fishing limits and phosphorus controls; and the population of the burrowing mayfly, historically the dominant benthic invertebrate in the lakes, rebounded between 1990 and 2001. On the other hand, the available data suggest that wetland-dependent birds have been static or declining, nonnative fish dominate prey fish in most areas, and invasive zebra mussels have decimated native freshwater mussel communities.

Less prominent initiatives have also fallen short of restoring resilient ecosystems. For example, Arizona's collaborative Upper San Pedro Partnership (USPP), formed in 1998, failed to prevent the San Pedro River from running dry for the first time in 2005. Although the partnership has generated and disseminated an impressive amount of information to stakeholders, and local governments have taken several steps to reduce water consumption in the region, there has been no consensus on the ultimate issue, which is growth control. According to Saliba and Jacobs (2008), "Perhaps the largest criticism leveled against the USPP is its inability to make difficult decisions regarding growth.…The politics and economics of growth in Arizona make this conversation very difficult."

Scholars investigating marine EBM have also turned up pallid results. Pitcher et al. (2009) reviewed the role of EBM in fisheries management around the world and found that, while many countries had adopted EBM principles, few had actually taken the steps to achieve effective implementation. Only a handful of countries in the developed world were clearly moving toward EBM, and many European countries garnered dismal ratings. The authors concluded that "Whilst the late nineties also saw the blossoming of 'Oceans' approaches aimed at developing and applying EBM principles to multiple sectors in multistakeholder processes, the gradual pace of these reforms and their perceived expense has meant that few have been implemented" (Pitcher et al. 2009). Arkema et al. (2006) investigated forty-nine management plans for eight large marine and coastal ecosystems to assess the extent to which managers were actually practicing EBM. They, too, found that implementation of EBM principles was lagging. Moreover, management objectives included more detailed criteria relevant to human activity than ecological criteria; many of the plans focused on objectives that promoted commercial and recreational uses, such as maintaining public access and rebuilding depleted fisheries.

Other studies support claims about the propensity of collaborative planning, in particular, to yield environmentally risk-tolerant solutions. Pralle (2006), for example, conducted a detailed analysis of the Quincy Library Group (QLG), and found that the decision by key activists to plan collaboratively in a local forum led planners to redefine the central problem facing the ecosystem as forest fires, rather than excessive logging. Doing so defused conflict and allowed for a solution that gave something to everyone without necessarily addressing the root cause of the region's environmental degradation. Pralle noted that the focus on process disarmed environmental challengers, who found it difficult to combat the "overwhelmingly positive characterization" of local, collaborative decision making. She observed that "In a world of polarized interest groups and partisan gridlock, policymakers may be more than willing to settle for outward signs of consensus rather than true political compromises" (Pralle 2006).

Similarly, in a study of two collaborative projects within the Everglades restoration in South Florida, Frank (2009) found that collaboration was better at resolving conflict than at problem solving, which was a subordinate objective. She concluded that:

> Since collaborative processes did not significantly change power relations, collaborative outputs and the political capital upon which they depended were largely transient. Collaboration produced a delicate balancing act of aligned interests in keeping with the rhetoric of win–win and sustainability. Collaborative recommendations appeared highly integrated, yet under the surface there were strategic motivations and shallow commitments. The agreements began to unravel when system dynamics of technical shortcomings changed the conditions upon which the agreements depended. Combine this with the long-range dominance of economic interests, and the result was poorer implementation performance for environmental plan features.

Other scholars, however, have identified more encouraging examples of EBM that incontrovertibly have resulted in environmental improvements. These projects tend to feature most of the characteristics posited by political scientist Ostrom (1990) and her colleagues as essential to effective local, collaborative management of common pool resources. In particular, appropriators have the following characteristics: (1) they believe they will be harmed if they do not adopt rules to govern use of the resource; (2) they are affected in similar ways by the proposed rules; (3) they value the continued use of this common property resource (discount rates are low); (4) they face low information, transformation, and enforcement costs; (5) they share generalized norms of reciprocity and trust; and (6) they constitute a relatively small and stable group. In addition, the target resource is in sufficiently good shape that efforts to protect it will confer benefits, there are valid and reliable indicators of system health, the flow of resources is relatively predictable, and the system is sufficiently small to allow knowledge of external boundaries and internal microenvironments (Ostrom 2001). Such conditions rarely hold at the larger scales that are typical of EBM.

Furthermore, in many cases, evaluators who discern positive results have relied on the testimony of participants, rather than on actual evidence of ecological improvement. For example, Steven Yaffee (2002) reported on 105 EBM-like partnerships throughout the United States. According to surveys of those initiatives, a majority had not only produced better relationships and greater awareness of the ecosystem, but also improved scientific understanding, ecological restoration, increased native species populations, and improvements in "overall ecosystem integrity." Similarly, in a survey by the GAO (2008) of seven collaborative initiatives, participants claimed they had improved natural resource conditions—although none had collected any data to demonstrate their claimed impact(s) at a landscape scale. Alternatively, analysts have evaluated EBM on the basis of process, but not outcomes. Tallis et al. (2010), for instance, proffer two case studies of "successful" EBM, one in Raja Ampat, Indonesia, and the other in Puget Sound, Washington. Neither purported to show ecological improvements, however.

Conclusions

Overall, the evidence suggests that EBM, although widely embraced in theory, does not necessarily result in ecological restoration in practice. In cases where EBM has been fully implemented, landscape-scale planning has yielded discernible environmental benefits, but the effects of collaboration with stakeholders and flexible, adaptive implementation are more ambiguous. This is not entirely surprising: proponents of EBM often gloss over the potential trade-offs among environmental, economic, and social considerations, particularly in the short run. They assume that long-term thinking, and a related preoccupation with ecological sustainability, will somehow emerge from a collaborative process. For this to happen, however, participants must adopt a view that healthy, functioning ecosystems are essential to human well-being. They must embrace a land ethic and eschew a short-run economic point of view.

Such transformation is unlikely under any circumstances but, counterintuitively, appears to be made more probable by the exercise of political leadership and regulatory leverage. That is why, as legal scholar Bradley Karkkainen (2002) observes, in the United States the federal government plays a critical role in EBM. More generally, he notes that productive collaboration is most likely when the most powerful actors have their backs against the wall—usually as a result of a stringent federal law that is likely to be enforced. Mobilization and litigation by environmental advocates can also generate the kind of "pervasive, persistent, and profound uncertainty, and the associated recognition of mutual dependence" necessary to bring about shifts in the balance of power that precede productive deliberation (Cohen and Rogers 2003).

Proponents of sustainability science anticipate that new modeling tools will do a better job of forecasting trajectories, assessing complex risks, and laying bare the potential trade-offs between short-run economic development and long-run environmental sustainability. The link between a shift toward more holistic, integrative science and more environmentally protective governance is not straightforward, however. We will also need insight into human motivation, and better mechanisms for transforming social priorities, to bring about the kinds of behavior changes required for society to become genuinely sustainable.

References

Amy DJ (1990) Environmental dispute resolution: the promise and the pitfalls. In: Norman JV, Kraft ME (eds) Environmental policy in the 1990s. CQ Press, Washington

Andrews CJ (2002) Humble analysis: the practice of joint fact-finding. Praeger, Westport

Arkema KK, Abramson SC, Dewsbury BM (2006) Marine ecosystem-based management: from characterization to implementation. Front Ecol Environ 4:525–532

Axelrod RM (1984) The evolution of cooperation. Basic Books, New York

Beatley T, Manning K (1997) The ecology of place: planning for the environment, economy, and community. Island Press, Washington

Beattie M (1996) An ecosystem approach to fish and wildlife conservation. Ecol Appl 6:696–699

Beierle TC, Cayford J (2002) Democracy in practice: public participation in environmental decisions. Resources for the Future, Washington
Berkes F (1999) Sacred ecology. Taylor and Francis, Philadelphia
Blumenthal D, Jannink J-L (2000) A classification of collaborative management methods. Ecol Soc 4:13. http://www.consecol.org/vol4/iss2/art13/
Botkin DB (1990) Discordant harmonies: A new ecology for the twenty-first century. Oxford University Press, New York
Brick P, Weber EP (2001) Will the rain follow the plow? Unearthing a new environmental movement. In: Brick P, Snow D, Van De Wetering S (eds) Across the great divide: explorations in collaborative conservation and the american west. Island Press, Washington
Browman HI, Cury PM, Hilborn R et al (2004) Perspectives on ecosystem-based approaches to the management of marine resources. Mar Ecol Prog Ser 274:269–303
Brunner RD, Steelman DA, Co-Juell L et al (2005) Adaptive governance: integrating science, policy, and decision making. Columbia University Press, New York
Callicott JB (2000) Harmony between man and land: Aldo Leopold and the foundations of ecosystem management. J For 98:4–13
Cestero B (1999) Beyond the hundredth meeting: a field guide to collaborative conservation on the west's public lands. Sonoran Institute, Tucson
Christensen NL, Bartuska AM, Carpenter S et al (1996) The report of the Ecological Society of America committee on the scientific basis for ecosystem management. Ecol Appl 6:665–691
Clark JR (1999) The ecosystem approach from a practical point of view. Conserv Biol 13:679–681
Coggins GC (2001) Of californicators, quislings, and crazies: some perils of devolved collaboration. In: Brick P, Snow D, and Ven de Wetering S (eds) Across the great divide: explorations in collaborative conservation and the American west. Island Press, Washington
Cohen J, Rogers J (2003) Power and reason. In: Fung A, Olin Wright E (eds) Deepening democracy: Institutional innovations in empowered participatory governance. Verso, New York
Cortner HJ, Moote MA (1999) The politics of ecosystem management. Island Press, Washington
Coglianese G (2001) Is consensus an appropriate basis for regulatory policy? In: Orts EW, Deketelaere K (eds) Environmental contracts. Kluwer Law International, Boston
Dombeck MP (1996) Thinking like a mountain: BLM's approach to ecosystem management. Ecol Appl 6:699–702
Dryzek J (1990) Discursive democracy: politics, policy, and political science. Cambridge University Press, New York
Duane TP (1997) Community participation in ecosystem management. Ecol Law Quart 24:771–797
Eckersley R (2002) Environmental pragmatism, ecocentrism, and deliberative democracy: between problem-solving and fundamental critique. In: Minteer B, Taylor BP (eds) Democracy and the claims of nature: critical perspectives for a new century. Rowman and Littlefield, Landham
Ernst HR (2003) Chesapeake Bay blues: science, politics, and the struggle to save the bay. Rowman and Littlefield, Landham
Esty DC, Chertow MR (1997) Thinking ecologically: an introduction. In: Marion R, Chertow MR, Esty DC (eds) Thinking ecologically. Yale University Press, New Haven
Fiorino DJ (2004) Flexibility. In: Durant RF, Fiorino DJ, O'Leary R (eds) Environmental governance reconsidered: challenges, choices, and opportunities. MIT Press, Cambridge
Fischer F (2000) Citizens, experts, and the environment. Duke University Press, Durham
Fitzsimmons AK (1999) Defending illusions: federal protection of ecosystems. Rowman and Littlefield Publishers Inc, Landham
Frank K (2009) The role of collaboration in Everglades restoration (Unpublished dissertation). Georgia Institute of Technology, Atlanta
Franklin J (1997) Ecosystem management: an overview. In: Haney A, Boyce MS (eds) Ecosystem management: applications for sustainable forest and wildlife resources. Yale University Press, New Haven

Freeman J (1997) Collaborative governance in the administrative state. UCLA Law Rev 45:1–98
Glick D, Clark TW (1998) Overcoming boundaries: the greater Yellowstone ecosystem. In: Meffe G, Nielsen LA, Knight RL, Schenborn DA (eds) Ecosystem management: adaptive, community-based conservation. Island Press, Washington
Goldstein B (1991) The struggle over ecosystem management at Yellowstone. Bioscience 42:183–187
Gordon J, Coppock J (1997) Ecosystem management and economic development. In: Chertow MR, Esty DC (eds) Thinking ecologically. Yale University Press, New Haven
Graham M (1999) The morning after earth day. Brookings Institution Press, Washington
Grumbine RE (1994) What is ecosystem management? Conserv Biol 8:27–38
Haeuber R (1996) Setting the environmental policy agenda: the case of ecosystem management. Nat Resour J 36:1–28
Haeuber R, Franklin J (1996) Perspectives on ecosystem management. Ecol Appl 6:692–693
Holling CS (ed) (1978) Adaptive environmental assessment and management. Wiley, New York
Holling CS (1995) What barriers? What bridges? In: Gunderson LH, Holling CS, Light SS (eds) Barriers and bridges to the renewal of ecosystems and institutions. Columbia University Press, New York
Holling CS (1996) Surprise for science, resilience for ecosystems, and incentives for people. Ecol Appl 6:733–735
Holling CS, Meffe G (1996) Command and control and the pathology of natural resource management. Conserv Biol 10:328–337
Horton T (2003) Turning the tide: saving the Chesapeake Bay. Island Press, Washington
Innes JE (1996) Planning through consensus building. J Am Plann Assoc 62:460–472
Innes JE, Booher D (1999) Consensus building and complex adaptive systems: a framework for evaluating collaborative planning. J Am Plann Assoc 65:412–423
Interagency Ecosystem Management Task Force (1995) The ecosystem approach: healthy ecosystems and sustainable economies, vol 1. Report of The Interagency Ecosystem Management Task Force. http://www.denix.osd.mil/nr/upload/ecosystem1.html
Johnson BL (1999) Introduction to the special feature: adaptive management—scientifically sound, socially challenged? Conser Ecol 3:10. http://www.ecologyandsociety.org/vol3/iss1/art10
Karkkainen BC (2002) Collaborative ecosystem governance: scale, complexity, and dynamism. Virginia Environ Law J 21:189–243
Keiter RB (1991) An introduction to the ecosystem management debate. In: Keiter RB, Boyce MS (eds) The greater Yellowstone ecosystem: redefining America's wilderness heritage. Yale University Press, New Haven
Keiter RB (1998) Ecosystems and the law: toward an integrated approach. Ecol Appl 8:332–341
Keiter RB (2003) Keeping faith with nature: ecosystems, democracy, and America's public lands. Yale University Press, New Haven
Lackey RT (1998) Seven pillars of ecosystem management. Landscape Urban Plan 40:21–30
Lamont A (2006) Policy characterization of ecosystem management. Environ Monit Assess 113:5–18
Layzer JA (2008) Natural experiments: ecosystem-based management and the environment. MIT Press, Cambridge
Layzer JA (2012) Ecosystem-based management in the Chesapeake Bay. In: The environmental case, 3rd edn. CQ Press, Washington
Lee KN (1993) Compass and gyroscope: integrating science and politics for the environment. Island Press, Washington, DC
Leslie HM, McLeod KL (2007) Confronting the challenges of implementing marine ecosystem-based management. Front Ecol Environ 5:540–548
Levin SA, Clark WC (eds) (2010) Toward a science of sustainability. Airlie Center, Warrenton
Lowi TJ (1999) Frontyard propaganda. Boston Rev 24:17–18
Ludwg D, Hilborn R, Walters C (1993) Uncertainty, resource exploitation, and conservation: lessons from history. Science 260:36

McCloskey M (1996) The skeptic: collaboration has its limits. High Country News, May 13
McLeod KL et al (2005) Scientific consensus statement on marine ecosystem-based management. Accessed 28 March 2012 http://www.compassonline.org/sites/all/files/document_files/EBM_Consensus_Statement_v12.pdf
Meffe GK, Carroll CR (1994) Principles of conservation biology. Sinauer Associates, Sunderland
Meffe GK, Nielsen LA, Knight RL, Schenborn DA (2002) Ecosystem management: adaptive, community-based conservation. Island Press, Washington
Millennium Ecosystem Assessment (MEA) (2006) Ecosystems and human well-being. Island Press, Washington, DC
Murphy D (1999) Case study. In: Johnson KN, Swanson F, Herring M et al (eds) Bioregional assessments. Island Press, Washington
National Academy of Public Administration (NAPA) (1995) Setting priorities, getting results: a new direction for the U.S. Environmental Protection Agency. National Academy of Public Administration, Washington
Noss RF, Cooperrider AY (1994) Saving nature's legacy. Island Press, Washington
Ostrom E (1990) Governing the commons: the evolution of institutions for collective action. Cambridge University Press, New York
Ostrom E (2001) Reformulating the commons. In: Burger J, Ostrom E, Norgaard R (eds) Protection the commons: a framework for resource management in the Americas. Island Press, Washington
Ozawa C (1991) Recasting science: consensual procedures in public policy making. Westview Press, Boulder, CO
Perry DA, Amaranthus MP (1997) Disturbance, recovery, and stability. In: Kohm KA, Franklin J (eds) Creating a forestry for the 21st century. Island Press, Washington
Peterson MN, Nils M, Peterson MJ et al (2005) Conservation and the myths of consensus. Conserv Biol 19:762–767
Pew Oceans Commission (2003) America's living oceans: charting a course for sea change. Pew Oceans Commission, Arlington
Pickett STA, Ostfeld RC (1995) The shifting paradigm in ecology. In: Knight RL, Bates SF (eds) A new century for natural resources management. Island Press, Washington
Pirot J-Y, Meynell P-J, Elder D (2000) Ecosystem management: lessons from around the world. World Conservation Union, Cambridge
Pitcher TJ, Kalikoski D, Short K et al (2009) An evaluation of progress in implementing ecosystem-based management in fisheries in 33 countries. Mar Policy 33:223–232
Pralle SB (2006) Branching out, digging in. Environmental advocacy and agenda setting. Georgetown University Press, Washington
President's Council on Sustainable Development (PCSD) (1996) Sustainable America: new consensus for prosperity, opportunity, and a healthy environment for the future. US Government Printing Office, Washington
Rosenberg AA, McLeod KL (2005) Implementing ecosystem-based approaches to management for the conservation of ecosystem services. Mar Ecol Progr Ser 300:270–274
Rosenberg AA, Sandifer PA (2009) What do managers need? In: McLeod K, Leslie H (eds) Ecosystem-based management for the oceans. Island Press, Washington
Ruckelshaus M, Klinger T, Knowlton N, DeMaster DP (2008) Marine ecosystem-based management in practice: scientific and governance challenges. Bioscience 58:53–63
Saliba G, Jacobs KL (2008) Saving the San Pedro River. Environment 50:30–43
Salwasser H (1998) Ecosystem management: a new perspective for national forest and grasslands. In: Aley J et al (eds) Ecosystem management. Taylor and Francis, Philadelphia
Shear H (2006) The Great Lakes, an ecosystem rehabilitated, but still under threat. Environ Monit Assess 113:199–225
Society of American Foresters (1993) Sustaining long-term forest health and productivity. Society of American Foresters, Bethesda
Stahl A (2001) Ownership, accountability, and collaboration. In: Brock P, Snow D, Van de Wetering S (eds) Across the great divide: explorations in collaborative conservation and the American west. Island Press, Washington

Stankey GH, Bormann BT, Ryan C et al (2003) Adaptive management and the northwest forest plan. Forestry 101:40–46
Stanley TR Jr (1995) Ecosystem management and the arrogance of humanism. Conserv Biol 9:255–262
Steinzor RI (2000) The corruption of civic environmentalism. Environ Law Reporter 30:10909–10921
Susskind L, Cruikshank J (1987) Breaking the impasse: consensual approaches to resolving public disputes. Basic Books, New York
Szaro RC, Sexton WT, Malone CR (1998) The emergence of ecosystem management as a tool for meeting people's needs and sustaining ecosystems. Landscape Urban Plan 40:1–7
Tallis H, Levin PS, Ruckelshaus M et al (2010) The many faces of ecosystem-based management: making the process work today in real places. Mar Policy 34:340–348
Tarlock AD (2003) Slouching toward eden: the eco-pragmatic challenges of ecosystem revival. Minn Law Rev 87:1173–1208
Thomas JW (1996) Forest service perspectives on ecosystem management. Ecol Appl 6:703–705
United Nations Convention on Biological Diversity (UNCB) (2003) Ecosystem approach: further elaboration, guidelines for implementation and relationship with sustainable forest management. Report of the expert meeting on the ecosystem approach, UNEP, Montreal
United Nations Development Programme (UNDP), United National Environmental Programme, World Bank, and World Resources Institute (2003) A guide to world resources 2002–2004—decisions for the earth—balance, voice, and power. World Resources Institute, Washington
United Nations Environmental Programme (UNEP) and Global programme of action for the protection of the marine environment from land-based activities (GPA) (2006) Ecosystem-based management: markers for assessing progress. UNEP/GPA, The Hague, The Netherlands. http://gpa.unep.org/documents/ecosystem-based_management.english.pdf
US Commission on Ocean Policy (2004) An ocean blueprint for the 21st century. US Commission on Ocean Policy, Washington
US Environmental Protection Agency (USEPA) (1994) The new generation of environmental protection. US EPA, Washington
US General Accounting Office (GAO) (1994) Ecosystem management: additional actions needed to adequately test a promising approach. August, GAO/RCED-94-111
US General Accounting Office (GAO) (2008) Opportunities exist to enhance federal participation in collaborative efforts to reduce conflicts and improve natural resource conditions. February, GAO-08-262
Wallace MG, Cortner HJ, Moote MA et al (1996) Moving toward ecosystem management: examining a change in philosophy for resource management. J Polit Ecol 3:1–36
Walters C (1997) Challenges in adaptive management of riparian and coastal ecosystems. Conser Ecol 1:1. http://www.consecol.org/vol1/iss2/art1
Weber EP (2003) Bringing society back. Grassroots ecosystem management, accountability, and sustainable communities. MIT Press, Cambridge
Western Governors' Association (1998) Principles for environmental management in the west. Resolution 98-001, Denver
Wondollek JM, Yaffee SL (2000) Making collaboration work: lessons from innovation in natural resource management. Island Press, Washington
Wood CA (1994) Ecosystem management: achieving the new land ethic. Renew Res J 12:6–12
Yaffee SL (1999) Three faces of ecosystem management. Conserv Biol 13:713–725
Yaffee SL (2002) Ecosystem management in policy and practice. In: Meffe G, Nielsen LA, Knight RL et al (eds) Ecosystem management: adaptive, community-based conservation. Island Press, Washington

Sustainable Coastal Margins: Challenges of Tempo and Mode for the Policy Domain

Ronald C. Baird

Abstract The urgent need for coastal management of tempo and mode for global population growth and urbanization is generally underappreciated. That growth is inevitable in the coming two decades. With a population of 311 million, the United States must absorb one to two million people per year in coastal watersheds, yet maintain a sustainable and built environment. Degree of success is largely dependent on the policy domain–political will and performance of governance. The current compartmentalized governance structure has been inadequate in meeting environmental goals, and the structure is unlikely to change in the next decade. Strategic and targeted approaches that account for the social, economic, and environmental realities of urban space in a sustainability context are needed for framing contemporary management strategies. That is, confronting reality, thinking strategically, and changing the way institutions are managed and their degree of connectivity. Strategic guidelines are advanced as a blueprint for creating practical, sustainability-based frameworks for performance enhancement. Operational imperatives include pragmatism, prioritization, alignment, understanding, anticipation, context, and implementation.

Keywords Coastal urbanization • Sustainability frameworks • Policy and performance

Introduction

Considerable thought has been given in recent years concerning sustainability in the context of today's rapidly expanding populations and resource consumptive societies (e.g., NRC 1999; Kennedy 2002; Hall and Day 2009). Likewise, frameworks for

R.C. Baird (✉)
Rhode Island Sea Grant, University of Rhode Island,
South Ferry Road, Narragansett, RI, 02882, USA
e-mail: rbaird@gso.uri.edu

M.P. Weinstein and R.E. Turner (eds.), *Sustainability Science: The Emerging Paradigm and the Urban Environment*, DOI 10.1007/978-1-4614-3188-6_10,

science and technology to support transitions to sustainability have also been articulated (e.g., Kates et al. 2001; Clark and Dickson 2003; Holdren 2008). As a consequence, the notion of sustainability has become the central focus and organizing principle of contemporary environmental policy. The concept has great value in that it is understandable to broad audiences and captures the essential management objectives in addressing today's very real issues of resource utilization, environmental degradation, limitations, and the human condition. The problem comes in the translation of a global concept of reconciling human needs with environment into one of practical application in context-specific public policy and management interventions.

The reality is that society will continually have to address the central issue of maintenance of living standards, environments, and quality of life, one that is fraught with human values, conflicting objectives, economic reality, and enormous environmental complexity (Roe and van Eeten 2001; Weinstein and Reed 2005). As Albert Speer, in an excellent treatise on sustainable cities put it, "a sustainably planned and built environment is a must—not a nice to have" (Gaines and Jager 2009). When applied to coastal environments, there are implications of great significance for coastal stewardship—the central focus of this paper. Sustainability implies maintenance of environment in an acceptable state or condition. Finally, and this is critical, sustainability is forever, in that time horizons for maintaining such states are indefinite in the face of natural and anthropomorphic driven changes.

The application of sustainability concepts to contemporary, coastal management contexts remains an immense challenge at virtually all spatial scales, global to local (USCOP 2004). The reality for practical application is that societal response to environmental issues rests with a complex of socioeconomic and governance interrelationships that constitutes the domain of public policy. Appropriate decisions on courses and principles of action must be made and then translated by a complex institutional infrastructure into public policy that elicits effective responses from society at large. How well society addresses sustainability challenges is in great measure a function of the performance of our institutions of governance (Orbach 2002; Baird 2005a; Layzer 2008). Institutional performance can be judged by determining how well environmental goals and standards are being met and are coastal ecosystems at scales of concern, within or trending toward acceptable states in a sustainability sense.

Coastal margins are of immense economic importance to society, as are associated environments (e.g., Beach 2002; PEW 2003; Pendleton 2008). Such environments have long been subject to human-related stresses, yet those impacts have increased dramatically in recent decades, resulting in substantial changes to associated ecosystems (e.g., USCOP 2004; Lotze et al. 2006). What has changed in those decades is tempo and mode in the size and distribution of human populations in coastal watersheds. That distribution is primarily urban. By urban, I mean simply having the characteristics of cities, including suburban areas, and not to specific political boundaries. Current demographic momentum is expected to continue at significant rates for several decades and beyond with profound implications for the domains of public policy and coastal management (Crossett et al. 2004; MEA 2005). It is the magnitude of these changes, the time lines involved, and the environmental challenges they portend that are generally underappreciated, even for those immersed daily in management and sustainability issues.

The purpose here is to review some well-known facts relative to tempo and mode of population increase and the environmental challenges this portends for coastal margins. The next decade or so will be a time of critical transition for humanity in a sustainability sense (Holdren 2008; Hall and Day 2009; O'Neill et al. 2010). This is a short time frame for our current domain of public policy, where time periods to pass legislation, settle litigation, develop regulations, and restore degraded ecosystems are measured in years. The objectives are to provide a framework for thinking strategically and realistically about a fast changing world, and stimulate thinking about practical and effective approaches in improving performance of our institutions of governance. The need is strategic, targeted approaches that take into account the qualities, characteristics, and dynamics of urban societies and the social/economic/environmental realities of urban dominated space as essentials to framing coastal management strategies global to local.

Sustainability in Application

Conceptually there are limits, many well articulated in the literature, to human consumption in a finite world of natural capital (e.g., Rees 2012). The reality, however, is that transitions to sustainable states and defining what sets of conditions are suitable for human well-being as sustainability goals are enormously complex, poorly understood, involve many dimensions, but clearly are dependent on the behavior of human societies (e.g., Sterman 2012). The first order of business then is to develop effective approaches to resource management based on sustainability concepts. Sustainability implies maintenance of environmental conditions consistent with human well-being. In a practical sense, there are no absolutes, just collective judgments made at a given time as to conditions suitable for society. It is this central precept of defined conditions that provides the conceptual basis and operational definition of sustainability that can now be applied in practical management contexts.

A second reality is that we are faced with rapidly changing conditions fueled by demographic momentum, development, and expanding economies. These linked systems are insufficiently understood. There are feedback loops, time delays, and multiple effects on environment. Consequently, to be effective, management approaches must be highly responsive, adaptive, and flexible in addressing issues and generating/applying new knowledge/technology. Operationally, the old axiom attributed to Congressman Jim Cooper of Tennessee applies. "If you can't measure it, you can't manage it, and if you don't measure it, you don't deserve to manage it." A second axiom is equally true. If the measurement or indicator has no context, you still cannot manage it. Operationally, then, measurements must be related to environmental conditions or state and that state deemed by current understanding and policy to be acceptable or unacceptable in a sustainability sense.

In concept, institutions of governance are responsible for defining sustainability goals and indicators of acceptable conditions for natural capital management based on current knowledge and public policy. Institutional performance can be judged on how well the defined/desired conditions are being maintained, restored, and adapted to new understanding. In practice, this is a highly complex, value laden and dynamic approach of many dimensions under rapidly changing conditions. Nonetheless, the approach provides a sound conceptual basis for developing effective management applications in a sustainability framework, as well as targets for implementation.

Tempo and Mode

Tempo. The term simply refers to the rate or pace of change in the size of human populations. Tempo can be applied in a spatial context, such as jurisdictional, coastal, or urbanized space. Population is the principal driver of the multitude of interconnected systems that constitute human drivers of environmental change and therefore is directly related to issues of sustainability. Population is a widely measured variable at many spatial scales relevant to management decision-making. A multitude of socioeconomic, environmental, and spatial variables can be expressed as a function of tempo, even though mechanisms are poorly understood. These are discussed in more detail in the sections that follow. The important point is that population is an essential indicator and proxy for a host of drivers of environmental change. Tempo provides a focal point for scientific inquiry, predictive capacity, and decision-making for sustainability frameworks. The impetus to socioeconomic and environmental variables from tempo is defined here as demographic momentum.

Concerning current tempo, the human global population will shortly reach seven billion; an astounding number given that world population in 1930 was estimated at two billion. There are people living today who have seen world population more than triple. Moreover, we continue to add 70+ million people per year, and high rates are predicted for the next several decades. Much of this growth is in developing countries where population momentum and sustainability issues will be particularly acute, yet capacity and governance problematic (Kunzig 2011).

For the United States, today's population is approaching 311 million and adding some 3.3 million per year. Over 52% of the population resides in coastal counties. Population in coastal counties increased by 46% from 1970 to 2010 and is expected to grow by another 7–8 million people by 2015—a rate of about 1.5 million people per annum. Such rates are expected to continue throughout the coming decades and beyond (Crossett et al. 2004; NOAA 2010). That is equivalent to adding the population of metro Philadelphia to our coastal margins every year.

Similar patterns are manifest at regional and local scales. Since 1985, the six-state Chesapeake Bay Watershed has added 4 million people to the 13 million already there. The Watershed is currently adding 157,000 people per annum, a rate expected to continue for another 2 decades (EPA 2007). The State of New Jersey, where 90% of the population lives in coastal counties, has the nation's highest population density. The state is expected to grow by 20–30% in the next 2 decades (NJDEP 2009; Hasse and Lathrop 2008). At yet smaller scales, New Jersey's Barnegat Bay Watershed has grown 170% in population since the Clean Water Act was passed in the early 1970s to about 570,000 people today. Current estimates predict adding another 150,000 people by 2020 (Moore and Bates 2010). Tempo along the world's coastal margins is a sobering reality. The environmental and socioeconomic consequences of that momentum must be a primary and urgent driver in crafting practical, effective approaches to coastal management at all relevant spatial scales.

Mode. As used here, mode refers to the manner in which human populations are distributed, in this case in urbanized environments. In the United States, the urban percentage of population is 80% and urban space dominates coastal watersheds. Urban space and environmental footprints are currently expanding in concert with tempo. Contemporary human society can be for all intents and purposes considered an urban civilization (Rees 2012). Consequently, any strategic approach to engagement of coastal sustainability issues needs to incorporate the social, economic, and environmental attributes of urbanized areas as a first principle in framing management strategies.

It is evident that human populations are undergoing one of the most profound and rapid changes in distribution in our evolutionary history. In just two generations, the number of humans living in urban environments is expected to increase by 60% to fully 80% of the global population. That is an additional four billion people living in cities (McGranahan and Satterthwaite 2003; McDonald 2008). In the United States, about 80% of the population now lives in urban environments, and the majority of the country's major urban areas lie in coastal and Great Lakes watersheds (Grimm et al. 2008a). The nation's top 20 coastal cities are on track to add 32 million people between 2000 and 2030 while expanding their urban footprints by 46% (McGrath 2000).

The process of urbanization involves population size and density, location, and spatial dimensions driven by a complex dynamic of multiple interrelated factors that are the subject of intense study (e.g., Bettencourt et al. 2007; Batty 2008; Gaines and Jager 2009). By way of example, of the world's 25 largest so-called mega cities (population now or expected by 2020 to exceed ten million inhabitants), only two are located in the United States, and 19 of the 25 are in coastal locations. About 80% of all cities with populations exceeding eight million inhabitants are coastal (Marshall 2005). Spatial extent presents a very different picture. Of the 25 largest cities in the world ranked by areal extent, 16 are under US jurisdiction, and all of these are located in coastal watersheds (Gaines and Jager 2009). This is a graphic illustration of the propensity of US cities to develop new land at high rates per capita compared to elsewhere. Such sprawl is particularly acute in coastal margins (Beach 2002).

Fig. 1 shows another important pattern of the urban mode in the United States, namely the emergence of highly urbanized corridors—the megalopolis (Grimm

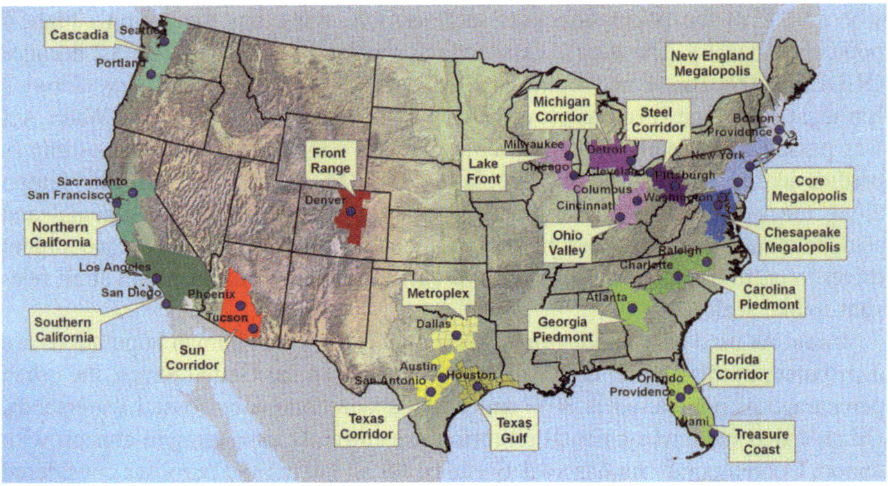

Fig. 1 Map of megapolitan areas overlaying on the topography of the continental USA. *Source:* after Grimm et al. (2008a)

et al. 2008a). This phenomenon is prevalent worldwide (Liu and Diamond 2005; Montgomery 2008). Note the number of megalopolis areas in Fig. 1 located in coastal or Great Lakes watersheds. The State of New Jersey (Fig. 2) is connected to two major metroplexes at either end of the state, and over half of the total land area is predicted to now be classified as developed (Hasse and Lathrop 2008). It could be argued that New Jersey is part of a single urban complex stretching from Boston to Washington, DC. This phenomenon of connectivity in the growth of urban development along coastal margins is critical in environmental management contexts. To reiterate, coastal management strategies must address the urban mode and put urbanization into a context of time, space, people, and dynamic process.

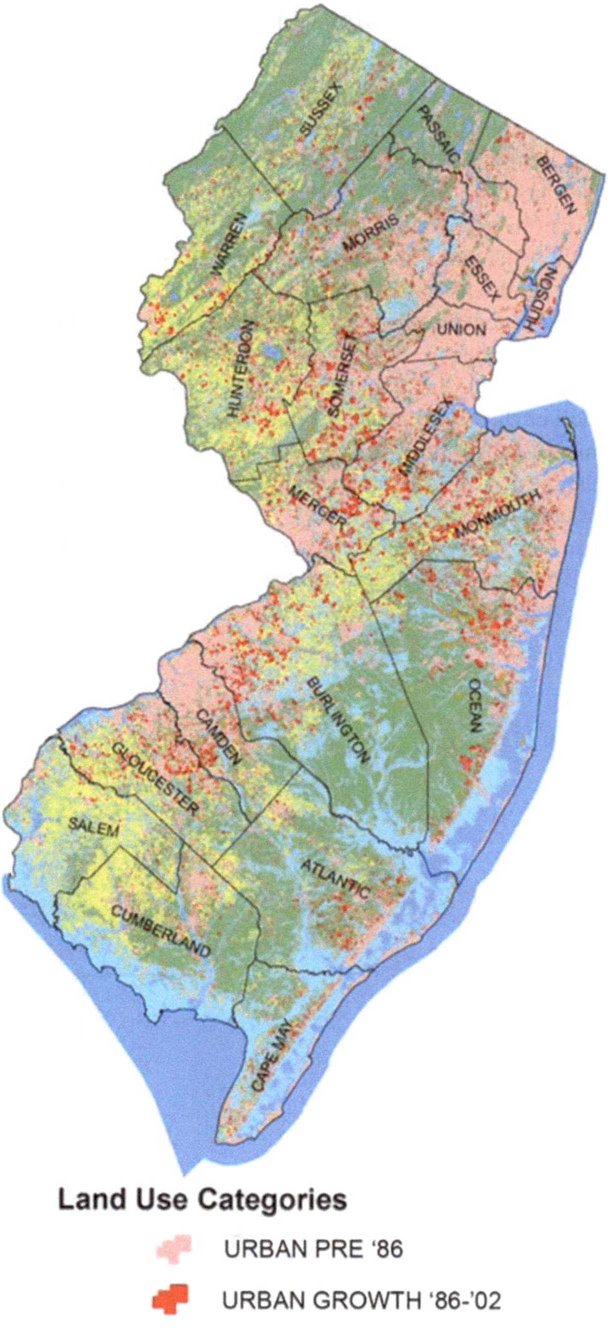

Fig. 2 Patterns of urban development depicted for the coastal state of New Jersey. Note presence of urbanized corridors and jurisdictional complexity. *Source:* after Hasse and Lathrop (2008)

Dynamic Systems and Coastal Margins

Domains of interaction. In practical terms, managing for sustainability is first and foremost a dynamic process that responds to constant change and continually reevaluates both sustainability goals and implementation strategies in response to new knowledge and changed conditions. Institutional performance is judged on how well sustainability goals are being met. Strategic frameworks and resulting actions for meeting such goals then involve interactions with highly interconnected, dynamic systems of urban space, human drivers of environmental condition, and dynamics of coastal ecosystems and natural resources. Spatial context can vary from global to small scales.

Fig. 3 is a conceptual diagram to illustrate the interactions of these domains. The domains will be discussed in more detail in later sections. Coastal management lies in the domain of institutions of governance, that collective network of human institutions that set policy, create new knowledge, provide resources for building capacity, and fund management agencies. The domain also involves educational, legal, legislative, and enforcement elements and relationships with the private sector (Baird 2005b). This institutional network has the responsibility for measuring indicators, creating strategic frameworks, and implementing actions to achieve sustainability goals.

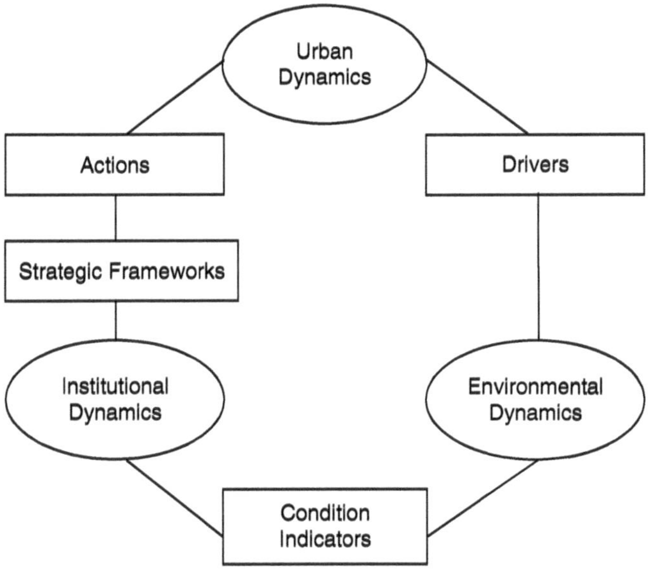

Fig. 3 Conceptual diagram of interactions of principal domains of coastal resource management. The system is highly dynamic, interconnected, and illustrates the critical roles of measurement, strategic frameworks, and management actions

Urban dynamics. The challenge, from a coastal management perspective, is in restoring and maintaining acceptable environmental conditions as defined by institutions of governance in the face of the momentum of urbanization. With over one million km^2 of urban area expected to be added globally in the next two decades, understanding the relationships of cities to multiple environmental stressors is paramount (McDonald 2008). The cumulative impacts of coastal urbanization on ecosystems are highly complex and occur at many spatial scales (Bettencourt et al. 2007; Grimm et al. 2008b; McDonald 2008). Knowledge about the relationships of urban growth to consumption of and impacts on natural resources must be rapidly assimilated and translated into management actions (Baird 2009). Many diverse properties of cities (e.g., wealth, pollution, and infrastructure) have been shown to scale with population size and density and urban form, the overall spatial pattern of development. These properties in turn affect resource use and ecosystem services.

While the mechanisms behind these relationships are incompletely understood, many can be tied to spatial metrics and associated census data. For instance, auto use and energy consumption are nonlinear functions of population size and density. Such relationships with population can be characterized at scales from regional to local and can be very useful as key indicators of socioeconomic and environmental impacts for planning, prediction, and management decision-making. Greater knowledge and understanding of socioeconomic and environmental parameters critical to quality of life at spatial scales appropriate to decision-making is key to developing new and more comprehensive approaches to integrated coastal zone management. The following are examples of known relationships useful in better understanding how urban processes relate to coastal resource management.

1. The ecological footprint of coastal cities far exceeds their actual dimensions (e.g., air quality, food, transportation, water, pollution).
2. Urban form affects resource use in that past patterns tend to persist because of infrastructure (e.g., roads) that creates a physical imprint that persists for years.
3. Variables that affect coastal ecosystems that exceed per capita urban population growth rates include vehicle miles, land development, impervious surfaces, urban runoff, consumption (food, water, energy), production (trash, fertilizers, wastewater, pollutants).
4. Important variables that lag per capita growth are infrastructure efficiency (roads, wastewater treatment, and energy distribution) and management response to ecosystem problems.
5. Contemporary urban complexes (e.g., New Jersey, Shanghai) now exist as part of larger scale multijurisdictional, regional urban corridors that exhibit high degrees of connectivity and commonality in dynamics, form, economics, and ecological impacts.

Institutional performance and coastal margins. We have made progress in the policy domain both globally and domestically particularly with the development of ecosystem based and spatial approaches to management (e.g., Leslie and McLeod 2007; Sivas and Caldwell 2008) and the advent of valuing ecosystem services and natural resource values (Freeman 2003; Tallis et al. 2008). This includes sustainability science, legislation, international policy forums, councils, and NGO/private sector engagement, plus new technology, stakeholder involvement, and conflict resolution. Likewise, the human dimensions, such as core values, political will, cultural complexity, management structure and dynamics, information flow, and human resource development, are areas of active study. Indeed, institutions of governance can be thought of as systems and networks in their own right. Recent studies have examined the multitude of issues and problems facing mankind in addressing sustainability goals, including knowledge bases required (Parris and Kates 2003; Holdren 2008; Levin and Clark 2010).

The central problem for coastal margins is how to make sustainability goals operational, and then translate these into practical management applications and performance metrics for current institutions of governance. We continue to experience ecosystem degradation, deplete resources, and fail to meet goals for coastal margins, strong indicators of inadequacy of current institutional performance. The reasons are many and varied and are the subject of other chapters in this volume (Layzer 2012; Sterman 2012). The following few examples illustrate the complex nature, context, and scale of the problem:

1. Recent analysis indicates that no marine ecosystem is unaffected by human activities. The intensity varies spatially, and 41% of systems examined globally are strongly affected by human-related drivers (Halpern et al. 2008).
2. For estuaries and coastal seas, over 65% of sea grass and wetland habitat has been destroyed worldwide, and many formally abundant species have been severely depleted (Lotze et al. 2006).
3. Dead zones from nutrient runoff in coastal seas are spreading in extent globally (Diaz and Rosenberg 2008).
4. For the metropolitan area of greater Jakarta, population 20 million and growing, only about 3% of sewage is treated, while traffic is responsible for 70% of nitrous oxides and particulates in the air (Marshall 2005).
5. In the United States, the size of anoxic zones has doubled in 22 years and includes Gulf of Mexico, Chesapeake Bay, and Lake Erie; 1/3 of saltwater fish tested had at least one chemical contaminant above benchmarks for human health; greater than 60% of coastal rivers are degraded by nutrient runoff and half of assessed estuaries are classified as impaired (Heinz 2008; EPA 2009).
6. At smaller scales yet, water quality goals for Chesapeake Bay have not been met in spite of 26 years of regulation and many millions of dollars spent on research, while in Barnegat Bay, dissolved oxygen levels are below federal standards and jellyfish now plague the upper Bay (EPA 2007; Moore and Bates 2010).

Thus, major indicators of environmental stress along coastal margins abound, and the situation is much more serious in developing countries. The reality is performance of current institutions of governance in the United States and globally has been inadequate in maintaining or restoring coastal habitats to acceptable states as currently defined. The question is what can we do about it in a complex world driven by tempo and mode? There are no pat answers, yet a sustainable environment is a necessity. In this country alone, performance enhancement strategies will have to address adding one to two million additional people per year in coastal watersheds for at least several decades.

Progress will have to come primarily from reducing impediments to the collective performance of current institutions. That means a focus on practical, workable solutions, best practices, and the cumulative contributions of countless individual steps.

On Building Strategic Frameworks

Confronting reality. Good performance rests on three essential steps—confronting reality, understanding the big picture, and thinking about the practical details of what to do about it (Bossidy and Charan 2004). The process starts with an understanding of current conditions, future trends, and systems characteristics. The following are examples of contemporary reality institutions of governance must confront:

1. Increases in population, urbanization, and intensity of human activity in coastal margins are inevitable over the next 2 decades and the environmental consequences must be dealt with in a management context.
2. Human institutions of governance, political systems, and cultural proclivities have not evolved to operate under such time horizons. Today there is considerable time lag between onset of ecological degradation and management response; it takes time to pass a law, build infrastructure, settle litigation, and improve technology (Baird 2009).
3. The current multijurisdictional, compartmentalized and hierarchical management structure will not radically change in the next decade. Today's management framework involves a broad array of agencies, jurisdictions, and policies. By way of example, in coastal New Jersey over 20 federal agencies and 140+ laws affect coastal management. The State's Department of Environmental Protection lists 66 separate programs and units, while some 245 coastal municipalities have authority over land use and zoning (NJDEP 2009; Kennedy 2009).
4. Human drivers of ecosystem stress involve context (e.g., socioeconomic, geographic, and cultural), space, time, and dynamics. The spatial dimensions and rates of human activity in urbanizing landscapes are increasing while spatial dimensions for coastal resources that provide services essential to human well-being are finite. From the perspective of sustainability, maintaining and/or restoring coastal ecosystems to acceptable states in the future must involve reduction of the per capita contribution to ecosystem stress and/or rates of resource consumption in the face of tempo and mode.
5. It is political will that determines in large measure society's response to environmental issues. The political system provides resources to the complex bureaucracies that manage coastal resources; sets policy through legislation, litigation, and regulation; and provides enforcement and supports the complex of institutions

and corporations that provide the backbone of science and technology essential to support sustainability transitions.

6. Economic systems and economic strength determine in large measure society's capacity and performance capability in addressing environmental challenges at scales global to local. GDP, costs and incentive structures, tax rates and land use policies, risks and insurance are important determinants of cause, effect, and solution potential for environmental problems. Pressures from tempo will require economic growth rates of 3% or greater in the United States to avoid high unemployment rates, now a worldwide problem (Elliot 2010). Signs of financial exhaustion in public budgets are evident. Understanding of socioeconomic variables and systems is essential to resolving sustainability issues (Pendleton 2008; Julia and Duchin 2007; Lumenello 2008).

Thinking strategically. This is an approach to problem solving—not the formal process of strategic planning. It may be viewed as the practice of devising and employing plans and actions for setting and achieving performance goals once these are understood and defined. It involves pragmatism, defining problems and feasible solutions, then deciding on a course of action to address identified issues. That means making operational such global concepts as sustainability, ecosystem state, and adaptive management, then applying them to specific contexts and spatial scales (Gaines and Jager 2009). This process involves utilizing a broad array of approaches, engaging many constituencies, blending scientific understanding with the realities and limits of current institutions in order to enhance management performance. In practice, good strategic thinking is truly an art form utilizing science, knowledge of human behavior, leadership, and experience in devising and implementing effective actions (Baird 2005a).

Given the realities of slow evolutionary change rates in the structure of contemporary institutions of governance the primary strategic focus need be on changing the way they are managed and their degree of connectivity. This involves commonality of goals, objectives, and core values. It also involves engagement of the political systems that provide resources, set policy/regulate, and resolve conflicts. In the United States, the major sources of financial and human resources, as well as dictates of policy, reside at the federal level. The inability of existing federal institutions to adequately address pressing coastal issues and the difficulty of bureaucratic structures to change behavior in the face of rapidly changing conditions must be a prime target for strategic focus (Rassam 2006).

Strategically, performance enhancement starts with a common understanding of the primary goals of coastal management—that is, restoration and maintenance of acceptable ecosystem states as defined by institutions of governance. The second is accepting that current performance has been inadequate in a dynamic and fast changing world where rapid response is an imperative. Once this is understood, then management at every level in the policy domain need determine the primary impediments to performance in their sphere of influence. The next step is crafting practical, workable, and actionable performance enhancement frameworks.

Foundations for frameworks. Strategic thinking must lead to effective action that is based on smart, creative thinking, reliable information, defining objectives, taking responsibility, and evaluating performance. The goal is effective approaches and a culture of implementation. Action agendas are by necessity specific in context (e.g., spatially). Eight broadly applicable conceptual guidelines for creating performance enhancement frameworks are summarized below:

1. There has been little fundamental change in the federal agency structure and related legislative oversight concerned with coastal management in over two decades. There have been calls for consolidation and new structure for environmental agencies, but little of substance has been done (Schaefer et al. 2008). There is now an interagency National Ocean Council to enhance coordination and articulate national priorities. For instance, new mandates have been developed for research and marine spatial planning (Obama 2010). The issue remains that a more systematic approach to implementation of policy among and within management agencies at spatial dimensions of application, including better coordination and pooling of effort, promises significant performance enhancement. That means more efficient, coordinated, and decentralized organizations must be coupled with concepts of management by objective (Drucker 2008). Management by objective means setting specific measurable environmental objectives at multiple spatial contexts and measuring progress on the timeliness and effectiveness of actions taken in addressing those specific goals.
2. Performance by necessity rests on timely, evidence-based assessments, sound indicators, and feedback whereby critical information (data with a purpose) is rapidly assimilated and broadly distributed. Given the complexity of problems and need for rapid response, there is no substitute for knowledge of environmental condition, trends, and their relationship to human populations, urban/land/ocean spatial use characteristics, and sustainability goals. Resources for monitoring and assessment are chronically underfunded and underutilized in a world of scarce resources. Concentration on this sector, including technology, use of heuristics, proxies (such as population-related parameters), logic-based predictive models, and data management protocols, promises significant performance advantages. Measurements, monitoring, and assessment are the necessary foundations of management capacity and performance.
3. We need to better understand contemporary driving forces that influence political will. The political process impacts decision-making, be it legislation, litigation, conflict resolution, sustainability goals, or capacity to perform through resource distribution. It is the array of laws and regulations especially at the national and state level that provide the legal framework for the application of management interventions. Cities are centers of political power. Engage the political process in constructive ways, and advance the art of outreach and information transfer to political leadership and key stakeholders in creating understanding for policy development. Current events can be highly emotive in today's world in the sense of arousing intense feelings among segments of a population. These episodes of emotional contagion (rapid spreading of emotion in populace) can rapidly change political will, leading to episodes of new legislation and resource availability for the environment. Emotional contagion can provide great opportunities to advance performance agendas, build partnerships, and shift paradigms (Gladwell 2002).

4. Tempo and mode have vastly increased performance challenges arising from the problems of home rule and local/regional management complexities. Not only is there institutional and jurisdictional complexity, but urbanized regions also exhibit major cultural and associated value/political differences over small spatial scales (Schlesinger 1991; Gaines and Jager 2009). Management agencies concerned with coastal management, especially at federal and state levels, often have little direct jurisdictional authority over development, patterns of growth, infrastructure, enforcement (a significant, highly relevant and underappreciated performance drag), and service provision in urban municipalities, yet such patterns have major economic and environmental consequences. The issues of Chesapeake and Barnegat Bays are cases in point, including the common problem of shifting baselines for management objectives (Duarte et al. 2009). Jurisdictional alignment around common sustainability goals is essential for performance enhancement agendas. Increase engagement among jurisdictional authorities and promote innovations that increase coordination, joint planning, information exchange, and improved decision support.
5. Cities and populations need be related to effects on associated ecosystems and to the landscape (Angermeier et al. 2004; Weinstein 2008). Cities are highly dynamic, complex, and difficult to manage. Cities operate through major cycles such as waste, water, energy, and transportation. These can be related to infrastructure efficiency and environmental stressors in time/space and quantified. Urbanization occurs over long time horizons, and once constructed, is difficult and expensive to change (Fig. 4). Relate where possible urban processes

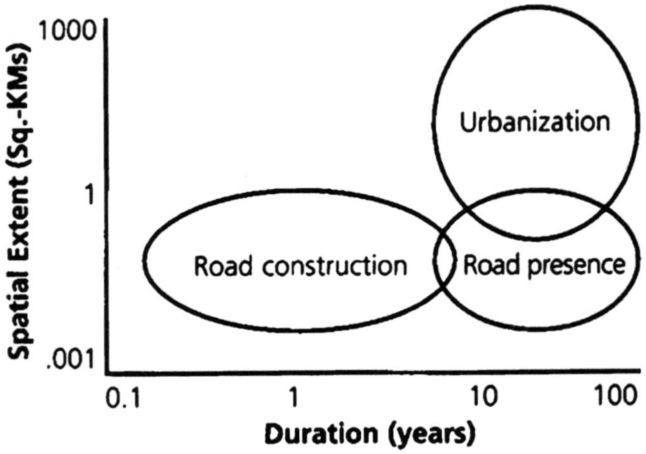

Fig. 4 Temporal and spatial extent of environmental impacts due to three major phases of urban growth: road construction, road presence, and built environment. Note axes are on a logarithmic scale. *Source:* after Angermeier et al. (2004)

to population (size, density, per capita). Understand the sustainability consequences of urban processes and concentrate management effort on regulating new development and land uses. Determine how development can most benignly grow and the eco-services trade-offs with increasing development.

6. Understand that performance is dependent on good science and policy, especially nationally, where most of the resources for research agendas are managed and national policy determined (e.g., Bloch 1996; Levin and Clark 2010). Know what managers need in various management contexts to enhance performance (Rosenberg and Sandifer 2009). Policy must be supported by resource concentration on objectives. Concerning national policy, the need is more focus on sustainability-related research. That means less fragmentation and more coordination (share and pool resources) and focus of research on common problems. More emphasis need be placed on research with a purpose, less on the basic vs. applied dichotomy. Assessments to identify research priorities for spatially specific areas such as watersheds, urban space, and regional ecosystems help clarify science objectives (Lester et al. 2010). Reduce time from knowledge creation to its use in application. Focus on the interface between research and practitioners (Palmer 2009). Add practitioners and stakeholders as part of the research review and policy process (Byron et al. 2011). Promote partnerships among agencies, universities, and industry. Interdisciplinary research approaches need to better engage the many practitioner elements of modern development (e.g., urban planners, landscape ecologists, coastal managers, and infrastructure-related expertise). Increase investment in spatially (regional, local) and process specific research.
7. Economic factors are major drivers of human activity and therefore stressors to coastal environments. Economies, particularly the public economy, impact options for action agendas. For instance, only 1/3 of nutrient loads to Chesapeake Bay are derived from developed lands, yet require 2/3s of remediation costs (EPA 2007). The replacement rate of US public infrastructure is lagging rates of obsolescence (Petroski 2009). Valuations, tax structure, zoning, and regulatory interventions relate to performance. These interrelationships need to be better understood and incorporated into performance frameworks. Economic understanding is central to good strategic frameworks. The tension between continued economic growth and development and achieving/maintaining sustainability targets will intensify with demographic momentum.
8. Finally, there are operational imperatives essential to strategic frameworks. Emphasize pragmatism, problem definition, and feasible solutions, realizing that planning and implementation is a continuous process. Stress the importance of innovations that reduce response times to emerging problems. Focus on anticipation, scenario creation, and predictive models for decision support. Get critical information in understandable formats to decision makers when needed. Base performance on evidence and indicators, and promote practices known to be effective and why. Improve communication and information transfer across jurisdictions and stakeholders. Performance is all about context—spatial scale, time, and place. Local scales are particularly critical to reducing environmental stress and resolving

conflicts (e.g., Layzer 2008). Create a mind-set of urgency, rapid assimilation, innovation, and use of new technologies and methods. Above all, continually prioritize actions, resources, and policy on the most important problems.

Summary and Conclusions

Exploitation of land and water resources is already greater than most of us realize, and demand fueled by tempo and urbanization promises much greater pressure on coastal ecosystem functions and services. We know urban form lasts for years, and maintaining acceptable environments rests on getting it right the first time with new development—a daunting and expensive challenge. Humans have a strong sense of place and culture. Cities have unique spatial characteristics. Conflict resolution and management success will hinge on relating problems to spatially specific contexts, especially at the local level. While sustainability is essential to modern urban society, socioeconomic issues in the United States generally outweigh environmental issues for the urban voting public. A poll taken in Washington DC found environment ranked eighth among concerns of city residents (Harper 2004). That is cause for concern and a target for our attention.

As has been stressed throughout, management actions must be applied in specific spatial contexts. Contemporary approaches for coastal management such as ecosystem-based management are inherently spatial (Layzer 2012). Coastal management is all about spatial context. Promising new developments now entering the lexicon of coastal management involve coastal marine spatial planning (CMSP), where multiple complex databases can be expressed spatially. Humans can rapidly assimilate spatial information, greatly improving our ability to get information to decision makers and resolve jurisdictional alignment issues (Crowder et al. 2006; Dennison et al. 2007). CMSP has now become a national priority and agencies like the US National Oceanic and Atmospheric Administration (NOAA) are adopting spatial protocols. We are in our infancy with regard to developing a spatial language for sustainability and management of coastal watersheds. There is emerging a world of possibilities for spatially explicit research agendas, trend analysis, decision support tools, and predictive modeling—all with great promise for performance enhancement.

Sustainability thresholds are no longer theoretical constructs, but today's reality. We are closer on a per capita basis than ever before in our evolutionary history to such thresholds and now have the challenge of shaping landscapes and ecosystems for human welfare (Kareiva et al. 2007). Issues of carrying capacity and world poverty abound (Butler 2004). Fig. 5 is a graphic illustration of tipping points, sustainability thresholds, and the socioeconomic consequences of remediation. From 1900 to about 1975 water levels dropped about 10 m in the Dead Sea. At that point the usage slope increased and in only 15 years water levels dropped another 10 m, indicative of a tipping point being crossed. By 1990 it was clear that a sustainability threshold had been crossed, yet growth-based usage continued and from 2000 to the present, well over ten additional meters were lost. Addressing the problem now

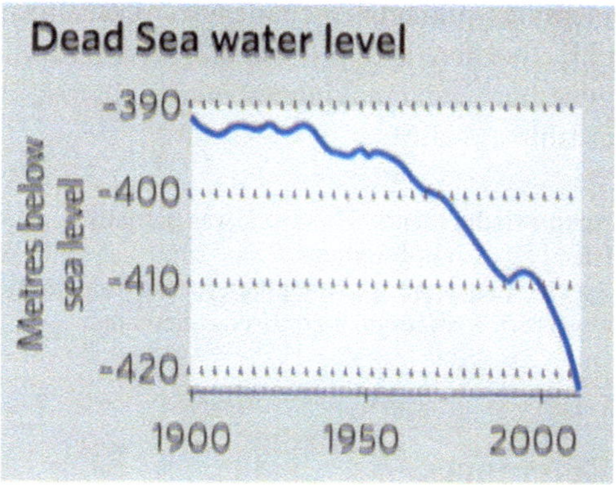

Fig. 5 Depiction of "tipping point" phenomenon for water levels in the Dead Sea. Note rapid decline after 1980. New development has had increasingly larger impacts on water levels. *Source:* reprinted by permission from Macmillan Publishers Ltd: Nature Glausiusz (2010), copyright 2010

involves international dialog, changes in usage patterns, severe socioeconomic consequences, and large remediation costs with uncertain outcomes (Glausiusz 2010). The real dilemma of tipping points is that small increments in growth at one spatial scale can lead to major costs for a far greater area and population in order to maintain/restore acceptable ecosystem states under current policies. Many other examples can be cited (e.g., Oczkowski and Nixon 2008; Sterman 2012). Note the relationships of population, consumption, socioeconomics, slow response, and environmental consequences.

The good news is that marine ecosystems and fisheries have been surprisingly resilient to many anthropogenic impacts, and evidence is mounting that mitigation/regulation can lead to varying degrees of recovery. Land, water, and other finite resources are more problematic. The next decade will present many economic, legal, and policy dilemmas and reevaluation of trade-offs in decision-making. The reality is one of urgency in dealing with the complexity, knowledge gaps, and process understanding and their relationships to population and urbanization. There is no substitute for leadership and effective decision-making at all levels in a complex world of tempo and mode. President G.W. Bush, in discussing decision-making, makes the following points of fundamental relevance to sustainable coastal margins (Bush 2010). Decisions must be made based on best available knowledge that are practical, purposeful, and about getting results. One must learn from decisions what was effective or not, and why. Most importantly, in a world of uncertainties, many decisions are difficult, knowledge is incomplete, and outcomes uncertain. Strong arguments can be made for alternatives, yet inaction is not an option. Decide, then set clear objectives, evaluate outcomes, and adjust.

Much has been written about our coastal management shortcomings, global to local. We simply must take action and responsibility to reshape thinking and current approaches of our institutions of governance at all levels. The old adage "The future is now" is an apt description. That is not just a higher order problem. It must be addressed in the trenches of every day politics, policy, and practical application with spatial specificity and a mind-set of management by objective. Faced with the seemingly intractable challenges of urbanization and growth in coastal margins, the objective here is to provide a blueprint for more proactive, integrated, and adaptive governance and making a spatial and jurisdictionally nested system of coastal stewardship a reality.

Acknowledgements This work was partially supported by the Universities of Hawaii and Rhode Island Sea Grant Programs. Thanks to Michael P. Weinstein for his support and encouragement. Special thanks to Lisa Humphrey for her numerous contributions to this effort and to two anonymous reviewers for their constructive comments.

References

Angermeier PL, Wheeler AP, Rosenberger AE (2004) A conceptual framework for assessing impacts of roads on aquatic biota. Fisheries 29:19–29

NOAA (National Oceanic and Atmospheric Administration) (2010) State of the coast report. http://stateofthecoast.noaa.gov/. Accessed 6 July 2010

Baird RC (2005a) On sustainability, estuaries, and ecosystem restoration: the art of the practical. Restor Ecol 13:154–158

Baird RC (2005b) The human dimension in ecosystem management: institutional performance and the Sea Grant paradigm. In: Hennessey T, Sutinen J (eds) Sustaining large marine ecosystems: the human dimension. Elsevier, New York, NY

Baird RC (2009) Coastal urbanization: the challenge of management lag. Manag Environ Qual Int J 20:371–382

Batty M (2008) The size, scale, and shape of cities. Science 319:769–771

Beach D (2002) Coastal sprawl: the effects of urban design on aquatic ecosystems in the United States. Pew Oceans Commission, Arlington, VA

Bettencourt LMA, Lobo J, Helbing D et al (2007) Growth, innovation, scaling, and the pace of life in cities. PNAS 104:7301–7306

Bloch E (1996) Cooperation, competition, and science policy. Science 273:1779

Bossidy L, Charan R (2004) Confronting reality: doing what matters to get things right. Crown Business, New York, NY

Bush GW (2010) Decision points. Random House, New York, NY

Butler CD (2004) Human carrying capacity and human health. PLoS Med 1:192–194

Byron C, Bengtson D, Costa-Pierce B et al (2011) Integrating science into management: ecological carrying capacity of bivalve shellfish aquaculture. Mar Pol 35:363–370

Clark WC, Dickson NM (2003) Sustainability science: the emerging research program. PNAS 100:8059–8061

Crossett KM, Culliton TJ et al (2004) Population trends along the coastal United States: 1980–2008. National Oceanic and Atmospheric Administration, National Ocean Service, Washington, DC

Crowder LB, Osherenko G, Young OR et al (2006) Resolving mismatches in US ocean governance. Science 313:617–618

Dennison WC, Lookingbill TR, Carruthers TJB et al (2007) An eye-opening approach to developing and communicating integrated environmental assessments. Front Ecol Environ 5:307–314

Diaz RJ, Rosenberg R (2008) Spreading dead zones and consequences for marine ecosystems. Science 321:926–929

Drucker PF (2008) The essential Drucker. HarperCollins, New York, NY

Duarte CM, Conley DJ, Carstensen J et al (2009) Return to Neverland: shifting baselines affect eutrophication. Estuaries Coasts 32:29–36

Elliot M (2010) The US isn't alone: high unemployment rates are a new global reality. Fortune 162:36

EPA (Environmental Protection Agency) (2007) Development growth outpacing progress in watershed effort to restore the Chesapeake Bay. EPA Report 2007-P00031. Washington, DC
EPA (Environmental Protection Agency) (2009) Bay barometer: a health and restoration assessment of the Chesapeake Bay and watershed in 2008. EPA-903-R-09-001. Washington, DC
Freeman AM (2003) The measurement of environmental and resource values, 2nd edn. Resources for the Future Press, Washington, DC
Gaines J, Jager S (2009) Albert Speer & partners: a manifesto for sustainable cities: think local, act global. Prestel, London, UK
Gladwell M (2002) The tipping point: how little things can make a big difference. Little Brown and Company, New York, NY
Glausiusz J (2010) Environmental science: new life for the Dead Sea? Nature 464:1118–1120
Grimm NB, Foster D, Groffman P et al (2008a) The changing landscape: ecosystem responses to urbanization and pollution across climatic and societal gradients. Front Ecol Environ 6:264–272
Grimm NB, Faeth SH, Golubiewski NE et al (2008b) Global change and the ecology of cities. Science 319:756–760
Hall CAS, Day JW Jr (2009) Revisiting the limits to growth after peak oil. Am Sci 97:230–237
Halpern BS, Walbridge S, Selkoe KA et al (2008) A global map of human impact on marine ecosystems. Science 319:948–952
Harper J (2004) Environment lags in poll of concerns; 35% Americans worry about it. Washington Times, 20 April
Hasse J, Lathrop RG (2008) Tracking New Jersey's dynamic landscape: urban growth and open space loss 1986-1995-2002. Rowan University and Center for Remote Sensing & Spatial Analysis, Rutgers University. http://crssa.rutgers.edu/projects/lc/download/urbangrowth86_95_02/HasseLathrop_njluc_executive_summary_7_14_08.pdf. Accessed 28 Jan 2011
Heinz (H. John Heinz Third Center for Science, Economics, and the Environment) (2008) The state of the Nation's ecosystems 2008: measuring the lands, waters and living resources of the United States. The Heinz Center, Washington, DC
Holdren JP (2008) Science and technology for sustainable well being. Science 319:424–434
Julia R, Duchin F (2007) World trade as an adjustment mechanism of agriculture to climate. Clim Chang 82:393–409
Kareiva P, Watts S, McDonald R et al (2007) Domesticated nature: shaping landscapes and ecosystems for human welfare. Science 316:1866–1869
Kates RW, Clark WC, Corell R et al (2001) Sustainability science. Science 292:641–642
Kennedy D (2002) Sustainability problems, science and solutions. In: Kennedy D (ed) Recommendations for achieving sustainable communities, science and solutions. National Council for Science and the Environment, Washington, DC
Kennedy SM (2009) Coastal resource protection: a review of select New Jersey regulatory & planning tools. White paper for the Monmouth University Urban Coast Institute. Monmouth, NJ http://www.monmouth.edu/urban_coast_institute/articles/SusanKennedyCoastal.ProtectionWhitePaper.pdf. Accessed 28 Jan 2011
Kunzig R (2011) Population 7 billion: find out why you shouldn't panic-at least not yet. Natl Geogr 219:32–69
Layzer J (2008) Natural experiments: ecosystem-based management and the environment. The MIT Press, Cambridge, MA
Layzer J (2012) The purpose and politics of ecosystem based management. In: Weinstein MP (ed) Sustainability science. Springer, New York
Leslie HM, McLeod KL (2007) Confronting the challenges of implementing marine ecosystem-based management. Front Ecol Environ 5:540–548
Lester SE, McLeod KL, Tallis H et al (2010) Science in support of ecosystem-based management for the US West Coast and beyond. Biol Conserv 143:576–587
Levin SA, Clark WC (eds) (2010) Toward a science of sustainability report from toward a science of sustainability conference. Princeton University, Princeton, NJ
Liu J, Diamond J (2005) China's environment in a globalizing world. Nature 435:1179–1186

Lotze HK, Lenihan HS, Bourque BJ et al (2006) Depletion, degradation, and recovery potential of estuaries and coastal seas. Science 312:1806–1809
Lumenello S (2008) Keynote address—Jeffrey Sachs: "common wealth: economics for a crowded planet". Alumni Quarterly Colloquy. Harvard University, Cambridge, pp 2–4
Marshall J (2005) Megacity, mega mess. Nature 437:312–314
McDonald RI (2008) Global urbanization: can ecologists identify a sustainable way forward? Front Ecol Environ 6:99–104
McGranahan G, Satterthwaite D (2003) Urban centers: an assessment of sustainability. Annu Rev Environ Resour 28:243–274
McGrath D (2000) Predicting urban sprawl in top 20 U.S. coastal cities. Illinois-Indiana Sea Grant, The Helm Fall, pp 1–6
MEA (Millennium Ecosystem Assessment) (2005) Ecosystems and human well being: current state and trends. Island Press, Washington, DC
Montgomery MR (2008) The urban transformation of the developing world. Science 319: 761–764
Moore K, Bates TB (2010) Barnegat Bay under stress. The Sunday Press, Asbury Park, NJ
National Research Council (NRC) (1999) Our common journey, a transition toward sustainability. National Academy Press, Washington, DC
NJDEP (New Jersey Department of Environmental Protection) (2009) Coastal management program. http://www.nj.gov/dep/cmp/. Accessed May 2009
O'Neill DW, Dietz R, Jones N (2010) Enough is enough: ideas for a sustainable economy in a world of finite resources. The report of the Steady State Economy Conference, Leeds, UK. Center for the Advancement of the Steady State Economy (Arlington, VA, USA) and Economic Justice for All (Leeds, UK)
Obama B (2010) National Ocean Policy. Executive order, July 19, 2010. http://www.whitehouse.gov/the-press-office/executive-order-stewardship-ocean-our-coasts-and-great-lakes. Accessed 27 Jan 2011
Oczkowski A, Nixon S (2008) Increasing nutrient concentrations and fall of a coastal fishery; a review of data from the Nile Delta, Egypt. Estuar Coast Shelf Sci 77:309–319
Orbach MK (2002) Beyond freedom of the seas: ocean policy for the third millennium. National Academy of Sciences, Washington, DC
Palmer MA (2009) Reforming watershed restoration: science in need of application and applications in need of science. Estuaries Coasts 32:1–17
Parris TM, Kates RW (2003) Characterizing a sustainability transition: goals, targets, trends and driving forces. Proc Natl Acad Sci USA 100:8068–8073
Pendleton LH (2008) The economic and market value of coasts and estuaries: what's at stake? Coastal Ocean Values Press, North Sandwich, NH
Petroski H (2009) Infrastructure. Am Sci 97:370–374
PEW Ocean Commission Report (2003) America's living oceans: charting a course for sea change. A report to the nation. Pew Oceans Commission, Arlington, VA
Rassam G (2006) The ocean action plan: what is it? Where is it going? Fisheries 31:556
Rees WF (2012) Cities as dissipative structures: global change and the vulnerability of urban civilization. In: Weinstein MP (ed) Sustainability science. Springer, New York
Roe E, van Eeten M (2001) Threshold-based resource management: a framework for comprehensive ecosystem management. Environ Manage 27:195–214
Rosenberg AA, Sandifer PA (2009) What do managers need? In: McLeod KL, Leslie HM (eds) Ecosystem-based management for the oceans. Island Press, Washington, DC
Schaefer M, Baker DJ, Gibbons JH et al (2008) An earth systems science agency. Science 321:44–45
Schlesinger AM Jr (1991) The disuniting of America. Whittle Direct Books, Knoxville
Sivas DA, Caldwell MR (2008) A new vision for California ocean governance: comprehensive ecosystem-based marine zoning. Stanford Environ Law J 27:209–279

Sterman J (2012) Sustaining sustainability: creating a systems science in a fragmented academy and polarized world. In: Weinstein MP (ed) Sustainability science. Springer, New York

Tallis H, Kareiva P, Marvier M, Chang A (2008) An ecosystem services framework to support both practical conservation and economic development. PNAS 105:9457–9464

USCOP (U.S. Commission on Ocean Policy) (2004) An ocean blueprint for the 21st century. Final report, Washington, DC

Weinstein MP (2008) Ecological restoration and coastal management: placing people in the landscape. J Appl Ecol 45:296–304

Weinstein MP, Reed DJ (2005) Sustainable coastal development: the dual mandate and a recommendation for "commerce managed areas". Restor Ecol 13:174–182

Fishery and Forest Transitions to Sustainability: A Comparative Analysis

Bonnie J. McCay and Thomas K. Rudel

Abstract What are the causes of and sources of resistance to transitions from depleting, damaging trends to conserving and restoring trends in the use and management of natural resources? This is a central question in sustainability science, which we address by discussing "forest transition theory," one well-established area of analysis, and proposing "fisheries transition theory" for another. The general question is whether such transitions take place, their timing, and evident causes. Forest transition theory developed around the questions of how factors such as industrialization and urbanization affect forest cover and what situations encourage turnarounds in forest cover, from deforestation to forestation. We point to similarities and differences in the factors that appear to be involved in the recovery of depleted fish stocks as a first step toward a comparable theory concerning fisheries transitions to sustainability.

Keywords Forest transition • Fishery transition • Fisheries sustainability

Introduction

With concern mounting over an array of global environmental problems, "sustainability science" emerged in the 1990s to focus on human processes that repair, restore, or stabilize imperiled environmental resources like fish, forests, and the global climate (Kates et al. 2001). Scientists began to talk about "transitions to sustainability" (Speth 1992; National Research Council 1999; Lebel 2005; Fischer-Kowalski and Rotmans 2009; Rock et al. 2009), whereby humans put their

B.J. McCay (✉) • T.K. Rudel
Department of Human Ecology, Rutgers University,
55 Dudley Road, New Brunswick, NJ, 08901, USA
e-mail: Mccay@aesop.rutgers.edu; rudel@aesop.rutgers.edu

M.P. Weinstein and R.E. Turner (eds.), *Sustainability Science: The Emerging Paradigm and the Urban Environment*, DOI 10.1007/978-1-4614-3188-6_11,

exploitation of natural resources on a more sustainable footing by transforming how they use these resources. In this context, the search for driving forces behind shifts toward sustainability became a high priority for investigators.

To date considerable attention has been directed toward "forest transitions," documenting the restoration of degraded forests and the emergence of sustainable forest exploitation practices, and analyzing forces that drive these transitions. Most analyses identify changes in human incentives as the most important force driving these transitions (Baird 2012). These assertions no doubt reflect empirical realities, at least in the case of the forest transition, but to some extent these analytic tendencies may reflect the systems approach or social scientific training of most of the people who have studied these transitions as well as technical capabilities to measure changes. Differences between the sustainability transition that characterizes one natural resource, forests for example, and sustainability transitions that characterize other natural resources like fisheries have not received much attention. In this context it seems useful to examine less frequently analyzed fisheries transitions, a matter that is timely given mounting evidence that at least some major marine fisheries have been restored to viable levels of biomass and sustainable levels of harvest (Hilborn 2007; Worm et al. 2009). The analysis begins with a general description of the trajectory and underlying causes for forest transitions over the past two centuries. Then, we describe the empirical patterns in an emerging fisheries transition. The chapter concludes with a comparison of the patterns and processes that characterize the two transitions.

Forest Transitions

By forest transition is meant a long-term shift in patterns of land cover change from forest depletion to forest recovery. If a country undergoes a forest transition, this means most generally a shift from net losses in forest cover during a period of agricultural and settlement expansion to net gains in forest cover during a subsequent period of urbanization and urban economic growth. As implied by this definition, a point of inflection occurs in forest transitions in which long-term declines give way to long-term gains in the extent of forest cover (Fig. 1) (Rudel et al. 2005). In effect, the term "forest transition" is intellectual shorthand for a historical generalization about changes in the relationship between human societies and forests. Because forest transitions in most instances enhance the delivery of important environmental services like carbon sequestration, water purification, pollination, and nutrient recycling, they represent an important type of sustainability transition.

Major changes in human societies have periodically led to large-scale changes in forest cover. For example, the Black Death in fourteenth century Europe led to a considerable expansion of forests into the depopulated regions. With the onset of industrialization in the early nineteenth century a sequence of changes in forest cover–society relations began to recur, first in western Europe, then in North America, and most recently in Eurasia and selected countries in the Global South. With urbanization and industrialization demand for wood increases at the same time that the numbers of factory jobs grow. At the same time farmers become better acquainted with the agricultural potential of their lands. Faced with higher prices for wood, diminishing supplies of labor as farm workers depart for urban jobs, and an increased appreciation for differentials in soil fertility, farmers concentrate agricultural production on their most fertile lands and return their remaining lands to forests (Mather and Needle 1998).

The transitions occur in variable ways across nations. In the wealthier countries (Europe and the Americas) most forest regrowth has occurred spontaneously, after scarce labor leaves the land. In more labor abundant and poorer places in Africa, the Middle East, and Asia, smallholders plant many of the trees and plantations that contribute to the increments in forest cover (Rudel et al. 2005). In the East Asian cases where governments have had a tradition of intervening actively in markets to promote economic development, states have created significant economic incentives to expedite the conversion of marginal agricultural lands into forest plantations (Mather 2007). In the East African cases householders have led the way in planting fruit trees around their houses, creating what might be called "fruit forests" in densely populated zones (Rudel 2010). A social movement organized around the iconic figure of Wangari Maathai, a 2004 Nobel laureate, aims to plant a billion trees; it has spurred the planting of household orchards and kitchen forests in some locales (see www.unep.org/billiontreecampaign).

Temporal as well as geographical differences have characterized forest transitions. Perhaps most notably, the point of inflection (Fig. 1) for forest transitions has come to take place earlier in the process (Rudel et al. 2005). For example, forest cover did not begin to increase in Scotland in the early twentieth century until forest cover had declined to only 6% of the entire land area. In contrast in Costa

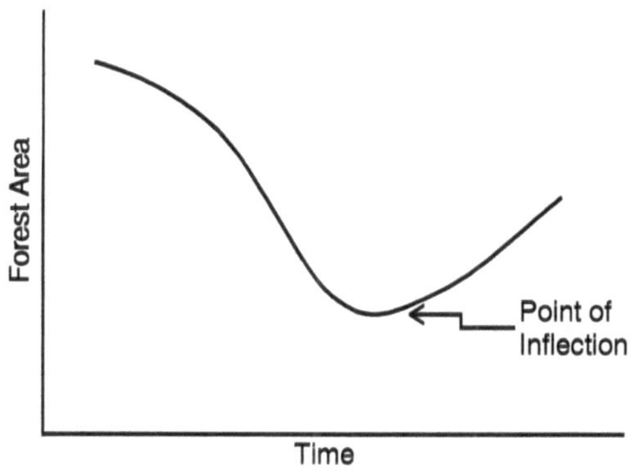

Fig. 1 A forest transition with the point of inflection

Rica forest cover began, after a long period of decline, to increase during the 1990s when the country still had around 30% of its land area in forest. One partial explanation for the change in these "turning points" is the recent creation of extensive park systems in tropical countries. These preservation efforts have prevented further declines in the extent of forests until demographic and economic trends began to promote forest recovery. Comparable park protections for forests did not exist when the nineteenth and early twentieth century forest transitions began, so in those periods the extent of forests declined even further before recovery began.

The historical changes from the earliest to the most recent forest transitions suggest that forest transitions, like economic development efforts (Gerschenkron 1962), exhibit latecomer effects. As people begin to recognize these recurring patterns of change in forests, they evaluate the changes and, if judged positively, these historical shifts in forest cover become the aim of government policies, foundation projects, and social movements. Arguably, these conscious efforts to induce the transitions speed them up, so the most recent transitions occur more quickly than the first transitions (Rudel and Hooper 2005).

A critical eye on forest transitions raises important questions about the extent of the environmental benefits and in particular the likelihood that a kind of "environmental leakage" occurs in which forest growth in one place (or country) comes at the cost of forest losses in other places. In other words as European countries increased their forest cover and reduced the extent of their agricultural lands did they begin to import more wood and agricultural products from other countries that had to deforest lands to provide these products for European consumers? If extensive leakage characterizes a forest transition, it could undo at a global scale the environmental benefits that appear at some national scales. Alternatively, if the shift in agricultural lands from one country to another country involves a shift toward more productive agricultural lands and an associated reduction in cultivated lands, then one might be able to make the case that a global forest transition has occurred. A recent empirical investigation of this question from 1961 to 2007 in 12 countries, eight of which experienced forest transitions, demonstrates that leakage in other countries offset 22% of the forest gains in the countries experiencing forest transitions. More disturbingly, the magnitude of the offsetting leakage effects has increased to 50% over the past decade (Meyfroidt et al. 2010).

Fishery Transitions

Can we discern comparable transitions in the world's fisheries? Are there points of inflection, within which long-term declines give way to long-term gains in the abundance or productivity of marine (and freshwater) living resources, which may be taken as the equivalent of forest cover? Are these points occurring earlier or later in the process? What forms do the transitions take? Restoration of wild stocks? Replacement of wild production with farmed production? How might they be related to the creation of protected areas? Are there other "latecomer" measures? Are cases of successful fish stock restoration counterbalanced by overexploitation of other stocks, the kind of environmental leakage or externality noted for some forest transitions? The following discussion is a preliminary effort to address these questions which, to our knowledge, have not been raised in comparison with forest transitions and in the context of sustainability science.

The dominant narrative about wild marine fisheries continues to be one of long-term and gradual as well as short-term and precipitous decline and a disturbing trend toward impoverished marine ecosystems as larger predators at higher trophic levels are removed or fished down, leading to complex ecological changes in marine systems that some say is a process of "fishing down the food web" (Pauly et al. 1988). There is a concurrent process of harvesting species lower in the food chain, the "forage fishes" and the shellfish that rely on phyto- and zooplankton and provide essential functions for the ecosystem. Although considerable debate and uncertainty remain about these processes, the standard view is that if a transition is underway, it is going in the wrong direction: from what may have been sustainable fisheries some time ago to overharvested fisheries and highly vulnerable and stressed ecosystems (Worm et al. 2006). A well-known and dramatic instance is that of Atlantic cod (*Gadus morhua*) on both sides of the Atlantic ocean since 1950 (Fig. 2).

What are the signs of transition in the other direction, toward regeneration of fish stocks and a resumption or emergence of more sustainable human uses? At a global scale, the situation is complex and can be interpreted in different ways. Global fish

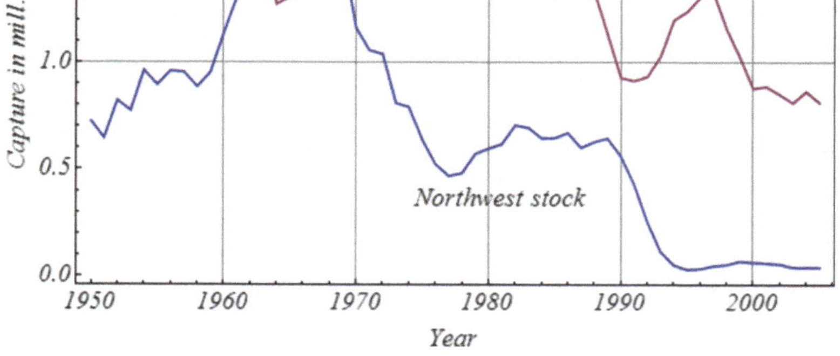

Fig. 2 Atlantic cod biomass estimates, Northwest and Northeast stocks, 1950–2005. *Source*: FAO Fishery Statistics programme (FIGIS Online), Atlantic cod capture 1950–2005; available at http://en.wikipedia.org/wiki/File:Atlantic_cod_capture_1950_2005.png

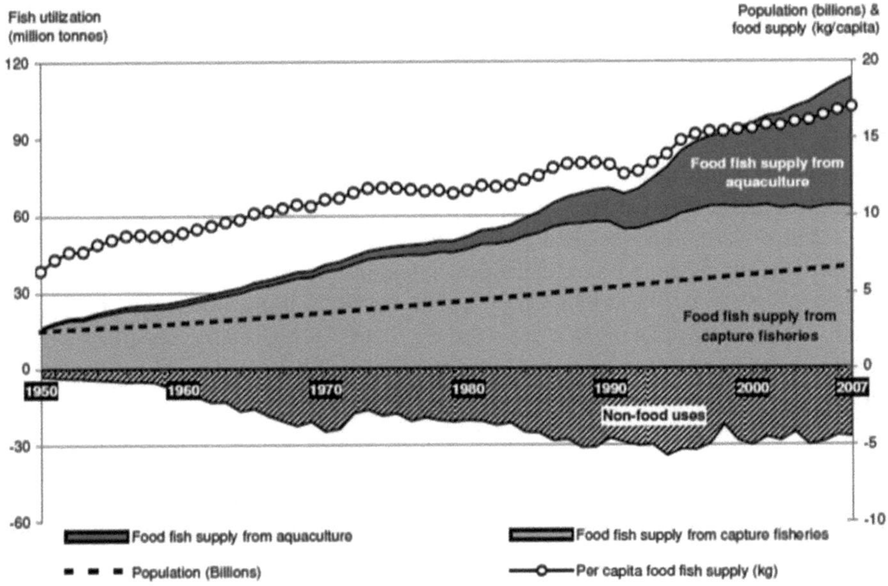

Fig. 3 World fish utilization and supply, 1960–2007. *Source*: FAO (2009)

production continued to increase well throughout the twentieth century, according to the Fisheries and Agricultural Organization of the United Nations, the major source of global-scale data. However, as shown in Fig. 3, it had begun to level off by 2000.

In 2007, overall catches of coastal fishes and of "demersal" or bottom fishes had dropped by 6% compared to 2006; catches of small pelagics (like the herrings and anchovetas) grew by 2.5%; and tuna and tuna-like species began to decrease slightly after an increasing trend period which had led to a historical catch peak in 2006 (FAO 2009). The data may signal not only further depletion but also cases where there is indeed a transition toward more sustainable fisheries taking place. The 2009 review of global fisheries between 1960 and 2007 showed that catches of some classes of marine organisms remained stable or improved. Marine crustaceans and shell mollusks remained stable, and catches of cephalopods (squids, cuttlefishes, octopuses) rose, continuing a strong upward trend, 35% over 5 years (FAO 2009:xvi), a trend that increased through 2008 (FAO 2010, Fig. 4).

These data should be treated with caution. From another perspective, the situation has greatly worsened over time, masked by high catches for some species. The FAO estimated that the proportion of different fish stocks (defined in terms of species and area) that are considered to be either depleted, overexploited, or in the process of recovering from depletion had risen from 10% in 1975 to 32% in 2008 (FAO 2010). However, the process of recovery is important, the signal of transition, to which we will turn after a discussion of the role of fish farming in global seafood production.

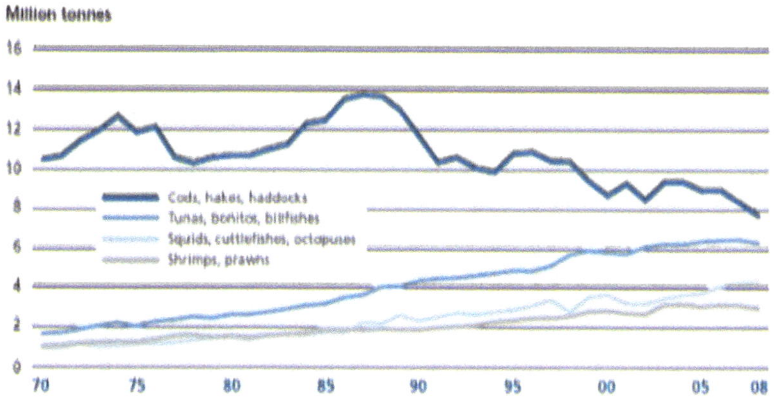

Fig. 4 Catch trends by marine species groups, 1970–2008. *Source*: FAO (2010:16)

Fisheries Transitions: Fish Farming

Using nations as the jurisdictional unit (somewhat problematic given the transnational and international nature of many fish populations), we can say that a considerable number of countries have entered into a fisheries transition due to the expansion of fish farming, which is similar to the expansion of tree-planting and husbandry in forested systems. An ever-increasing portion of global fish production comes from fish farming (Fig. 3) (FAO 2009). As of 2007, the increase of total production of fish (including crustaceans and mollusks) had continued, but increase was almost entirely from aquaculture. The production of wild fish and shellfish ("capture production") stayed fairly level since 2001, whereas aquaculture production continued strong growth, about 6.5% per year, and as of 2007 had reached about 50.3 million tons, compared with 90 million for capture production. The trend continued through 2008, the latest reported date (FAO 2010).

Is this sign of significant transition in the world's fisheries? Following measurement conventions for forest transitions, a fisheries transition is defined as taking place when the total stock of fish in a jurisdiction begins to increase, irrespective of what that stock is comprised of. This stock of fish can include farmed fish in pens and ponds as well as wild fish in the oceans, just as a forest transition can come from planted as well as wild trees. This is problematic if the negative effects of tree and fish farming on biodiversity are at issue, but it may be justified from the perspective of sustainable production of food and fiber.

In some regions of the world, particularly Southeast Asia and Asia, coastal fish-farming ventures have long histories and continue to expand. Some types of fish farming are mainly for local consumption (e.g., milkfish [*Chanos chanos*] in the Philippines); others are for high-end luxury and export markets (e.g., shrimp).

Salmon farming is very big business in temperate environments, having expanded from native homes of salmonids in the North Atlantic and North Pacific rivers and oceans to South American fjords. Shellfish culture is significant as well, and new farmed products appear in supermarkets; for example, tilapia and hybrid striped bass are now staples in U.S. supermarkets, though rarely seen 2 decades ago, and branzini (*Dicentrarchus labrax*), from Mediterranean fish farms, is a popular newcomer in the northeast USA.

The rise of fish farming can be viewed as a positive transition, whereby fish farming is compensating for decline in wild fishes and growth in demand for seafood given rising human populations. This interpretation is reasonable in some cases and not in others. Given the diversity of fish farming, aquaculture can be paradoxically both a solution and a contributing factor to the collapse of fisheries stocks. For species such as carp and mollusks, the culture of which involves little transformation of fragile habitats and which are herbivorous or filter feeders, the net contribution to global fish supplies is substantial. However, many farmed animals, like salmon and tuna, are carnivores that require fish as food. This adds to demand for small pelagic "forage" fish (Naylor and Burke 2005), and the rise of demand for fishmeal for aquaculture feeds in countries like China is of concern.

Shrimp is another problematic example. It is a major farmed export crop, mainly grown in brackish coastal ponds. Wild shrimp harvesting is costly to marine ecosystems because of the often large and destructive by-catch involved, reflecting difficulty avoiding juvenile and other vulnerable individuals when harvesting fish using fine-meshed nets (comparable perhaps to clear-cutting forests in consequences for biodiversity). However, shrimp aquaculture has problems, too. One is the demand on wild shrimp stocks (and associated by-catch) of the harvest of "seed" shrimp for farms because of the high costs and technical challenges of breeding shrimp in controlled facilities; another is loss of coastal ecosystem services due to alterations of coastal habitats for the ponds, and a third is the need for high protein feeds. These features of shrimp farming pose threats to marine ecosystems that may not be compensated for by reliance on farming for final production.

Production data for salmon and shrimp, and likely for other species as farming for them expands, indicate that farming supplements rather than substitutes for fishing (Goldburg and Naylor 2005). Furthermore, most mariculture systems are not fully enclosed; they are open to the sea, one way or the other, allowing for exchange of nutrients, pollutants, disease organisms, and genes with the wild system. Accordingly, in some places and from some perspectives, farming seafood does not reduce pressure on wild stocks and may actually maintain or increase pressures on them. Calls for more "ecological" approaches to fish farming, especially in the open ocean (e.g., Goldburg and Naylor 2005) are just that; a transition to such has not yet occurred.

Fisheries Transitions: Leaving the Seas

A major kind of transition identified for forests comes about as a side effect or unintended consequence of people shifting to other activities or places, due largely to urbanization and industrialization. It complements the creation of protected areas that force people to reduce their forest activities. Little is known about this kind of transition for fisheries. No comparable effort has been made to examine systematically the movement of people in and out of marine fisheries occupations, due to the pull of other activities and places, and the possible restorative effects on fish stocks and marine ecosystems. More predominant are accounts of intensified exploitation over time, particularly recent efforts to document prehistorical and historical effects of human activities on marine ecosystems, some of which highlight how profoundly those ecosystems have changed in a negative sense (Jackson et al. 2001; Sandin et al. 2008).

Glimmers and suggestions for future research can be found in historical accounts that show intensified or changing fishing patterns in relation to population migration to coastal areas, changes in other sectors such as agriculture or industry, and the exigencies of conflict. One example is the closure of the herring fishery of the North Sea during World Wars I (Parmanne 1999) and II because of military activities in the waters; it is often referred to as an exemplary case of how the cessation of fishing can lead to fish stock recovery. The history of the herring fishery shows many points of time in the more distant past—indeed, throughout the seventeenth century and during the Napoleonic Era—when maritime warfare and privateering resulted in lower productivity (Poulsen 2006), a possible example of inadvertent transition, when lowered effort corrected for harvesting levels that may have led to depletion. Furthermore, shifts in the distribution of power among mercantilist networks could result in loss of markets, also leading to downturns in fishing activity, seen in the eighteenth century for North Sea herring. Social changes affecting diets also play upon fisheries one way and the other. The demand for salted herring led to increased exploitation generally in the seventeenth to nineteenth centuries, but there were times of declining per capita consumption (1650–1750) that tempered demand. Furthermore, changes in the distribution and abundance of herring occurred for "natural" causes, as for example in the period 1815–1850 (Poulsen 2006). Market factors, wars, and environmental changes also seem to have been dominant causes for change in the productivity of other important fisheries, such as the Newfoundland salt cod fishery (Rose 2007).

Were these transitions toward sustainability? At least on a fairly large scale, it seems as if Thomas Huxley was at least 80% right at the 1883 National Fisheries Exposition when he dismissed as "unscientific" the complaints of fishermen about how other fishing boats were destroying the stocks. He stated that "[a]ny tendency to over-fishing will meet with its natural check....This check will always come into operation long before anything like permanent exhaustion has occurred" (quoted in Kurlansky 1998:121–122). Many of the world's fish stocks have proved amazingly resilient as long as the level of fishing mortality stayed relatively low, and thus what might look at a historical distance like transition could very well be simply shift from one more-or-less sustainable system of resource exploitation to another. Future research should, however, look at fisheries for slow-to-reproduce whales and other marine mammals, slow-growing and long-lived fisheries, and endemic and highly specialized tropical species, which would be more vulnerable to overexploitation and could, in theory, yield accounts of positive transition, although the literature available to us at this point in time seems to indicate otherwise (Jackson et al. 2001; Schipper et al. 2008). Future research could also examine shifts in cultural factors, ranging from food proscriptions to consumer tastes and bioethics, as they have affected fisheries transitions. The linkages may not be obvious: a decline in the availability of "bush meat" (e.g., large primates) in some tropical regions, linked to large-scale conservation measures, appears to have increased demand for fish, adding to pressures on West African fish stocks (Brashares et al. 2004).

More generally, research has yet to be done on the question of how urbanization, industrialization, patterns of employment and unemployment, and so forth have affected marine fisheries. The overall pattern of rising coastal populations does not bode well for coastal fisheries, given increased demand and likely increases in effort. A growing literature on poverty and fisheries (e.g., Béné et al. 2007) shows close and mutual interaction between fisheries decline and poverty, decline being both cause and effect of poverty. Economic development could both alleviate poverty and create more alternatives to fishing, helping to reverse downward trends in coastal fish stocks. However, all of this is predicated on close associations between fishing effort and local populations and economies, whereas the reality in many parts of the world is that fishing effort in a region can have as much to do with conditions elsewhere, the "leakage" discussed in the forest transition literature. Thus, for example, *The Ecologist* recently reported activists' claims that increased fish "piracy" or illegality and overfishing are destroying the livelihoods of coastal fishing groups in Africa in order to provide fish that compensates for declining fish stocks in European Union waters (Ecologist 2011).

Fishery Transitions: Protected Areas and Regulatory Interventions

In many tropical countries, the creation of forest preserves and national parks helped to slow down the depletion of forests, enabling the "inflection point" or transition to begin when substantial forested areas remained. In contrast the creation of marine-protected areas has been of minor scale until very recently, representing less than 0.7 or 0.8% of the world's oceans by 2010 (World Database on Protected Areas 2009), and very little of that fully protected from harvesting. Of course, there are other sources of closure, including those created for military purposes or because of public health concerns due to pollution, and the effects of these on fish stocks should be examined. The creation in 2006 of extremely large marine reserves in tropical and subtropical waters, including the 139,000 miles^{-2} (361,400 km^{-2})

Papahanaumokuakea Marine National Monument in the northwest Hawaiian Islands and comparably large areas in the Mariana Islands, American Samoa, and Guam may be similar to the case in forest transition. However, some of those areas are in waters that have not suffered extreme degrees of depletion as yet.

The more evident fishery recoveries or transitions are associated with intentional interventions by the state, reflecting the evolution of "resource management" expertise and missions on the part of governments. In relation to the forest transition literature, government interventions into marine fisheries management would be considered "latecomer" efforts, building upon prior experiences and the evolution of scientific and administrative capacities. The science and practice of marine fisheries management took shape at the end of the nineteenth century and in the first half of the twentieth century (Smith 1994) but its effectiveness was constrained by the geopolitics of the oceans, whereby all but narrow coastal bands of the sea were open access international waters, subject only to difficult-to-enforce international management regimes. Fisheries that occur farther offshore, especially those that take place beyond the 200 nautical mile (370 km) zone of extended national jurisdiction are far more prone to "tragedies of the unmanaged commons" (Hardin 1994) due to the great difficulties involved in international environmental management, given the sovereignty of individual nations and weakness of international institutions (Young 1994).

In the mid-1970s the United Nations Conference on the Law of the Sea led to claims of 200-mile limits of exclusive economic or fisheries jurisdiction throughout the world (Burke 1994), enabling the expansion of science-based management to encompass a huge portion of the species-rich coastal fishing grounds of the world. Capacity for management varied, of course, favoring the wealthier nations and regions of the world, leaving the poorer ones to rely on international aid and on local, customary, and informal management regimes (McGoodwin 1990; Dyer and McGoodwin 1994). North Sea herring offers a case showing the emergence of more rapid responses with time and experience. After World War II the fishery expanded rapidly, with technological changes in fish finding and harvesting and increased demand from recovering European economies. In the 1970s the stock declined precipitously but nothing was done until 1978, when the fishery was closed to harvest just before the herring stocks might have become extinct. Recovery was very slow. In the mid-1990s decline occurred again due to high fishing mortality, but this time management actions were adopted early enough to enable the fish stock to recover within a few years (Simmonds 2007).

As noted earlier, globally some of the crustacean fisheries have shown marked increase in recent years, and we would be remiss in ignoring the account of the American lobster (*Homarus americanus*), which appears to be as extreme a case of positive transition as the Atlantic cod was one of depletion. Fig. 5 shows ups and downs through the early 1940s, followed by sustained catches and a recent boom in catches.

Downturns in catches in the State of Maine, accounting for most of the Gulf of Maine fishery, led to legislature and industry acceptance of more stringent management measures (Acheson 2003, 1997), which may have played a role in the transition to a strong fishery. Tempering this account, like so many others, are

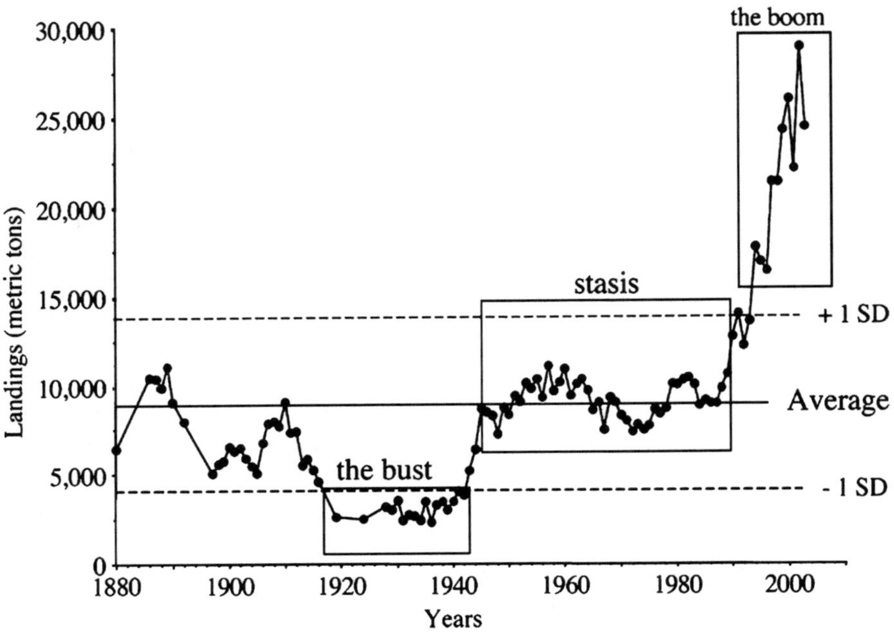

Fig. 5 Gulf of Maine lobster landings, 1880–2008

high levels of uncertainty bolstered by recognition of complexity in marine coastal ecosystems. The American lobster "transition" could be reinterpreted as the outcome of a variety of cumulative and synergistic effects involving sea urchins, cod, herring, and human interactions with each. For example, the rise of a sea urchin harvest alleviated sea urchin pressure on kelp beds, which recovered and provided habitat for lobsters and other species. Overfishing of cod and other large predators removed predator pressure on young lobsters, and loss of cod fishing opportunities led many fishermen to concentrate instead on lobster (Acheson and Steneck 1997).

Less well understood but recognized are the roles of environmental factors in the timing and extent of recovery of fish stocks that governments seek to manage by restraints on fishing. The extent and timing of recovery of North Sea herring stocks and the cod stocks of Newfoundland have been diminished and slowed by environmental factors that may include changing water temperatures and predation (Hislop 1996; Rose 2007; May 2009), although it is difficult to pinpoint the precise causes.

In the introduction to this chapter, we referred to the possibility that transitions may be affected by biophysical features of the systems involved. It does seem that fisheries transitions are harder to assess than forest transitions. At the simplest level, they are harder to see, as fish remain underwater until caught and fishing activities often take place far from points of observation. Consequently, in marine fisheries, transitions to more sustainable systems can take a long time to be visible or may not be visible at all. They also may be short-lived. In multi-dimensional highly motile fluid media, fish populations have high levels of variability and unpredictability.

The good news can be unexpected as well as hard to explain, making it difficult to assess whether management was the cause, a problem magnified by the common pool characteristics of fisheries relative to forests, the latter being more easily divisible as well as observed. Something that appears to be on a downward trend can suddenly emerge as a strong resource, as happened in the 1960s with the so-called "gadoid burst" in the North Sea (Simmonds 2007). Similarly, salmon fisheries are notorious for variability and unpredictability; for example, the 2010 Fraser River salmon fishery may be the best in almost a century, following upon 3 years of serious decline (Bernton 2010). This poses another order of difficulty for fisheries transitions: how can one evaluate whether or not a regulatory intervention works and convince people that it is worthwhile to support it when there are so many "unknowns" at play? The term "faith-based" has been applied to criticize some applications of fisheries science (Hilborn 2006), but it is also true that much of fisheries management must be accepted on the basis of faith that somehow it should work.

Fishery Transitions: Incentives and Property Rights

Our introduction noted the recognition of human incentives as significant driving forces in forest transitions. This is true for fisheries as well but may work out differently. Open-access conditions are widely understood to create dis-incentives for conservation and stewardship, instead promoting competitive behavior for short-term goals. Although the creation of 200-mile limits enabled coastal states to claim territories at sea, in general it is logistically difficult to create boundaries around marine resources, a first step toward controlling access and creating more exclusive property rights, which provide incentives inherent in ownership and enable the use of market forces in regulation. The development of bounded units of control over marine resources is also restricted in marine systems by the high mobility of both prey (especially the finfishes) and predator (the fishers), as well as the variety of habitats needed to support the various life-history stages of many if not most fish species. Consequently, there has been relatively little use of the tools of property rights for fishing grounds and fishes comparable to those of land and tree tenure (Fortmann and Bruce 1988), and this constitutes a major difference between forests and fisheries. However, the situation is changing through both the privatization of fishing rights and strengthening of local community engagement in fisheries management.

Recognition of the dis-incentives for stewardship inherent in a situation of few property rights led to the creation of quasi-property rights in fisheries from about 1980 to the present in major industrialized nations, closely linked ideologically with neoliberalism. With economic rather than biological goals in mind but recognition of the interactions between them, economists have long argued and worked for the implementation of individual transferable quotas (ITQs) in fisheries management (Squires et al. 1995; Grafton et al. 2006). ITQs are analogous to cap-and-trade programs in emissions control: rights to percentages of an overall quota are assigned to entities (individuals, businesses) who are able to trade them in order to achieve efficiencies while accepting restrictions on amounts overall. The economic benefits of ITQs are self-evident, as they have corrected tendencies in fisheries toward over-capitalization. It is less obvious that they assist in transitions toward sustainability.

Examining a large number of cases worldwide, Costello et al. concluded that fisheries collapse was much less likely with ITQs than without (Costello et al. 2008; Costello and Kaffine 2008), lending credence to the notion that this form of privatization too could help with the transition to more sustainable fisheries. However, other studies that examined a wider range of variables concluded that such gains were in effect dependent on the workings of government regulation combined with property-related incentives (Branch 2009; Chu 2009; Essington 2010).

More simply, without a quota that has scientific and political legitimacy and credibility, as well as monitoring and enforcement inputs from government, ITQs cannot be effective either for economic or biological purposes, and privatization comes at significant social and administrative costs (McCay 1995; Young and McCay 1995). However, there are other forms of property rights that may play a role in fisheries transitions whereby social groups of various dimensions and kinds claim territorial or other rights and are engaged in regulating their fisheries, often in concert with government agencies, in forms of community-based resource management that have counterparts in community forestry movements throughout the world.

The American lobster fishery transition (Fig. 5) owes success in some part to the informal territories established and defended by local groups of lobster fishers; their territorial actions combine with other industry-driven measures such as minimum and maximum sizes and avoidance of egg-bearing lobsters to provide the basis for what appears to be high sustainable fishing, including a history of transition from hard times (Acheson 1997, 2003; Acheson and Steneck 1997). Informal territoriality has been complemented in recent years by regional management zones to help implement fisher-led controls over effort in the fisheries.

Transition to sustainable fishing along a remote coast of northwestern Mexico also depends on local community engagement and property rights. Small cooperatives of the Vizcaino Peninsula of Baja California, Mexico, hold official long-term concessions for valuable species, mainly abalone and lobster, in demarcated territories off their shores. Although the state is involved in the management regime, its success has as much to do with the incentives and capacities for monitoring and enforcement made possible by the concessions, as well as trial-and-error experience with near collapse (Ponce-Díaz et al. 2009). Similar efforts in Chile have led to nationwide restructuring of coastal benthic fisheries, and as observed in other cases, increased awareness of and capacity for the kinds of changes in fishing activity that can support transition (Schumann 2007). There are many other instances of community-based property rights contributing to sustainable fisheries (Pinkerton and Weinstein 1995), including coastal Japan. In

Japan as elsewhere, the outcomes are not predictable given the uncertainties inherent in marine ecosystem dynamics as well as human socioeconomic systems (Takahashi et al. 2006).

Large-Scale Fisheries Transitions: US Fisheries

It is possible to identify specific fisheries around the world that show signs of transition from overfished to healthier status. The European Union nations have a long and worthy history of fisheries science and management, and some successes, but transitions to sustainability have been plagued by political difficulties within the union, stemming in part from a requirement about quantitative sharing of catches among nations, the so-called Common Fisheries Policy (Daw and Gray 2005). With an equally long and substantial history of fisheries science, the United States has emerged as a much stronger case of fisheries transition, less hampered by jurisdictional issues despite evolving a complex system of regional and state-federal governance since the 200-mile limit of 1977.

Evidence for a large-scale transition, not just for individual species, comes from the 2010 government report on the status of the marine fish stocks under jurisdiction of the federal management system (3–200 n.m.). The National Marine Fisheries Service uses a "fish stock sustainability index" (FSSI) to measure performance of 230 key fish stocks; it increases as stock assessments improve, "overfishing" is ended, and "overfished" stocks are rebuilt to the level deemed to provide maximum sustainable yield (MSY) the legal goal.[1] The index increased greatly from 2000 to 2009 (Fig. 6), a 60% increase in 9 years (National Marine Fisheries Service 2010).

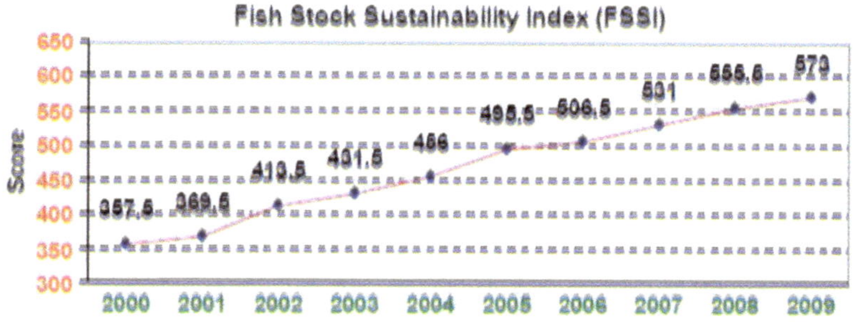

Fig. 6 Fish stock sustainability index (FSSI), United States, 2000–2009

[1] More information about the FSSI can be found at: http://www.nmfs.noaa.gov/sfa/statusoffisheries/SOSmain.htm.

"Overfishing" and "overfished" are key technical terms for indicators that require action. A stock that is subject to overfishing has a fishing mortality (harvest) rate above the level that provides for the maximum sustainable yield. A stock that is overfished has a biomass level below its prescribed biological threshold (NMFS 2010:3).

Put another way, only 23% of the 203 fish stocks—or fish stock complexes—for which indicators are available were overfished in 2009. There were, however, 317 fish stocks, or stock complexes, for which assessments had not been done or been possible, and there was substantial regional variation. Some regions, like the Mid-Atlantic, show virtually no overfished stocks and impressive records of restoration from overfished status; others have many overfished stocks and fewer cases of improvement. In contrast, regions like the Pacific coast have numerous overfished stocks but none currently being fished as levels that are deemed overfishing, and therefore can be construed as in the process of restoration.

As might be expected, given the international nature of most of the fishing that takes place for them, the "highly migratory" tunas, marlins, sharks, etc. that are the responsibility of the federal government are in poor condition by these standards, both "overfished" and subject to "overfishing." Also in trouble by these measures are subtropical complexes of the South Atlantic, Caribbean, and Gulf of Mexico, as well as many of the groundfish of New England.

Despite such problems, the United States appears to be at a major "inflection point," moving toward overall restoration of fish stocks. This conclusion is supported by a recent widely publicized interview with Dr. Steven Murawski, a highly respected fisheries scientist. He was quoted as declaring that as of 2011 no U.S.-managed stocks would be defined as overfished:

"The projected end of overfishing comes during a turbulent fishing year that's seen New England fishermen switch to a radically new management system. But scientist Steve Murawski said that for the first time in written fishing history, which goes back to 1900, "As far as we know, we've hit the right levels, which is a milestone. And this isn't just a decadal milestone, this is a century phenomenon," said Murawski, who retired recently as chief scientist at the National Oceanic and Atmospheric Administration's Fisheries Service" (Anonymous 2011).

Fisheries Transition at Larger Scales: Environmental Leakage

Environmental "leakage," where strong controls over fishing effort in one place come at the cost of intensified harvesting pressure in other places, is evident and likely quite high in fisheries as it is in forest situations. The United States, for example, is increasingly dependent on seafood imports, and not only from neighbors like Canada but also increasingly on seafood harvested by Asian nations, inside and outside their territorial waters. The sustainability of their practices calls for greater scrutiny, as does the wider problem of illegal, unreported, and unidentified (IUU) fishing, which is a major impediment to the achievement of sustainable world fisheries, respecting neither national boundaries nor international attempts to manage high sea resources (FAO 2010). The EU's dependence on imports has also increased, as has its direct and indirect engagement in fisheries off the coasts of other nations, particularly Africa, as noted earlier.

Other "leakage" takes place through the activities of predatory fishing fleets and buyers, "roving bandits" that by-pass or violate the rules developed locally or act too quickly for the formation of effective local responses. This is true in fisheries as it is in logging. It has been enhanced not only by the globalization of markets but also by the transnational or highly fluid nature of the harvesters, as seen in the recent history of both sea urchin and live aquarium fish trades (Berkes et al. 2006).

Fisheries Transition at Global Scales: Eco-certification

The use of eco-certification of fisheries to identify and encourage the continuation of sustainable fisheries or the end of unsustainable ones is one counter to IUU, "roving bandit," and other global fishery problems. It builds upon rising consumer concerns about ecological sustainability and health, and as such is an "ecological modernization" strategy. The Marine Stewardship Council (MSC) is a "latecomer" institution modeled on earlier forest eco-certification programs. The MSC became an international nonprofit organization in 1999 and began relying on third party firms to assess marine fisheries against environmental and institutional criteria for sustainability. In 2009 it was estimated that fisheries at one or another stage of assessment accounted for about 7% of the world's annual wild harvest (MSC 2009), and the percentage has grown since. As of 2010 there were 96 certified fisheries as well as 131 undergoing assessment by a third party and untold others undergoing pre-assessment within the MSC (www.msc.org), as well as uncounted others within competing eco-certification organizations.

Certification has been criticized as "green-washing," favoring already successful, large-scale capitalist ventures. Similarly to the experience in forest certification, all but one of the first 42 fisheries to be certified were from developed nations (the "North" including Australia). Moreover, some were at a very large industrial as well as geographic scale such as the Alaskan pollock fisheries, although others were much more localized and less capital-intensive such as the Scottish Loch Torridon creel fishery for the lobster-like nephrops (*Nephrops norvegicus*). An exception to the developed nations trend was the Baja California, Mexico, red rock lobster (*Panulirus interuptus*) fishery, the first artisanal-scale and developing fishery to be certified (certification is a costly process), helped considerably by efforts of the World Wildlife Fund as well as the fishery's own concession-based system of co-management between local fishing cooperatives and government, referred to earlier in this chapter (McCay and Weisman 2007).

The MSC is frequently criticized for being too lenient in its standards in the effort to expand the scope of certification (Greenpeace USA 2009), but the assessments are carried out by independent bodies that employ highly qualified scientists, and the criteria are stringent. Therefore, although there are acknowledged problems with certification, it seems reasonable to use certification as one indicator that a fishery is being managed sustainably. It is a possible indicator of transition when combined with information about the longer history of the fishery.

The scope of MSC certification is slowly expanding, aided by food retailers seeking to serve consumers concerned about the sources of their food, and other sources and forms of certification are emerging. The MSC system not only recognizes sustainable practices and viable systems of management but encourages and requires improvement, and thus bears watching. Certification of aquacultured fisheries is also expanding.

Discussion and Conclusions

Fisheries transitions might be expected to occur more often and earlier in the coastal fisheries, which are usually within the boundaries of national jurisdiction, as compared with offshore and highly migratory fisheries. The coastal fisheries often have long histories of exploitation, which provide the possibility for episodes of trouble and trial-and-error responses that may lead to sustainable outcomes. Offshore fisheries are troubled by challenges in international management as well as particulars such as the technical difficulties involved in monitoring what is happening far from shore and strategies such as using flags-of-convenience to avoid stringent government regulation.

Fishery transitions in recent times have been facilitated by the capacity to create and enforce boundaries and some forms of property rights, but the state and its role in fisheries transitions is important. Strong states, with political commitment to fisheries, access to scientific knowledge that allows reasonable decisions to be made in the face of the high degree of uncertainty typical of marine systems, and fairly good institutions for effective participation of resource users in knowledge creation, rule-making and monitoring, seem critical to fishery transitions in recent decades. Tools for improving the sustainability of marine ecosystem services, such as regulating fisheries and creating truly effective marine-protected areas (a matter we have not had space to address here), are meaningless without strong financial, political, and legal support on the one hand, and effective outreach to and inclusion of affected groups on the other.

Finally, there is hope. The United States is experiencing a transition toward more sustainable coastal fisheries. Transition is happening elsewhere as well. A recent analysis of trends by a prestigious group of scientists found that "After a long history of overexploitation, increasing efforts to restore marine ecosystems and rebuild fisheries are under way" (Worm et al. 2009). In half of the ten ecosystems examined, the average exploitation rate had recently declined; in seven of them the exploitation rate was estimated to be at or below the rate predicted to achieve MSY. At an individual fish stock level, the results were less sanguine: 63% of the fish stocks assessed require rebuilding and some are on the verge of collapse. Well-known regulatory tools are available, but two major issues remain: international fleets and the lack of alternatives to fishing in many poorer regions.

What are the key differences if any between fisheries and forests that seem to affect the likelihood, timing, form of transition to sustainable and recovered condi-

Table 1 Characteristics of fishery and forest transitions

Characteristics	Fishery transitions	Forest transitions
Historical onset	Late twentieth century, early twenty-first century	Early twentieth century
Biological reserves	Small-scale creation and spread, 1990s—<0.8% gazetted, 2010	Extensive creation, 1950–2000, ca. 10% land surface
Deliberate cultivation	27% of the world's foodfish production in 2007 came from aquaculture	~5% of world's forests in 2000
Role of the state	Wild fisheries: public resources, extensive formal role of the state, variable effectiveness; international fisheries issue; recent development of privatized and community-based management to complement state power; privatized property rights slow to emerge	Extensive recently in Asia, less so elsewhere. Private property and communal property extensive in many domains, with much remaining as state or public property
Leakage	Appreciable and growing, United States, European reliance on imports from Asia, etc.; illegal and unreported fisheries; scale thought to be major impediment to global sustainability	Appreciable, growing in extent, but not so much as to nullify all environmental benefits of forest transitions

tions? Table 1 sums up some of the contrasts we see between fishery and forest transitions which may be used to guide further exploration of the topic. The historical onset of fishery transitions is clearly much later than that of forest transitions, and there is cross-over, the "latecomer" effect operating in the application of science-based resource management and measures such as protected areas and eco-certification. Nation-states often became involved in forest management earlier, and the skills, missions, tools used eventually surfaced in fisheries management in the late nineteenth and early twentieth centuries. The transition tool of setting aside large areas for protection and restoration, as biological reserves or protected areas is found in both cases, again, later for fisheries than for forests, and at a far smaller scale for fisheries. The transition tool of deliberate cultivation—planting trees, farming fish, and shellfish—has a very long history in fisheries as in forestry, but in fisheries it has expanded greatly in the past 2 decades, particularly in the last 10 years, now accounting for about 27% of the world's foodfish production (FAO 2009).

A key difference is that fisheries—at least wild capture fisheries—are less prone than are forests to being treated as private property, which means that they have longer histories of being treated as open-access or informally-enclosed systems. The greater difficulty imposing exclusive and long-term property rights in fisheries in contrast with forests, whether state, private, or communal, also means that eventual transition depends not only on the abilities of local communities and user groups to control activities—facilitated by the ability to create boundaries or exclusive fishing rights—but also on the role of the state, which retains responsibility for public property. The difficulties are logistical, but they have also been moral and political, as progressive tightening of access to land-based resources increased the value of having freer access to the seas, and many nations have enshrined the free rights of citizens to the seas in their charters (this also poses challenges to protected area management).

Another key point of difference is that a large component of the world's fisheries takes place beyond or straddles national boundaries, subject only to informal user-group and weak international jurisdiction. Forests are always within the boundaries of a nation-state, although the effectiveness of that state in extending its controls and exerting some regulatory protection varies greatly in time and space. Furthermore, forested ecosystems often straddle the boundaries between two or more nations, and, like distant oceans, can be effectively beyond successful monitoring and enforcement. In both instances, environmental "leakage" is a serious issue, whereby the benefits of transition in one place may be outweighed or even nullified by the costs of extraction and other activities elsewhere. For both forests and fisheries, we offer cautious optimism.

The fluidity and multi-dimensionality of the marine environment, the high variability of marine ecological communities and oceanographic dynamics, the large scope and taxonomic complexity of the world's oceans (Worm et al. 2009) bear mention. It may be that marine ecosystems have been buffered somewhat from human activities until recently in part because of remoteness of some from human settlement and activity and in part because some of them—e.g., the northern temperate systems and upwelling systems, as well as species such as lobster—have shown tremendous productivity and resilience. Consequently, the transitions for ocean fisheries—movement of people to other activities, setting aside areas and species for special protection, adoption of effective regulations on harvesting—have been later than those for forests.

For both fisheries and forests, we offer cautious optimism. The major goal of this article was to generate interest in furthering the comparison and, in particular, examining the conditions for successful transitions to sustainability. We have taken a very simple approach, asking whether there are signs and signals of transition in fisheries, as in forests, of transition to more sustainable systems. A next step is to carefully analyze causes and correlates of transitions and failures of transition. This topic is now also one concerning vulnerabilities and resilience in the face of climate change and the many ecological, social, and political ramifications of climate change. Consequently, the idea of transition, while still valuable, may have to yield way to "transformability" (Walker et al. 2004), just as the old natural resource notions of "maximum sustained yield" so central to both forest and fisheries management must be replaced with more adaptive and precautionary ways of coping with dynamic natural systems.

Acknowledgments This work is based on numerous projects of the authors, supported by the New Jersey Sea Grant College Program, the National Science Foundation, and the New Jersey Agricultural Experiment Station. We are grateful to the editors and to anonymous reviewers for useful suggestions.

References

Acheson JM (1997) The politics of managing the Maine lobster industry: 1860 to the present. Human Ecol 25:3–27

Acheson JM (2003) Capturing the commons: devising institutions to manage the Maine lobster industry. University Press of New England, Hanover, NH

Acheson JM, Steneck RS (1997) Bust and then boom in the Maine lobster industry: perspectives of fishers and biologists. N Am J Fish Manage 17:826–847

Anonymous (2011) Is overfishing ended? Top US scientist says yes. Associated Press. http://www.google.com/hostednews/ap/article/ALeqM5jlaWeLkS0t__O2_fChxd-Ty6w3qA?docId=f808be18d5cc48d58ab923de17d7a604. Accessed 9 Jan 2011

Baird RC (2012) Sustainable coastal margins: challenges of tempo and mode for the policy domain. In: Weinstein MP, Eugene Turner R (eds) Sustainability science: the emerging paradigm and the urban environment. Springer, New York

Béné C, Macfadyne G, Allison EH (2007) Increasing the contribution of small-scale fisheries to poverty alleviation and food security. FAO Fisheries Technical Paper 9251056641, Food and Agriculture Organisation of the United Nations, Rome

Berkes F, Hughes TP, Steneck RS et al (2006) Globalization, roving bandits, and marine resources. Science 311:1557–1558

Bernton H (2010) Fraser River whopper sockeye salmon run even bigger. Seattle Times, 1 Sept 2010. http://seattletimes.nwsource.com/html/localnews/2012777922_salmon02m.html. Accessed 5 September 2010

Branch TA (2009) How do individual transferable quotas affect marine ecosystems? Fish Fisher 10:39–57

Brashares JS, Arcese P, Sam MK et al (2004) Bushmeat hunting, wildlife declines, and fish supply in West Africa. Science 306:1180–1183

Burke WT (1994) The new international law of fisheries; UNCLOS 1982 and beyond. Clarendon Press, Oxford

Chu C (2009) Thirty years later: the global growth of ITQs and their influence on stock status in marine fisheries. Fish Fisher 10:217–230

Costello CJ, Kaffine D (2008) Natural resource use with limited-tenure property rights. J Environ Econ Manage 55:20–36

Costello CJ, Gaines SD, Lynham J (2008) Can catch shares prevent fisheries collapse? Science 321:1678–1681

Daw T, Gray T (2005) Fisheries science and sustainability in international policy: a study of failure in the European Union's Common Fisheries Policy. Mar Policy 29:189–197

Dyer C, McGoodwin J (1994) Folk management in the world's fisheries: lessons for modern fisheries management. University Press of Colorado, Boulder

Ecologist (2011) Illegal European fishing contributing to poverty and piracy in Africa. The Ecologist, 20 April 2011. http://www.theecologist.org/News/news_analysis/852417/illegal_european_fishing_contributing_to_poverty_and_piracy_in_africa.html. Accessed 9 January 2012

Essington TE (2010) Ecological indicators display reduced variation in North American catch share fisheries. Proc Natl Acad Sci 107:754–759

FAO (2009) FAO yearbook: fisheries and aquaculture statistics 2007. Food and Agriculture Organization of the United Nations, Rome. ftp://ftp.fao.org/docrep/fao/012/i1013t/i1013t.pdf. Accessed 11 October 2010

FAO (2010) The state of world fisheries and aquaculture 2010. Food and Agriculture Organisation, Rome

Fischer-Kowalski M, Rotmans J (2009) Conceptualizing, observing, and influencing social-ecological transitions. Ecol Soc 14, http://www.ecologyandsociety.org/issues/. Accessed 7 June 2010

Fortmann L, Bruce JW (1988) Whose trees? Proprietary dimensions of forestry, Rural studies series. Westview Press, Boulder

Gerschenkron A (1962) Economic backwardness in historical perspective: a book of essays. Harvard University Press, Cambridge
Goldburg R, Naylor RL (2005) Future seascapes, fishing, and fish farming. Front Ecol Environ 3:21–28
Grafton RQ, Arnason R, Bjørndal T et al (2006) Incentive-based approaches to sustainable fisheries. Can J Fish Aquat Sci 63:699–710
Greenpeace USA (2009) Assessment of the Marine Stewardship Council (MSC) fisheries certification program. http://www.greenpeace.org/usa/en/media-center/reports/assessment-of-the-msc/. Accessed 20 July 2009
Hardin G (1994) The tragedy of the unmanaged commons. Trends Ecol Evol 9:199
Hilborn RW (2006) Faith-based fisheries. Fisheries 31:545–555
Hilborn R (2007) Reinterpreting the state of fisheries and their management. Ecosystems 10:1362–1369
Hislop JRG (1996) Changes in North Sea gadoid stocks. ICES J Mar Sci 53:1146–1156
Jackson JBC, Kirby MX, Berger WH et al (2001) Historical overfishing and the recent collapse of coastal ecosystems. Science 293:629–637
Kates RW, Clark WC, Corell R et al (2001) Sustainability science. Science 292:641–642
Kurlansky M (1998) Cod: a biography of the fish that changed the world. Penguin Books, London
Lebel L (2005) Transitions to sustainability in production-consumption systems. J Ind Ecol 9:11–13
Marine Stewardship Council (2009) Net benefits: the first ten years of MSC certified sustainable fisheries. http://www.msc.org/business-support/net-benefits. Accessed 19 August 2010
Mather A (2007) Recent Asian forest transitions in relation to forest transition theory. Int Forest Rev 9:491–502
Mather A, Needle C (1998) The forest transition: a theoretical basis. Area 30:117–124
May A (2009) The collapse of the Northern Cod. Newfoundland Quart 102. http://www.mun.ca/harriscentre/reports/nlquarterly/NQ_article_Vol_102_No_2.pdf. Accessed 9 June 2010
McCay BJ (1995) Social and ecological implications of ITQs: an overview. Ocean Coast Manage 28:3–22
McCay BJ, Weisman W (2007) Greening the red rock lobster: ecolabeling and control in a Mexican fishery. In Paper presented to annual meeting of the American Anthropological Association, November 27 to December 2, 2007, Washington, DC (Session anthropology and sustainability: issues, outcomes and experiences in Mexican fisheries and coastal communities, organized by Maria Cruz Torres)
McGoodwin JR (1990) Crisis in the world's fisheries: people, problems, and policies. Stanford University Press, Stanford
Meyfroidt P, Rudel TK, Lambin EF (2010) Forest transitions, trade, and the displacement of land use. Proc Natl Acad Sci 107:20917–20922. doi:10.1073/pnas.1014773107
National Marine Fisheries Service (2010) 2009 status of U.S. fisheries. 2009 report to congress, U.S. Department of Commerce. National Oceanic and Atmospheric Administration. National Marine Fisheries Service, Silver Spring, MD. http://www.nmfs.noaa.gov/sfa/statusoffisheries/sos_full28_press.pdf. Accessed 2 September 2011
National Research Council (1999) Our common journey: a transition to sustainability. National Academies Press, Washington, DC
Naylor RL, Burke M (2005) Aquaculture and ocean resources: raising tigers of the sea. Ann Rev Environ Resour 30:185–218
Parmanne R (1999) Herring fishery in the Bothnian Sea (southern Gulf of Bothnia) and the North Sea: similarities and differences. Boreal Environ Res 3:321–328
Pauly D, Christensen V, Dalsgaard A, Froese R, Torres L (1988) Fishing down marine food webs. Science 279:860–863
Pinkerton E, Weinstein M (1995) Fisheries that work: sustainability through community-based management. The David Suzuki Foundation, Vancouver

Ponce-Díaz G, Weisman W, McCay BJ (2009) Co-responsabilidad y participación en el manejo de pesquerías en México: lecciones de Baja California Sur. Pesca Conserv 1:1–9
Poulsen B (2006) Historical exploitation of North Sea herring stocks—an environmental history of the Dutch herring fisheries, c. 1600–1860. Dissertation, Southern Denmark. English summary. http://hmapcoml.org/news/documents/HistoricalexploitationofNorthSeaherringstocks EngSumBoP.pdf. Accessed 26 June 2010
Rock M, Murphy JT, Rasiah R et al (2009) A hard slog, not a leapfrog: globalization and sustainability transitions in developing Asia. Technol Forecast Soc Change 76:241–255
Rose G (2007) Cod: the ecological history of the North Atlantic fisheries. Breakwater Books, St. Johns
Rudel TK (2010) The human ecology of tropical regrowth. J Sustain Forest (in press)
Rudel TK, Hooper L (2005) Is the pace of social change accelerating?: latecomers, common languages, and rapid historical declines in fertility. Int J Comp Sociol 46:275–296
Rudel TK, Coomes OT, Moran E, Achard F, Angelsen A, Xu J, Lambin E (2005) Forest transitions: towards a global understanding of land use change. Global Environ Change 15:23–31
Sandin SA, Smith JE, DeMartini EE et al (2008) Baselines and degradation of coral reefs in the Northern Line Islands. PLoS One 3:e1548
Schipper J, Chanson JS, Chiozza F et al (2008) The status of the world's land and marine mammals: diversity, threat, and knowledge. Science 322:225–230
Schumann S (2007) Co-management and "consciousness": fishers' assimilation of management principles in Chile. Mar Policy 31:101–111
Simmonds EJ (2007) Comparison of two periods of North Sea herring stock management: success, failure, and monetary value. ICES J Mar Sci 64:686–692
Smith TD (1994) Scaling fisheries: the science of measuring the effects of fishing, 1855–1955. Cambridge University Press, New York
Speth JG (1992) The transition to a sustainable society. Proc Natl Acad Sci 89:870–872
Squires D, Kirkley J, Tisdell CA (1995) Individual transferable quotas as a fisheries management tool. Rev Fish Sci 3:141–169
Takahashi SB, McCay BJ, Baba O (2006) The good, the bad, or the ugly? Advantages and challenges of Japanese coastal fisheries management. Bull Mar Sci 78:575–591
Walker BH, Holling CS, Carpenter SR, Kinzig A (2004) Resilience, adaptability and transformability in social-ecological systems. Ecol Soc 9:5. http://www.ecologyandsociety.org/vol9/iss2/art5. Accessed 11 June 2010
World Database on Protected Areas (2009) World database on protected areas. United Nations Environment Program. Explore the world database on marine protected areas. http://www.wdpa-marine.org/#/countries/about. Accessed 7 June 2010
Worm B, Barbier EB, Beaumont N et al (2006) Impacts of biodiversity loss on ocean ecosystem services. Science 314:787–790
Worm B, Hilborn R, Baum JK et al (2009) Rebuilding global fisheries. Science 325:578–585
Young OR (1994) International governance: protecting the environment in a stateless society. Cornell University Press, Ithaca
Young MD, McCay BJ (1995) Building equity, stewardship, and resilience into market-based property rights systems. In: Hanna S, Munasinghe M (eds) Property rights and the environment: social and ecological issues. The World Bank, Washington, DC

Part IV
The Ecology of Cities

The global phenomenon of human migration to cities, combined with the observation that most megacities (populations exceeding 10,000,000) are situated on coasts and estuaries, makes the ecology of cities an open frontier for sustainability science research, education, and outreach. This theme will focus on the emergence of urban coastal research integrated with the social sciences and humanities.

There is growing opportunity to extend and integrate knowledge of the urban center combining an ecological lens with social and economic understanding at different scales governing the flows and cycles of critical resources. Moreover, Pickett and Cadenasso (2006) note that all ecosystems inhabited by humans should be "modeled to include individuals as well as the social aggregations they generate or influence." They suggest further that "it is perfectly reasonable to incorporate such factors and processes into ecosystem" and pose the questions:

1. How do the spatial structures of socioeconomic, ecological, and physical features of an urban area relate to one another and how do they change with time?
2. What are the fluxes of energy, matter, human built capital, and social capital in an urban system; how do they relate to one another; and how do they change over the long term?
3. How can people develop and use an understanding of the metropolis as an ecological system to improve the quality of their environment and to reduce pollution to downstream air, watersheds, and coastal environs?

In short, what are the institutional arrangements, constraints, and opportunities out there that test our mettle as scientists (natural, social, and economic)?

The concept of urban ecosystems as regions (landscapes) also lends credibility to the boundary between planning and ecology. It recognizes that the human populace must have "access and responsibility for the distributed resources and amenities upon which they depend" (Pickett and Cadenasso 2006). The authors note further, "watersheds, floodplains, large natural areas, connecting corridors, green buffers, green water management infrastructure, and recreational parks and playgrounds are all parts of the kind of patches that must be planned for, appropriately arranged and phased in as urban regions grow and change."

While the knowledge of nature in cities sets the foundation for addressing ecological processes, it is not sufficient for understanding how those processes ultimately become a function of the feedback dynamics associated with interactions among social, ecological, and economic drivers. Like it or not, humans are here to stay, and we need to "wedge sustainability" into our own consciousness, as well as engaging the public and decision-makers to this reality (Coontz 2007).

References

Pickett S, Cadenasso M (2006) Advancing urban ecological studies: frameworks, concepts, and results from the Baltimore ecosystem study. Austral Ecol 31:114–125

Coontz R (2007) Wedging sustainability into the public consciousness. Science 315:1068–1069

Cities as Dissipative Structures: Global Change and the Vulnerability of Urban Civilization

William E. Rees

Abstract Techno-industrial society and modern cities as presently conceived are inherently unsustainable. This conclusion flows from the energy and material dynamics of growing cities interpreted in light of the second law of thermodynamics. In second law terms, cities are self-organizing, far-from-equilibrium dissipative structures whose "self-organization" is utterly dependent on access to abundant energy and material resources. Cities are also open, growing, dependent subsystems of the materially-closed nongrowing ecosphere—they produce themselves and grow by feeding on energy and matter extracted from their host ecosystems. Indeed, high-income consumer cities are concentrated nodes of material consumption and waste production that parasitize large areas of productive ecosystems and waste sinks lying far outside the cities. The latter constitute the cities' true "ecological footprints." In effect, thermodynamic law dictates that cities can increase their own local structure and complexity (negentropy) only by increasing the disorder and randomness (entropy) in their host system, the ecosphere. The problem is that anthropogenic degradation now exceeds ecospheric regeneration and threatens to undermine the very urban civilization causing it. To achieve sustainability, global society must rebalance production and consumption, abandon the growth ethic, relocalize our economies and increase urban-regional self-reliance, all of which fly in the face of prevailing global development ideology.

Keywords Sustainable cities • Ecological footprint • Alternatives • Second law of thermodynamics

W.E. Rees (✉)
School of Community and Regional Planning, University of British Columbia,
6333 Memorial Road, Vancouver, BC V6T 1Z2, Canada
e-mail: wrees@exchange.ubc.ca

M.P. Weinstein and R.E. Turner (eds.), *Sustainability Science: The Emerging Paradigm and the Urban Environment*, DOI 10.1007/978-1-4614-3188-6_12,

H. sapiens: Unsustainable by Nature and Nurture

Sometime in 2009 the number of people living in urban regions reached 3.4 billion, equal to 50% of the total human population. Already the greatest mass-migration of people in history, the pace of urbanization is accelerating—the United Nations projects that urban populations will increase by an additional 2.9 billion the next 4 decades (UN 2009). In other words, the world's cities are expected to add more people in the coming 40 years than had accumulated on Earth in the entire history of *H. sapiens* up until 1957! Urban population growth will exceed the anticipated increase in total human numbers (2.3 billion) as cities absorb the equivalent of global growth while drawing significantly from rural populations. All such projections, of course, assume a more or less "business-as-usual" scenario that the prevailing global cultural narrative of progress-with-growth will continue to play out smoothly and without interruption. They ignore potential potholes on the road to urban utopia stemming from the increasing impact of human economic demands on critical biophysical life-support systems.

This paper poses an alternative interpretation of the prospects for the world's cities, great and small. The future will not unfold as a steadily improving extension of the past; the uninterrupted urbanization of humanity cannot be taken for granted. Cities are facing ecologically uncertain times. Indeed, the very expansion of cities increases their exposure to hazards associated with self-induced global ecological change.

This outcome is not due to any fault of cities per se (cities actually have significant potential sustainability advantages over lower-density settlement patterns). Rather, my starting premise is that our entire resource-intensive industrial culture is inherently unsustainable. Indeed, I have argued elsewhere that unsustainability is an *inevitable emergent property* of the systemic interaction between contemporary global society and the ecosphere and that the symptoms are steadily worsening (Rees 2008, 2010). This implies that any scenarios or plans based on simple projections of prevailing trends are likely to founder.

The proximal roots of this conundrum spring from today's global development paradigm. The growth-oriented beliefs, values, and assumptions underpinning contemporary economic models and consequential "environmental" behavior are fundamentally at odds with the biophysical laws and dynamics governing vital ecosystems and geophysical systems. It is, therefore, difficult to envision any politically acceptable reform of the prevailing paradigm that would produce a sustainable relationship between the modern human enterprise and nature. As matters stand, the global human enterprise is in a state of overshoot. Aggregate energy and material consumption and waste production exceed the regenerative and waste assimilation capacities of the ecosphere. This happens despite the fact that over three billion people—almost half the human family—remain in poverty living on less than $(US) 2.50 per day.

The distal roots of unsustainability are deeply intractable. The ecologically dysfunctional behavior of modern society is shaped by the complex interplay of both innate (i.e., genetically influenced) predispositions and socially-constructed cultural narratives. It does not help that both sets of drivers operate mostly subconsciously. People are generally unaware of the quiet "whisperings" of their genes (or deny their existence) (Barrash 1979). Nor are they conscious that they are acting out of various socially-constructed beliefs and ideologies acquired automatically simply by growing up in a particular culture (Grant 1998; Rees 2002). Human (un) sustainability is a truly "wicked problem."

The biological contribution to this conundrum derives, in part, from the most basic of Darwinian survival strategies: unless constrained by negative feedback, *H. sapiens*, like all other species, will expand numerically to fill all accessible habitats and consume all available resources (Ponting 1991; Fowler and Hobbs 2003). Of course, in the case of humans, "availability" is defined by contemporary technology and there can be little doubt that human technological ingenuity accounts for much of our species' unparalleled success in the material struggle for existence (Rees 2010).

The socio-cultural element is somewhat more complex, in part because there is actually a second layer of nature–nurture interaction at play. On the one hand, humans are obviously capable of acquiring virtually *any* set of cultural norms. On the other, they are naturally predisposed to adopting the beliefs, narratives, and behavioral norms of the *particular* culture in which they are raised. (Shared meaning is essential for group cohesion and for personal and cultural identity.) Thus, while different societies subscribe to markedly different foundational myths, religious doctrines and grand cultural narratives (not to mention disciplinary paradigms and political ideologies), the "social construction" of such unifying cultural norms is a universal property of human societies (Berger and Luckmann 1966).

This observation is no mere curiosity—it is shared myths, models, and narratives that give shape and direction to a society's economic and political life and determine how its members "act out" in the natural world (Grant 1998). The critical point here is that, while the need for socially-constructed meaning is yet another vessel cast from our genes, *the specific contents of our shared "stories" are determined by socio-cultural context.*

Some social scientists may question the role of innate predispositions in human population dynamics, economic expansionism, and tendency to over-consume. Such doubters, however, are obliged to provide an alternative explanation of the fact that: (1) released by technology from negative feedback (disease, starvation, resource shortages), the human population exploded super-exponentially beginning in the nineteenth century and is only beginning to level off; (2) humans have come to have the widest geographical range of any mammalian species (no habitable landmass remains unsettled by humans and even the most uninhabitable of Earthly environments, save the bottom of the sea, have semi-permanent human outposts); (3) even the popular press has been forced to acknowledge evidence of "peak oil," "peak phosphorus"—indeed, "peak *everything*" (Heinberg 2007). It is worth noting here the world's increasing anxiety that some nations, particularly China are stockpiling or withholding from markets various minerals essential to today's high-tech electronic lifestyles. To bring matters closer to home, anyone concerned about fiscal deficits, the national debt, or even his/her own credit card (im)balances has direct empirical evidence of the human propensity to use up available resources—and then some! All such observations are consistent with Fowler's and Hobbs' (2003) conclusion that, in material terms, *H. sapiens* is a true outlier species. Modern *H. sapiens*' demands on our supportive ecosystems typically dwarf those of 95 ecologically similar species by one or two orders of magnitude![1]

There will be less questioning of the facts that humans routinely "socially construct" their effective realities (Berger and Luckmann 1966) and that global society has spent the past half-century purposefully constructing an economic reality based on perpetual material growth. The governing elites of the market democracies have persuaded or cajoled virtually the entire world into adopting a common myth of uncommon power. All major national governments and mainstream international agencies are united in a vision of global development and poverty alleviation centered on unlimited economic expansion fueled by open markets and more liberalized trade (Rees 2002). Supporting (or rationalizing) this core narrative is a critical secondary myth: that human welfare can be all but equated with ever-increasing material well-being (i.e., perpetual income growth).

In short, the world community is currently enthralled with a global development myth that echoes the expansionist "whisperings" of our genes. And who would deny that the expansionist vision has been the principal dynamic lending "shape and direction" to economic political life everywhere for at least the past half century? It is this vision that provides the context for our assessment of prospects for cities in the twenty-first century.

The City as Biophysical Entity

The word "City" means different things to different people. Wikipedia defines the city as "a relatively large and permanent settlement" and, somewhat circularly, as "an urban settlement with a large population." These are fairly typical of definitions by people who think in geographic or demographic terms. Then again, architects and urban designers think of cities as areas dominated by elements of the "built environment"; anthropologists and sociologists see the city as a hotbed of artistic, cultural and political activity; engineers regard cities as intensely concentrated nodes in a global network of transportation, communication and related infrastructure, and Jacobs (1969) famously described cities as the "engines of national economic growth."

[1] Even preagricultural humans significantly altered energy and material flows through ecosystems by virtue of large per capita energy demands and group living. This necessarily affected biodiversity. As Diamond (1992) writes: "For every area of the world that paleontologists have studied and that humans first reached within the last fifty thousand years, human arrival approximately coincided with massive prehistoric extinctions." Pimm et al. (1995) estimate "that with only Stone Age technology, the Polynesians exterminated >2,000 bird species, some ~15% of the world total."

There is no disputing these and other definitions of the city—the city is all of these things. The problem is that even the most inclusive definitions are incomplete, whether in urban reference works or standard dictionaries. What is missing from most traditional popular and academic descriptions is any appreciation of the city as a complex *ecological* entity subject to thermodynamic and other biophysical laws.

This omission is perversely illogical. Cities increasingly help to define the human ecological niche. They comprise the *major habitat of the dominant species on the planet* and make unmatched biophysical demands on the ecosphere. In short, cities have become a critical structural, functional, and spatial expression of human ecological reality. It is the more remarkable, therefore, that the very concept of cities as ecological entities remains below most people's cognitive radar.

The reasons for this perceptual gap are no doubt complex but certainly related to modern humanity's general sense of alienation from nature. So-called "Cartesian dualism" has created a nearly impermeable psychological barrier between all things human and the natural world. To make matters worse, modern technology and urbanization itself tend to widen the gap and foster the illusion that humanity is decoupling from nature. Even well-educated people do not generally think of themselves as ecological or even biological entities until forced to by health or related "environmental" problems. By extension, modern urbanites see cities and the urban world as polar opposites to wilderness and the natural world; they have been forced to contemplate the biophysics of cities only by global change.

Cities: Quintessential "Dissipative Structures"

Perhaps the most fruitful way to reassert the city's de facto connectedness to the rest of nature is through far-from-equilibrium thermodynamic theory. The starting point for this approach is the second law of thermodynamics, i.e., the entropy law.

In its simplest form, the second law states that any spontaneous change in an isolated system—a system that can exchange neither energy nor material with its environment—increases the system's "entropy." This is a technical way of stating that things naturally tend to wear out and run down. With each successive change, an isolated system loses potential; it becomes more randomly structured and disordered, energy dissipates, concentrations disperse, and gradients disappear. Eventually, the system reaches "thermodynamic equilibrium," which is a state of maximum local entropy in which nothing further can happen.

Of course, many complex real-world systems, ranging from newborn infants to cities, to the entire ecosphere, are neither isolated nor sliding toward equilibrium. The ecosphere, for example, is a highly-ordered self-organizing system of mind-boggling complexity, multi-layered structure and steep gradients as represented by millions of distinct species, differentiated matter, and accumulated biomass. Over geological time, its biodiversity, systemic complexity, and energy/material flows have been increasing—i.e., the ecosphere has been moving ever *further* from equilibrium. Indeed, this phenomenon may well be the measure of life.

As Prigogine (1997) asserts, "distance from equilibrium becomes an essential parameter in describing nature, much like temperature [is] in [standard] equilibrium thermodynamics."

Scientists long thought that living systems were exempt from the second law because they *gain* in structural mass and functional complexity over time. This is not the case—all systems are subject to the same processes of entropic decay. (There are no known violations of the second law.) The paradox dissolves only when we recognize that all living systems, from cellular organelles to entire ecosystems, and from cities to the ecosphere, are *open* systems that freely exchange energy and matter with their host "environments."

Most critically, some systems biologists now describe living systems as existing in overlapping nested hierarchies in which each component system ("holon") is contained by the next level up and itself comprises a chain of linked subsystems at lower levels. This organizational form is the basis for "SOHO" (*s*elf-*o*rganizing *h*olarchic *o*pen) systems theory (see Kay and Regier 2000). Each subsystem or holon in the hierarchy grows, develops, and maintains itself by extracting usable energy and material (negentropy) from its host "environment" one level up. It processes this energy/matter internally to produce and maintain its own structure/function and exports degraded energy and material wastes (entropy) back into its host. In short, living organisms maintain their *local* level of organization as far-from-equilibrium systems at the expense of increasing *global* entropy, and particularly at the expense of the entropy of their immediate host system (Schneider and Kay 1994, 1995).[2] They are called "dissipative structures" because all self-organizing systems survive by continuously degrading and dissipating available energy and matter (Prigogine 1997).

What has this to do with humans and where do cities fit into this discussion? Like the ecosphere, cities are self-organizing far-from-equilibrium dissipative structures. Cities, however (indeed, the entire human enterprise), are open, growing, dependent subsystems of the materially closed, nongrowing finite ecosphere. Thus, while the ecosphere evolves and maintains itself by feeding on an extra-terrestrial source of energy, and by continuously recycling matter, cities grow and maintain themselves by feeding on the rest of the ecosphere and ejecting their wastes back into it. In other words, cities can grow and increase their internal order (negentropy) only by

[2] Renegade economist Georgescu-Roegan (1971a, b) was among the first to understand the implications of the second law for the human economy. Starting from the fact that all economic activity must draw low entropy resources out of nature and dump useless high entropy waste back in, he reasoned first that "…in a finite space there can be only a finite amount of low entropy and, second, that low entropy continuously and irrevocably dwindles away." He further speculated that since modern humans are unlikely to practice restraint in their use of resources, nature and human nature may combine to ensure that "…the destiny of man is to have a short but fiery, exciting, and extravagant life…" (Georgescu-Roegen 1975). This view naturally remains controversial with opponents relying on resource substitutions and human technological ingenuity to defeat such second-law pessimism.

Table 1 A "second-law" comparison of human-less and humanized ecosystems

Ecosystems without humans	Humanized ecosystems
Grow and evolve by assimilating, degrading, and dissipating high-grade solar energy (exergy) through photosynthesis and evapotranspiration	Dedicated to economic processes involving the extraction, processing and consumptive degradation of fossil exergy and other material "resources" that have accumulated in the ecosphere, including biomass and nonhuman species
Anabolic processes (bio-production) exceed catabolic processes (respiration and dissipation)	Catabolism (destructive dissipation) exceeds anabolism (production of artifacts and manufactured capital)
Available energy and matter (biomass and other resource gradients) accumulate; species proliferate, ecosystems differentiate, complexity increases	Resource stocks are depleted and dissipated; biodiversity and structural complexity decline; ecosystems unravel and simplify
Waste heat dissipates off-earth; the entropy of the solar system and the universe increases	Waste heat dissipates off-earth; material wastes (often toxic) accumulate in the ecosphere; ecosystems lose functional integrity; the entropy of ecosphere (and ultimately the universe) increases

"disordering" the ecosphere and increasing global entropy. Table 1 includes a comparison of unexploited ecosystems with "humanized" ecosystems that are structured and redirected to supporting urban culture.

There is nothing inherently unsustainable about either the SOHO hierarchy or cities themselves, despite the dysfunctional relationship between contemporary urban culture and the ecosphere. Ecosystems can self-produce indefinitely empowered by solar energy. Anabolism marginally exceeds catabolism in the nonhumanized ecosphere, and so biomass accumulates and natural sinks (the recycling process) are capacious. In short, net primary production by producer species (mostly green plants) has always been more than adequate to sustain the world's entire complement of consumer organisms, including preindustrial humans. Critical problems have emerged only recently with the constantly rising demand and the excessive scale of the human enterprise. Far-from-equilibrium thermodynamic theory thus provides a simple double-barreled criterion for global sustainability: the human enterprise must not persistently consume more than nature produces, nor generate more waste than nature can assimilate (with a generous allowance for the thousands of other consumer species with whom we share the planet).

That said, the dynamics of perpetual throughput growth obviously contain the seeds of potential pathology. The increasingly urban human enterprise is thermodynamically positioned within the SOHO hierarchy to consume and dissipate the ecosphere from the inside out (Rees 1999). Indeed, the accelerating pace of human-induced global ecological change suggests that humanity is already a dangerously parasite on its planetary host. It does not help that the spatial separation of city-dwellers from the distant ecosystems fueling their consumer lifestyles blinds them to the worst eco-thermodynamic consequences of over-consumption and excessive pollution.[3] Wealthy urbanites may experience a world of burgeoning cities and expanding economies, but it is also a world of depleting resources, degraded landscapes, declining biodiversity, dying oceans, green-house gas accumulation, and climate change. Because of the sheer scale of human demand, wealthy cities are now as much the engines of global ecological decay as they are the "engines of national economic growth" (Table 1; Box 1).

Box 1 Dissipating the Ecosphere

The human enterprise is a global thermodynamic system that produces itself and grows by converting nature and its products into more humans and their cultural artifacts. To fuel this process, the world economy necessarily dissipates prodigious quantities of energy and matter. According to the US Energy Information Administration, the world consumed an astonishing 495×10^{15} Btu (522×10^{18} J) of energy in 2007 (EIA 2010). Liquid fuels, mostly petroleum, accounted for the largest share, equivalent to about 35% of the total, or 173×10^{15} Btu (183×10^{18} J). (In 2007, the world extracted an average of 85,899,740 barrels of oil daily for a total of 31.4 billion barrels that year.) Coal and natural gas together account for most of the rest of global consumption and about half of total energy use 245×10^{15} Btu (258.3×10^{18} J).

Only a small fraction of the material flows and high-quality (low-entropy) energy processed through the economy are actually incorporated into useful goods and services because of unavoidable thermodynamic and mechanical inefficiencies and such factors as the declining quality of mineral ores. The economic process dissipates a vastly larger quantity of energy and material back into supportive ecosystems as low-grade heat and useless waste, including climate-forcing carbon dioxide. The combustion of petroleum products alone dissipated over 11 billion metric tons (tonnes) of carbon dioxide—containing three billion tonnes of carbon—into the atmosphere in 2007.

Raw energy, of course, is merely a means to an end. Humans use energy to access, transport, and process all the other resources required or demanded by modern civilization. The total material waste-load is enormous. The World Resources Institute analyzed some of world's richest, but most efficient countries, and found that material waste output from economic production ranges from 11 tons per person per year in Japan to 25 tons per person per year in the United States. When so-called "hidden flows" are included—flows resulting

(continued)

[3] For example, as much as a third of pollution and landscape degradation in China is attributable to manufacturing for blissfully unaware consumers in the wealthy cities of North America, Europe, etc.

Box 1 (continued)

from economic activity that do not actually enter the production process, such as soil erosion, mining overburden, and earth moved during construction—then the total annual waste output increases to 21 tons per person in Japan and 86 tons per person in the United States (WRI 2000). That was already 86,000 kg (198,598 lb) every year for every man, woman, and child in the United States in the 1990s; per capita solid, liquid, and gaseous discharges are still increasing in the United States and almost everywhere else. The bottom line? The aggregate displacement of materials makes humanity the single most significant biological and geophysical force changing (mostly degrading) the planet. Growing the human enterprise necessarily increases global net entropy.

Cities as Ecosystems

Modern humans have been forced to seriously consider their role as ecological agents only by the deepening "environmental crisis." An important by-product of this awakening is a growing interest in the notion of cities as ecosystems (e.g., Register 2006). The curious student will discover that there is even a relatively new journal *Urban Ecosystems*.

The concept of "urban ecosystem" remains ill-defined and ambiguous, as might be expected this early in the transition. To some analysts, humans are only incidental to the real story. Most natural scientists studying "urban ecosystems," cast the city mainly as an unnatural habitat for *other* species. To them, the urban ecosystem consists of the assemblage of nonhuman organisms in the city and the purpose of their inquiries is to determine how these species have adapted to the structural and chemical characteristics of the "built environment." Thus, the majority of papers in *Urban Ecosystems* focus on the impacts of urbanization on nonhuman plants and animals, or on remnant "natural" ecosystems within the city.

This emphasis on other species' adaptations to an "alien" urban environment is perfectly legitimate, but it also poses an absurdist paradox. After all, it is *H. sapiens* that created "the city" and this most human of artifacts is clearly the most structurally remarkable habitat of any animal species on the planet. Moreover, the economic activities to satisfy the material demands of wealthy urbanites have become the major drivers of global ecological change. Surely, these realities should refocus attention on humans as the dominant ecological actor in all the world's cities. Not to recognize *H. sapiens* as the keystone species of the urban ecosystem is a major cognitive lapse (and yet another consequence of Cartesian dualism, which fosters the perceptual separation of people and all things human from the natural world.

On the other hand, those who do acknowledge that humans are ecological beings would be wrong if they think of the "the city" per se as comprising a human ecosystem. It is doubtlessly true that every city is a complex system (or perhaps "complex of systems") and that cities represent an ecologically critical *component* of the human ecosystem. Even if everyone lived in town, however, this would not qualify "the city," as presently conceived, as a functional human "ecosystem."

We will briefly explore the ecosystem concept itself to understand this second paradox. Odum (1971) emphasized that ecosystems should be characterized more in terms of structural properties and functional relationships than by gross morphology. He defined "ecosystem" as any biotic community interacting with its physical environment such that the flow and dissipation of energy results in a defined trophic (feeding) structure, the emergence of biodiversity, and characteristic material cycles between the living and nonliving components. In short, a functionally complete ecosystem consists of a self-sustaining assemblage of living species existing in complementary relationships with each other and their physical environs. Thermodynamically speaking, ecosystems are complex, quasi-independent, self-organizing, far-from-equilibrium dissipative structures that continuously (re)generate themselves from simple inorganic chemicals (water, carbon dioxide, nitrates, phosphates, and a few trace minerals) that are continuously recycled, using high-quality available energy (exergy) from the sun.

Clearly, by these definitions, no modern city qualifies as a complete human ecosystem. To achieve this standard, a city would have to include a sufficient complement of producer organisms (green plants), macro-consumers (animals, including humans), micro-consumers (bacteria and fungi), and abiotic factors to support its human population indefinitely. As matters stand, some defining elements are missing from cities altogether or are insufficiently abundant to maintain the system's functional integrity. Consider, for example, that the largest and functionally most important components of urbanites' supportive ecosystems—the assemblage of organisms producing their food, fiber and oxygen; and most of the micro-consumers that complete their nutrient cycles—are found mostly in so-called rural "environments" scattered all over the planet.

Significantly, the distancing of urban consumers from the productive components of their de facto ecosystems (distant agricultural and forest lands) inhibits the replenishment of soil organic matter and the recycling of phosphorus, nitrogen, and other nutrients and contained in household and human wastes. These are typically washed out to sea (where they contribute to the emergence of anoxic marine "dead zones"). In effect, urbanization transforms local, integrated, cyclical human ecological production systems into global, horizontally disintegrated, unidirectional, throughput systems (Rees 1997). Ironically, the resultant continuous leakage of nutrients from farmland in shipments of food to cities threatens to undermine organic agriculture even as its products gain ground in the urban marketplace.

To summarize, in spatial, structural, and functional terms, cities as presently conceived are, at best, incomplete "heterotrophic" ecosystems. They are nodes of intense energy/material consumption and waste generation entirely dependent for their survival on the productive and assimilative capacities of complementary producer ecosystems located at great distance from the cities themselves (see Box 2). Urbanization thus creates a dramatic shift in city-dwellers' spatial/psychological relationships to the land, but *there is no corresponding change in eco-functional relationships*. Indeed, far-from-decoupling people from nature, urbanization generally increases their per capita "load" on cities' supportive ecosystems. Failure to understand the basic facts of urban *human* ecology will doom our quest for sustainability and increase cities' vulnerability to global ecological change.

Box 2 The City as Human Feedlot

On a crude but illustratively useful level, "the city" is the human analogue of a livestock feedlot (Rees 2003). Industrial feedlots, like cities, are densely populated by a single macro-consumer species such as cattle (or pigs, or chickens). However, the fields that produce fodder for feedlot animals may be hundreds of kilometers distant from the feedlot operation. Also missing from feedlots are adequate populations of micro-consuming decomposers. Thus, like cities, feedlots have separated the functionally inseparable, short-circuiting even the possibility of within-system decomposition and nutrient recycling. Feedlots thus generate vast quantities of manure containing vital nutrients that are often not re-deposited on range or cropland for nutrient recycling, but rather are disposed of inappropriately, contaminating the atmosphere, soils, and surface and subsurface waters at great distance and over large areas. Sound familiar? Since feedlots are, in effect, subsystems of the human urban industrial system, and constructed from the same paradigmatic framing, it is not surprising that they are eco-structurally similar to cities.

The Human Ecological Footprint

If cities constitute only a small fraction of the total urban human ecosystem, then it seems logical to ask "just how big is the whole?" Specifically, the first question of urban ecology should be "what is the total productive ecosystem area, inside and outside the political boundaries of the city that is effectively appropriated to sustain its human population?"

This kind of question is answerable using ecological footprint analysis (Rees 1992; Wackernagel and Rees 1996). An extension of energy and material flows assessment (Haberl et al. 2004), eco-footprint analysis (EFA) starts from a comprehensive inventory of the annualized energy and material flows generated by the subject population to feed, clothe, house, transport and otherwise maintain and grow itself. EF analysts also quantify the flows of critical wastes, particularly carbon dioxide. The next step is to convert each significant material flow into the corresponding area of productive land and water ecosystems required to produce (or assimilate) it. Thus, the ecological footprint of a specified population is formally defined as:

> The area of land and water ecosystems required, on a continuous basis, to produce the resources that the population consumes and to assimilate (some of) the wastes that the population produces, wherever on Earth the relevant land/water is located (Rees 2006).

The size of a population's eco-footprint (EF) generally depends on four factors: population size; average material standard of living (i.e., the EF reflects consumption); productivity of the land/water base; and the efficiency of resource harvesting, processing, and use at the time of the analysis. In other words, EFA provides an area-based snap-shot of a study population's demand on the ecosphere at the time of the analysis—technological change can affect any of these factors and would be reflected in subsequent EF studies.[4]

Whether its members are aware of it or not, every population imposes an "ecological footprint" on the planet. A moment's reflection reveals that population EFs constitute mutually exclusive appropriations of nature—the bio-capacity used by one person or population is not available for use by another. Bottom line: *all human sub-groups are competing for access to the finite area of productive ecosystems on earth.*

A population's ecological footprint represents much of the biocapacity or "natural capital" required to meet that population's consumptive and assimilative demands. However, EFA obviously does not capture all human ecological impacts. Various wastes such as ozone-depleting chemicals or the toxic chemical residues accumulating in our food chain, for example, cannot be converted into a corresponding ecosystem area. Also, the method assumes that land and waterscape are being used sustainably which is obviously not the case. In short, EFA provides a *conservative* estimate of the human load on ecosystems.

Table 2 ranks a selection of high- to low-income countries according to their *per capita* EFs. Domestic bio-capacities and overshoot factors are also given. To facilitate comparison, national EF and bio-capacity data have been converted to their equivalents in standardized global hectare (gha; hectares of global average productivity).

[4] For fuller details of the method, including inclusions, exceptions and limitations, see Rees (2006), WWF (2008, 2010), and www.footprintnetwork.org/atlas.

Table 2 The eco-footprints, bio-capacities, and overshoot factors of selected nations (estimated from 2007 data in WWF 2010)

Country	GDP per capita (in 2010 international dollars)[a]	*Per capita* eco-footprint (gha)	*Per capita* domestic bio-capacity (gha)	Overshoot factor
World	10,700	2.7	1.8	1.5
United States	47,284	8.0	3.8	2.1
Canada	39,057	7.0	15.0	0.5
The Netherlands	40,765	6.1	1.1	5.5
France	34,077	5.0	2.7	1.9
United Kingdom	34,920	4.8	1.5	3.2
Malaysia	14,670	4.8	2.6	1.8
Japan	33,805	4.7	0.7	6.7
Hungary	18,738	3.0	2.3	1.3
Mexico	14,430	3.0	3.5	0.9
Brazil	11,239	2.9	8.9	0.3
Thailand	9,187	2.3	0.6	3.8
China	7,519	2.2	1.0	2.2
Indonesia	4,394	1.2	1.4	0.9
India	3,339	0.9	0.5	1.8
Malawi	827	0.8	0.8	1.0
Bangladesh	1,572	0.6	0.4	1.5

[a]US dollar equivalent purchasing power in the United States (data from IMF 2011)

Wealthy, mainly urban, consumers clearly impose a larger average load on the ecosphere than do the mainly rural population. The citizens of wasteful high-income countries, like the United States, Canada and the Netherlands, have average EFs ranging from 6 to over 10 gha, or up to 18 times larger than the EFs of the mainly rural citizens of the world's poorest countries such as Afghanistan and Bangladesh. European countries and Japan typically have *per capita* EFs in the 4–6 gha range. China is fairly representative of the emerging economies which show rapidly growing EFs currently in the range of 1.5–3 gha. These data reflect the growing global income gap; the richest 20% of the human family consume more than 75% of the world's income; the poorest 20% subsist on just 1.5% of the world's income (Shah 2010).

The final column of Table 2 shows each country's "overshoot factor." The overshoot factor is the ratio of the national average eco-footprint to *per capita* domestic bio-capacity. Countries with overshoot factors larger than one impose a greater burden on the ecosphere than could be supported by their domestic ecosystems. That is, these countries are at least partially dependent on trade and on exploitation of the global commons to maintain their current lifestyles. The Netherlands and Japan, for example, use the ecological services of 5.5 and 6.7 times as much productive land/water elsewhere in the world than is contained within their respective borders. All countries in overshoot are running "ecological deficits" with the rest of the world.

Those few countries with overshoot ratios less than 1 (e.g., Brazil, Canada) seem to have ecological surpluses and could therefore still theoretically live within their "natural incomes." But they only *seem* to have surpluses because any extra bio-capacity is generally being traded away (or appropriated by pollution) to cover the ecological deficits of other countries. The agricultural, forestry, fisheries, and carbon sink surpluses of Canada, for example, serve a large export market. One consequence is that production for trade contributes proportionately to the ongoing degradation of that nation's soils, forests, and fish stocks (Kissinger and Rees 2009).

Ominously, the world as a whole is in a state of overshoot (Table 2). Human demand exceeds Earth's regenerative capacity by about 50% (WWF 2010). We are living, in part, by depleting and dissipating as waste, the enormous stocks of potentially renewable natural capital (fish, forests, soils, etc.) that have accumulated in ecosystems over millions of years. Continuing in this vein, we can interpret the eco-footprint in thermodynamic terms, as the minimal area of natural photosynthetic "solar collector" needed to regenerate the biomass and chemical energy equivalents of the useful resources and fossil energy that any study population consumes and dissipates.

The Ecological Reach of Global Cities

As might be expected, the eco-footprints of cities are virtually all a deficit. Urban populations are almost totally dependent on rural peoples, ecosystems, and life-support processes increasingly scattered all over the planet (Rees 1992, 2003; Girardet 2004; Wackernagel et al. 2006; Newman and Jennings 2008). Of course, the urban–rural relationship temporarily benefits both parties—cities provide markets for the products of the countryside and technologies that can improve rural living, for example. But, while rural populations have survived historically without cities, the ecological dependence of modern urbanites on "the hinterland" is absolute. *There can be no urban sustainability without rural sustainability.*

So, just how heavily does a typical modern city tread on the global countryside? Despite methodological and data-quality differences,[5] urban eco-footprint studies invariably show that the EFs of high-income cities exceed their geographic or political areas by two to three orders of magnitude. For example:

- Assuming they are typical Canadians (per-capita EF = 7.0 gha), the 600,000 citizens of my home town, Vancouver, effectively occupy an ecosystem area

[5] Urban eco-footprints are more difficult to estimate than national EFs. National EFs are based on data routinely collected by government statistical agencies and international (e.g., UN) organizations, but no such agencies monitor trade across municipal boundaries. In the absence of original local data-gathering capacity, city EFs can be based on the national per capita estimate adjusted for local variations in income, energy sources, lifestyles, etc.

outside the municipal boundaries that is 368 times larger than the city's 11,400 ha. Even the metropolitan population of 2.2 million, living at lower average densities, has an extraterritorial eco-footprint 55 times larger than the metropolitan region's 2,787 km^2.

- Folke et al. (1997) estimated that the 29 largest cities of Europe's Baltic region require the bio-capacity of forest, agricultural, marine, and wetland ecosystems that is 565–1,130 times larger than the area of the cities themselves.
- Warren-Rhodes and Koenig (2001) showed that the almost seven million people of Hong Kong (EF = 5.0–7.2 ha/capita) have a total eco-footprint of 332,150–478,300 km^2. Thus, Hong Kong's eco-footprint is 303 times the total land area of the Hong Kong Special Administrative Region (1,097 km^2) and 3,020 times the built-up area of the city (110 km^2).
- With a population of 33 million and assuming the Japanese average per capita EF of about 4.7 gha, the Tokyo metropolitan region has a total eco-footprint of 155,100,000 gha. However, the entire domestic bio-capacity of Japan is only about 89,000,000 gha (2007 data).[6] In short, metropolitan Tokyo alone, with only 26% of Japan's population, lives on an area of productive ecosystems 1.7 times larger than the nation's entire terrestrial bio-capacity!

These data show, first, that in energy and material terms, contemporary cities are entropic black holes sweeping up the productivity of a vastly larger and increasingly global resource hinterland (and, necessarily, spewing an equivalent quantity of waste back into it). Second, a footprint analysis provides a conservative estimate of the scale of true (i.e., complete) human urban ecosystems. Both the consumptive built-up core and the vastly more extensive supportive countryside are essential components of the whole—is there any logical reason why the lifeless parking lot at the mall is more legitimately a part of the urban ecosystem than are the farm- and forest-lands dedicated to sustaining the city's human inhabitants? Third, an EFA underscores the fact that no city or country can achieve sustainability on its own if it is part of a SOHO hierarchy operating unsustainably. Vancouver, Hong Kong, and Tokyo might strive to become exemplars of compact urban design and sustainable lifestyles but, if the extra-urban ecosystems upon which they are dependent fail, so would our model cities. Given the increasing global entanglement of nations, the best any national or urban subsystem can attain independently is a state of quasi-sustainability. "Quasi-sustainable" describes that level of economic activity and energy/material consumption per capita which, if extended to the entire system, would result in global sustainability (Rees 2009). In 2010, quasi-sustainability implied a per capita eco-footprint of about 1.8 gha which represents an equitable share of global bio-capacity (Table 2). Using my own country as an example, Canadians would have to take steps to

[6] The terrestrial area of Japan is only about 37,770,000 ha but Japan's terrestrial ecosystems are considerably more productive than the world average. This increases the country's bio-capacity to about 89,000,000 gha.

reduce their average eco-footprints by 74% (from 7.0 to 1.8 gha per capita) to meet the quasi-sustainability standard! Bangladeshis, on the other hand could *increase* their material consumption by 200% (both cases assume stable populations and sustainable use of natural capital).

Global Change and the Vulnerability of Cities

The EFA suggests a fourth and more ominous set of implications for the future of cities. Trade and accelerating globalization have historically assured the abundant supplies and uninterrupted flows of the energy and other material resources required to grow the modern metropolis. But this raises an awkward question in an era of global change: just how secure is any city of millions, or even a relative "town" of 100,000, if resource scarcity, shifting climate, or geopolitical unrest threaten to cut it off from vital supplies? Several accelerating trends driven by population growth and explosive urbanization suggest that this is no idle question. Here are four examples:

Land and soil degradation. The Food and Agriculture Organization estimates that feeding an anticipated population of 9.1 billion people "would require a 70% increase in global food production between 2005–2007 and 2050. Production in the developing countries would need to almost double" (OECD-FAO 2010). While there is significant debate over prospects for agriculture, achieving increases of this magnitude may be difficult. Erosion rates resulting from the conventional tillage-based agriculture still dominant in the world today average an unsustainable 10–100 times greater than rates of soil production, erosion under native vegetation, and long-term geological erosion (Montgomery 2007). One study suggests topsoil is currently being eroded 16–300 times faster than it can be replaced (Barrow 1991). Severe and very severe land degradation due to agricultural activities affects 1.2 billion ha (about 35% of all severely degraded land). The total area presently under arable use is only slightly greater than this at about 1.4 billion ha (FAO 2000). About 300 million hectares of cultivated land—enough to feed almost all of Europe—has been lost to production, and we are still losing five to seven million hectares annually (FAO 2000; SDIS 2004). So far, part of the impact of land and soil degradation on production is unnoticed because we have managed to substitute fossil fuel for depleted soils and landscape degradation—but that may be about to change.

Peak oil and the costs of energy. As the "second law" reminds us, available energy is prerequisite for everything. The first permanent human settlements can be traced to the dawn of agriculture and surpluses of food energy; modern mega-cities are very much the product of abundant cheap fossil energy. The explosion of human populations and urban accumulations of manufactured capital would not have been possible without fossil fuels; raising the urban human enterprise ever-further from equilibrium requires increasingly prodigious quantities of energy (Box 1). Oil is also the major driver of the green revolution. Mechanization, diesel-powered irrigation, the capacity to double-crop, and agro-chemicals (fertilizers and pesticides) made from oil and natural gas account for 79–96% of the increased yields of wheat, rice, and maize production since 1967 (Conforti and Giampietro 1997; Cassman 1999).

For all these reasons, various analysts have argued that the peaking of global crude output could pose a greater challenge to geopolitical stability and urban security than any other factor (Duncan and Youngquist 1999; Laherrere 2003). It is significant then, that in November 2010, the International Energy Agency (IEA) acknowledged for the first time that conventional oil production has *already* peaked—in 2006. Remarkably, and with no explanation, the agency goes on to suggest in its central "New Policies Scenario" that oil supplies will continue to increase and still be the mainstay of the global economy in 2035, although more than half of the anticipated supply will come from oil fields "yet to be developed or found" (IEA 2010)! Meanwhile, it is not entirely a coincidence that as 2011 and economic recovery get under way, oil prices are rising toward $US 90 a barrel and world food prices are again approaching the disastrous high reached in 2008. Of course, with every up-tick in food prices millions more people join the billion already calorically malnourished. If current trends continue, a return to food export restrictions and price riots is likely. In addition to shrinking the supply and exploding the price of food, "peak oil" could have an enormous impact on urban transportation, urban form/structure, and the future size of cities. Wither urban security? Duncan (2001) argues that in the absence of viable substitutes for petroleum, the life expectancy of industrial civilization is only about 100 years; i.e., from 1930 to 2030.

Global climate change. Some analysts see human-induced climate change (in part, a product of fossil fuel combustion and the entropic dissipation of carbon dioxide into the atmosphere) as the greatest threat to urban civilization. The concentration of atmospheric carbon dioxide reached 390 ppm at the end of 2010, which was 39% above the preindustrial level of 280 ppm, and the rate of increase is rising. Other greenhouse gases are climbing at a faster pace. Even modest resultant increases in the global heat balance could permanently change the spatial coincidence of suitable climate regimes and arable soils, thus negatively affecting food production. It would also alter the historic availability and spatial distribution of water that have long-shaped human settlement patterns and the geography of development, in general. By some accounts, climate change could bring the world to the edge of anarchy (Schwartz and Randall 2003; CSIS 2007). Washington's Center for Strategic and International Studies suggests that human-induced climate change could end peaceful global integration as various nations contract inwardly to conserve what they need—or expand outwardly to take what they need—for survival (more on this below). In the event of "severe climate change," corresponding to an average increase in global temperature of 2.6°C by 2040 (now deemed to be increasingly likely), major nonlinear changes in biophysical systems will give rise to major nonlinear socio-political events. Shifting climate will force internal and cross-border migrations as people abandon areas where food and water are scarce. People will also flee rising seas and areas devastated by increasingly frequent droughts, floods, and severe storms. Dramatic increases in migration combined with food, energy, and water shortages will impose great pressure on the internal cohesion of nations. War is likely and nuclear war is possible (CSIS 2007).

Before dismissing such disaster scenarios as alarmist, consider just some of the extreme weather events of 2010 (WMO 2010). To provide context, note that the January to October period was the warmest in the instrumental record, and that 2010 as a whole is in a virtual three-way tie with 1989 and 2005 as the warmest year since 1850. Warming was especially strong in Africa, parts of Asia, and parts of the Arctic including Arctic Canada with average temperatures running 1–3°C above the long-term mean. Seventeen countries recorded their highest-ever record temperatures. Here are some of the more notable extreme weather events of 2010:

- Several parts of Eurasia recorded exceptional heat-waves with the most extreme heat centered over western Russia. The July mean temperatures in Moscow were 7.6°C above normal, making it the city's hottest month by more than 2°C. The city reached 30°C or above on 33 consecutive days (compared to 0 days above 30°C in the summer of 2009) and a new record high of 38.2°C was set on 29 July. About 11,000 excess deaths in Moscow alone were attributed to the extreme heat. The Russian grain crop was reduced by a third, prompting the government to restrict exports. This had an immediate effect on global food prices and, therefore, on import-dependent populations.
- Parts of the Northern Hemisphere had an abnormally cold winter. Central Russia experienced the greatest temperature anomalies (4°C below normal), but the most unusual conditions were on the western periphery of Europe, where Ireland and Scotland both experienced their coldest winter since 1962–1963. The United States had its coldest winter since 1984–1985.
- While western Russia baked, an exceptional monsoon brought Pakistan its worst flooding in history. Over 20 million people were displaced, more than 1,500 people were killed, and large areas of Pakistan's agricultural land were inundated. In terms of numbers affected, the UN rated the flood as the greatest humanitarian crisis in recent history. By contrast, north-eastern India and Bangladesh "enjoyed" their driest monsoon season since 1994.
- Other regions also suffered heavy rains and flooding in 2010. In Indonesia, at least double the normal monthly rainfall fell each month from June to October in Java, the islands east of Java, and in southern Sulawesi. The May–October period was the wettest on record for northern Australia with rainfall 152% above normal. Averaged over Australia, the spring was the wettest on record. (Australia's travails continued into the New Year with massive flooding in Queensland covering an area of the size of France and Germany combined, displacing hundreds of thousands, and destroying much of the state's grain crop.) Germany had its wettest August on record; Bursa (Turkey) had its wettest January–October on record (1,152 mm, 132% above normal); in South America, November brought Colombia its most severe floods in more than 30 years.
- Other regions suffered severe drought. Yunnan and Guizhou provinces in China had their lowest rainfalls on record during the period from September 2009 to

mid-March 2010 with totals 30–80% below normal. Pakistan experienced drought in the early months of 2010 before the onset of record monsoon floods. An unusually dry July–September in north-western Brazil sharply reduced stream-flow in many parts of the Amazon basin. The Rio Negro, a major Amazon tributary, fell to its lowest level on record. The UK had its driest January–June period since 1929 (and the list goes on—see WMO 2010 for details).

Recent results from climate scientists suggest that such extreme events may well become the norm as GHG emissions accelerate. Anderson and Bows (2008) argue that "an optimistic interpretation of the current framing of climate change implies that stabilization [of green house gases] much below 650 ppmv CO_2e [carbon dioxide equivalents] is improbable." This is partly because, in order to stabilize at 650 ppmv CO_2e, the majority of OECD nations will soon have to begin decarbonizing at rates in excess of 6% per year and there is virtually no possibility of this happening (even though it would simultaneously help to address looming petroleum shortage). Note that atmospheric GHG concentrations of 650 ppmv CO_2e imply a catastrophic 4°C increase in mean global temperature, cf., the mere 2.6°C increase assumed in CSIS's already horrific "severe climate change" scenario. Other models suggest that a 4°C warming would be sufficient to convert much of the United States, Southern Europe, China, India, Africa, and South America into uninhabitable wastelands (see Vince 2009), displacing hundreds of millions of people and jeopardizing prospects for maintaining any form of civilization.

Schwartz and Randall (2003) set out to "imagine the unthinkable" in studying the security-related implications of shifting global climate. They created a seemingly extreme climate change scenario based on fairly abrupt shifts in historic weather patterns which "although not the most likely, is plausible." Given the accelerating pace of change, one wonders whether events have elevated the probabilities from merely plausible toward "possible" or even "likely."

Can Urban Societies Adapt? (Do We Have a Choice?)

> [Thermodynamics]...holds the supreme position among the laws of nature... If your theory is found to be against the Second Law of Thermodynamics, I can give you no hope; there is nothing for it but to collapse in deepest humiliation (Eddington 1929).

Urban civilization necessarily exploits its supportive ecosystems to produce itself and grow. The consumption by human enterprise, however, has now surpassed the regenerative capacity of the ecosphere. Continued growth in energy and material throughput will inevitably accelerate the pace of fisheries collapse, biodiversity loss, land and water degradation, resource scarcity, marine eutrophication (ocean dead zones), greenhouse gas accumulation, climate change, etc. In short, humans are destructuring and dissipating critical resource ecosystems, polluting most others, and disrupting life-support functions essential to our own survival. For all the wishful thinking of politicians, growth economists, and other techno-optimists, there is no escape from the second law of thermodynamics—beyond a certain point, further growth in the human enterprise can occur *only* at the expense of the entropic dissipation of the ecosphere.

Resource scarcity and degraded landscapes have always led to civil strife and war. Global civilization is, therefore, in double jeopardy, for nothing is more vulnerable to rapid ecological change and potential geopolitical chaos than world's cities.

Humans self-describe themselves as having certain unique qualities, including high intelligence, the ability to reason and think logically, the capacity for forward planning, and a sense of compassion for other people and even other species. It is clear, however, that other factors hold sway when debate turns to policies to ensure sustainability. Political decision making, particularly at the national and global levels, is governed by instincts rooted in primitive tribalism, suspicion of others, narrow short-term self-interest, the felt need to protect one's socio-political status, and unwavering allegiance to continued material economic growth, all of which are expressed through competitive economic and military saber-rattling. This ensures that the age-old "march of folly" by governments (Tuchman 1984) tramples reason underfoot in the crusade for an ecologically secure future. One need look no further than the failed climate negotiations at Copenhagen in 2009, and Cancun in 2010, for recent proof that the pace of ecological change consistently outpaces the political response.

Not surprisingly, both public and private sector urban planning applications are also falling short of the mark. Such popular "sustainability" strategies as smart growth, the "new urbanism," green buildings, and urban densification create the illusion of progress while doing little to reduce our ecological footprints (Rees 2009). Usually, they do not even establish sustainability targets or establish monitoring programs, and none focuses directly on personal consumption habits.

These are critical lapses—wealthy nations should be taking steps to shrink their *per capita* material consumption and carbon emissions by up to 80%. If a given sustainability project or proposal does not result in significant, permanent, and absolute reductions in material throughputs, then it is part of the problem.

On the positive side, a growing number of mainly civil society organizations, such as Bioregional[7] and the Transition Movement (Hopkins 2008) do recognize the need for reducing energy and material use.

It seems fair to ask in the circumstances, what humans might achieve if inclusive intelligence and logic were allowed to play a more substantial role in the quest for local and global sustainability. For example, might we be able to reorganize cities better to avoid global change and adapt to unavoidable change when it comes? A comprehensive discussion of policy responses is beyond the scope of this paper, but I can provide the skeleton of one possible resilience-based scenario.

[7] See http://www.bioregional.com/ and http://www.oneplanetliving.org/index.html.

Planning for Urban Resilience: The Eco-Regional City State

> The climate crisis won't be solved by changing light bulbs and inflating your tires more, planting a tree and driving a little less. It's going to require a truly fundamental shift in how we build our cities and live in them (Register 2009).

To begin, we must agree on the factual basis of our dilemma. The data are clear that ecological degradation, including climate change, is accelerating. The sustainability-related problems of both first and third world cities are well documented (e.g., Marcotullio and McGranahan 2007; Martine et al. 2008). Regrettably, the official responses have tended to side-step the data and ignore scientific understanding. Sustainability policies and poverty reduction based on economic efficiency and technological substitutions have, if anything, exacerbated the problem by lowing costs and otherwise encouraging consumption and material growth *especially in rich countries that do not need it.* Meanwhile, the chronically impoverished are left unsustainably even further behind (Shah 2010). Solving these problems requires that we rewrite our dominant cultural narrative.

As a first step, society must come to accept that the era of unconstrained population and economic growth on a finite (and considerably diminished) planet will come to an end, either because we end rationally it or because nature ends it chaotically. Even climate scientists Anderson and Bows (2008) suggested that achieving the necessary rates of emissions reductions would require "a planned economic recession." Second, we must accept that gross socio-economic inequity (as reflected in EFA) is morally unsustainable and, if unattended, will undermine efforts to achieve ecological sustainability. Third, the global community should come to realize that for perhaps the first time in history, individual and national self-interest have converged with humanity's collective self-interest. Remember, no nation can achieve sustainability in isolation—the first-class suites on the Titanic sank just as quickly and as deeply as the steerage cabins.

How might the forgoing fundamental precepts affect our conceptualization of and planning for sustainable cities? Some ecologically sensitive goals/guidelines for sustainable urbanism include:

- Establishing an appropriate special scale for "urban" governance. Complexity theory suggests that, while regional-scale human community-associated ecosystems and watersheds are theoretically manageable, larger systems are less so (predictability gives way to uncertainty and surprise at larger scales).
- Implementing a rigorous population policy. The human population should be controlled safely below each planning region's average carrying capacity, and with due consideration of the global context.
- Initiating steady-state economic planning to maintain energy and material throughput within the regenerative capacity of supportive ecosystems.

This, in turn, requires:

- Taxation and income policies supportive of greater social equity and life-quality—e.g., appropriate social safety nets to assist needy families through the transition

to a sustainable economy. Gross inequity undermines population health and is politically destabilizing (Wilkinson and Pickett 2009).
- Self-governance at the urban-regional level, including control over as much as possible of the local land and resource base.
- Mechanisms to collectively manage land, ecosystems, and other resources vital to sustaining the community. (This will require stinting customary private property rights).
- Policies to enhance self-reliance. Planning regions need not necessarily be materially self-sufficient, but ideally they should also not become dependent on trade for vital supplies. A reasonable guiding principle might be: Trade if necessary, but do not necessarily trade.

In ideal circumstances (i.e., if we could start over knowing what we do now), we would design human settlements as bioregional urban-centered city-states. (Existing provinces and "states" are redundant in an urban world.) The overarching ecological goal of such transformative urban planning would be to reconstruct the nature–city relationship, and to convert it from a host-parasite to a mutual partnership.

Consistent with the previous guidelines, essential elements of eco-city planning include:

- Redefining the role of globalization and relocalizing regional economies.
- Reconceiving "the city system" as a complete human ecosystem (this is the ultimate in bio-mimicry).
- Consolidating as much as possible of the city's de facto "eco-footprint" (its supportive hinterland) within the natural eco-region surrounding the urban core.
- Supporting the population as much as possible on regional resources and ecosystems, thus reducing reliance on trade goods. Citizens would have an incentive to husband their local ecosystem's sustainably because their lives depend on it.
- Rethinking and reengineering "the city," to convert it from a resource-depleting throughput system, to a self-sustaining circular-flows ecosystem. For example, animal and human domestic wastes would be treated and recycled on the eco-region's farm-and-forest lands, improving soil quality, reducing the need for artificial fertilizers, and simultaneously reducing ground and surface water contamination.
- Densification of the urban core in ways that spatially reintegrate workplaces with living space. The goal would be to facilitate the city taking full advantage of the economies of scale and agglomeration economies that confer on well-designed high-density cities a substantial "sustainability multiplier" (Rees 2003) (e.g., reduced per capita demand for occupied space and transportation; greater potential for recycling, reuse and remanufacturing, electricity co-generation and district heating/cooling; expanded opportunities for co-housing, tool-sharing, and other activities that reduce material demand).
- Implementing economic and social planning policies that facilitate reducing residents' ecological footprints to a globally equitable 1.8 ha per capita. This would achieve "quasi-sustainability" and is technically possible, while actually improving individual and community well-being (Rees 2010).

- Creating multiple-redundant energy/water/food sources and related redundant infrastructure that are not fossil fuel dependent. In particular, sustainable cities should buffer water and food supplies from short-term climate vagaries (e.g., crop failures, drought, flooding) by engineering substantial storage capacity.

In this context, it is encouraging to see the emergence of handbooks for urban sustainability explicitly oriented to treating cities as ecosystems (Register 2006; Newman and Jennings 2008). What the world needs now is a number of strategically-placed demonstration projects of the eco-city state concept at work.

Epilogue

I am under no illusion that anything like the forgoing proposal will be quickly implemented anywhere. To accomplish this feat requires a complete revisioning of contemporary society's conceptualization of cities, flies in the teeth of a prevailing world development paradigm based on efficient material growth through global economic integration, and demands a total break from the premises and behaviors that dominate geopolitics.[8] Nevertheless, the human enterprise is in peril and urgent responses are necessary. Since the transition to a sustainable urban society would take decades in the best of circumstances, there may be some virtue in adding radical ideas to the mix of potential solutions under debate.

Meanwhile, in the absence of a globally-coordinated adaptive strategy for urban survival, some countries are reacting to the crisis in novel, if potentially counterproductive ways. For example, in response to land scarcity, the run-up in food prices in 2008 and restrictions imposed by some governments on food exports, many overpopulated import dependent countries (e.g., China, Saudi Arabia, Libya, South Korea) began acquiring large tracts of foreign land in mostly poor, developing countries to sustain populations back home. The International Food Policy Research Institute (IFPRI) claims that foreign investors sought or secured between 37 and 15–20 million ha (49 million ac) of farmland in developing countries between 2006 and mid-2009 (cited in The Economist 2009). Cotula et al. (2009) documented a total of 2,492,684 ha (6,156,930 ac) of approved land acquisitions from 2004 to early 2009 in just five African countries, Ethiopia, Ghana, Madagascar, Mali, and Sudan.

In the short term, these land purchases may seem to be a rational response to overpopulation, land shortages, food-insecurity and growing resource competition. It even reflects a growing sense that greater self-reliance (a major goal of my

[8] One must also ask how we could accommodate urbanized nations that are deeply in ecological debt. Remember Tokyo and Japan? Globalization has enabled that country to grow and run such a large ecological deficit that it would be impossible to create a viable bioregional "City State of Tokyo" even if we could ignore Japan's other 95 million inhabitants and had access to the entire country! (Other mega-cities and their home states face similar problems).

eco-city state proposal) may not be such a bad idea after all.[9] Distant land-lease arrangements, however, are ultimately doomed to failure if they merely facilitate additional growth in the importing country (which cancels any initial benefit), or if the now even-more-dependent country is cut off from external sources by permanent climate change, international strife, or protests by citizens of the host country (who may ultimately discover they need their domestic lands for their own purposes).[10]

What is the "bottom line?" We live in interesting times that demand dramatic approaches to the problems of city and urban vulnerabilities. It is at least possible that debating proposals for urban sustainability will produce a hybrid solution that actually works. If we succeed, society will have once again succeeded in "complexifying" in response to an unprecedented challenge, and ensuring the survival of global civilization at least until the next challenge comes along.

It is also conceivable that the challenge is insurmountable—that it defies our institutional capacity to respond, that people are simply so disillusioned by the failures of their governing elites that no effective response is possible, or that any response we do make is overwhelmed by the pace of change. In these circumstances, the implosion of global civilization is a real possibility (which event would at least serve to validate Tainter's observation that "what is perhaps most intriguing in the evolution of human societies is the regularity with which the pattern of increasing complexity is interrupted by collapse…" (Tainter 1988, 1995)). Indeed, periodic collapse may be a natural phenomenon of overmature and brittle systems that we can at best delay, but not avoid indefinitely (see Gunderson and Holling 2002). Whatever the merits of this hypothesis, this is clearly no time to repress truly out-of-the-box solutions to modern *H. sapiens*' truly out-of-the-box predicament.

References

Anderson K, Bows A (2008) Reframing the climate change challenge in light of post-2000 emission trends. Philos Transact A Math Phys Eng Sci 366(1882):3863–3882. doi:10.1098/rsta.2008.0138

Barrash D (1979) The Whisperings within—evolution and the origin of human nature. Penguin Books, New York

Barrow C (1991) Land degradation. Cambridge University Press, Cambridge

Berger PL, Luckmann T (1966) The social construction of reality. Doubleday, Garden City

Cassman KG (1999) Ecological intensification of cereal production systems: yield potential, soil quality, and precision agriculture. Proc Natl Acad Sci 96:5952–5959

[9] That said, the example of Tokyo illustrates that some form of trade will be necessary even as the balance tilts toward greater self-reliance.

[10] There is already a precedent. In April 2009, a military-backed popular uprising toppled the government of Madagascar when the citizenry learned their leaders had entered a long-term lease agreement for much of their nation's farmland with the Daewoo Corporation of South Korea. Madagascar's new government canceled the agreement.

Conforti P, Giampietro M (1997) Fossil energy use in agriculture: an international comparison. Agric Ecosyst Environ 65:231–243

Cotula L, Vermeulen S, Leonard R et al (2009) Land grab or development opportunity? Agricultural investment and international land deals in Africa. IIED/FAO/IFAD, London/Rome. ISBN 978-1-84369-741-1

CSIS (2007) The age of consequences: the foreign policy and national security implications of climate change. Center for Strategic and International Studies, Washington. http://www.csis.org/media/csis/pubs/071105_ageofconsequences.pdf

Diamond J (1992) The third chimpanzee. Harper Collins Publishers, New York

Duncan RC (2001) World energy production, population growth, and the road to the Olduvai Gorge. Popul Environ 22:503–522

Duncan RC, Youngquist W (1999) Encircling the peak of world oil production. Nat Resour Res 8:219–232

Eddington AS (1929) The nature of the physical world. Macmillan Company, New York ([re] Published by Kessinger Publishing, 2005)

EIA (2010) International energy outlook 2010. Energy Information Administration, US Department of Energy

FAO (2000) Land resource potential and constraints at regional and country levels. Food and Agriculture Organization (United Nations), Rome, Italy

Folke C, Jansson A-M, Larsson J et al (1997) Ecosystem appropriation by cities. Ambio 26:167–172

Fowler C, Hobbs L (2003) Is humanity sustainable? Proc R Soc Lond B 270(1533):2579–2583

Georgescu-Roegan N (1971a) The entropy law and the economic problem. Distinguished lecture series no 1. University of Alabama, Department of Economics, Alabama

Georgescu-Roegan N (1971b) The entropy law and the economic process. Harvard University Press, Cambridge

Georgescu-Roegen N (1975) Energy and economic myths. South Econ J 41:347–381

Girardet H (2004) The metabolism of cities. In: Wheeler S, Beatley T (eds) Sustainable urban development reader. Routledge, London

Grant C (1998) Myths we live by. University of Ottawa Press, Ottawa

Gunderson LH, Holling CS (eds) (2002) Panarchy: understanding transformations in human and natural systems. Island Press, Washington DC

Haberl H, Fischer-Kowalski M, Krausmann J et al (2004) Progress toward sustainability? What the conceptual framework of material and energy flow accounting (MEFA) can offer. Land Use Policy 21:199–213

Heinberg R (2007) Peak everything—waking up to the century of declines. New Society Publishers, Gabriola Island

Hopkins R (2008) The transition handbook: from oil dependency to local resilience. Green Books, Totnes

IEA (2010) World energy outlook 2010. International Energy Agency, Paris. Executive summary at http://www.iea.org/Textbase/npsum/weo2010sum.pdf

IMF (2011) World economic outlook database (April 2011). International Monetary Fund, Washington, DC

Jacobs J (1969) The economy of cities. Vintage Books, New York

Kay J, Regier H (2000) Uncertainty, complexity, and ecological integrity. In: Crabbé P, Holland A et al (eds) Implementing ecological integrity: restoring regional and global environment and human health (NATO science series IV: earth and environmental sciences), vol 1. Kluwer Academic Publishers, Dordrecht, The Netherlands, pp 121–156

Kissinger M, Rees WE (2009) Footprints on the prairies: degradation and sustainability of Canadian agricultural land in a globalizing world. Ecol Econ 68:2309–2315

Laherrere J (2003) Forecast of oil and gas supply to 2050. In: Paper presented to 'Petrotech 2003—New Delhi'. http://www.hubbertpeak.com/laherrere/Petrotech090103.pdf

Marcotullio P, McGranahan G (eds) (2007) Scaling urban environmental challenges—from local to global and back. Earthscan, London

Martine G, McGranahan G, Montgomery M, Fernández-Castilla R (eds) (2008) The new global frontier—urbanization poverty and environment in the 21st century. Earthscan, London
Montgomery D (2007) Soil erosion and agricultural sustainability. Proc Natl Acad Sci 104:13268–13272. http://www.pnas.org/cgi/doi/10.1073/pnas.0611508104
Newman P, Jennings I (2008) Cities as sustainable ecosystems. Island Press, Washington, DC
Odum EP (1971) Fundamentals of ecology. WB Saunders, Philadelphia
OECD-FAO (2010) Agricultural outlook 2010–2019. Organization for Economic Cooperation and Development, Paris
Pimm SL, Russell GJ et al (1995) The future of biodiversity. Science 296:347–350
Ponting C (1991) A green history of the world. Sinclair-Stevenson, London
Prigogine I (1997) The end of certainty: time, chaos and the new laws of nature. The Free Press, New York
Rees WE (1992) Ecological footprints and appropriated carrying capacity: what urban economics leaves out. Environ Urban 4:120–130
Rees WE (1997) Is 'sustainable city' an oxymoron? Local Environ 2:303–310
Rees WE (1999) Consuming the earth: the biophysics of sustainability. Ecol Econ 29:23–27
Rees WE (2002) Globalization and sustainability: conflict or convergence? Bull Sci Technol Soc 22:249–268
Rees WE (2003) Understanding urban ecosystems: an ecological economics perspective. In: Berkowitz A et al (eds) Understanding urban ecosystems. Springer, New York
Rees WE (2006) Ecological footprints and biocapacity: essential elements in sustainability assessment. In: Dewulf J, Van Langenhove H (eds) Renewables-based technology: sustainability assessment. Wiley, Chichester
Rees WE (2008) Human nature, eco-footprints and environmental injustice. Local Environ 13:685–701. http://dx.doi.org/10.1080/13549830802475609
Rees WE (2009) The ecological crisis and self-delusion: implications for the building sector. Build Res Inform 37:300–311. doi:10.1080/09613210902781470
Rees WE (2010) What's blocking sustainability? Human nature, cognition and denial. Sustain Sci Pract Pol 6:13–25. http://sspp.proquest.com/archives/vol6iss2/1001-012.rees.html
Register R (2006) Ecocities: rebuilding cities in balance with nature (revised edition). New Society Publishers, Gabriola Island
Register R (2009) Cities can save the earth. Foreign policy in focus. http://www.fpif.org/articles/cities_can_save_the_earth. Accessed 12 May 2009
Schneider ED, Kay JJ (1994) Complexity and thermodynamics: toward a new ecology. Futures 26:626–647
Schneider ED, Kay JJ (1995) Order from disorder: the thermodynamics of complexity in biology. In: Murphy MP, O'Neill LAJ (eds) What is life: the next fifty years. Cambridge University Press, Cambridge
Schwartz P, Randall D (2003) An abrupt climate change scenario and its implications for United States national security. Report prepared for the US Defense Department, Washington. http://www.climate.org/PDF/clim_change_scenario.pdf
SDIS (2004) Disappearing land: soil degradation. Sustainable Development Information Service, Global Trends (World Resources Institute), Washington, DC
Shah A (2010) Poverty facts and stats (citing World Bank Development Indicators 2008). http://www.globalissues.org/article/26/poverty-facts-and-stats
Tainter J (1988) The collapse of complex societies. Cambridge University Press, Cambridge
Tainter J (1995) Sustainability of complex societies. Futures 27:397–404
The Economist (2009) Outsourcing's Third Wave (21 May 2009)
Tuchman B (1984) The march of folly. Alfred Knopf, New York
UN (2009) World urbanization prospects: the 2009 revision. United Nations Department of Economic and Social Affairs/Population Division, United Nations, NY
Vince G (2009) Surviving in a warmer world. New Sci 201:29–33
Wackernagel M, Rees WE (1996) Our ecological footprint: reducing human impact on the earth. New Society Publishers, Gabriola Island

Wackernagel M, Kitzes J et al (2006) The ecological footprint of cities and regions: comparing resource availability with resource demand. Environ Urban 18:103–112. doi:10.1177/0956247806063978

Warren-Rhodes K, Koenig A (2001) Ecosystem appropriation by Hong Kong and its implications for sustainable development. Ecol Econ 39:347–359

Wilkinson R, Pickett K (2009) The spirit level—why greater equality makes societies stronger. Bloomsbury Press, New York

WMO (2010) 2010 in the top three warmest years, 2001–2010 warmest 10-year period. Press Release No 904 (2 Oct 2010). World Meteorological Organization, Geneva http://www.wmo.int/pages/mediacentre/press_releases/pr_904_en.html

WRI (2000) Weight of nations: material outflows from industrial economies. World Resources Institute, Washington, DC

WWF (2008) Living planet report 2008. World Wide Fund for Nature, Gland, Switzerland

WWF (2010) Living planet report 2010. World Wide Fund for Nature, Gland, Switzerland

A Mathematical Description of Urban Metabolism

Christopher Kennedy

Abstract A comprehensive, yet simple, model of the urban metabolism is described using approximately 25 closed-form equations. The equations represent essential interrelationships between the major components of metabolism—materials, water, nutrients, energy, and contaminants. The model expresses the role of infrastructure in the urban metabolism through parameters such as per capita floor space and the density of transportation infrastructure which, as part of a city's material stock, influence the flows of energy and/or water flows through the city. The density of transportation infrastructure is found to be a potentially universal parameter, with a value of 0.10 km ha^{-1}, which is relatively invariant between cities. The model also includes other parameters which, although having more variation, are independent of climate, city size, population, and other unique characteristics of cities. These other parameters include: material intensities, per capita floor space, intensity of water use for cooling, leakage rates for water distribution systems, heating and cooling intensities of buildings, and utilization rates for transportation infrastructure.

Keywords Urban metabolism • Material stock and dynamics • Infrastructure • Sustainability

Introduction

The notion of the urban metabolism is central to understanding and addressing many of the issues of resource flows and environmental impacts associated with cities, i.e., the sustainability of cities. Analogous to the metabolism of organisms, "cities transform raw materials, fuel, and water into the built environment, human

C. Kennedy (✉)
Department of Civil Engineering, University of Toronto,
35 St. George Street, Toronto, ON M5S 1A4, Canada
e-mail: christopher.kennedy@utoronto.ca

M.P. Weinstein and R.E. Turner (eds.), *Sustainability Science: The Emerging Paradigm and the Urban Environment*, DOI 10.1007/978-1-4614-3188-6_13,

biomass and waste" (Decker et al. 2000). Following Wolman's (1965) application to a hypothetical US city of one million people, about 15–20 comprehensive urban metabolism studies of actual cities have been undertaken worldwide (Kennedy et al. 2011). Such studies have quantified the flows of materials, water, nutrients, and energy through cities, including associated contaminant emissions.

In many cases, the urban metabolism concept has been used to underpin an accounting framework, rather than as a mathematical model per se. This point needs clarification. A study of urban metabolism essentially involves application of the method of material flow analysis (MFA) to cities (Baccini and Bader 1996; Brunner and Rechberger 2004). MFA researchers have developed substantial mathematical models, e.g., employing differential equations and matrix algebra, to describe the flows of one or more materials, sometimes with associated energy flows, through countries, regions, and occasionally cities. With the possible exception of studies of Vienna and a region of the Swiss Lowlands (Baccini 1987; Hendricks et al. 2000), however, mathematical MFA models have not been applied in a comprehensive analysis of urban metabolism. As a result, the studies of urban metabolism that have sought to quantify all or most of the material, water, nutrient, energy, and contaminant flows through cities have mainly become large data collection exercises.

A wide variety of mathematical models have been used in the detailed study of individual components of urban metabolism (Fung and Kennedy 2005). If we move beyond the aggregated (non-spatial) MFA models, we find that there is, for example, a large class of spatially disaggregated land-use/transportation models (Wegener 1995) that have the potential to simulate transportation energy use and emissions. Other detailed studies have modeled the spatial distribution of nutrients in urban watersheds (Costanza et al. 2002), or residential green house emitting activities (VandeWeghe and Kennedy 2007). Some of the large, complex spatially disaggregated models such as UrbanSim (Wadell 2002) are perhaps progressing to a point where they can mathematically model many aspects of urban metabolism.

The more modest objective of this paper is to describe the urban metabolism in a comprehensive, yet mathematically simple form. Essentially, the urban metabolism will be described by approximately 25 closed-form equations. While aiming at simplicity, the mathematical model needs to be sophisticated enough to:

1. Represent essential interrelationships between the various components of metabolism—materials, water, nutrients, energy, and contaminants—especially with regard to how growth in the material stock; i.e., infrastructure, impacts the other components.
2. Develop metabolism parameters in such a way that it is possible to separate the effects of city size, climate, population density, and other unique physical characteristics from remaining parameters that reflect the intensity of material and energy use. Such mathematical form may be useful for either fairly comparing cities accounting for differences in unique characteristics or, in the absence of reliable data, help with estimating the components of metabolism for cities with incomplete data.
3. Assist with developing consistent data standards for urban metabolism studies.

Although not undertaken here, the model has a potentially wide range of applications in efforts to develop sustainable cities. These include scenario development and long-term planning, in which the model could support estimation of greenhouse gas emissions, or urban ecological footprints. The parameters of the model may also be used to make comparisons between cities, thereby promoting learning of sustainable city strategies. Given its simplicity, one of the particular strengths of the model is the ease with which it might be combined with other macrourban models; e.g., macroeconomic models, to assess or guide urban sustainability. The focus of this paper, though, is on the mathematical foundations.

The mathematical model is developed using the systems boundary shown in Fig. 1. The urban system is defined so as to include peri-urban areas, aquifers below the city, and the urban atmospheric boundary layer above. Fig. 1 broadly depicts the annual inflows, I, annual outflows, O, and the storage, S, of the main components of the urban metabolism. Also included are annual internal fluxes, Q, e.g., between surface and subsurface parts of the urban system, and the internal production, P, of food by urban agriculture. More detailed description of the metabolism is added in the subsequent equations.

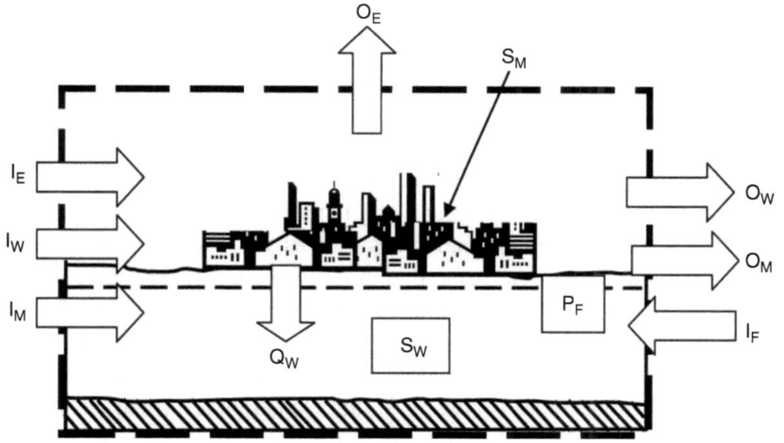

Fig. 1 Urban systems boundary broadly showing inputs (I), outputs (O), internal flows (Q), storage (S), and production (P) of materials (M), water (W), energy (E), and food (F). Greater detail is described by the mathematical equations developed in the text

Materials

Material Stock

The stock of materials in a city is an appropriate starting point to describe the metabolism because the quantities and properties of buildings and other infrastructure substantially determine other components of the metabolism. A subcomponent of the material stock, for all materials of type *m*, in structures, *s*, may be expressed by $S^s_{\mathrm{M},m}$. The total material stock is then:

$$S_{\mathrm{M}} = \sum_s \sum_m S^s_{\mathrm{M},m} \quad (1)$$

The range of material types considered could vary from a few to many, depending on the issue at hand. A concise categorization scheme based on the EUROSTAT (2001) methodology considers four types of materials: biomass, metals, non-metallic minerals, and fossil fuels.

The structure is included as a superscript recognizing that most of the anthropogenically placed materials in cities are bound up in the buildings and physical infrastructure systems. At the broadest level, there are essentially two types of urban structures: buildings and linear infrastructure, and I shall just develop an example equation of each type. A more detailed classification for structures would be: residential buildings, commercial/institutional buildings, industrial buildings, transportation infrastructure, drinking water infrastructure, and wastewater/sewer infrastructure. The determinants of the material stock may be quite different for various structures.

The stock of materials in residential buildings, rb, may be expressed by:

$$S^{\mathrm{rb}}_{\mathrm{M},m} = P \cdot f^{\mathrm{rb}} \cdot i^{\mathrm{rb}}_{\mathrm{M},m} \quad (2)$$

where

P is the population of the city or urban region; f^{rb} is the per capita floor space for residential buildings [m^2 person^{-1}]; and $i^{\mathrm{rb}}_{\mathrm{M},m}$ is material intensity [tons m^{-2} of floor space]. Floor space per capita is sometimes used as a measure of standard of living. Average values vary from 8 m^2 per capita in African cities to 34.5 m^2 per capita for cities in industrialized countries (Flood 1997). The material intensity of residential construction can vary depending on the material used. Bergsdal et al. (2007), for example, report intensities of ~0.6 $\mathrm{t\,m}^{-2}$ for concrete and 0.11 $\mathrm{t\,m}^{-2}$ for wood in Norwegian residential homes (Table 1). In the Netherlands, where wood construction is less common, Müller (2006) reports an intensity of 2 $\mathrm{t\,m}^{-2}$ for concrete. The parameters of (2) might thus be expected to vary between cities in different countries.

Table 1 Material intensity of residential construction

Material and location	Year	Intensity ($\mathrm{t\,m}^{-2}$)	Source
Concrete in Dutch housing construction	2000	~2.0	Müller (2006)
Concrete in Norwegian housing construction	2000	~0.6	Bergsdal et al. (2007)
Wood in Norwegian housing construction	2000	0.11	Bergsdal et al. (2007)

An equation similar to (2), i.e., based on per capita floor space, can be used for commercial/institutional buildings and possibly industrial buildings. Alternatively, it may be useful to define the ratio of materials in other buildings relative to residential buildings.

The stock of material in linear transportation infrastructure may be given by:

$$S_{M,m}^{ti} = A \cdot \rho_{ti} \cdot i_{M,m}^{ti} \quad (3)$$

where

A is the area of the city [km²]; ρ_{ti} is the density of transportation infrastructure [km km⁻²]; and $i_{M,m}^{ti}$ is material intensity [t km⁻¹ of linear transport infrastructure]. Clearly, greater detail can be added here, e.g., distinguishing between road and rail infrastructure. There is also a hierarchy in urban road systems from local roads, to main arterials roads, to urban expressways. These differ significantly in their densities and material intensities as shown in Table 2.

Table 2 Material intensity of linear infrastructure

Category	Material Intensity
Roads (Finland; Saari et al. 2007)	
Connecting road (7.1 m wide)	2.97 Mt km^{-1}
Regional road (7.1 m wide)	2.22 Mt km^{-1}
Class I main road (10.0 m wide)	5.93 Mt km^{-1}
Motorways (2×11.75 m wide)	30.55 Mt km^{-1}
Storm drainage (Taipei, Huang and Hsu 2003)	
Arterial pipes	6,490 t km^{-1} sand/gravel 998 t km^{-1}cement
Branch pipes	1300 t km^{-1} sand/gravel 200 t km^{-1}cement
Sewerage (Taipei, Huang and Hsu 2003)	
Arterial pipes	6,430 m^3 km^{-1}sand/gravel 1,446 t km^{-1}cement
Branch pipes	480 m^3 km^{-1}sand/gravel 122 t km^{-1}cement
Flood prevention (Taipei, Huang and Hsu 2003)	
Dike/levee	1,35,000 t km^{-1} sand/gravel 20,770 t km^{-1}cement
Bank protection	90,000 t km^{-1} sand/gravel 13,850 t km^{-1}cement

Urban transportation infrastructure also includes non-linear elements. The stock of materials in structures such as bridges and stations might, however, be averaged in with the linear transportation elements. Other structures such as airports and main railway stations can be included as special forms of institutional buildings.

The stock of materials in other linear infrastructure systems may be expressed in a similar form to (3). Indeed, since water and sewer systems generally underlie roads, the densities of such systems might be approximately the same. Further data on material intensities for water and sewer infrastructure are given in Table 2.

Material Dynamics

Material stocks, S_M, in cities are generally increasing (Brunner and Rechberger 2001; Kennedy et al. 2007) clearly indicating that the inputs, I_M, exceed the outputs, O_M, in the material budget equation:

$$\frac{\mathrm{d}S_{\mathrm{M}}}{\mathrm{d}t} = I_{\mathrm{M}} - O_{\mathrm{M}} \tag{4}$$

Dynamic MFA models have been developed to solve the budget equation with applications to materials such as concrete and timber for residential buildings in Switzerland (Müller et al. 2004), Holland (Müller 2006), and Norway (Bergsdal et al. 2007). Outflows are dependent on the distribution of material life time in stock, L_M, and the historical inflow I_M (Müller 2006):

$$O_m(t) = \int_{t_0}^{t} L_{\mathrm{M}}(t,t')I_{\mathrm{M}}(t')d^{t'} \tag{5}$$

Current and future material inflows are essentially then determined by changes in the stock required to provide for changes in population, changes in level of service, and to replace outflows.

The interaction between population, level of service, and material use are well described by Müller (2006). The main/key effects can also be seen by differentiating (2) and (3):

$$\frac{\mathrm{d}S^{\mathrm{rb}}_{\mathrm{M},m}}{\mathrm{d}t} = \frac{\mathrm{d}P}{\mathrm{d}t} f^{\mathrm{rb}} i^{\mathrm{rb}}_{\mathrm{M},m} + P\frac{\mathrm{d}f^{\mathrm{rb}}}{\mathrm{d}t} i^{\mathrm{rb}}_{\mathrm{M},m} + Pf^{\mathrm{rb}}\frac{\mathrm{d}i^{\mathrm{rb}}_{\mathrm{M},m}}{\mathrm{d}t} \tag{6}$$

$$\frac{\mathrm{d}S^{\mathrm{ti}}_{\mathrm{M},m}}{\mathrm{d}t} = i^{\mathrm{ti}}_{\mathrm{M},m}\rho_I \frac{\mathrm{d}A}{\mathrm{d}t} \tag{7}$$

Equation (6) shows that increases in materials in the residential building stock can occur due to changes in per capita floor space and material intensity, as well as population growth. Change in per capita floor space has been significant. From the 1940s/1950s to 2002, the average single detached house in the United States increased from 1,100 ft^2 (102 m^2) to 2,340 ft^2 (218 m^2). Moreover, while house floor space more than doubled, the shrinking of family sizes resulted in living area per family member increasing by a factor of three (Wilson and Boehland 2005). Less is known about how material intensity may have changed with the increases in floor area.

Equation (7) shows that the accumulation of materials in linear infrastructure systems depends on the density of infrastructure, its material intensity, and the rate at which the urban area, A, is increasing over time. Data for 22 world cities from 1980 to 1990 show that urbanized areas were increasing at rates varying between

Table 3 Rates of change of urbanized areas (1980–1990) and density of transportation infrastructure (1990) for 22 cities. Calculations based on data from Kenworthy et al. (1999)

	Min.	Max.	Mean	St. Dev.
Urbanized area, 1980 (ha)	11,695	3,87,951	1,06,002	1,04,276
Urbanized area, 1990 (ha)	12,872	4,10,380	1,21,250	1,13,331
Rate of change of urbanized area (ha year^{-1})	27	3,851	1,524	1,316
Rate of change of urbanized area (% year^{1})	0.19%	3.75%	1.60%	1.03%
Density of transportation infrastructure, 1,990 (km ha^{-1})	0.075	0.157	0.104	0.022

The cities are: Amsterdam, Brisbane, Brussels, Calgary, Canberra, Chicago, Detroit, Edmonton, Frankfurt, Hamburg, Houston, Jakarta, London, Melbourne, Perth, Portland, San Diego, San Francisco, Singapore, Sydney, Vancouver, and Vienna

0.19 and 3.75% per year (Table 3). An analysis of the same dataset indicates that the density of transportation infrastructure is relatively consistent between cities, with a mean of 0.104 km ha^{-1} and a standard deviation of 0.022 km ha^{-1}. These values are for 1990, but were found to be essentially the same for 1980. Material intensity has been assumed to be constant in differentiating (3), but this assumption could be relaxed if this parameter is changing over time.

Consumables

Beyond the large quantities of construction materials, which become bound up in urban infrastructure systems, there is a wide range of other material goods which have relatively short life spans in the urban metabolism. The list is vast and includes items such as food (discussed in the next section), paper, packaging, electronic goods, and household furnishings. Some of these enter cities and are deposited as solid waste within weeks; others may remain in the metabolism stream for decades. Browne et al. (2009) compared the relative rates of inflows and outflows in the urban metabolism for consumables and capital equipment in different economic sectors. The material budget (4) can be used to model any particular material good, component or substance in the urban metabolism.

Water

Although the flow of water through a city can, in principle, also be understood as just another type of material flow, there are good reasons for developing specific equations for water. First, the inflows and outflows are far greater than any other materials in magnitude (Wolman 1965; Kennedy et al. 2007). Second, they include both natural and anthropogenic components. Third, while changes in storage of water can be significant over the duration of rainfall events, they are often relatively insignificant in the context of annual flows; i.e., there is no change in storage from year to year. One exception to this last point, however, has to do with the groundwater levels in aquifers below cities. Thus, water has some unique characteristics within the urban metabolism.

Natural Water Balance

The natural water balance for a city can be described by:

$$I_{\mathrm{W,precip}} + I_{\mathrm{W,pipe}} + I_{\mathrm{W,sw}} + I_{\mathrm{W,gw}} = O_{\mathrm{W,evap}} + O_{\mathrm{W,out}} + \Delta S_{\mathrm{W}} \tag{8}$$

where

$I_{\mathrm{W,precip}}$ is natural inflow from precipitation, dew, and hoar frost; $I_{\mathrm{W,pipe}}$ is water piped into the city;

$I_{\mathrm{W,sw}}$ is the net surface water flow into the city, e.g, accounting for rivers flowing through;

$I_{\mathrm{W,gw}}$ is the net ground water flow into city aquifers;

$O_{\mathrm{W,evap}}$ is evapotranspiration;

$O_{\mathrm{W,out}}$ is water piped out of cities, where it has not been expressed within the net surface water flow term above. This term would apply when the outflow from wastewater or stormwater systems is outside of city boundaries;

ΔS_{W} is the annual change in water stored within the city. This term may be close to zero unless there are changes to groundwater level below the city or changes in the amount of water stored in surface reservoirs within city boundaries.

Anthropogenic Water Use

The water used for human consumption in cities may come from several of the inflow terms in (8): (1) it could be captured in city reservoirs following precipitation, (2) be pumped into the city, or (3) be withdrawn from rivers, lakes or groundwater sources within the city. From a consumption perspective, the annual anthropogenic water use Q_{W} in cities has two components:

$$Q_{\mathrm{W}} = Q_{\mathrm{W,D}} + Q_{\mathrm{W,L}} \tag{9}$$

where

$Q_{\mathrm{W,D}}$ is water demanded for end use and $Q_{\mathrm{W,L}}$ is losses in the water distribution system (including leaks within buildings). The water demanded for end-use typically has a constant base component, mainly for indoor use, $Q_{\mathrm{W,D}}^{\mathrm{base}}$, and a seasonal component, which reflects the increases in water for outdoor use during hotter days. Thus,

$$Q_{\mathrm{W,D}} = Q_{\mathrm{W,D}}^{\mathrm{base}} + CDD \cdot i_{\mathrm{W}}^{\mathrm{cooling}} \tag{10}$$

where
CDD is cooling degree days for the city [°C.d] and $i_W^{cooling}$ is the intensity of water use for cooling [m^3°C^{-1}].

There is some variation in the amounts of water used by cities. Reviews of urban metabolism studies, mainly in developed cities, show that the annual urban water use varies between 50 and 200 t per capita (Kennedy et al. 2007). The relative sizes of base and seasonal components might also vary between cities. For example, in Toronto, the base demand of approximately 480 ML in 2001 was 90% of the total demand of 535 ML (Sahely and Kennedy 2007, Sadiq 2003). Given an average CDD of 370°C.d year^{-1} (over 18°C; from http://www.degreedays.net), this indicates that $i_W^{cooling}$ is close to 0.15 ML°C.d for Toronto.

Losses from the water distribution system can be related to the density of linear infrastructure by:

$$Q_{W,L} + A \cdot \rho_{ti} \cdot l \tag{11}$$

where
l is the annual leakage rate per length of linear infrastructure, including losses from pipe breaks. A range of complex factors underlie water losses and pipe breaks, including material properties, soil properties, age, and operational factors (Kleiner and Rajani 2002).

The annual volume of wastewater treated, $Q_{WWT,}$ for a city is:

$$Q_{WWT} = Q_{WWE} + Q_{WWF} + Q_{INF} \tag{12}$$

The volume of wastewater generated Q_{WWE} is what enters the sewer system (sanitary and/or combined, as applicable) from end users. This volume is usually less than the volume of clean water supplied, due to consumptive losses, i.e., human consumption, and watering of gardens. Consumptive losses in Toronto, for example, are of the order 20–25% (Sahely and Kennedy 2007). For the remaining terms in equation (11), Q_{WWF} is the annual volume of wet weather flow that enters the wastewater treatment plant via the combined sewer system or inflow and infiltration; and Q_{INF} is the annual volume of water, known as base infiltration, which enters the sewer systems during dry weather.

Urban Aquifers

A budget equation for the subsurface system can be developed based on the change from virgin conditions (Zhang and Kennedy 2006, Bredehoeft et al. 1982). In absence of the city, the groundwater budget equation is:

$$Q_{W,R0} + I_{W,gw} = Q_{W,D0} \tag{13}$$

where

$Q_{W,R0}$ is natural recharge and $Q_{W,D0}$ is the natural surface discharge. The net natural groundwater flux (as given for (8)) would also be that under virgin conditions.

As the city develops, surface features changes and ground water is pumped from the aquifers, the equilibrium condition is disturbed and natural recharge and discharge change. The incremental ground water budget equation is then

$$\Delta S_{W,gw} = \Delta Q_{W,R0} + Q_{W,ar} + \Delta I_{W,gw} - \Delta Q_{W,D0} - Q_{W,gwpump} \tag{14}$$

where

$\Delta S_{W,gw}$ is the rate of change in ground water storage in the system [L^3 T^{-1}];
$\Delta Q_{W,R0}$ is the change in natural recharge from virgin conditions [L^3 T^{-1}];
$\Delta Q_{W,D0}$ is the change in natural discharge from virgin conditions [L^3 T^{-1}];
$Q_{W,gwpump}$ is the net pumping rate of the city [L^3 T^{-1}];
$\Delta I_{W,gw}$ is the change in net groundwater inflow [L^3 T^{-1}].

The other term in (14), $Q_{W,ar}$ is net anthropogenic urban recharge [L^3 T^{-1}], which includes: leakage from water mains, net leakage from sewerage and stormwater systems in sewered areas; returning groundwater by irrigation and gardening; infiltration by artificial recharge; seepage through septic systems in unsewered areas; and seepage from industrial facilities. An example demonstration of (14) to the aquifers below Beijing is given by Zhang and Kennedy (2006).

Energy

Energy flow is the ultimate measure of the urban metabolism. The magnitude of energy flows for heating and cooling are typically related to climate, but other components of urban energy use can be linked back to the shape and form of a city, reflected by its infrastructure systems and hence material stocks. Energy flow is also similar to water in the metabolism in that it has natural and anthropogenic components, and typically has negligible change in net storage from a year-to-year perspective.

Anthropogenic Energy Use

The anthropogenic energy consumed by a city, I_E, can be expressed by:

$$I_E = I_{E,buildings} + I_{E,transport} + I_{E,industry} + I_{E,construction} + I_{E,water\ pumping} + I_{E,waste} \tag{15}$$

Equation (15) has three major components: the energy used by buildings; transportation and industry; and three typically minor components: energy use for construction, water supply (pumping and treatment), and waste management. Energy used for industrial processes is likely quite industry specific, so is not expanded further here.

The energy used in buildings can be further expressed as:

$$I_{\text{E,buildings}} = I_{\text{E,heating}} + I_{\text{E,cooling}} + I_{\text{E,light and appl.}} + I_{\text{E,water heating}} \tag{16}$$

with components for space heating, cooling, lighting and appliances, and water heating.

The heating and cooling energy for buildings can be given by:

$$I_{\text{E,heating}} = \sum_{\text{building type}} HDD \cdot i_{\text{E,heating}} \cdot P \cdot f \tag{17}$$

$$I_{\text{E,cooling}} = \sum_{\text{building type}} CDD \cdot i_{\text{E,heating}} \cdot P \cdot f \cdot cp \tag{18}$$

where

HDD = heating degree days [°C.d];
CDD = cooling degree days [°C.d];
$i_{\text{E,heating}}$ = heating intensity [$\text{J.m}^{-2}.{}^{\circ}\text{C}^{-1}.\text{d}^{-1}$];
$i_{\text{E,cooling}}$ = cooling intensity [$\text{J.m}^{-2}.{}^{\circ}\text{C}^{-1}.\text{d}^{-1}$].

Equations (17) and (18) link building energy use back to material stocks (2) through the floor area per person, *f*.

The effect of climate is captured by the heating and cooling degree days, with the heating and cooling intensity values providing a measure of the thermal properties of the building stock. A study of ten global city regions by Kennedy et al. (2009) has shown that heating degree days correlate well ($R^2 = 0.70$) with energy used for heating and industrial use in cities. Cooling degree days is perhaps not so good at explaining energy used for cooling on cities, because humidity is also an important factor. Glaeser and Khan (2008) showed, however, that household electricity use in US cities rises with average July temperatures. Using data for residential homes for Boston and St. Louis from Wilson and Boehland (2005), example values for heating intensity and cooling intensity are estimated to be of the order 0.05 MJ m^{-2} ($^{\circ}$F.day)$^{-1}$ year^{-1}, and 01 m^{-2} ($^{\circ}$F.day)$^{-1}$ year^{-1}, respectively (Table 4).

Table 4 Calculation of heating and cooling intensities for representative residential homes in Boston and St. Louis (based on energy modeling data in Table 3 of Wilson and Boehland 2005; HDD and CDD from National Climatic Data Centre 2001 http://ggweather.com/ccd)

City	House size (m^2)	Heating energy (MJ)	Cooling energy (MJ)	Heating intensity (MJ m^{-2} HDD year^{-1})	Cooling intensity (MJ m^{-2} HDD year^{-})
Boston (HDD:	279 g	77,015	20,045	0.049	0.093
5,630°F day;	139 g	36,925	13,715	0.047	0.127
CDD: 777°F day)	139 p	50,640	12,660	0.065	0.117
St. Louis (HDD:	279 g	64,355	30,595	0.049	0.070
5,630°F.day;	139 g	30,595	21,100	0.046	0.097
CDD: 777°F.day)	139 p	42,200	22,155	0.064	0.102

p = poor; g = good

Transportation energy use can be subdivided into surface passenger, surface freight aviation, and marine components:

$$I_{\mathrm{E,transport}} = I_{\mathrm{E,passenger}} + I_{\mathrm{E,freight}} + I_{\mathrm{E,aviation}} + I_{\mathrm{E,marine}} \tag{19}$$

The first of these components will be considered further. Energy used for surface passenger transportation can be expressed by:

$$I_{\mathrm{E,passenger}} \sum_{\mathrm{mode}} \frac{1}{\rho_{\mathrm{p}}} \cdot P \cdot \rho_{\mathrm{I}} \cdot h \cdot \varepsilon \tag{20}$$

where

ρ_{P} = average population density [km^{-2}];
ρ_{I} = density of transportation infrastructure [km km^{-2}];
h = utilization of infrastructure [veh-km km^{-1}];
ε = fuel efficiency [J veh-km^{-1}].

Equation (20) is useful in several respects. First it recognizes the reciprocal relationship between transportation energy use and population density (Newman and Kenworthy 1989). Second the product of the first four terms within the summation equals vehicle-kilometers traveled (VKT), which is sometimes used as a sustainability indictor for urban transportation. Third, there is a link back to the material stock (3) through the density of transportation infrastructure.

The usefulness of (20) to explain variation in urban transportation energy use between cities has been explained elsewhere (Roux et al. 2010), but a brief analysis can be given using data for the same 22 cities as in Table 3. As previously explained, the density of transportation infrastructure, ρ_I, has a relatively low standard deviation. There is greater variability, however, in the utilization of infrastructure, h, by automobiles, which for the same set of cities has a mean of 1,630,000 [km km^{-1}] and a standard deviation of 630,000. The fuel efficiency, ε, for automobiles has a mean of 4.9 MJ veh-km^{-1} and a standard deviation of 1.3 MJ veh-km^{-1}. Thus, most of the variation in per capita energy use by automobiles in cities can be explained by road utilization rates and vehicle efficiency, as well as population density, ρ_p.

Surface Urban Energy Balance

A further equation which describes energy in the urban metabolism is that for the surface energy balance (Douglas 1983):

$$\mathrm{I}_{E,S} + I_{E,F} + I_{E,I} = O_{E,L} + O_{E,G} + O_{E,E} \tag{21}$$

where

$\mathrm{I}_{E,S}$ = Rate of arrival of radiant energy from the sun
$I_{E,F}$ = Rate of generation of heat due to combustion and dissipation in machinery
$I_{E,I}$ = Rate of heat arrival from the earth's interior
$O_{E,L}$ = Rate of loss of heat by evapotranspiration
$O_{E,G}$ = Rate of loss of heat by conduction to soil, buildings, roads, etc.
$O_{E,E}$ = Rate of loss of heat by radiation

Food and Nutrients

An equation for the flow of food in the urban metabolism can be adapted from equations developed by Codoban and Kennedy (2008) for neighborhood metabolism. These include the mass of food and consumed water that are used in human metabolism, as well as waste products from the food stream.

$$I_F + P_F + I_{W,Kit} = O_{F,RetFW} + O_{F,ResFW} + O_{F,Met} + O_{F,S} \tag{22}$$

where

I_F = mass of food and packaged drinks imported to the city;
P_F = mass of food and packaged drinks produced in the city, for internal consumption;
$I_{W,Kit}$ = mass of kitchen water used during food preparation or drunk during meals;
$O_{F,RetFW}$ = mass of retail food waste produced by grocery stores and restaurants;
$O_{F,ResFW}$ = mass of residential food waste going to landfill, compost, or organic waste collection;
$O_{F,Met}$ = mass of carbon and water lost via respiration and transpiration in residents metabolism;
$O_{F,S}$ = mass of feces and urine exported to sewerage system.

This equation for food in the urban metabolism can also be expressed in terms of energy content:

$$I_{\mathrm{E,Food}} + P_{\mathrm{E,Food}} = O_{\mathrm{E,RetFW}} + O_{\mathrm{E,ResFW}} + O_{\mathrm{E,Met}} + O_{\mathrm{E,S}} \tag{23}$$

where

$I_{\mathrm{E,Food}}$ = energy in food and packaged drinks imported to the city;

$P_{\mathrm{E,Food}}$ = energy in food and packaged drinks produced in the city, for internal consumption;

$O_{\mathrm{E,RetFW}}$ = energy in retail food waste produced by grocery stores and restaurants;

$O_{\mathrm{E,ResFW}}$ = energy in residential food waste going to landfill, compost, or organic waste collection;

$O_{\mathrm{E,Met}}$ = energy used in residents metabolism;

$O_{\mathrm{E,S}}$ = energy in feces and urine exported to sewerage system.

Food is clearly a special type of material in the urban metabolism, because of its provision of human sustenance. Rather than study the entire mass flows of food, some researchers have alternatively focused on specific nutrients in the urban metabolism, which can also have inorganic forms; see for example Nilson (1995), Baker et al. (2001) or Forkes (2007).

Urban Contaminants and Wastes

Many types of contaminants or wastes are produced within the urban metabolism—and can mix between streams of the metabolism before being emitted in solid, liquid, or gaseous forms. Air contaminants are derived in particular from energy use, via combustion of fossil fuels, depending on the technologies employed. Aqueous contaminants arrive from the entry of materials and nutrients into the water system, varying in composition and concentration between stormwater and wastewater. Solid waste is similarly produced by the material and nutrient streams. Other contaminants, e.g., organic solvents or polybrominated flame retardants, can partition in complex ways between different streams.

No further equations to describe contaminant flows will be given here. This is in part due to the complexity and wide variation in the type of contaminants, but also because the fundamental streams of metabolism that gives rise to contaminants have already been described; the contaminants are flows of materials, water, and nutrients as described by the equations above. While these equations provide a basis from which to mathematically describe contaminants flows, further parameters may be required for several reasons. First, the technologies used for treatment, separation, or recycling play a role in determining contaminant or waste levels. Second, for some contaminants, particularly those in stormwater, the surface areas of buildings, roads, and other surfaces are important for determining loadings (Hodge and Diamond 2010). Surface areas have not been reflected in the equations above, but have been considered in an urban metabolism context by Deilmann (2009).

An example of how one class of contaminant—greenhouse gas emissions—relate to the urban metabolism can be given. Table 5 shows the type of data required in conducting a GHG inventory for a city. This data includes quantities of energy consumed, waste generated, and where applicable materials produced within the city—all of which can be determined using the equations, or notation, in this paper.

Table 5 Components of urban metabolism that are required for the inventorying of GHG emissions for cities and local communities (Table 2, from Kennedy et al. 2011)

Components of urban metabolism	Preferred units
Total electricity consumption	GWh
Consumption of heating and industrial fuels by each fuel type (e.g., natural gas, fuel oils, coal, LPG—includes fuels used in combined heat and power plants).	TJ for each fuel type
Total consumption of ground transportation fuels (gasoline, diesel, other) based on sales data.	Million liters for each fuel type
Volume of jet fuel loaded onto planes at airports within the boundary of the city/urban region.	Million liters
Volume of marine fuel loaded onto vessels at the city's port (if applicable).	Million liters
Tonnage and composition of landfill waste (% food, garden, paper, wood, textiles, industrial, other/inert) from all sectors; and percentage of landfill methane that is captured	t and %
Tonnage of solid waste incinerated (if applicable)	t
Masses of steel, cement, and other materials or chemicals produced in the city causing non-energy-related industrial process emissions.	t

Conclusions

This paper has broadly provided a mathematical model of urban metabolism, along with example data for pertinent model parameters. Its main contribution is an integration of mathematical descriptions of urban processes from different fields (e.g., MFA, hydrology, transportation) into a common mathematical form. Of particular interest are parameters such as the per capita floor space, *f*, and the density of transportation infrastructure, ρ_I which are first defined in relation to the material balance of the city, but then contribute to the description of energy and/or water flows in the city. These parameters help to reflect the important role that infrastructure serves in determining the metabolism of cities.

The density of transportation infrastructure is of further interest because the preliminary analysis suggests that it is relatively consistent between cities. For a sample of 22 cities, the density of transportation infrastructure had a mean of 0.104 km ha^{-1} with a relatively small standard deviation of just 0.022 km ha^{-1}. This suggests that it may be a universal parameter for all cities.

The model also includes other parameters which, while having more variation, are independent of the climate, city size, population, and other unique characteristics of cities. These parameters include: material intensities, per capita floor space, intensity of water use for cooling, leakage rates for water distribution systems, heating and cooling intensities of buildings, and utilization rates for transportation infrastructure. A few of these parameters may possibly be new, or are at least rarely used.

Acknowledgement The author thanks the anonymous reviewers of this work.

References

Baccini P (1987) A city's metabolism: towards the sustainable development of urban systems. J Urban Technol 4:27–39

Baccini P, Bader H-P (1996) Regionlar stoffhaushalt: erfassung, bewertung and steuerung. Spektrum, Heidelberg

Baker LA, Hope D, Xu Y et al (2001) Nitrogen balance for the central Arizona-Phoenix (CAP) ecosystem. Ecosystems 4:582–602

Bergsdal H, Brattebø H, Bohne RA et al (2007) Dynamic material flow analysis for Norway's dwelling stock. Build Res Inform 35:557–570

Bredehoeft JD, Papadopulos SS, Cooper HH Jr (1982) Groundwater—the water-budget myth. Scientific basis of water resource management, studies in geophysics. National Research Council, National Academy Press, Washington

Browne D, O'Regan B, Moles R (2009) Assessment of total urban metabolism and metabolic inefficiency in an Irish city-region. Waste Manage 29:2765–2771

Brunner PH, Rechberger H (2001) Anthropogenic metabolism and environmental legacies. In: Munn T (ed) Encyclopedia of global environmental change, vol 3. Willey, UK

Brunner PH, Rechberger H (2004) Practical handbook of material flow analysis. CRC Press, Boca Raton

Codoban N, Kennedy CA (2008) The metabolism of neighbourhoods. ASCE, J Urban Plan and Devel 134:21–31

Costanza R, Voinov A, Boumans R et al (2002) Integrated ecological economic modeling of the Patuxent River watershed, Maryland. Ecol Monogr 72:203–231

Decker H, Elliott S, Smith FA et al (2000) Energy and material flow through the urban ecosystem. Ann Rev Energ Envir 25:685–740

Deilmann C. (2009) Urban metabolism and the surface of the city. In: German annual of spatial research and policy, guiding principles for spatial development in Germany. Springer Berlin Heidelberg

Douglas I (1983) The urban environment. Edward Arnold, London

EUROSTAT (2001) Economy-wide material flow accounts and derived indicators. A methodological guide. Statistical Office of the European Union, Luxembourg, Belgium

Flood J (1997) Urban and housing indicators. Urban Stud 34:1635–1665

Forkes J (2007) Nitrogen balance for the urban food metabolism of Toronto, Canada. Resour Conser Recy 52:74–94

Fung M, Kennedy CA (2005) An integrated macroeconomic model for assessing urban sustainability. Environ Plann B 32:639–656

Glaeser EL, Khan ME (2008) The greenness of cities: carbon dioxide emissions and urban development. National Bureau of Economic Research, Working Paper 14238. Aug 2008

Hendriks C, Obernosterer R, Müller D et al (2000) Material flow analysis: a tool to support environmental policy decision making. Case studies on the city of Vienna and the Swiss lowlands. Local Environ 5:311–328

Hodge E, Diamond M (2010) Sources, fate and effects of contaminant emissions in urban areas (chapter 6). In: Harrad S (ed) Persistent organic pollutants. Wiley, Chichester, West Sussex

Huang S-L, Hsu W-L (2003) Material flow analysis and emergy evaluation of Taipei's urban construction. Landscape Urban Plan 63:61–74
Kennedy CA, Cuddihy J, Engel YJ (2007) The changing metabolism of cities. J Indust Ecol 11:43–59
Kennedy CA, Steinberger J, Gasson B et al (2009) Greenhouse gas emissions from global cities. Environ Sci Tech 43:7297–7302
Kennedy CA, Pincetl S, Bunje P (2011) The study of urban metabolism and its applications to urban planning and design. J Environ Pollut 159(8–9):1965–1973
Kenworthy JR, Laube P, Newman P et al (1999) An international sourcebook of automobile dependence in cities, 1960-1990. University Press of Colorado, Boulder
Kleiner Y, Rajani B (2002) Forecasting variations and trends in watermain breaks. J Infrastr Syst 8(4):122–131
Müller D (2006) Stock dynamics for forecasting material flows—case study for residential housing in The Netherlands. Ecol Econ 59:142–156
Müller D, Bader H-P, Baccini P (2004) Long-term coordination of timber production and consumption using a dynamic material and energy flow analysis. J Ind Ecol 8:65–87
Newman P, Kenworthy J (1989) Gasoline consumption and cities: a comparison of US cities with a global survey. J Amer Plan Assoc 55:24–37
Nilson J (1995) A phosphorus budget for a Swedish municipality. J Environ Manag 45:243–253
Roux S, Derrible S, Kennedy CA (2010) Urban transportation infrastructure and its influence on energy use. Unpublished material
Sadiq WA (2003) Impact of climate change on the water demand of Toronto. Unpublished MA Sci Thesis. Department of Civil Engineering, University of Toronto
Saari A, Lettenmeier M, Pusenius K et al (2007) Influence of vehicle type and road category on natural resource consumption in road transport. Transp Res, Part D 12:23–32
Sahely HR, Kennedy CA (2007) Integrated systems flow model for quantifying environmental and economic sustainability indicators: case study of the city of Toronto urban water system. ASCE J Water Resour Plan Manag 133:550–559
VandeWeghe J, Kennedy CA (2007) A spatial analysis of residential greenhouse gas emissions in the Toronto census metropolitan area. J Ind Ecol 11:133–144
Waddell P (2002) UrbanSim: modeling urban development for land use, transportation and environmental planning. J Amer Plan Assoc 68:297–314
Wegener M (1995) Operational urban models: state of the art. J Amer Plan Assoc 60:17–29
Wilson A, Boehland J (2005) Small is beautiful: U.S. house size, resource use and the environment. J Ind Ecol 9:277–287
Wolman A (1965) The metabolism of cities. Sci Am 213:179–190
Zhang L, Kennedy CA (2006) Determination of sustainable yield in urban groundwater systems: Beijing, China. ASCE J Hydrol Eng 11:21–28

Urbanization, Local Government, and Planning for Sustainability

Robert W. Taylor

Abstract Local governments are in a unique position to manifest and implement the practices of sustainability. They possess a decision-making apparatus that allows sustainability practices to be readily implemented; they are the institutions closest to the people and whose decisions reflect on developing the holistic health of the community, meaning that the goals of equity, economy and environmental quality must all be satisfied equally; and they are the institutions that are most directly accountable to the people. This chapter discusses the origins and principles of sustainability planning for cities, various strategies for implementation, and concludes by providing a case study of a sustainability plan for the city of Manila, Philippines.

Keywords Urban sustainability • Sustainability metrics • Urban planning • Manila, Philippines

Introduction

Today, nearly 50% of the world's population lives in cities. By 2030, this percentage will increase to 60% and cities of the developing world are expected to absorb 95% of this growth as a result of rural to urban migration, transformation of rural settlements into urban ones, and natural population increase (Fig. 1). Although comprising only 3% of the earth's land area, cities consume 75% of global energy, create 80% of global greenhouse gas emissions, and intensely concentrate industry, people, materials and energy (Davis 2008; Schulz 2002).

R.W. Taylor (✉)
Earth and Environmental Studies, Montclair State University, Montclair, NJ 07093, USA
e-mail: taylorr@mail.montclair.edu

M.P. Weinstein and R.E. Turner (eds.), *Sustainability Science: The Emerging Paradigm and the Urban Environment*, DOI 10.1007/978-1-4614-3188-6_14,

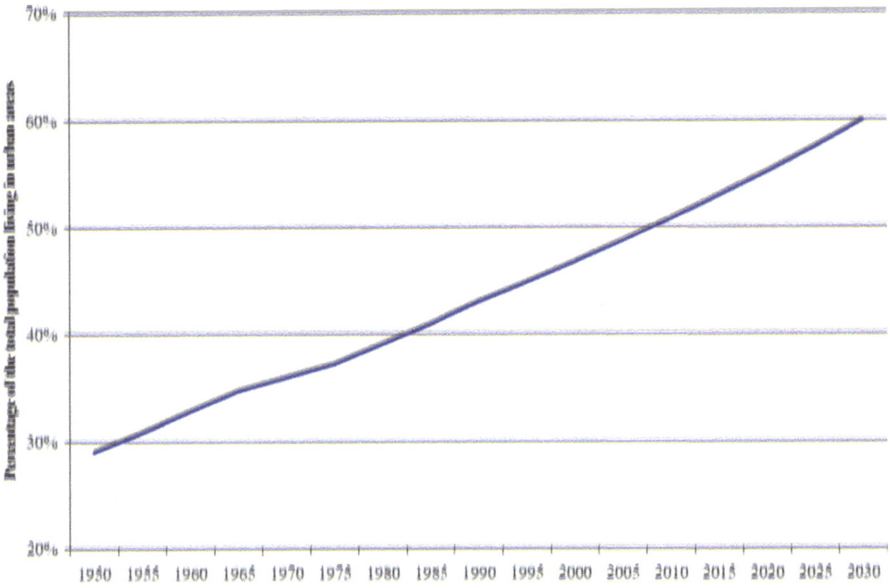

Fig. 1 Growth of global urban population 1950–2030. *Source*: Urban Settlement: Data, Measures, and Trends, Harvard School of Public Health, December 2007

Asia alone will add an additional 1.5% people to its urban regions by 2020 (ADB 2001). This wave of urban expansion, lead by China and India, is restoring Asia to the global prominence that it enjoyed before the European and North American industrial revolution. Nearly 2.5 billion Asians will live in cities by 2025, accounting for almost 54% of the world's urban population (Dobbs and Sankhe 2010).

Without proper management planning, cities constitute a major environmental hazard. Yet, cities also possess great opportunities for sustainability. Their compact settlement pattern provides economies of scale that can encourage resource and energy efficiency. With people living closer together, public transit can be encouraged, critical infrastructure such as sewers, roads, and electricity can be minimized resulting in more efficient land and material use. Also, cities encourage innovation and resource efficiency. Building construction, renewable energy advances, and innovative waste management can be successfully adopted. In New York City, for example, global greenhouse gases per capita are just one-third of the US average (Glaeser 2009; Owen 2009).

Sustainability planning is a relatively new approach to local government management that seeks to integrate urban planning with the principles and practices of sustainability. It is not designed to replace urban master plans since the objectives and approaches taken by sustainability planning are different. First, sustainability planning recognizes the unique characteristics of the local community in crafting sustainable strategies. Second, it utilizes considerable capacity building through stakeholder engagement, often assessing local stakeholder needs and desires through an inclusive "collaborative" approach. In this process, it seeks advice on existing and potential sustainability initiatives from a wide variety of local stakeholders. Third, it uses sustainability issues mapping as a tool to assess the major environmental, social and economic risks to the local community, including adaptive responses to climate change and natural disasters. Fourth, it delineates practical and cost-effective programs for sustainability, those that can be easily implemented, while recognizing the need for long-term strategies. Fifth, it discusses "best practices" in sustainability and utilizes a sustainability plan as an effective reporting document. Sixth, it advocates a funding or financial strategy that can lead to successful implementation. And lastly, it provides a mechanism for continuity and monitoring of sustainability strategies (Evans and Sunback 2005; Leitman 1999; Newman and Jennings 2008; Moore 2010; Taylor and Carandang, 2010).

Sustainability and Cities

The notion of sustainability has emerged as a major advocacy to deal with issues of environmental degradation. The publication of *Our Common Future* by the World Commission on Environment and Development defined sustainable development or sustainability as "development that meets the needs of the present without compromising the ability of future generations to meet their own needs." (WCED 1987) Sustainable Development became central to the United Nations Conference on Environment and Development, the Earth Summit, at Rio de Janeiro in 1992. The profile of cities was raised through Agenda 21, which recognized the importance of urbanization to sustainability and called for a Local Agenda 21 lead by local authorities (Ravetz 2000; Leitman 1999). Sustainability was further linked to urbanization through a key United Nations Report issued in early 1996 that laid out the blueprint for the Habitat II City Summit held later that year in Istanbul (UNCHS 1996). This report further tied the principles of Agenda 21 to urbanization, particularly the problems associated with the rapid urbanization of mega-cities in developing countries. A second United Nations Report was issued in 2001 that broadened the scope of sustainability to include the social, political and economic impacts of globalization and a greater emphasis on issues of poverty and social justice (UNCHS 2001).

The urban implications of sustainability have been widely debated by various scholars (Romaya and Radoki 2002). Some have emphasized an urban systems approach that seeks to determine the levels of socioeconomic, demographic, and technological output that can reinforce an urban system and maintain its long-term survival (Nijkamp 1990). Others have sought to use it as an approach to replace traditional modes of urban planning with sustainable urban management (Weiland and Hilty 1998). In this idea, urban sustainability includes the "minimization in the use of nonrenewable resources, the achievement of the sustainable use of renewable resources, staying within the absorptive capacity of local and global waste absorption limits and meeting basic human needs" (Choguill 1993). Others have sought to distance the concept of urban sustainability from sustainable development whereby urban sustainability means "strong sustainability," emphasizing ecological objectives over economic ones (Satterwaite 1997). And lastly, urban sustainability has meant to some a strictly social dimension, incorporating notions of equity, empowerment, accessibility, and participation (Basiago 1999).

Because they are heterotrophic, meaning that they are ecosystems that do not capture sufficient energy to meet their own needs, some argue that cities can never be sustainable. Most sustainable ecosystems are autotrophic, ones that capture sufficient energy for their needs. Sustainability planning for cities, however, centers on strategies, policies and programs by which cities can become more "photosynthetic," or closer to autotrophic ecosystems (McDonough and Braugart 2002). Also, sustainability planning is largely a response to and a modification of modernist and postmodernist planning styles. It holds at its core an ecological worldview. Modernist planning, instead, relies on a rational, comprehensive view of urban development that emphasizes reliance on the efficiency of technological solutions. The horizontal development of twentieth century cities is often the result of the extensive use of one such technology, the automobile, to provide maximum mobility and metropolitan reach. Postmodernist planning tends to emphasize pluralism and localized cultural traditions. Decision-making models for modernist planning are unitary and for postmodernist planning decentralized. Although sustainability planning incorporates some of the characteristics of both these approaches, it holds a distinctly ecological worldview. While recognizing pluralism, sustainability planning is centered on systems-thinking or the interconnection of people, values, things, events and resource use (Sterman 2012). As a planning mode, it uses communication and education to help evolve public understanding; advocacy planning to achieve shared goals; and incentives and mandates to implement agreed upon strategies (Wheeler 2004).

The basic premise of this paper is that cities provide solutions for a sustainable world, not problems. They possess the three major attributes necessary for successful sustainability: financial capital, human capital, and social capital. Two basic models for urban sustainability can be delineated. The first is the "Urban Metabolism" model (Fig. 2) that recognizes how resources and materials flow through an urban system and leads to both human productivity and waste outputs (see Kennedy 2012). The model also utilizes an eco-footprint measurement instrument as a way to determine successful or failed processes and public policies. A second model specifically dedicated to sustainable urban planning emphasizes the concept of the tensions of the "dual mandate" (Fig. 3) (Roe and van Eeten 2001).

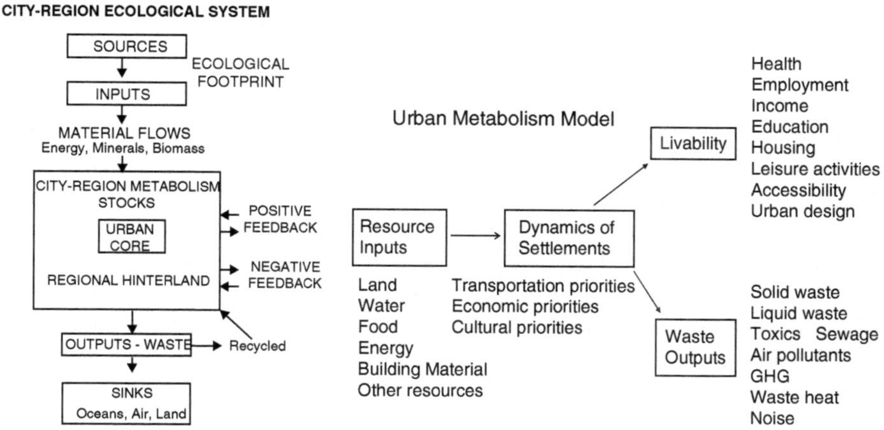

Fig. 2 Urban metabolism model. *Source*: Roe and van Eeten (2001)

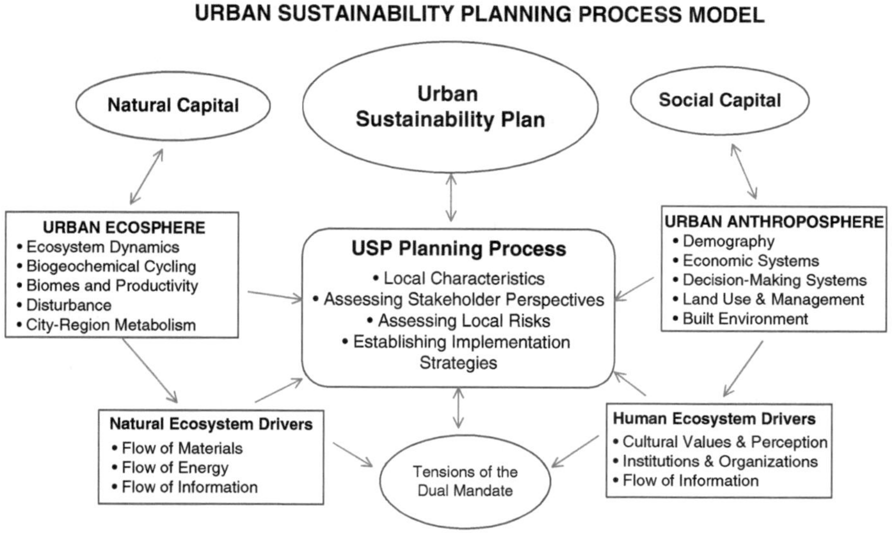

Fig. 3 Urban sustainability planning process model

These are the tensions or conflict points between the intersection of human ecosystem drivers and natural ecosystem drivers. For instance, all human modifications of the environment produce environmental disturbance. The questions of trade-off and whether benefits are sufficient to accommodate costs need to be determined, assessed, and communicated through a collaborative, consensual process. A successful sustainability plan must satisfy or accommodate numerous drivers and constituencies.

Approaches to Sustainability Planning for Cities

A number of existing sustainability planning approaches can provide suitable guides for cities. There are four basic approaches to urban sustainability planning (Table 1): vision-based, strategy-based, performance-based, and stakeholder-based. Vision-based approaches tend to emphasize guidelines, benchmarks, measures of city quality, and best practices in sustainability. An example of this approach is the Melbourne Principles that established ten objectives which cities must meet to progress toward sustainability. Other examples include the technical assistance by ICLEI (Local Governments for Sustainability), SustainLane's ranking of US cities based upon their capacity to attain prescribed measures of city quality, the American Society of Landscape Architects' "Sustainability Sites Initiative", and the Center for Natural Resources Defense Council's "City Grid—Smarter Cities."

Table 1 Approaches to urban sustainability planning

Approach	Characteristics	Examples
Vision-based	Vision statements Technical assistance Establishing guidelines and benchmarks Measures of city quality Best practices and innovative solutions	Melbourne Principles ICLEI Technical Guidelines "Sustainable Sites Initiative" American Society of Landscape Architects "SustainLane" US City Rankings "City Grid—Smarter Cities" Center for Natural Resources Defense Council
Strategy-based	Management plan Policy and strategy plan City-government directives	City of Steamboat Springs "Sustainability Management Plan" City of Rockville, "Strategy for a Sustainable Rockville" Ontario, Canada Sustainable Design Guidelines and Toolkit Makati Environmental Management Plan
Performance-based	Point-based standard Mostly voluntary Public image enhancement	Florida Green Local Government Standard Sustainable Jersey LEED for Neighborhoods
Stakeholder-based	Inclusive stakeholder engagement Educational and awareness driven Grassroots based	Aiming for Sustainability in Manila—De La Salle University Enhancing Urban Sustainability in Clearwater, Florida, University of South Florida

A second approach is strategy-based which emphasizes management plans and is often commissioned by city governments. This approach tends to provide specific strategy plans that deal with immediate problems facing cities or establish a detailed planning process by which cities can undertake sustainability planning. Examples of the former are the City of Steamboat Springs, Colorado's "Sustainability Management Plan," and the City of Rockville, Illinois' "Strategy of a Sustainable Rockville." An example of the latter is the Association of Municipalities of Ontario's

(AMO) "sustainability toolkit" for Ontario, Canada municipalities (Fig. 4; Bendle Group 2006; Ontario Model 2007; Makati Model 2009).

Tool 1: Making the Case for Sustainability
Tool 2: Structuring a Sustainability Planning Process
Tool 3: Defining Sustainability
Tool 4: Capacity Building
Tool 5: Assembling Baseline Data
Tool 6: Creating a Vision and Identifying Priorities
Tool 7: Preparing an Integrated Community Sustainability Plan
Tool 8: An Option to ICSPS: Adopting an Adaptive Management/Learn-by-Doing Approach
Tool 9: Engaging the Stakeholders
Tool 10: Adopting Sustainability Indicators, Performance Monitoring and Evaluation
Tool 11: Institutionalizing Sustainability
Tool 12: Securing Funding/Other Support for Sustainability Initiatives
Tool 13: Making the Links and Creating Value: Capital Investment Plans & Sustainability

Fig. 4 Ontario, Canada sustainability toolkit. *Source*: http://pubs.pembina.org/reports, 2008–2009

What makes the AMO toolkit an effective planning tool is that the Canadian Federal Gas Tax provides monies for investment in infrastructure that "achieves cleaner air, water and lower greenhouse gases". With this incentive in place, Ontario communities can make use of the toolkit to develop a sustainability plan that can leverage infrastructure investment.

A third approach to sustainability planning emphasizes performance-based planning models. This approach measures sustainability practices very precisely through a point-based system. Examples are LEED for Neighborhoods (www.usgbc.org/leed/nd); Sustainable Jersey (www.sustainablejersey.com); and the Florida Green Local Government Standard (www.foridagreenbuilding.org/local-governments). This approach is undertaken voluntarily by the city and certified by an outside third party. The benefits of undergoing this elaborate and detailed process are to generate a certified public awareness or eco-labeling of the sustainability practices of these cities. It is the approach that generates the most public image enhancement for a city.

The fourth approach is stakeholder-based. While all sustainability planning approaches recognize the importance of stakeholder engagement, few utilize it as a basic strategy to provide community education and awareness on sustainability practices and to engage the community is actually becoming a part of the plan. Examples are the "Enhancing Urban Sustainability in Clearwater, Florida" plan of the University of South Florida and the De La Salle University plan for Manila, Philippines, *Aiming for Sustainability in Manila* (Taylor and Carandang 2010, 2011). The stakeholder-based approach holds that in sustainability planning, the collaborative process is often the most important part of the plan. Even if strategies are not implemented, the process can stand alone as a plan. The process of gathering information from the community on perceptions, goals, problems, and engaging the community through workshops and other public forums are basic to achieving sustainability.

Lastly, sustainability planning ultimately depends on strategies for implementation. There are three basic types: voluntary, market incentive, and regulatory (Table 2). Most sustainability planning is voluntary with implementation strategies based on community education or public image enhancement. Often the plan is the major vehicle for community awareness of sustainability practices and potential projects in the community. Also, the creation of performance-based plans usually results in the implementation of sustainability practices. A second implementation strategy uses market-based incentives. This is based on a secure funding source, usually derived from government funding or large corporate or foundation donors. An example of a successful incentive that links the plan to implementation in the Ontario model where dedicated gas taxes are used to incentivize municipalities to develop sustainable infrastructure improvement projects. The third and last implementation strategy is regulatory. Examples are the use of zoning ordinances and building codes to insure sustainable practices such as energy and water conservation, and laws to maintain open space standards and reduce impervious land use.

Table 2 Implementation strategies for urban sustainability planning

Implementation strategy	Characteristics
Voluntary-based	Established and financed by stakeholder groups; inclusive, educational, and emphasis on wider view of the civic community
Financial incentive-based	Funded through various established funding mechanisms, such as a tax dedicated to certain performance criteria, i.e. Ontario Program
Legislative-based	Based upon specific regulatory requirements that establish criteria for zoning and building codes; also used as a tool to encourage fast-tracking of development projects

Source: Taylor (2010)

Can Cities Be Made Sustainable?

The scholarly literature on the belief that cities are really sustainable is mixed. For instance, Bugliarello (2006) discusses the virtues of cities as places to implement sustainability. He notes that cities concentrate human population, resource and material use, and economic activity. They exhibit certain advantages: their compactness, creativity, and diversity of design can promote equitable and just distribution of amenities and resources; the degree and ease of contact and mobility can contribute to a more livable habitat; and integrated mixed use communities and high-density urban living can shrink per capita ecological footprints by reducing energy and material needs. Further, he describes the paradigm of the city as biological, social and machine, complex in nature and involving three basic components and their interaction with the environment. It is biological in sense that it encompasses humans and other species that together strive to a balance while sharing of same resources, exposed to heightened exposure to microbiological threats in dense urban environments. Organizations, businesses, the city government, ethnic and informal social groups, and families form the social component. At the same time, presence of structures, vehicles, and other artifacts can be said to represent the machine component of cities (Justus and Taylor 2011).

Wood (2007) is representative of writers who believe that not all cities can become sustainable. He notes that although cities in the developing world face great challenges—climate change, loss of biodiversity, and land degradation, they still continue to emphasize traditional unsustainable designs. He believes that developing countries lack the political will and are unable to consider sustainable solutions. Sometimes, sustainable issues are perceived as secondary to more "urgent" needs such as HIV/AIDs and drought and food insecurity (not seen as related to sustainability). Also, developing countries often heavily prioritize economic growth ahead of other concerns, i.e., environmental considerations.

Urban sustainability is often an idea or concept neither well understood nor supported by majority populations in both developed and developing countries. There needs to be a discourse change to one that encourages greater reciprocal opportunity and perceived mutual advantage for all "eco aware" citizens. Synergistic urban lifestyles that are desirable, attainable, maintainable and reproducible—better known as "meta design planning"—is needed by today's high dense cities. "Eco" cities can be developed through a strong consensus that is inclusive of the business community, consumers, politicians, educators, bankers, and developers. Through collaboration, we can thus envision future designs, coherent and multilayered in nature, between wider groups of professionals working with urban planners and designers to create future sustainable cities (Justus and Taylor 2011).

One of the best arguments for sustainability planning for cities has been made by university professor and community planner Wheeler (2004). He discusses how cities can move to sustainability by emphasizing compact urban design, preserve open space, adopt alternatives to automobile use, and implement building codes that emphasize energy conservation and efficiency. He indicates that sustainability is indeed achievable in cities. One way is through environmental planning where cities create management plans for watersheds to protect and restore ecosystems. Another way is by reducing consumption levels and waste generation, and by adopting eco-friendly designs that reduce packaging and emphasize local production. Also, cities that reduce energy consumption and invest in alternative sources are more likely to achieve sustainability.

Wheeler notes four ways that cities can move toward sustainability. First, as mentioned above, to foster environmental planning that emphasizes watershed management. Second, to create a land use policy that preserves agricultural land on urban peripheries, maintains open space and greenways within cities, and advocates compact mixed use development that reduces vehicular use and promotes a sense of place and community. Third, to design transportation plans aimed at reducing vehicle dependency by improved transport alternatives, e.g., pedestrian and bike paths, new public transit options, car share programs, and revised street designs. And fourth, to increase both affordable housing and energy efficient buildings through codes that emphasize solar energy use and energy conservation.

An argument that is skeptical of cities to achieve sustainability is presented in the work of urban geographer Ooi (2009) of Nanyang Technological University. She shows that the challenges of rapid urbanization in emerging Asian economies is making it difficult for these cities to meet the basics of sanitation, water supply and housing, not to mention the more lofty goals of sustainability advocates. Professor Ooi points out the many challenges of cities in developing countries to include sustainability into their development agendas. She discusses how rampant urbanization has created resource depletion, air and water quality pollution, traffic congestion, poor sanitation and lack of housing. Extensive migration of rural people to urban areas in search of jobs has caused the mushrooming of informal settlements in cities with poor sanitation and health issues. Also, the lack of integrative or multi-sector policy at all levels of government compounds the challenges that these cities face.

Sustainable Cities in the United States and Britain

In the United States, sustainability planning for cities has largely targeted adaptation policies for climate change. In the absence of strong federal policy to address climate change, local governments have taken on the responsibility to deal with the issue and have often used sustainability plans as their main policy and planning instrument. A "Living Cities" survey taken in 2009 found that four out of five cities surveyed reported that sustainability was among their top five priorities (www.livingcities.org). It is also found that over 75 cities have, or will have soon, a detailed plan on how to reduce greenhouse gases, generally calling for a 10–20% reduction within the next 5–10 years. But their report also noted that cities were slower at addressing sustainability issues such as expanding mass transit, promoting green jobs, and improving the energy efficiency of existing building stock.

Portney Kent (2005) contends that at least 41 cities in the United States have begun sustainable city programs that are believed to effectively improve their livability. He highlights programs such as smart growth, increased bicycle ridership, integrated pest management, urban gardens, composting, local energy generation, and recycling and waste reuse. These programs are designed to move a city toward sustainability. Cities such as Seattle, Washington and Portland, Oregon have been recognized as leaders in urban sustainability. These cities, like all cities that encourage sustainability, have a common theme. It is the role of public participation at the grass-roots level and recognition of broader civil society, particularly Non-Government Organizations (NGOs) in helping to shape and implement city programs. Fundamental program elements vary from city to city, but successful programs incorporate benchmarks that measure progress toward sustainability over time (Justus and Taylor 2011).

Chicago is perhaps one of the best examples of an American city moving toward a concrete program of adapting to climate change, one of the key objectives of urban sustainability. They have used thermal radar to spot the city's hottest spots so that pavement can be removed and vegetation placed on urban rooftops (Kaufman 2011). Their program of climate change adaptation revolves around the belief that Chicago will experience much hotter summers and 35% more precipitation in winter and spring and 20% less rain in summer and fall. This will significantly alter Chicago's micro-climate and increase its vulnerability to ecological and financial risks. As a solution, Chicago is preparing new street design which utilizes permeable pavement, underground urban storm water runoff storage tanks, drought-resistant plants, and *Spartina* grasses for filtering pollutants and salts used on winter roads (Fig. 5).

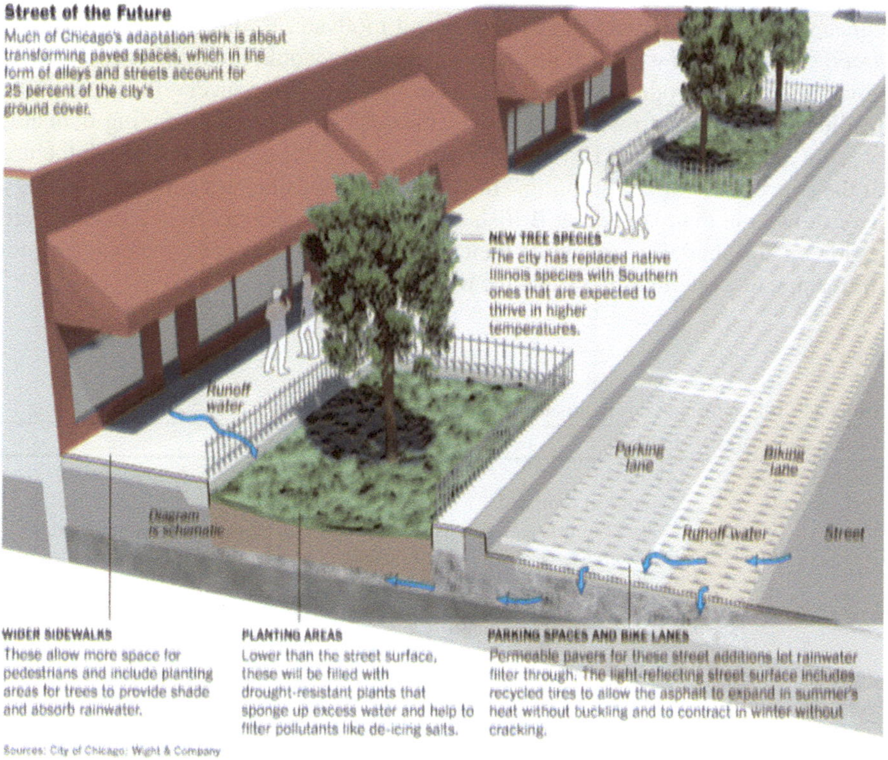

Fig. 5 Chicago's Climate Adaptation Plan (CAP). *Source*: Retrieved from *New York Times*, May 23, 2011

Throughout the world, scholars and activists are seeking ways to measure and determine whether their country's cities are moving toward sustainability. In the United Kingdom, the Manchester City Region estimates that the cost of doing nothing about climate change will cost the North-West region of England about four billion USD in 10 years. It seeks through its long-term strategic plan to alter its historic industrial heritage and to move Manchester from red brick to green brick. David Aeron-Thomas et al. in *Sustainable Cities Index: Ranking the Largest 20 British Cities* (www.forumforthefuture.org) have measured the progress toward sustainability of the largest British cities across three broad areas: environmental performance, quality of life, and "future proofing"; i.e., how will these cities address key future issues. They developed a rating system that is based on 13 indicators across these three broad areas—a set of indicators that were derived from publically available data and where local governments have the power to improve the sustainability of their cities (Table 3). Hence, their measurement system is both viable and meets a common-sense approach to evaluating urban sustainability. An interesting methodological conclusion was to place less emphasis on surveys and more on developing indicators based on quantifiable data from public sources. Also, they moved away from the concept of measuring green business, a vague and difficult concept for local governments to access and motivate, to the notion of an urban green economy, and what that urban green economy would look like and how local governments can produce a set of policies to enable it. They successfully integrated resilience thinking—to what extent are cities able to adapt to changing demands brought about by external factors—into their sustainability criteria through the use of the concept of "future proofing." In all, while establishing a viable methodology they showed in their study that British cities have moved away from a north–south divide on sustainability and while their ecological footprint is decreasing, it is not decreasing at a rate that is sustainable over the long term.

Table 3 Criteria for measuring urban sustainability in British cities

Sustainability area	Indicator	Criteria for measurement
Environmental Performance	Air quality	Annual mean background concentrations of nitrogen oxides
	Ecological footprint	Hectare of land per capita for services, food, housing, transportation, and consumables per capita
	Household waste	Household waste collected per capita
	Biodiversity	Percentage of local biodiversity sites that have undergone conservation management in the last 5 years
Quality of life	Life expectancy from birth	Measure of health and longevity
	Green spaces	Measurement of green space per 100,000 population
	Transport	Number of minutes spent per person per month traveling to city services
	Unemployment	Number of benefit claimants as a percentage of working age population
	Education	Percentage of working age population with education or equivalent
Future proofing	Local Government commitment to climate change	A measurement of existing or future commitment to climate adaptation or mitigation strategies
	Economy	Number of business start-ups per 100,000 people
	Recycling	Percentage of household waste recycled or composted
	Food	Number of plots of land in local food production

Source: Retrieved and modified from www.forumforthfuture.org

Sustainability Planning for Cities in Emerging Economies

While sustainability planning is relatively advanced in the developed economies of Europe and North America, it is a new and somewhat untested planning regime for emerging economies. As stated previously, most future urbanization will occur in Asia and other emerging regions of the world (Fig. 6; Global Compact Cities Program 2010). These regions bear similar characteristics in terms of higher amounts of environmental degradation, lower per capita incomes, higher rates of poverty and greater exposure to the effects of climate change. But, a case can be made that sustainability planning will produce more significant benefits with less cost to the cities of these emerging economies. The key areas that sustainability planning can provide support are disaster management, climate change, environmental degradation and natural resource loss, poverty, and effective service delivery.

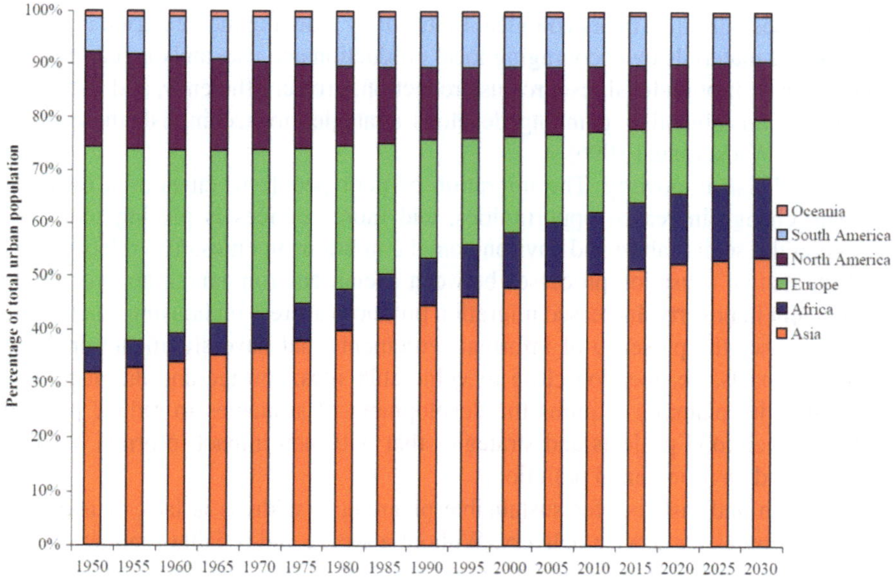

Fig. 6 Future urbanization will occur mostly in developing countries. *Source*: Urban Settlement: Data, Measures, and Trends, Harvard School of Public Health, December 2007

Climate change or weather-related risks are one of the major drivers for sustainability planning. The major vulnerabilities to cities are weather-related changes due to global warming including rising sea levels and intensity and frequency of rainfall. Since most major cities are located in coastal environments, sea-level rise has increased local storm surges and flooding events (Manila Bulletin 2010). Global temperature increase has resulted in temperature extremes that put vulnerable urban populations like the elderly and the poor at risk. Also, cities are affected by the "urban heat island effect," which is the process by which man-made surfaces and limited green spaces in cities produce higher temperatures than in surrounding rural and forested areas. This compounds the effects of global temperature increase in cities and requires adaptive planning solutions. Also, the effects of climate change places undue stress on rural and coastal environments that result in increased migration to the city, causing social tension and environmental problems.

There are three broad risk management strategies for climate change: protection, mitigation, and adaptive strategies. Protection strategies aim at developing expensive solutions, e.g., seawalls. Mitigation strategies are directed at larger national strategies, e.g., replacement of fossil fuel dependence in electricity productions to renewable energy. And lastly, adaptive strategies that are smaller, more cost-effective, and are most likely to be implemented at the local community level, such as natural drainage systems, green building programs, etc. Sustainability planning for cities tends to emphasize adaptive strategies for climate change risk (Claudio 2010).

A second risk is environmental degradation and natural resource loss. Sustainability planning views the urban system as an ecosystem. In an ecosystems model, cities minimize their ecological footprint by reducing their external inputs of energy and materials by encouraging local production and use, and decreasing output of waste through local resource use reduction, greater efficiency, and reuse and recycling. Sustainability planning develops strategies to accomplish these goals (Rees and Wackernagel 1996).

A third risk is poverty. The migration of poor people to cities as a result of climate change increased opportunities, and more services is placing increased stress on the social fabric and environmental conditions in cities. The relationships and interactive causation processes between social, environmental, and economic are evident in poverty. Increased migration produces a stress on housing availability that increases the presence of informal settlements that have significant environmental impacts, i.e. location close to vulnerable sites, sewage and drainage, etc. Sustainability planning requires the development of strategies to reduce poverty through livelihood projects and strategies that will then impact informal housing location and environmental pollution.

A fourth and last risk for sustainable planning to address is lack of adequate revenue sources for service delivery. Without a secure and reliable funding or revenue source, sustainable strategies cannot succeed. Hence, the development of an effective sustainable management plan for local governments requires the introduction of creative financial mechanisms and instruments and the integration of finance into the other sectors of the plan, i.e. housing, economy, equity, and environment.

Sustainability Planning: Manila, Philippines Case Study

An example of the potential effectiveness and some of the difficulties in engaging in sustainability planning in emerging economies is demonstrated through the development of a sustainability management plan for the city of Manila, Philippines (Taylor 2005; Taylor and Carandang 2010, 2011). The city of Manila (Fig. 7) has

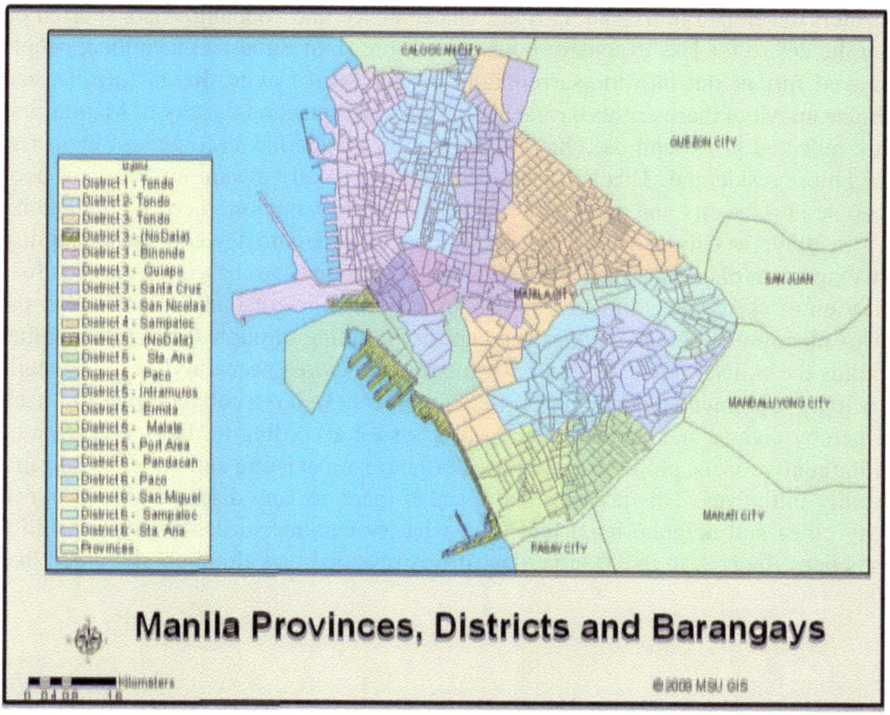

Fig. 7 GIS map of City of Manila and Metro region. *Source*: Amy Ferdinand, MSU

the highest population density of any major city in the world, which theoretically can be a positive characteristic associated with sustainability if used as tool for public transportation. Also, the city of Manila resides largely at sea-level and the Philippines as a region has major exposure to the effects of climate change. According to the 2007 census, Manila has a population of close to 1.7 million occupying a land area of only 38.6 km^2. It comprises one of the 17 local government units of Metro Manila with a population of roughly 12 million. The city of Manila is also the capital of the Philippines.

The city of Manila is the old central business district and cultural center of the metropolis. It consists of 16 geographic and six legislative districts (Fig. 7). There are 897 barangays, the smallest unit of city government, which roughly coincides with a neighborhood unit and has its own political designation. Located at the confluence of the Pasig River and Manila Bay, Manila developed first along the river and waterfront area and spread outward from the geographical districts of Escolta, Quiapo, and Santa Cruz. The Spanish heritage is manifested in Fort Santiago and Intramuros, the old walled city, which has the feel and ambiance of an old European medieval city. The areas around Roxas Boulevard in Ermita and Malate bear testament to the American period and the Burnham Plan, one of the first examples of sustainability planning in the Philippines (Taylor 1994). In the last decades of the twentieth century, Metropolitan Manila expanded outward as new population centers developed in Makati, Ortigas, Quezon City, and Alabang. Also, spurred on by the desire for less expensive housing in more rural–suburban locations, people moved further out into the surrounding provinces of Cavite, Rizal, Laguna, and Bulacan. All of these events have had a significant impact on the city of Manila, and are reflected in its land use changes. Presently, Manila has become less industrial and more residential. This has been the result of a steady inward migration of poor people into the city and the natural increase of an existing low-income population.

In 2009, the City of Manila engaged a local university to develop a sustainability management plan. The plan, a set of recommendations, was based on a broad cross section of stakeholder inputs (Fig. 8) based on a series of interactive workshops. This planning approach emphasized capacity building through social learning that builds the skills, knowledge, and institutional structures necessary for sustainability to be implemented (Friedman 1987; Putnam 1993). It is a collaborative approach whereby diverse stakeholder groups are assessed according to their perception of community assets, problems, and for information on existing and potential sustainability initiatives. This process can reveal if there are any discrepancies between city elites and neighborhood locals on what constitutes successful sustainability practices. Indeed, in the Manila plan, it was revealed that although all agreed that

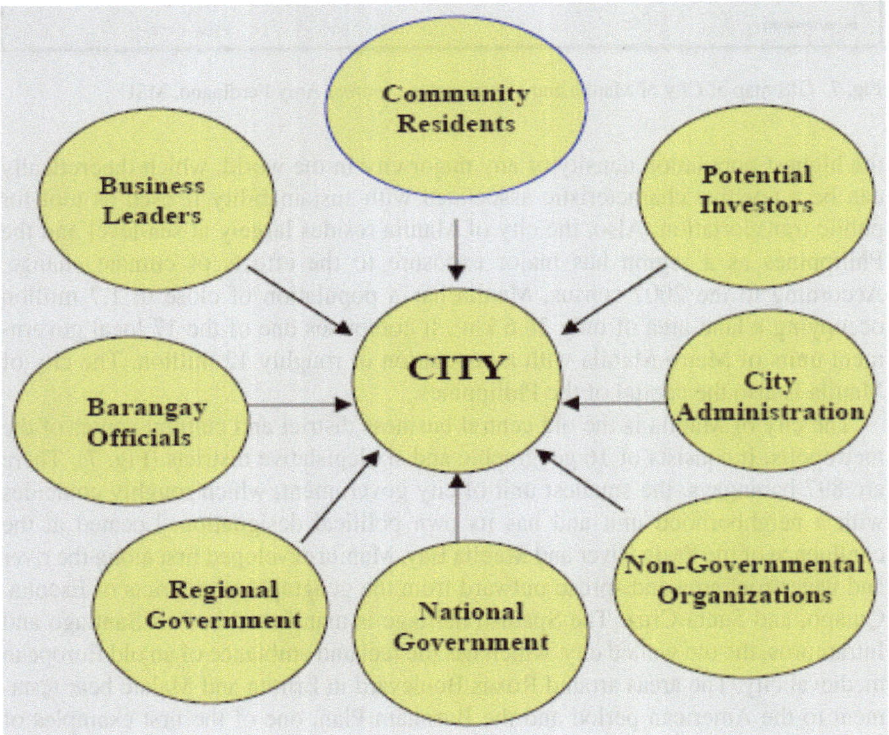

Fig. 8 Stakeholder map. *Source*: Taylor (2010)

actions needed to be taken to promote sustainability, there were major differences in priorities and vision. The locals tended to emphasize more "people-based" issues while the elites were more concerned about environmental and economic issues. This difference in perception and overall vision can be a significant deterrent for effective implementation of sustainability policies, practices, and projects.

A key component of any sustainability planning process is to map out the key challenges that a city confronts. This process takes on a number of forms, but basically consists of a matrix cross-referencing challenges analyzed against a number of categories such as issues, sources, and opportunities. Some of the major challenges facing cities are climate change, water budgets, land use, and air quality. An example of a Sustainability Issues Map (SIM) using this approach is provided in Table 4.

Table 4 Sustainability issues map (SIM) for cities

Challenges	Issues	Sources	Opportunities
Climate change	Reduction of greenhouse gas emissions, sea rise, and coastal vulnerability	Power utilities, business and household operations, transportation	Launch eco-efficiency efforts targeting energy use and related GHG emissions
Water	Water supply, water quality, wastewater treatment, storm water runoff, flooding	Business and household runoff, hospital waste, leaky supply pipes, overuse	Launch efforts to conserve, reuse water, and educate residents on these practices
Land	Waste management, brownfields, land use, urban design, built environment	Household and business consumption, subdivision development and sprawl	Launch programs to reuse land, adopt compact urban design, and reduce waste
Air	Air pollution, air quality	Transportation, power plants	Launch public transit program

Source: Taylor (2010)

The Manila Plan assessed the risks to land, air and water ecologies and produced a set of recommendations. For air ecology they were to: establish a carbon credit program, develop a green building program, develop an energy conservation program, develop a photovoltaic solar panel installation program, and establish a transit improvement program. For water ecology, it was to: continue the Pasig River Rehabilitation Program, establish waterfront buffer areas, implement natural drainage systems, create a rainwater harvesting program, and establish a water conservation program. And for land ecology, it was to: develop a land-recycling program for obsolete industrial sites, restore old buildings for adaptive reuse, establish a strategic implementation program for waste recycling, turn barangays into urban villages, and establish a greenways program.

The Manila Sustainability Plan was presented to the Mayor's office in 2009. It is presently being considered for implementation. The report advocates five strategies for successful sustainability programs, projects, and initiatives. First, it promotes the use of the Manila local governmental unit as the most effective organization to implement sustainability initiatives. They possess a decision-making apparatus whereby sustainability plans can be enforced; they are closest to the people for community consensus; and their success is determined by bringing environmental, economic, and social equity into equilibrium. The Local Government Code in the Philippines has devolved power to the local governments thereby providing them the necessary capacity to lead in sustainability. Second, the City needs to create an interdepartmental committee, an office, agency, or ombudsman whose sole responsibility is to implement sustainability initiatives. The stakeholder groups highly support this strategy and it can be combined with a sustainability educational campaign to educate people about energy and water use. Third, the City needs to collaborate with a local university to establish a "City of Manila Monitoring Station" that monitors social, economic, and environmental health utilizing a Geographical Information Mapping System at the barangay level. This community-based resource would be geared to work with the barangays to ameliorate their problems and initiate sustainability projects. Fourth, the city needs to stimulate and encourage the development of public–private partnerships that revolve around sustainability. These partnerships can be initiated by community-based cooperatives, NGOs, or private businesses, but need to work closely with barangays and City agencies. And lastly, Manila has very little land available for development. Hence, sustainability initiatives need to emphasize the reuse or recycling of land that is environmentally contaminated, under-utilized, or vacant. A major program needs to be implemented that identifies these sites, and proposes strategies to place them back into productive use, perhaps utilizing new technologies that emphasize resource efficiency. It is also important to recognize that local stakeholders advocate community designs that promote mixed-use development and the more intense use of land. These strategies are all designed to make their community more sustainable.

A key component to successful sustainability planning for any city but particularly for cities in emerging economies is to create a financing component to the Plan. In the Philippines many responsibilities and services have devolved from the national government to the cities, but revenue capacity has not often followed this decentralization of power. As a result, Philippine cities direct most of their national monetary transfers to well-established nondiscretionary city services, leaving little funding for sustainability projects or actions. Hence, one of the key goals of a sustainability plan is to seek out creative funding strategies. In the Philippines, creative financing strategies are local revenue bonding, development impact fees, transfer of development rights, special assessment districts and tax increment financing, real estate investment trusts, microfinance, and public–private partnerships.

A number of barriers to sustainability exist in cities like Manila. First, while there is a general recognition of the concept of "sustainability," there is little understanding of how the concept relates specifically to people and their communities. Second, there needs to be a steady supply of trained personnel to implement sustainability and a specific office that spearheads these initiatives. Third, there needs to be better communication between city officials and the neighborhoods as to what needs to be done. Often, there are contrasting visions of sustainability. And finally, many sustainability initiatives cannot withstand political change. Often successful programs initiated by one city administration are canceled in mid-stream by an incoming administration. Also, valuable data are lost or misfiled in governmental transition which affects the capacity of planners to construct important benchmark or performance indicators.

Conclusions

Sustainability planning is the next frontier for cities. Local governments or cities are the best organizations to manifest and implement sustainability. They possess a decision-making apparatus that allows the sustainability practice to be readily implemented; they are the institution that is closest to the people and whose decisions reflect on developing the holistic health of the community, meaning that the goals of equity, economy and environmental quality must all be satisfied equally; and they are the institution that is held most directly accountable to the people. While corporate sustainability is an admirable pursuit, corporations as a vehicle for sustainability are limited by the demand for short-term profitability. The local government sustainability plan is the best mechanism to advance sustainability as it constitutes a strategic planning approach with the following characteristics: a high level of stakeholder participation, short-term actions to solve priority issues, a long-term vision, and an ongoing monitoring system.

Western cities are a bit ahead of their sister cities in the developing world in adopting this new planning regime, although cities like Curitiba, Brazil and Bogota, Columbia are leaders in their own right. Yet, uniform cross-regional limitations to implementing sustainability in cities can be observed. The most pressing is that too much emphasis is placed on the initial plan and much less emphasis on the follow-up or in the process of implementation. This is generally a reflection of the inability for sustainability plans to establish their own revenue base. This needs to change for sustainability to be a meaningful pursuit. Cities must develop more creative financing sources for sustainability projects and political leaders need to have the political will to create the necessary incentives through zoning, bonding, building codes, impact fees, and land use plans.

References

ADB (Asian Development Bank) (2001) Asian environmental outlook 2001. ADB, Manila

Basiago A (1999) Economic, social, and environmental sustainability in development theory and urban planning practice. Environmental 19:145–161

Bugliarello G (2006) Urban sustainability: dilemmas, challenges and paradigms. Technol Soc 28:19–26

Choguill C (1993) Sustainable cities: urban policies for the future. Habitat Int 17:3
Claudio C (2010) Climate change adaptation. Business World, June 15
Davis C (2008) Monthly update: urbanization and environmental sustainability, Earthtrends, World Resources, February
Dobbs R, Sankhe S (2010) Comparing urbanization in China and India. McKinley Quarterly, July
Evans J, Sunback T (2005) Governing sustainable cities. Earthscan, London
Friedman J (1987) Planning in the public domain: from knowledge to action. Princeton University Press, Princeton
Global Compact Cities Programme (2010) Sustainable cities, vol 1. United National global compact, Executive summary, p 6
Justus F, Taylor R (2011) Introduction, Are cities sustainable? In: Taylor R (ed) Taking sides: clashing issues in sustainability. Mc-Graw Hill, New York
Kennedy C (2012) A mathematical description of urban metabolism. In: Weinstein MP, Eugene Turner R (eds) Sustainability science: the emerging paradigm and the urban environment. Springer, New York
Kent P (2005) Civic engagement and sustainable cities. Publ Admin Rev 65:579–591
Leitman J (1999) Sustaining cities. McGraw Hill, New York
Manila Bulletin (2010) (Environment and Nature B-10 Section), UN warns of mega-disasters due to extreme weather conditions. Tuesday, June 8
McDonough W, Braungart R (2002) Cradle to cradle: remaking the way we make things. North Point Press, New York
Moore S (ed) (2010) Pragmatic sustainability. Routledge, London
Newman P, Jennings I (2008) Cities as sustainable ecosystems. Island Press, Washington, DC
Nijkamp P (1990) Sustainability of urban systems. Averbury, Aldershot
Ooi GL (2009) Challenges of sustainability for Asian urbanization. Curr Opin Environ Sustain 1:187–191
Owen D (2009) The green metropolis. Penguin Group, New York
Putnam R (1993) Making democracy work: civic traditions in modern Italy. Princeton University Press, Princeton
Ravetz J (2000) City-region 2020. Earthscan, London
Rees W, Wackernagel M (1996) Urban ecological footprints: why cities cannot be sustainable and why they are a key to sustainability. Environ Impact Assess Rev 16:223–248
Roe E, van Eeten M (2001) Threshold-based resource management: a framework for comprehensive ecosystem management. Environ Manage 27:195–214
Romaya S, Radoki C (eds) (2002) Building urban settlements: approaches and case studies in the developing world. ITDG Publishing, London
Satterwaite D (1997) Sustainable cities or cities that contribute to sustainable development. Urban Stud 34:1667–1691
Schulz V (ed) (2002) Urbanization and transition to sustainability. International human dimensions program on global environmental change. FRG, Bonn
Sterman JD (2012) Sustaining sustainability: creating a systems science in a fragmented academy and polarized world. In: Weinstein MP, Eugene Turner R (eds) Sustainability science: the emerging paradigm and the urban environment. Springer, New York
Taylor R (1994) The best of Burnham: the Philippine plans. Plann Hist 16:14–17
Taylor R (2005) Beneficial urban sustainability in Asia: the metro-Manila Philippines experience. In: Feng H, Yu L, Soleki W (eds) Urban dimensions of environmental change—science, exposures, policies and technologies. Science Press, Montclair
Taylor R, Carandang J (2010) Sustainability planning for Philippine cities. De La Salle University, Yuchengco Center, Manila
Taylor R, Carandang J (2011) Sustainability and redevelopment in the City of Manila, Philippines. Human Ecol 23:14–21
United Nations Center for Human Settlements (UNCHS) (1996) An urbanizing world: global report on human settlement. Oxford University Press, New York

UNCHS (2001) The state of the world's cities report. Haitat, Nairobi
United States Asian Environmental Partnership (1999) Place-based public policy in southeast Asia: developing, managing, and innovating for sustainability. USAEP, Washington, DC
Wackernagel M, Rees W (1996) Our ecological footprint. New Society Publishers, Gabriola, British Columbia
WCED (1987) Our common future. Oxford University Press, Oxford
Weiland U, Hilty L (1998) Sustainable urban management: opportunities and risks of information technology. In: Hamm L, Muttagu S (eds) Sustainable development and the future of cities. ITDG Publishing, London
Wheeler S (2004) Planning for sustainability: creating livable, equitable and ecological communities. Routledge, London
Wood J (2007) Synergy city; planning for a high density, super-symbiotic society. Landscape Urban Plan 83:77–83

Internet Sources

Bendle Group (2006) Sustainable management plan: City of Steamboat Springs, Colorado. Steamboat Springs, Colorado, www.steamboatsprings.net; retrieved January 12, 2009
Glaeser E (2009) Help the environment, stay in the city, www.dcexaminer.com; retrieved January 8, 2008
Kaufman L (2011) A city prepares for a warm long-term forecast. New York Times, retrieved May 22, 2011
Living Cities (2009) Green cities: how urban sustainability efforts can and must drive America's climate change policies, www.livingcities.org; retrieved May 5, 2009
Makati Model (2009) Sustainable development in Makati: environmental management plan, http://cityriskpedia.com; retrieved May 8, 2009
Ontario Model (2007) Ontario community sustainability report, http://pubs.pembina.org/reports; retrieved May 10, 2009
Sustainable Cities Index (2010) www.forumforthefuture.org; retrieved July 12, 2011

Climate Change, Globalization, and the Double Exposure Challenge to Sustainability: Rolling the Dice in Coastal New Jersey

Robin Leichenko

Abstract Climate change and globalization present significant challenges for sustainability. Both processes enhance connections across space and time, such that actions taken in one part of the world have increasingly visible impacts in other parts of the world. The processes also magnify risks and uncertainties, exacerbate vulnerabilities, and undermine resilience to many types of shocks and stresses. This chapter explores how climate change and globalization are together influencing sustainability in urbanized coastal zones with particular emphasis on coastal New Jersey. While urban coastal zones have long confronted a multitude of development-related stresses including reductions in quantity and quality of freshwater flow into estuaries, destruction and degradation of wetlands, and dredging and development of harbor areas, climate change and globalization represent new and interconnected sources of stress. Under climate change, altered temperature regimes, shifts in the variability and seasonality of precipitation, increases in the frequency and magnitude of extreme events, and sea level rise are together transforming the environmental baseline of coastal areas. At the same time, processes of globalization are contributing to growth of coastal tourism, intensification of coastal property investment, expansion of port facilities and shipping traffic, and changes in the availability of public funds needed to manage these complex, coupled systems.

Keywords Coastal zones • Vulnerability • Economic impacts • Multiple stresses • Sustainable development

---

R. Leichenko (✉)
Department of Geography, Rutgers University, 54 Joyce Kilmer Avenue,
Piscataway, NJ 08854, USA
e-mail: robin.leichenko@rutgers.edu

M.P. Weinstein and R.E. Turner (eds.), *Sustainability Science: The Emerging Paradigm and the Urban Environment*, DOI 10.1007/978-1-4614-3188-6_15,

Introduction

Within the emerging field of sustainability science, there is general recognition that large-scale processes of environmental and economic change play a critical role in the sustainability of coupled human–environment systems (Adger 1999; Adger et al. 2009; Turner 2010; Wilbanks and Kates 2010). There is also growing interest in exploring how economic and environmental changes interact to influence sustainability efforts within cities and urbanizing areas (Seto et al. 2010; Ernstson et al. 2010; Leichenko et al. 2010). Yet relatively limited attention has been directed toward understanding how two of the most transformative processes of the current era, namely climate change and globalization, are jointly influencing sustainability within specific regions and sectors (e.g., O'Brien and Leichenko 2000; O'Brien et al. 2004; Eakin 2006; Leichenko and O'Brien 2008; Silva et al. 2009). Both processes provide new opportunities, but they also create new risks and uncertainties and exacerbate vulnerabilities to many types of shocks and stresses (Leichenko and O'Brien 2008). These new risks and uncertainties are compounded by spatial and temporal tele-connections created by both processes, such that actions taken in one part of the world or during one time period have increasingly visible impacts in other parts of the world or at other times (Adger et al. 2009; Leichenko and O'Brien 2008).

This chapter explores how climate change and globalization are together influencing sustainability in urbanized coastal zones, with particular emphasis on coastal New Jersey. While urban coastal zones have long confronted a multitude of development-related stresses including reductions in quantity and quality of freshwater flow into estuaries, destruction and degradation of wetlands, and dredging and development of harbors, climate change and globalization represent new and interconnected sources of stress. Under climate change, altered temperature regimes, shifts in the variability and seasonality of precipitation, increases in the frequency and magnitude of extreme events, and sea level rise are together transforming the environmental baseline of coastal areas. At the same time, processes of globalization are contributing to the growth of coastal tourism, intensification of coastal property investment, expansion of port facilities and shipping traffic, and changes in the availability of public funds needed to these manage these complex, coupled human–environment systems.

The "double exposure" framework developed by Leichenko and O'Brien (2008) is used to examine how climate change and globalization are creating new sustainability challenges in New Jersey's coastal zone. The double exposure framework provides a generalized approach for investigation of the interactions between global environmental and economic changes, focusing on ways that these interacting processes spread risk and vulnerability over both space and time. As such, the framework is reflective of broader efforts by sustainability scientists to enhance understandings of vulnerability and adaptation, including factors influencing the resilience of communities, regions, and socioecological systems to shocks and stresses associated with processes of global change (e.g., Turner et al. 2003; Eakin and Luers 2006; Polsky et al. 2007; Nelson et al. 2007; Berkes

2007; Acosta-Michlik et al. 2008; Eakin and Wehbe 2009; Eriksen and Silva 2009). Key questions emerging from this body of work include the mechanisms that create feedbacks between processes of economic and environmental change, how tele-connections create new risks and uncertainties across space and time, and how to identify strategies to promote adaptation and resilience under conditions of rapid socioeconomic change (Adger et al. 2009; Leichenko et al. 2010).

The next part of the chapter describes how climate change and globalization are contributing to rapid transformation of urbanized coastal zones throughout the world. The basic elements of the double exposure framework are used to describe how the framework may be applied to assess how interactions between the globalization and climate change affect sustainability across many types of coupled human–environment systems. Insights from the double exposure framework are then used to consider how globalization and climate change are together affecting coastal New Jersey. The chapter concludes by arguing that the emergence of new types of intersections and interactions between globalization and climate change create both opportunities and challenges for sustainability and calls for new research on combined responses to both processes.

Climate Change, Globalization, and Urbanized Coastal Zones

Coastal areas are home to a growing proportion of the world's population. Currently more than half of all humans live within 200 km (120 miles of a coast; UN 2006), and coastal population growth is substantially outpacing that of inland areas due to both in-migration from inland regions and high rates of natural population increase. Coastal populations include approximately 618 million people (i.e., nearly one tenth of the world's population) living in locations with an elevation of less than 10 m above sea level (McGranahan et al. 2007). While the residents of low-lying coastal areas are primarily concentrated in low and lower-middle income countries of Asia including large countries such as China, India, Bangladesh, Vietnam, Indonesia and Egypt, as well as in small island nations such as the Bahamas, Surinam and Guyana, they also include large numbers of residents in many higher income countries such as the Netherlands, Denmark, Japan, and the United States (McGranahan et al. 2007).

Coastal development is deeply interwoven with these global patterns of urbanization. Coastal areas are home to the majority of the world's largest cities. Among the 25 largest megacities—those with a population of over ten million people—16 are located in coastal areas. Among cities with populations over five million, 65 cities worldwide have some portion of their settlement area location in a coastal area with an elevation of 10 m or less (McGranahan et al. 2007).

Within the United States, coastal population patterns mirror those at the global level. Of the 60 cities within the United States with populations greater than 300,000, 20 have land area with elevations at or below 6 m and connectivity to the sea (Weiss et al. 2011). Six of the ten largest cities in the United States are among this group, including New York, Los Angeles, Houston, Philadelphia, San Diego, and San Jose. The population of US coastal cities is likely to increase substantially over the next several decades due to both natural increase and current patterns of population in-migration from other parts of United States and other parts of the world. As discussed next, international migration is an important facet of globalization in coastal areas.

Globalization and Coastal Zones

Globalization may be understood as a shift in scale of economic activity, such that markets for goods and services that used to be local and regional in scope now extend across international borders. Globalization is also commonly associated with "space–time" compression, whereby improvements in transport and communication technologies have reduced the time and costs associated with the movement of goods, people, and ideas across space, effectively "shrinking" the world (Harvey 1990; Leichenko and O'Brien 2008). Along with technological innovations, greater economic, political and cultural integration under globalization is also facilitated by policies advocating free market approaches to economic activity. Key metrics of globalization including rising levels of international trade and investment, financial flows, international migration, and homogenization of cultural practice and preferences across nations, all of which are facilitated by widespread access to television, increasing availability of high-speed internet connections, and proliferation of cell phones and related mobile technologies. While globalization is frequently associated with rising standards of living and expanded opportunities for individual consumption, it has also implicated rising levels of income inequality both within and across nations, and it has become a major driver of increased energy, material usage worldwide. These "negative externalities" of globalization present significant barriers to efforts to achieve a sustainability transition.

Activities in coastal zones both reflect and play an important role in many facets of globalization. International migration has dramatically increased the population of many coastal cities. Growth of international trade has meant expansion of ports and growth of manufacturing facilities in coastal zones. Over the past 2 decades, shipping ports have been upgraded and expanded in many coastal cities in order to handle larger container ships, a process which often entails significant change to coastal landscapes including armored shorelines, land reclamation from the sea, and dredging and deepening of harbors. In addition to new berthing areas for vessels, these new developments provide living space for port workers, storage space for unloaded materials, and land for expanded intermodal transportation to facilitate shipment of goods both for export and to inland destinations. Export processing zones for manufacturing goods now dominate coastal areas in southern China, Indonesia, the Philippines, and elsewhere.

Globalization has also contributed to intensification of coastal property development worldwide. The property construction boom that started in the late 1990s and ended with the onset of the global financial crisis in 2008 was facilitated by globalization of finance, which contributed to increased availability of low interest rate loans (i.e., "cheap" money). These loans helped spawn developments within many high amenity coastal areas (Leichenko and Solecki 2005; Leichenko and Solecki 2008). While much of the new development in coastal areas consists of residential properties, the globalization of tourism has also had a major influence on coastal landscapes in many countries. The tourism sector in many countries is increasingly dominated by large, multinational firms (Hjalager 2007). These firms cater to international tourists through construction of large hotel complexes and development of beaches and recreational facilities in warm weather coastal zones throughout the world.

Climate Change and Coastal Areas

The vulnerability of coastal areas is well-recognized within the climate change literature (Adger et al. 2005; McGranahan et al. 2007; Neumann et al. 2010; Nicholls et al. 2011; Weiss et al. 2011). Two of the most significant climate-related threats to coastal areas include sea level rise and increased frequency and magnitude of coastal storms. By the end of this century, global sea levels are projected to approach or exceed 1 m (compared to 1990 levels) (Overpeck and Weiss 2009; Vermeer and Rahmstorf 2009). Some of this increase is likely to be attributed to the thermal expansion of oceans, and some is linked to the melting of glaciers and ice on land (Alley et al. 2005). There is also growing recognition of the possibility of substantially larger increase in sea levels as the result of the melting of the Greenland and West Antarctic Ice Sheets (Nicholls et al. 2008). While the magnitude of the increase will be varying spatially, those populations living in low-lying coastal areas will be particularly vulnerable to the effects of sea level rise.

Climate change is also expected to influence the magnitude and frequency of coastal extreme weather events such as hurricanes and tropical cyclones (Knutson et al. 2010). These events, which already represent major threats to coastal areas in both tropical and mid-latitude regions, are likely to be exacerbated by increased storm surges associated with sea level rise. While wetlands, coral reefs, and coastal ecosystems often buffer coastal regions from such extremes, absorbing much of the energy and impact, land use changes in many regions have led to large-scale reductions in wetlands, beaches, and other protective features (Leichenko and O'Brien 2008).

Importantly, climate change will occur in the context of intense development and environmental degradation. In conjunction with sea level rise, another climate change-related threat in coastal areas concerns freshwater supply and water quality. Many coastal areas are already experiencing significant water shortages and water quality problems due to lack of freshwater supplies, salt-water intrusion in aquifers, inadequate reservoirs, and other constraints. Increased temperatures, reduced precipitation, sea level rise, or increased variability of rainfall water as the result of climate change are likely to exacerbate these and other water resource problems.

For many coastal regions, the impacts associated with both climate change and globalization are not limited to populations residing within and around these areas, but also reverberate to other areas as the result of various types of economic linkages and social and political connections. The next section describes a framework for assessing the overlapping impacts and reverberating effects of both climate change and globalization.

The Double Exposure Framework

Despite widespread recognition of the linkages between climatic change and economic activities, research on connections and interactions between climate change and globalization remains limited. The majority of research on climate change emerges from the physical or ecological sciences, with emphasis on dynamics of the atmosphere, hydrosphere, and biosphere. In contrast, studies of globalization tend to emphasize political, economic, and cultural phenomena such as liberalization of trade, formation of transnational commodity chains, and emergence of a global mass media. Although much globalization research addresses environmental issues, this literature does not typically consider how globalization influences or interacts with larger processes of climate change. As a consequence, critical linkages, feedbacks, and synergies between globalization and climate change often receive insufficient attention.

The double exposure framework (O'Brien and Leichenko 2000; Leichenko and O'Brien 2008) provides a general, conceptual approach for analysis of many types of interactions between environmental change and globalization. One important difference between the double exposure framework and the other sustainability approaches cited earlier is that other frameworks generally do not take into account the potential interactions between climate change and other global change processes, particularly the outcomes of multiple processes interacting across space and time. There is also relatively little recognition within other frameworks of how the two processes together transform the context in which people and places experience and respond to changes of many types. Many frameworks stress the importance of context for explaining differential outcomes, vulnerability and resilience, yet the frameworks seldom recognize the extent to which the context itself is dynamic, dramatically changing as the result of both global environmental change and globalization. Within the double exposure framework, changing contextual conditions may affect exposure and responses to future global change processes, resulting in new patterns of vulnerability and new challenges for sustainability and social and ecological resilience (Leichenko and O'Brien 2008).

The double exposure framework's point of departure is that multiple global change processes are occurring and interacting both simultaneously and sequentially creating either negative or positive outcomes for individuals, households, communities, firms, and industries. Within the framework, global environmental change and globalization manifest as either gradual or sudden changes (i.e., stressors or shocks) that have differential effects across a particular exposure frame. Depending on the focus of the research, an exposure frame might be a spatial, political or ecological region, an economic sector or a network of institutions. Exposure results in measurable outcomes, which may, in turn, affect the processes as well as the context in which future changes are experienced.

In each case, exposure to global change processes is influenced by the characteristics of the change (e.g., direction, rate, magnitude, intensity, and spatial extent) and by factors in the contextual environment (e.g., institutional, economic, social, political, biophysical, cultural, and technological conditions). Responses, which may include actions taken either in anticipation of or following from exposure, are conditioned by factors in the contextual environment, as well as by the individual attributes of each affected actor (e.g., education, values, beliefs, cognition). Outcomes depend on both the degree of exposure to each global process and on the actions taken by the affected individuals or other actors.

Processes of climate change and globalization are represented as partially overlapping triangles in the double exposure framework (Fig. 1). These processes manifest in a specific contextual environment, portrayed as an oval. The extent or magnitude of exposure to the processes is depicted as the intersection between the triangles and the oval. An arrow leading from the contextual environment to a square representing outcomes symbolizes responses to the processes. Outcomes are depicted as separate from the contextual environment to emphasize that any outcome reflects measurable conditions at a specific point in time.

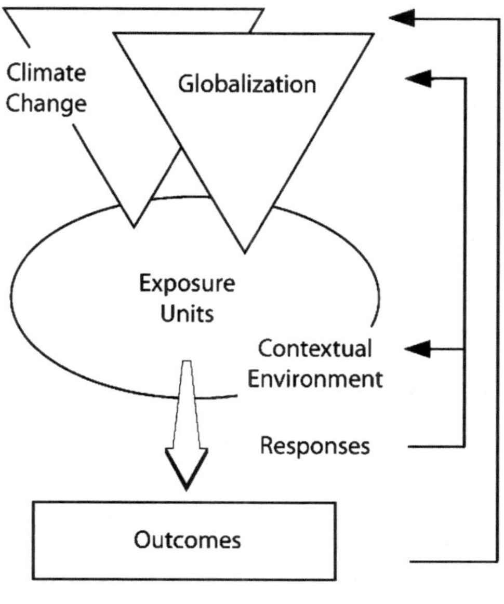

Fig. 1 A schematic of the double exposure framework (adapted from Leichenko and O'Brien 2008; Leichenko et al. 2010)

The framework incorporates dynamic linkages between the components. Processes may alter the contextual environment, responses may affect the processes, outcomes may affect responses, and so forth. Dynamics are also incorporated in the framework through recognition that processes and outcomes are often reflexive. Within the figure, the arrow leading from responses and outcomes back to the process triangles depicts these types of circular linkages, which are termed "feedbacks." Although the figure focuses on a single exposure frame, it is important to note that outcomes and responses that occur within one exposure frame may have widespread influence on other exposure frames across space and time.

The double exposure framework articulates a number of potential pathways of interaction between the two processes (Leichenko and O'Brien 2008). The pathway of *outcome double exposure* highlights overlapping impacts of both globalization and climate change on a particular exposure unit, whether it is a region, sector, system, or social group, showing how the combined effects of both processes often exacerbate existing patterns of spatial and social inequality and vulnerability. This pathway identifies what may be referred to as "double winners" and "double losers." The pathway of *context double exposure* shows how one process can influence the capacity to respond to shocks and stresses associated with the other process, often leading to negative outcomes. By incorporating the temporal dynamics of global change processes, context double exposure provides insights on how long-term resilience can be undermined by current changes to the contextual environment. The pathway of *feedback double exposure* demonstrates how the contextual changes, responses, and outcomes associated with either or both processes may contribute to drivers of the processes, thereby perpetuating cycles of double exposure and posing challenges to long-term sustainability. Feedback double exposure also emphasizes positive synergies between the two processes whereby actions made in response to one process may also contribute to solutions to the other process.

The double exposure framework provides a conceptual tool for investigating how processes of global change interact to affect sustainability within specific regions, sectors, and coupled human–environment systems. By emphasizing the dynamic interactions between processes, responses and outcomes, the framework aims to elicit new insights and research questions, beyond those associated with separate analyses of each global change process (Leichenko and O'Brien 2008). The pathways of double exposure are further illustrated in the next section through examination of coastal New Jersey.

Double Exposure and Coastal New Jersey

Coastal New Jersey extends from the state's Hudson River border with New York in the North to the Cape May Peninsula in the South. Processes of globalization within coastal New Jersey are perhaps most apparent in northern New Jersey. Northern New Jersey is home to the Newark-Elizabeth international port facilities, which are a major component of the Port of New York and New Jersey, the third largest international port by cargo tonnage and container volume in the United States (US Army Corp of Engineers 2010). Northern New Jersey also contains the high-rise office towers of Jersey City, which provide critical back office functions for New York City's global financial center, and Newark Liberty International Airport, a major hub for international travel. Evidence of earlier areas of global integration is notable in northern New Jersey, including historical rail terminals at Liberty State Park, the immigration museum at Ellis Island and the Statue of Liberty. Each of these historic and cultural landmarks is testament to processes of global integration during the nineteenth century, and each continues to attract a large and steady stream of US and international tourists.

Though less apparent, globalization's reach also extends to the sandy beaches and shores of central and southern New Jersey. The global visibility of the shore has increased recently with MTV's Jersey Shore™ television show. The show's broad global viewership has made its stars—"The Situation" and "Snooki"—as well as the "Jersey Shore" household names throughout the world. The cultural identity of the Jersey shore, as perpetuated through film and music, of course long precedes MTV: Atlantic City's international reputation as a "place to be seen" dates to the 1930s and the 1973 debut album of Bruce Springsteen, titled "Greetings from Asbury Park New Jersey" contained iconic songs about New Jersey that later became top hits both within the United States and internationally. Yet MTV's television show has brought renewed attention to and awareness of the Jersey shore, particularly for younger audiences from distant places.

Though less apparent than the cultural globalization of coastal New Jersey, the area's property markets are also deeply embedded in processes of financial globalization. As mentioned earlier, the property boom that was facilitated by globalization of finance during the past decade helped to accelerate the pace of property construction in high amenity, coastal regions (Leichenko et al. 2010). Ocean County, New Jersey, which located approximately 70 miles south of New York City, saw an increase in housing units between 2000 and 2010 of nearly 12%, nearly double the rate of increase for the state as a whole during the same period. The largest increases within Ocean County were concentrated in shore municipalities including Barnegat and Ocean Townships, where the number of housing units increased by 49.8 and 43.9%, respectively, between 2000 and 2010 (US Population Census 2010).

By the same token, the global financial crisis of 2008 and the economic recession that followed, both of which were also facilitated by financial globalization (Leichenko et al. 2010), affected the shore economy in many ways, including decline in tourism revenue, decline in value of residential and commercial properties, particularly those built as investment properties for second home buyers and summer rentals during the hey-day of the recent housing boom, loss of property tax revenue for municipalities, and loss of jobs, particular in the construction sector. By 2009, the number of construction jobs in Ocean County had declined by 22.5% in comparison with 5 years earlier (New Jersey Department of Labor and Workforce Development 2010).

While demographic, cultural and economic processes have been the driving forces for change in coastal New Jersey in recent years, climate change and associated sea level rise will play an increasingly important role in the future. As noted earlier, recent estimates suggest that sea levels may rise, may approach or exceed 1 m globally by the year 2100. For southern coastal New Jersey, the effects of global sea level rise are exacerbated by postglacial land subsidence that affects this portion of the US eastern seaboard, as well as local processes of subsidence due to groundwater extraction and other activities.

While coastal storms already occur on a regular basis in New Jersey, such events are likely to increase in frequency, duration, and magnitude as the result of climate change. Recent estimates for the New York metropolitan region suggest that by 2080, flood events that currently occur once every 100 years, on average, may occur once every 15–35 years (Horton et al. 2011). Sea level rise will exacerbate the effects of these events by creating higher storm surges that will contribute to coastal and back bay flooding. Other facets of climate change that will affect the flora, fauna and human populations of coastal New Jersey include more frequent droughts, and heat waves and changes in seasonality, including an earlier onset of Spring and more frost-free days overall. These changes will, in turn, affect water supply and water quality, ecosystem functions, habitat for migratory species, the length of the summer tourism season, and many other facets of shore life.

Double Exposure Challenges

Globalization and climate change are clearly transforming coastal New Jersey. The pathways of double exposure highlight some of the key challenges and opportunities that globalization and climate change are creating for the region.

Outcome Double Exposure

The pathway of outcome double exposure, which focuses on overlapping exposure to multiple processes of global change, can be illustrated via the example of New Jersey's coastal tourism sector. In 2008, tourism contributed an estimated $28 billion to the New Jersey economy. During that year, tourism also generated $7.7 billion in tax revenue, including $4.5 billion in local and state taxes (IHS Global Insight 2009). Tourism is the state's third largest private sector employer, accounting for more than 400,000 jobs. More than 60% of all tourism revenue in New Jersey can be attributed to the four counties of central and southern coastal New Jersey, including Monmouth, Ocean, Atlantic, and Cape May. Atlantic City, which, in recent years has hosted 30 million visitors annually, is estimated to account for about 33% of total tourism expenditure in the state; while the rest of the shore (Cape May, Ocean, Monmouth counties, and the remainder of Atlantic county) account for 28% (IHS Global Insight 2009).

The state's tourism sector has been significantly affected by the recession over the past 2 years. In particular, the counties of southern coastal New Jersey, including Atlantic and Cape May, are dominated by the tourism-based, leisure, and hospitality sectors. In Atlantic City, casino gaming revenue, which is generated through slot machines and gaming tables, decreased by 10% between 2009 and 2010, and is expected to decline another 10% during 2011 (Wittkowski 2011). Losses in gaming revenue, which have been exacerbated by regional competition from new casinos that have opened in neighboring states, particularly Delaware and Pennsylvania (Wittkowski 2011), have ultimately translated to 10,000 few jobs within the gaming sector in Atlantic City in 2010, as compared with 5 years earlier (Wittkowski 2010).

For Atlantic City and other shore communities loss of tourism revenue overall due to recession and regional competition has overlapped with a number of costly coastal storms including "Nor Ida" in November of 2009, which caused damage to coastal communities well in excess of $100 million (Associated Press 2009; Weaver et al. 2009), damaging Nor'easters in February and March 2010, and extreme snowfall events during December 2010 and January and February 2011. This overlap between the economic downturn and extreme storm events has created a situation of outcome double exposure for shore towns, which were financially strapped by the recession and at the same time must incur significant additional expenses for dune repair, beach replenishment, snow removal, and other infrastructure maintenance.

Context Double Exposure

The pathway of context double exposure explores transformations in the contextual environment as the result of either or both processes of global change. The interaction between the two processes may sometime create conditions that increase vulnerability or undermine adaptive capacity, thereby limiting future responses to shocks and stresses associated with one or both processes. Reductions in local public finances due to housing crisis and state reduction in funding for cities will not only affect the ability to pay for damages from increasingly frequent extreme events, but will also affect the ability of towns and cities to engage in adaptation planning and action that will be needed to respond to future climate change. In the case of Atlantic City, for example, adaptation planning is critical in order for the city to meet the sustainability challenges associated with climate change. Located on a barrier island with only one major causeway to the mainland, much of the city already falls within the 100-year floodplain. Plans for significant new investment of public funds for boardwalk restoration in Atlantic City, which are intended to help revitalize the city's gaming industrial and overall attractiveness for tourists, will need to be adjusted to take into account the altered conditions that result from climate change.

Feedback Double Exposure

The pathway of feedback double exposure emphasizes the relationships between the responses and outcomes of one crisis and the drivers of another. In coastal New Jersey, feedback double exposure is revealed through the connections between globalization-induced development of the housing sector and patterns of energy consumption. Much of the new development within Monmouth and Ocean Counties over the past two decades has drawn new homeowners who commute long distances, often by automobile, to jobs in Northern and Central New Jersey or New York. While this type of suburban expansion is an important driver of growth of transport and residential greenhouse gas emissions, the ongoing contraction of the housing sector amidst the economic slowdown may present an opportunity to alter this trend. Such a response, along the lines suggested by proponents of smart growth and green economy, might include an emphasis on construction of green buildings, expansion of alternative energy production, and water conservation (Leichenko et al. 2010). In order to be effective and sustainable at the local level, however, such efforts must incorporate adaptation planning in order to ensure resilience to future shocks and stresses.

Conclusions

Dramatic economic and environmental changes are underway in locales throughout the world. Individuals, households, communities, sectors, and regions will be confronted with impacts of these changes, whether it is through shifts in investments and sectoral upheavals, through alterations in sea level and changes in the frequency and magnitude of extreme weather events, or some combination of both. The double exposure framework provides sustainability scientists and others utilizing the approach a way to move beyond descriptive statements about the scope of change, and to more fully explain the processes and outcomes and to identify options that take advantage of new types of synergies between the processes. The three pathways of double exposure not only reveal vulnerabilities to multiple stresses, but also show ways to use the connections to make development more resilient to shocks and stresses of all types.

For coastal New Jersey, application of the double exposure framework reinforces the need for sustainability researchers to move beyond compartmentalized studies of either climate change or economic development, and look at how economic and environmental changes interact. The framework reveals that economic sectors such as tourism, though highly adaptable and accustomed to dynamic economic conditions, need to take steps to prepare for dramatic environmental shifts. In order to meet the sustainability challenge, a key direction for future work will be to identify ways to build upon the adaptive capacity and innovation potential of coastal industries and communities to promote resiliency of both coastal economies and socio-ecological systems.

Acknowledgments Research for this chapter was supported by a grant from the Rutgers University, Byrne Family First Year Seminar Program. I thank Rutgers students Dumebi Emetanjo and Jason Hanusey for research assistance, and I thank my colleague Briaval Holcomb for helpful discussions about the dynamics of coastal tourism in New Jersey. Portions of this chapter were adapted from an article by Leichenko et al. (2010).

References

Acosta-Michlik L, Kumar K, Klein RJT et al (2008) Application of fuzzy models to assess susceptibility to droughts from a socio-economic perspective. Reg Environ Change 8:150–161

Adger WN (1999) Social vulnerability to climate change and extremes in Coastal Vietnam. World Develop 27:249–269

Adger W, Hughes T, Folke C et al (2005) Social-ecological resilience to coastal disasters. Science 309:1036

Adger WN, Eakin H, Winkels A (2009) Nested and tele-connected vulnerabilities to environmental change. Front Ecol Environ 7:150–157

Alley RB, Clark P, Huybrechts P et al (2005) Ice-sheet and sea-level changes. Science 310:456–460

Associated Press (2009) East coast storm damage tops $100 billion. MSNBC.com. http://www.msnbc.msn.com/id/34016803/ns/weather/. Accessed 19 Nov 2009

Berkes F (2007) Understanding uncertainty and reducing vulnerability: lessons from resilience thinking. Nat Hazards 41:283–295

Eakin H (2006) Weathering risk in rural Mexico: climatic, institutional, and economic change. University of Arizona Press, Tucson

Eakin H, Luers A (2006) Assessing the vulnerability of social-environmental systems. Ann Rev Environ Resour 31:365–394

Eakin H, Wehbe M (2009) Linking local vulnerability to system sustainability in a resilience framework: two cases from Latin America. Clim Change 93:355–377

Eriksen S, Silva J (2009) The vulnerability context of a savanna area in Mozambique: household drought coping strategies and responses to economic change. Environ Sci Pol 12:33–52

Ernstson H, van der Leeuw S, Redman C et al (2010) Urban transitions: On urban resilience and human-dominated ecosystems. Ambio 39:531–545

Harvey D (1990) The condition of post-modernity: an enquiry into the origins of cultural change. Blackwell, Malden

Hjalager A (2007) Stages in the economic globalization of tourism. Ann Tour Res 34:437–457

Horton R, Bader D, DeGaetano A, Rosenzweig C (2011) "Climate Risks." In New York State ClimAID: Integrated Assessment for Effective Climate Change Adaptation Strategies in New York State, C. Rosenzweig, W. Solecki, A. DeGaetano, S. Hassol, P. Grabhorn, M. O'Grady, Eds., Annals of the NYAS, Vol. 1244, Issue 1, pp. 16–47. doi:10.1111/j.1749-6632.2011.06331.x. Accessed 12 March 2011

IHS Global Insight (2009) NJ tourism: holding its own during difficult times. 2009 New Jersey Governor's conference on tourism

Knutson T, McBride J, Chan J et al (2010) Tropical cyclones and climate change. Nat Geosci 3:157–163. doi:10.1038/ngeo779

Leichenko R, Solecki W (2005) Exporting the American dream: the globalization of suburban consumption landscapes. Reg Stud 39:241–253

Leichenko R, O'Brien K (2008) Environmental change and globalization: double exposures. Oxford University Press, New York

Leichenko R, Solecki W (2008) Consumption, inequity, and environmental justice: the making of new metropolitan landscapes in developing countries. Soc Nat Resour 21:611–624

Leichenko R, O'Brien K, Solecki W (2010) Climate change and the global financial crisis: a case of double exposure. Ann Assoc Am Geogr 100(4):963–972. doi:10.1080/00045608.2010.497340

McGranahan G, Balk D, Anderson B (2007) The rising tide: assessing the risks of climate change and human settlements in low elevation coastal zones. Environ Urban 19:17–37

Nelson D, Adger WN, Brown K (2007) Adaptation to environmental change: contributions of a resilience framework. Ann Rev Environ Resour 32:395–419

Neumann J, Hudgens D, Herter J et al (2010) Assessing sea-level rise impacts: a GIS-based framework and application to coastal New Jersey. Coast Manage 38:433–455

New Jersey Department of Labor and Workforce Development (2010) Central regional community fact book: Ocean County edition

Nicholls R, Tol R, Vafeidis A (2008) Global estimates of the impact of a collapse of the West Antarctic ice sheet: an application of FUND. Clim Chang 91:171–191

Nicholls R, Marinova N, Lowe JA et al (2011) Sea-level rise and its possible impacts given a 'beyond 4°C world' in the twenty-first century. Philos Trans R Soc A 369:161–181

O'Brien K, Leichenko R (2000) Double exposure: assessing the impacts of climate change within the context of economic globalization. Glob Environ Change 10:221–232

O'Brien K, Leichenko R, Kelkar U et al (2004) Mapping vulnerability to multiple stressors: climate change and globalization in India. Glob Environ Change 14:303–313

Overpeck J, Weiss J (2009) Projections of future sea level becoming more dire. Proc Natl Acad Sci 51:21461–21462

Polsky C, Neff R, Yarnal B (2007) Building comparable global change vulnerability assessments: the vulnerability scoping diagram. Glob Environ Change 17:472–485

Seto K, Sánchez-Rodríguez R, Fragkias M (2010) The new geography of contemporary urbanization and the environment. Ann Rev Environ Resour 35:167–194. doi:10.1146/annurev-environ-100809-125336

Silva JA, Eriksen S, Ombe ZA (2009) Double exposure in Mozambique's Limpopo River Basin. Geogr J 176:6–24

Turner B (2010) Vulnerability and resilience: coalescing or paralleling approaches for sustainability science? Glob Environ Change 20:570–576. doi:10.1016/j.gloenvcha.2010.07.003

Turner BL II, Kasperson RE, Matson PA et al (2003) A framework for vulnerability analysis in sustainability science. Proc Natl Acad Sci 100:8074–8079

United Nations (2006) Atlas of oceans. http://www.oceansatlas.org. Accessed 12 March 2011

US Army Corp of Engineers (2010) Waterborne Commerce Statistics Center. http://www.ndc.iwr.usace.army.mil//wcsc/wcsc.htm. Accessed 12 March 2011

US Census (2000) Census of population. US Department of Commerce, Bureau of the Census, Washington, DC

US Census (2010) Census of population. US Department of Commerce, Bureau of the Census, Washington, DC

Vermeer M, Rahmstorf S (2009) Global sea level linked to global temperature. Proc Natl Acad Sci 106:21527–21532

Weaver D, Barlos T, Miller, M (2009) N.J. shore towns calculate loss from coastal storm as officials head to region. PressofAtlanticCity.com http://www.pressofatlanticcity.com/news/press/cape_may/article_42e157b2-d2d3-11de-8e67-001cc4c03286.html. Accessed 12 March 2011

Weiss J, Overpeck J, Strauss B (2011) Implications of recent sea level rise science for low-elevation areas in coastal cities of the conterminous USA. Clim Chang 105:635–645. doi:10.1007/s10584-011-0024-x

Wilbanks T, Kates R (2010) Beyond adapting to climate change: embedding adaptation in responses to multiple threats and stresses. Ann Assoc Am Geogr 100:719–728

Wittkowski D (2010) Workers feel brunt of Atlantic City casino struggles as work force cut by 20 percent in five years. PressofAtlanticCity.com http://www.pressofatlanticcity.com/news/press/casinos_tourism/article_26bf755e-fffb-11df-9b54-001cc4c03286.html. Accessed 12 March 2011

Wittkowski D (2011) Atlantic City casino revenue falls nearly 10 percent in 2010, the fourth year of declines. PressofAtlanticCity.com http://www.pressofatlanticcity.com/news/breaking/article_02fa0fd6-1cf1-11e0-a05c-001cc4c03286.html. Accessed 12 March 2011

Sustainability Trajectories for Urban Waters

Richard Burroughs

Abstract Improving sustainability trajectories related to the biological health of urban waters requires enhancing the effectiveness of US nitrogen control programs for watersheds, cities, and ocean waters. A trajectory consists of identification of sustainability values, use of science to identify alternative solutions, selection of means for change, and assessment of results. Nitrogen, a limiting nutrient in most marine waters, contributes to algal blooms, declining levels of dissolved oxygen, and changes in biodiversity when it is present in bioavailable and excess amounts. Left unchecked nitrogen enrichment results in a regional trajectory trending away from biological sustainability. Its impacts have been observed on local, national, and global scales. The sustainability trajectory framework provides a novel way to view success or failure by clarifying values promoted and the means to reach them. Sequences of decisions related to nitrogen enrichment of New York Bight, Narragansett Bay, and Chesapeake Bay show that positive ecological trajectories rely upon the linkage of sustainability targets to authoritative governance techniques.

Keywords Sustainable development • Eutrophication • Sustainability trajectory • Coastal governance

Introduction

Urban water quality reflects development. Sustaining both environment and development requires full knowledge of the tradeoffs in play. The dimensions of sustainability include nature, life support, and human community as well as individual, social, and economic needs (NRC 1999). Creating a path for society based

R. Burroughs (✉)
Department of Marine Affairs, University of Rhode Island, Kingston, RI 02881, USA
e-mail: rburroughs@uri.edu

M.P. Weinstein and R.E. Turner (eds.), *Sustainability Science: The Emerging Paradigm and the Urban Environment*, DOI 10.1007/978-1-4614-3188-6_16,

on an understanding of these tradeoffs may be viewed as the policy process in action. The fundamental description and application of the policy process has been treated extensively (Lasswell 1971; Clark 2002; Birkland 2011; Burroughs 2011). In general the policy process is a way of viewing how an issue like urban water quality is formulated, assessed, and acted upon within society usually through government. Since environment/development problems are not resolved by one sequence of formulation, assessment and action, learning, subsequent reassessment is common practice. The results of repeated sequences of problem solving affect each of the dimensions of sustainability.

In explaining sustainability science, Kates et al. (2001) focus attention on interactions between society and nature as well as "society's capacity to guide those interactions along more sustainable trajectories." In the context adopted here, sustainability science provides the knowledge necessary to make informed tradeoffs in the policy process. Sustainable trajectories, proposed by Kates et al. (2001), would result when program implementation produces positive results along some combination of the dimensions identified by NRC (1999).

This chapter expands upon sustainability trajectories and applies them to urban waters and associated systems. In a positive sustainability trajectory, individuals in a geographic region make decisions and implement programs that assure continuity and improvement of one or more of the dimensions of sustainability. Trajectories call forth the sequence of information and actions that direct human behavior in ways that are beneficial for an identified objective or set of objectives. As will be explained in greater detail below, the actions that make up a sustainability trajectory may be viewed in sequence as: identification of a target and societal values related to it, engaging science to create feasible solutions, selecting and implementing authoritative means to meet targets, and assessing results (Fig. 1). In practice, sustainability decision-making consists of moving through the sequence of steps repeatedly as society grapples with both new information from sustainability science and new values, which collectively create new targets. A positive trajectory has the potential to restore biodiversity and ecosystem services while meeting individual needs, building human communities, and accumulating social capital.

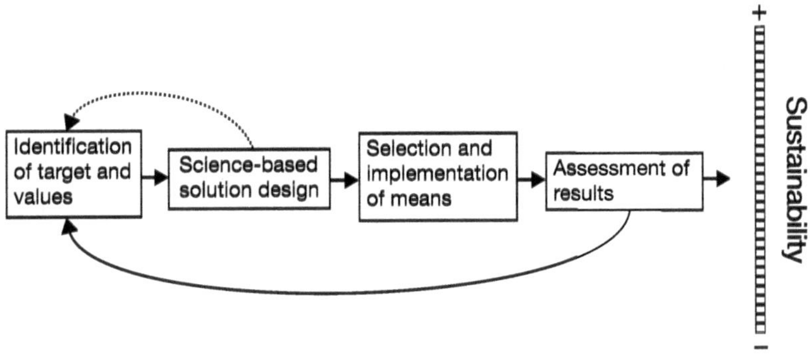

Fig. 1 Sustainability trajectory. Trajectories are cumulative results of iterative decision-making that result in enhanced or reduced sustainability of linked social and natural systems or discrete portions of them

The flow of nitrogen through agricultural, urban, and suburban environments has significant impacts on the development of regional economies and on the sustainability of nature and life support systems in coastal waters. The linkages between land and sea as well as social and natural systems are conclusively demonstrated by the Mississippi River, which delivers nitrogen from agricultural lands to an oxygen poor area of the Gulf of Mexico known as the "dead zone" (Turner and Rabalais 2003). As urban and suburban areas grow through human population increase, the flows of nitrogen in creation and consumption of food and also in air emissions, fertilizer, and other forms get directed to coastal waters and exact a greater toll on them. In this situation city metabolism, as originally described by Wolman (1965) and recently assessed across multiple urban areas (Kennedy et al. 2007; Kennedy 2012), can be managed to soften impacts on adjacent environments. City metabolism consists of all the materials and commodities required to meet the needs of the inhabitants. Nitrogen metabolism in an urban area focuses on resource consumption and waste generation related to nitrogen with specific consideration of impacts to marine waters. Poorly managed nitrogen flows to urban waters of the United States have resulted in eutrophication in many areas (Scavia and Bricker 2006). Thus, the flows of resources through urban systems can best be managed with full consideration of downstream impacts in coastal environments (Weinstein 2010). To do so requires the creation of forward-looking institutional structures that accommodate sustainability and urban development.

Positing a sustainability trajectory for nitrogen management related to linked urban–coastal systems in the northeastern United States sets the stage for retrospectively examining decision processes. The ultimate decisions and the means used to reach them comprise the institutional and organizational setting for sustainable development. To present and appraise sustainability trajectories for urban waters I will describe component parts (Fig. 1); assess anthropogenic sources and controls of nitrogen; describe factors that control trajectories in three locations; and assess causes for the trajectories observed.

Sustainability Trajectories

In defining sustainability science, Kates et al. (2001) noted the need to guide human–nature interactions along sustainable trajectories. For the purpose of this analysis redirecting human activities to pursue a more sustainable trajectory may be considered by identifying a target, supporting scientific analysis, selecting a means to reach the target, and assessing results prior to reinitiating this sequence (Fig. 1).

First, a sustainability target and the values surrounding it must be apparent. Targeting identifies the natural and social system or systems that are viewed to be unsustainable and establishes the predominant values that shape the desired result. In this example healthy coastal waters are the sustainability targets and, for the purpose of this analysis, other values are presumed subsidiary. Fischer et al. (2007)

envisioned the task of management as returning the biophysical, social, and economic system to a state that is sustainable. Nesting in this context requires that human societies and economies fit within the capacity of the earth to support life. This matter becomes more urgent when an ecological threshold is crossed (Lyytimaki and Hilden 2007). In coastal waters, species shifts and fish kills signal the penetration of thresholds. So, in the context of nitrogen flows to urban waters, eutrophication is one measure of the extent to which a threshold has been exceeded and as decisions are being made by the society they can be assessed with respect to the extent they restore nitrogen flows to more environmentally compatible levels. In concept, societal endorsement of the values surrounding a sustainability target, in this case healthy coastal waters, will inform multiple decisions.

Second, sustainability science equips society with a variety of means or solutions that are consistent with the target. Clark (2010) has characterized the main goal of sustainability science as discovering how to improve human wellbeing while recognizing its ultimate dependence on the natural environment. The challenge extends across multiple generations and to all of humanity. A transition toward sustainability will link goals with changes in human and environmental systems. For nitrogen, the sustainability goal of healthy coastal waters will require changes in human behavior such that nitrogen releases decline or their impacts become less burdensome. Sustainability science can identify technological as well as behavioral alternatives to meet this objective.

The third stage of a sustainability trajectory occurs when a governance technique is selected and implemented. It rests on the previous two steps. Sustainability science can provide alternate solutions to reach targets but options only become realities if adopted and effectively implemented. In this idealized sequence of considerations, sustainability science provides multiple pathways to reach the target. By comparing the predicted consequences of each approach, once implemented, with the predominant values of the affected communities, a solution may be selected. Solutions rest on technology change like the composting of sludge to biosolids or biological nitrogen removal (BNR) that reduces levels in effluent discharges. At a more fundamental level change in human diets, landscape utilization practices, and energy sources as well as consumption levels will become increasingly important to achieve sustainability goals. Key to all of these discussions is the political will to change institutions, which consist of the rules, expectations, and practices that a society uses to structure its affairs.

Finally, by evaluating the environmental and social results of repeated actions to control nitrogen one can determine whether the combined outcomes result in a positive sustainability trajectory in biological, social, or other terms. The ideal is a set of actions that improve the sustainability trajectory across multiple dimensions of the concept. Deviations from the ideal trajectory are noted for adjustment in subsequent sequences of decisions. With actions normed to the ideal trajectory, areas for improvement can be noted so that all concerned can evaluate subsequent decisions in the context of sustainability.

Nitrogen in Coastal Regions

During the twentieth century the global nitrogen cycle was substantially altered by the Haber-Bosch process to produce fertilizer, nitrogen fixation through crop selection, and fossil fuel consumption (Canfield et al. 2010). Much of this activity is related to the agricultural food chain (Jordan and Weller 1996). Since most people live in cities, urban environments are an important focal point for the nitrogen cycle. Cities are where most food is consumed. This role of nitrogen in city metabolism was recognized in considering sewage treatment in cities (Wolman 1965) and has since been assessed across several metropolitan regions (Kennedy et al. 2007). Since ecological footprints (Wackernagel et al. 2005) incorporate the land and water that a human population requires to produce resources and absorb wastes, the urban/suburban areas have both an upstream and a downstream component. The upstream consists primarily of the agricultural activities necessary to provide the food supply for the city, which may or may not occur in the same watershed. Downstream activities, the discharge of nitrogen, extend the footprint of the city to coastal waters. Assimilative capacity is the amount of nitrogen that can be tolerated without causing significant damage. Eutrophication indicates that assimilative capacities of coastal waters have been exceeded.

The transfer of reactive nitrogen from rivers to oceans has grown three- to fourfold due to human activity. Excess nitrogen unlocks rapid phytoplankton growth, which leads to respiration and decay, causing low oxygen in coastal waters. The resulting anoxic or hypoxic zones are marked by much lower levels of oxygen than would normally be expected (Rabalais and Gilbert 2009). When dissolved oxygen in the water column drops to less than 2 mg/L, the waters are referred to as hypoxic. When there is virtually no oxygen available, they are anoxic. These areas are known as dead zones because many animals commonly found in coastal waters cannot survive there.

In excess supply, nitrogen destabilizes coastal ecosystems in a variety of ways (Nixon 1995; Diaz and Rosenberg 2008). Animals living in and on the bottom, benthos, can be eliminated and secondary production is reduced which denies significant amounts of food energy for fisheries production. Oxygen decline in the water column coupled with sediment chemistry changes results in habitat compression for organisms that would normally be found in the water column or sediments. In some instances, coastal waters become anoxic and fish kills occur. As a result, understanding the nitrogen mass balance of a region with attention to anthropogenic inputs has become particularly important. Assessments of nitrogen metabolism in urban areas have been completed for Beijing (Han et al. 2011) and Stockholm (Jansson and Colding 2007) as well as Hong Kong, Phoenix, Bangkok, and Gavle, Sweden (Kennedy et al. 2007).

In the United States national surveys of eutrophication have been completed (Scavia and Bricker 2006), and the northeast part of the country has received detailed attention (Boyer et al. 2002; Howarth et al. 2002, 2006). The anthropogenic inputs to a watershed include atmospheric nitrogen deposition, fertilizer applications, fixation through agricultural practices, and the import or export of nitrogen in food, in its production, or in animal feed. Other components include river exports to coastal waters and the amount stored in the region. Within this larger context, much of the food imported to a coastal city is discharged to coastal waters through sewage treatment systems. Watersheds of the Northeast vary with respect to the sources of nitrogen inputs (Boyer et al. 2002). In far northern New England atmospheric deposition, a nonpoint source was dominant. In southern New England import of food contributed point source discharges from sewage treatment facilities. In the mid Atlantic region nonpoint, agricultural inputs dominate. A compilation by Alexander et al. (2001) shows Narragansett Bay and the Hudson-Raritan systems are dominated by point sources while the loads to the Chesapeake are spread across atmosphere, fertilizer, livestock, and nonagricultural nonpoint sources.

The sustainability of biological systems in the watershed, estuarine waters adjacent to the city, and adjacent ocean waters are in many instances affected by choices made concerning nitrogen. As such, these urban waters in close proximity to urban populations record, for better or for worse, the growth and development of cities. In addition since the input conditions for urban waters are determined upstream in the watershed, the ultimate conditions found in the estuary are often influenced by discharges at a distance from it. Furthermore, urban waters flow seaward. So, in sum nitrogen flows link the watershed, the estuarine waters of the city, and adjacent coastal marine waters. The sources of nitrogen, technological alternatives for controlling it, and impacts differ by geographic area. This means that constructing a sustainable trajectory with respect to nitrogen will utilize different governance techniques depending on local circumstances. This chapter examines sustainability trajectories in three different settings to understand whether iterative decision-making has created new behaviors or institutions to reduce deleterious impacts of nitrogen in urban settings.

New York Bight

For about a century, a location sited about 12 miles offshore in the New York Bight served as the repository for most sludge from the New York area. Sludge from sewage treatment consists of slurry of nitrogen-rich organic particles and water. Legislatively mandated increases in wastewater treatment created greater volumes of sludge (Burroughs 1988). New York practiced ocean dumping of this material for the longest time and ultimately delivered the largest amount of sludge to US waters. By 1990 ocean dumping of sludge in the United States had doubled from the early 1970s level to a new total of ten million wet tons per year, and almost all of it was dumped in New York Bight (Boesch et al. 2001).

In 1970, the President's Council on Environmental Quality had called for a ban on unregulated dumping of all materials and strict limitations on disposal of harmful wastes such as sludge (CEQ 1970). At the time ocean dumping was presumed to be a small portion of the ocean pollution problem. However, it was a specific source with an apparent easy solution, cessation of dumping. Guidelines for limitation and for elimination of ocean dumping emerged from the Convention of the Prevention of Marine Pollution by Dumping of Wastes and Other Matters or London Dumping Convention and coupled with the Marine Protection Research and Sanctuaries Act of 1972 set US policy in a direction toward the elimination of a variety ocean dumping practices (Fig. 2).

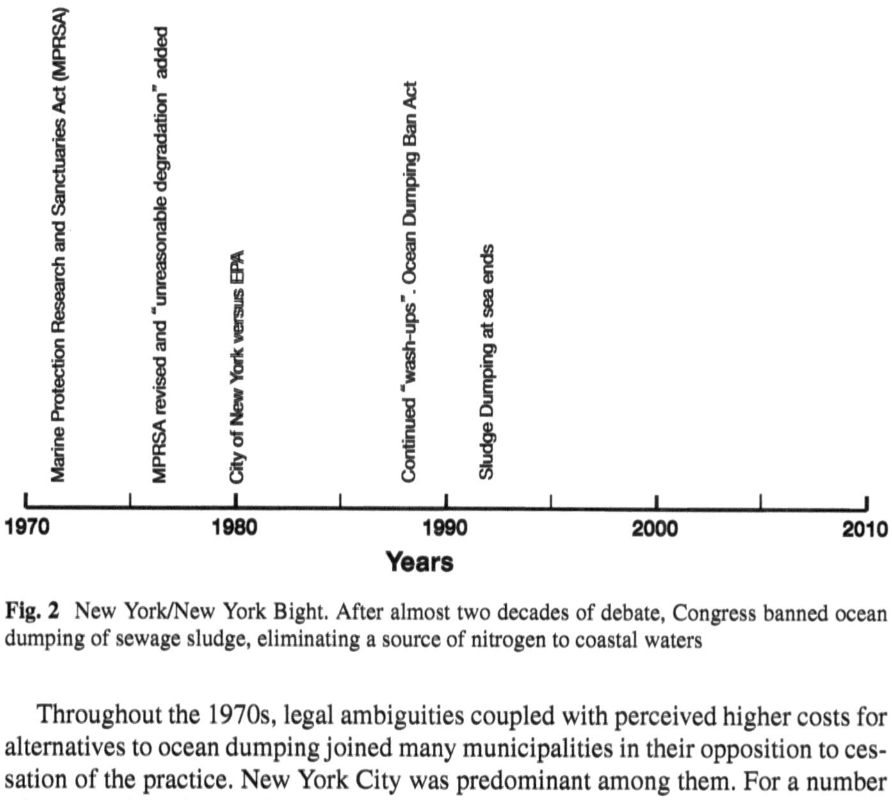

Fig. 2 New York/New York Bight. After almost two decades of debate, Congress banned ocean dumping of sewage sludge, eliminating a source of nitrogen to coastal waters

Throughout the 1970s, legal ambiguities coupled with perceived higher costs for alternatives to ocean dumping joined many municipalities in their opposition to cessation of the practice. New York City was predominant among them. For a number of years, values, law, and natural science circled ocean dumping without clear resolution. *Multimedia management* and assimilative capacity shaped these debates.

Multimedia management argued for a scientific assessment that would identify technical solutions and compare environmental impacts when sludge was placed on land, in the ocean, or burned with byproducts released to the air. Ideally by comparing impacts in each medium, one could select a solution that minimized total impacts. The US National Academy of Sciences laid out the possibilities as part of a review of US Environmental Protection Agency (USEPA) programs (NRC 1978). In addition to ocean dumping, two alternatives were assessed in detail. A second disposal alternative was incineration. However, the emissions from combustion have proven problematic, and, even with scrubbers, it was apparent many years ago that sludge incinerators would not meet air quality standards in many geographic areas (NRC 1978). The third option included various ways of using the material on the land. For many years, sludge has been disposed in landfills. This practice has opened the possibility for contamination of surface or groundwater. Careful design of approaches makes the practice viable, and many new opportunities for recycling the nutrients on land are developing.

After the 1977 revisions of ocean dumping law, USEPA interpreted it to prohibit dumping sludge only to find through a subsequent court decision (City of New York vs. EPA) that the situation was complex (Moore 1992). Ultimately the court found that EPA would have to determine if the proposed dumping would cause unreasonable degradation, whereas EPA had proceeded on the basis that sludge by its composition would be unacceptable. Furthermore, the court found that economic and environmental costs of alternatives to ocean dumping should also be weighed before reaching a decision. These arguments reintroduced some of the thinking that had been a part of the earlier NAS report on sludge. The comparative assessment of environmental impacts and costs awaited definitive resolution.

Concurrently discussion about *assimilative capacity*, the extent to which the ocean is an acceptable repository for certain wastes, augmented the multimedia analysis mentioned above. Assimilative capacity implies both an assessment of environmental impacts of the process and a determination that the biological impacts are acceptable to the society. The latter assessment is a value judgment, which attracted the participation of many interests within the society. Ultimately the National Advisory Committee on Oceans and Atmosphere (NACOA) supported continued dumping of sewage sludge in the ocean (NACOA 1981).

In 1984 USEPA designated the 106-mile sewage sludge dump site due East of Cape May, New Jersey (USEPA 1995). Disposal began at that deep water location in 1986 when the inshore site in New York Bight was phased out. Disposal of sludge continued until 1992.

Ultimately the uncertainties about intent were resolved in 1988. Action by the US Congress came after medical waste appeared on the shores of New York and New Jersey. Professionals realized that syringes and other materials were not a part of sewage sludge. However, the apparent level of citizen dismay about the abuse of the ocean conflated legalized ocean dumping of sewage sludge with illegal disposal of trash such as syringes. Pressure from tourism and environmental interests coupled with additional legal exchanges was ultimately resolved through unequivocal legislation. The US Ocean Dumping Ban Act of 1988 made it unlawful to dump sewage sludge in ocean waters after 1991 without paying a substantial fine. By June 30, 1992 New York, the last US city to ocean dump sludge, stopped the practice. By 2004, sludge generated at the major New York City sewage treatment facilities was heat dried for land application out of state, and other plants in the metropolitan region adopted composting (New York State Department of Environmental Conservation 2006).

Several actions have been taken by coastal cities and the regulatory agencies, primarily USEPA, in the decades since cessation of sludge dumping at sea. Eliminating the sea as a final resting place for sludge put further pressure on land at a time when air pollution regulations all but eliminated incineration. Conversion of sludge into fertilizer became prominent.

This was made possible by two important transitions. First, toxic contaminants in sludge have declined due to *pretreatment*, the requirement that industries remove contaminants before discharging to the sewers. Pretreatment dramatically lowered metals and synthetic organic chemicals in the sludge. Sludge that is too high in these chemicals is not desirable because organisms may assimilate toxins. Secondly, advances during the 1990s led to composting and other processing techniques that reduced pathogens after sludge was produced. As the levels of metals, synthetic organic chemicals, and pathogens decrease, sludge can increasingly be promoted as a biosolid with many potential applications. By the late 1990s, approximately 6.9 million tons of biosolids were produced annually in the United States, and about 60% were recycled on land. The USEPA projects that the total amount of biosolids will rise (USEPA 1999). Typical reuse includes land application, composting, and landfill cover.

Thus, the nitrogen flow formerly directed to New York Bight as sewage sludge has been stopped through federal legislation and most of the material is now dried and/or composted and applied to land outside of New York. While this action was undertaken quite apart from any overt consideration of sustainability, it illustrates an ability to alter human behavior in ways that benefit coastal waters and recycling. It culminates almost three decades of decisions during which both the underlying science and the values of society shifted. In the context of sustainability trajectories as defined here (Fig. 1), value and target shifts overwhelmed the more nuanced natural science arguments of NRC and NACOA. In effect Congress prioritized protection of coastal waters while at the same time air quality regulations limited incineration. Land application in various forms emerged as the alternative of choice, and the diversion of sludge away from New York Bight has resulted in more sustainable coastal waters and potentially more efficient fertilization of plants.

Upper Narragansett Bay, Rhode Island

Nitrogen flows to upper Narragansett Bay, Rhode Island through rivers and direct discharges by sewage treatment plants. While the circumstances are different—point source effluent in Rhode Island vs. ocean dumping of sludge in New York—each raises the question of the capacity and means of the society to make decisions that result in more sustainable coastal waters.

Wastewater treatment reduces the flow of nitrogen to coastal waters (Mueller and Anderson 1983; Tchobanoglous and Schroeder 1987; Laws 2000). When the wastewater enters the sewage treatment plant physical, chemical, and biological processes reduce pathogens and remove noxious materials. Physical separation reduces the levels of first inorganic and then organic particles in the waste stream. At the end of primary treatment, 50–60% of the suspended solids have been removed from the water. Because these particles, if released to coastal waters, would be degraded by bacteria, thereby consuming oxygen and releasing nitrogen in the water, successful primary treatment reduces some of the problem of hypoxia, or low oxygen, in coastal waters. Biochemical oxygen demand (BOD), the use of oxygen in degrading wastes, is reduced by approximately 35% after primary treatment.

Secondary treatment involves the breakdown of organic particles that remain in the effluent from primary treatment by creating conditions for bacteria to use the particles as a source of food. To accomplish this, the secondary treatment tank is inoculated with sludge containing the bacteria and provided with air to deliver oxygen for the bacteria. Under these conditions, the large supply of fine organic particles from primary treatment is degraded. After the bacteria grow, secondary sedimentation or clarification separates the microbial mass from the water by allowing the material to settle to the bottom of the tank, where it is drawn off as sludge. After primary and secondary treatment, 85–90% of the suspended solids have been removed along with a similar amount of the BOD (Mueller and Anderson 1983). Metals (e.g., copper, nickel, and cadmium) if present, most commonly reside with the organic particles. Therefore, collecting the particles as sludge can reduce the metal levels in the liquid waste stream by as much as 65%.

After secondary treatment and disinfection effluent waters discharged to the upper Bay still contain nitrogen. Adding nitrogen to marine waters can result in growth of algae and eutrophication. In upper Narragansett Bay summertime low dissolved oxygen in bottom waters has been linked with deleterious biological changes (Saarman et al. 2008; Deacutis 2008). Low oxygen events in bottom waters appear to be associated with weak neap tides when the water column is well stratified. These episodic occurrences are also influenced by earlier freshwater flows and are most common in the mid to late summer. One time series shows dips below 2.0 mg/L but no indication of anoxia (Saarman et al. 2008). Algal blooms and changes in benthic populations have been attributed to excess nutrient delivery to Bay waters (Deacutis 2008). Over time the form has shifted to dissolved inorganic nitrogen, which is readily assimilated by plants, and the amount of benthic regeneration has been documented (Fulweiler et al. 2010).

Since nitrogen delivery to the estuary can occur from multiple sources, and the control of each resulting from different management mechanisms, understanding sources and pathways are particularly important. When and how to insist on tertiary treatment of wastewater, one management approach, is a difficult and contentious activity (Fig. 3). It can also be very costly to implement. Tertiary treatment requires additional processing time and tank capacity because in the commonly used BNR process additional anoxic processing is necessary to liberate the nitrogen and release it as a gas to the atmosphere. Secondary treatment typically removes 10–30% of the nitrogen in wastewater whereas tertiary treatment can remove 80–95% (Mueller and Anderson 1983).

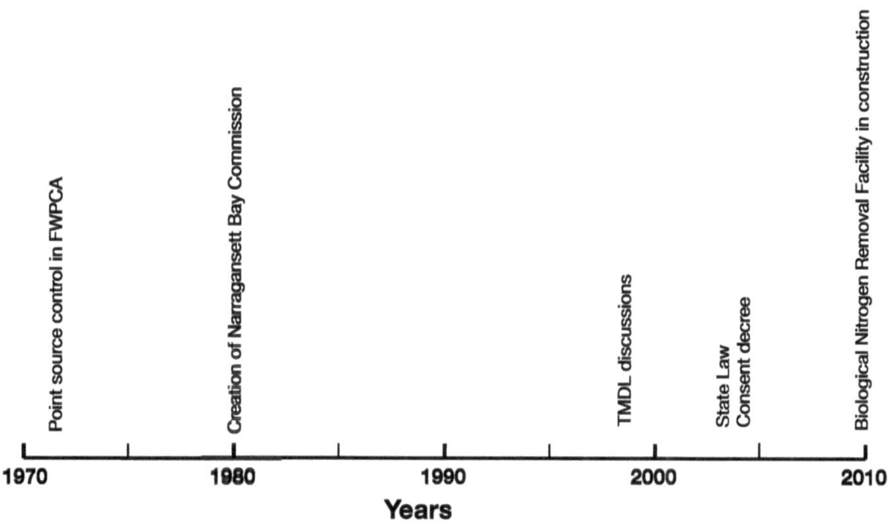

Fig. 3 Providence/Narragansett Bay, Rhode Island. Administrative actions invoking total maximum daily load (TMDL) coupled with state law ultimately required biological nitrogen removal to reduce loading to the estuary

The total maximum daily load (TMDL) provisions of the US Clean Water Act give the federal EPA and states considerable latitude in requiring additional treatment for point sources if a contaminant is deemed in excess in coastal waters (Houck 2002). The Act recognizes that legally mandated secondary treatment technology, while positive, may not succeed in raising ambient water quality to a level suitable for uses that citizens may expect and prefer. In short, the legally mandated technology-based approach does not assure all the uses people find desirable. To address this, section 303(d) of the US Clean Water Act requires identification of the contaminants responsible for low water quality. The process consists of identifying which low quality waters require attention, noting the load of the suspect contaminant that can be tolerated, and creating the means to reduce that load. In coastal waters, where nitrogen is the culprit, the TMDL approach is particularly important because of its purported role in establishing the total load and amounts of the contaminant that will be allowed from individual point and nonpoint sources. Unfortunately, a scientifically sound TMDL has not been completed for the upper Narragansett Bay nitrogen.

Nonetheless, after a gubernatorial commission reported its analysis in 2004 the Governor directed a removal of 40–50% of the nitrogen from treatment facilities discharging to the upper Bay. A state law mandated similar reductions. Nonpoint sources were not addressed through these actions. Through the Clean Water Act, the USEPA and state environmental agencies have directed the dischargers to reduce nitrogen concentrations to 5 mg/L in some cases. Additional reductions may be considered. This administrative authority stems from the periodic renewal of National Pollution Discharge Elimination System permits and reductions required by the federal program appear consistent with declarations by the state legislature and Governor.

Tertiary treatment can reduce the amount of nitrogen in treatment plant effluents. Narragansett Bay Commission plants will biologically convert nitrogen in the effluent to nitrogen gas through the BNR process (Water Environment Federation 2006).

Ultimately through the national pollution discharge permit renewal process, the two largest sewage treatment plants on Narragansett Bay adopted BNR to meet the lower nitrogen discharge limit. The sewage treatment plants in question contributed 28% of the total nitrogen load to the Bay in one recent compilation (Nixon et al. 2008). To the extent that nitrogen in effluent from these facilities is responsible for eutrophication of the upper Bay, this reduction in loading was deemed to ultimately result in higher water quality and hence a more sustainable estuarine system.

These administrative actions have two consequences on sustainability. First, the reduction in nitrogen should result in fewer hypoxic events and more sustainable ecosystems. Second, the costs of reducing nitrogen are spread across many urban poor in a local economy struggling to recover from a steep recession, which raises equity issues. In summary, critical institutional changes related to sustainability of biological systems in coastal waters came through administrative processes under the US Clean Water Act backed by state law. Implementation rests on BNR technology to be added at major sewage treatment plants.

Chesapeake Bay Watershed

In Chesapeake Bay, nonpoint sources of nitrogen dominate. They constitute a distinctly different management challenge in geographic extent and governance complexity. Success depends on collaboration among many political units. Addressing the problem requires altering agricultural and other practices that spread out over the land portions of a large watershed. Furthermore to be effective, actions must be coordinated. Regional solutions for watersheds and coastal areas occur when the involved parties agree on the need to resolve an environmental problem and believe that a comprehensive management system will ultimately benefit them. In this multi-jurisdictional setting, problems require new administrative structures. In fact, the ability to create and implement effective regional control processes is at the core of enhancing the sustainability of biological systems in the Bay.

For more than two centuries, states abutting the Chesapeake have negotiated regional arrangements to advance mutual interests. In a 1785 compact, Virginia gave free access to the Bay for ships headed to Maryland and both states stipulated that their citizens could navigate the Potomac (Capper et al. 1983). By the early twentieth century, pollution affecting oysters had been a significant concern for a number of years. The US Congress authorized a study, and with the assistance of health officials from both states, reports on the Potomac and the Bay were published in 1916. Years later in 1963 the Public Health Service initiated a project that spanned water quality issues for the Susquehanna River and the Bay. It noted the need for a basin-wide framework for water quality management decisions and called attention to enrichment of Bay waters as a source of the problems (Capper et al. 1983). The formation in 1940 and revision in 1970 of the Interstate Commission on the Potomac River Basin as well as the creation in 1970 of the Susquehanna River Basin Compact are precursors to the current Chesapeake Bay Commission.

In 1975 the US Congress directed the USEPA to conduct an extensive study of Chesapeake Bay, which initiated a series of actions concerning the water body (Fig. 4). At the time many felt that a better understanding of the natural system could establish a foundation for cleaning up the Bay. Contemporary scientific diagnoses identified nonpoint sources as the major cause of nitrogen flow to the Bay, and the work that led up to those conclusions began with the 7-year USEPA study (Boesch and Goldman 2009; Burroughs 2011).

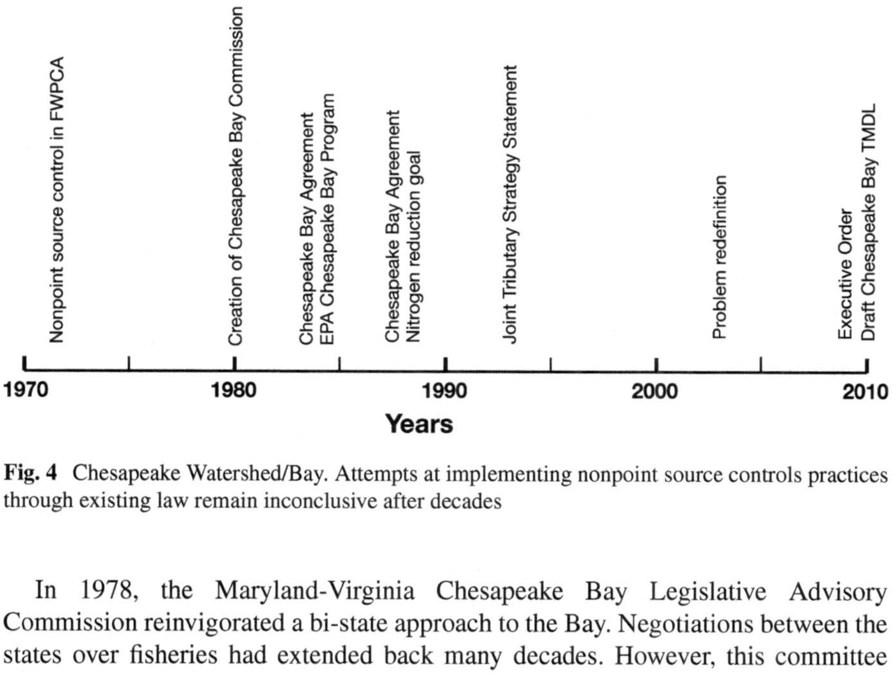

Fig. 4 Chesapeake Watershed/Bay. Attempts at implementing nonpoint source controls practices through existing law remain inconclusive after decades

In 1978, the Maryland-Virginia Chesapeake Bay Legislative Advisory Commission reinvigorated a bi-state approach to the Bay. Negotiations between the states over fisheries had extended back many decades. However, this committee signaled that although the mechanism might be similar, the objective was expanded to include environmental quality. The Legislative Advisory Commission recommended the formation of the Chesapeake Bay Commission, which was established in 1980. In 1985, Pennsylvania joined Virginia and Maryland on the Commission and with that change each state had five state legislators serving in addition to citizens and state government officials. The Commission advises the legislative branches in the states and serves as the legislative arm of the USEPA Chesapeake Bay Program. The Commission was and is the predominant regional management body for the bay.

In 1983 the governors of Maryland, Virginia, and Pennsylvania, the mayor of the District of Columbia, the chair of the Chesapeake Bay Commission, and the administrator of USEPA signed a Chesapeake Bay Agreement. Because recently completed scientific studies of the Bay had shown a decline in the living resources, observers felt that cooperation among EPA and the states was necessary to fully address pollutants entering the Bay. The agreement became the foundation of a new structure that allowed multiple governmental entities to collaboratively address bay/watershed environmental problems. The agreement established the Chesapeake Executive Council to oversee activities to improve water quality and living resources of the Bay. The 1983 agreement also created an implementation committee and called for a liaison office at the USEPA facility in Annapolis, Maryland (Chesapeake Bay Program). Following the 1983 agreement, numerous state initiatives were undertaken to manage land and resources in concert with the regional objectives (Costanza and Greer 1995).

Over time the Chesapeake Executive Council became committed to restoring and protecting the Bay and by 1987 selected eight goals to do so. At this point, the parties had begun the process of identifying activities to restore the Bay. Furthermore, by seeking to manage the impacts of human population growth and land development while restoring living resources and habitats, they firmly connected land and sea. Most dramatically they sought to implement "a basin-wide strategy to equitably achieve by the year 2000 at least a 40% reduction of the nitrogen and phosphorous entering the main stem of the Chesapeake Bay" as the means of reducing hypoxia (Chesapeake Bay Agreement 1987). This objective, which was not implemented, required a robust program to alter land use and reduce point discharges.

In 1987 the USEPA Chesapeake Bay Program was authorized through changes in the Clean Water Act, which assured annual consideration for federal funds. In specific, the law directed that EPA coordinate efforts to improve water quality through the Chesapeake Bay Program. This action by Congress further legitimized and empowered the solutions that the scientists and stakeholders from the Bay and its watersheds had selected. Grants to states would be a primary way of delivering the program.

In 1992, the Chesapeake Bay Agreement was amended to focus on tributaries to the Bay as a means of implementing the nutrient reductions proposed in 1987. Nitrogen reductions were to be monitored through the distribution of submerged aquatic vegetation. Tributary strategies instead of the TMDL approach in the Clean Water Act would identify the means to reduce nutrient and sediment loads to the Bay through use of best management practices (Hassett et al. 2005). Preferred practices included forest buffers, changes in agriculture, and retention/purification of storm water. In agriculture fertilization practices can be altered, cover crops can be planted, and planting may occur with minimal soil disturbance. Furthermore, signatories identified air deposition of nitrogen as an additional factor that should be considered.

A 2000 agreement affirmed previous goals and in some cases identified actions required to meet them. The signatories reemphasized land-use practices by setting goals for permanently preserving from development 20% of the land area in the watershed and by reducing sprawl by 30% through a variety of means. At that time Virginia, Maryland, Pennsylvania, and the District of Columbia were participants and brought with them innumerable county and local governments. Furthermore, an organizational entity that involved the states and the USEPA, the Chesapeake Executive Council, had been established. So at a minimum, coordinated management on a watershed/bay scale involved seven units at the federal level, four at the state level, two cross-state and federal coordinating entities, and innumerable county and local governments.

In 2003, the release of *Chesapeake Bay Blues* marked a shift to focusing on the politics of Bay restoration (Ernst 2003). Ernst (2003) reasoned that divided governments in a system structured to favor economic concerns would favor interest groups that apply financial and other resources to influence political deliberations. He targeted agricultural industries that are primary contributors of nonpoint source nitrogen to the Chesapeake. Given the circumstance where industrial groups were most powerful, he observed that environmental groups were unable to be effective which accounted for the impoverished condition of the Bay. In late 2008 a meeting of scientists, politicians, and policy analysts concluded that voluntary and collaborative structure of the Bay Program was inadequate (Ernst 2010). They advocated a regulatory program with enforceable measures for controlling nitrogen. The failed political structure was deemed to be at the core of the problem, and ultimately the ineffectual tributaries strategy of the 1990s was replaced by a return to the TMDL approach originally specified in the US Clean Water Act.

In 2009, President Obama ordered seven federal agencies to conduct their activities consistent with objectives for the watershed by establishing a Federal Leadership Committee for the Chesapeake Bay to oversee the development of programs and their implementation (Obama 2009). Chesapeake Bay policy has been redesigned several times, and the current TMDL requires an additional reduction of the 2009 nitrogen load (USEPA 2010). Point sources had been the principal focus for previous reductions in nitrogen. Technological solutions to reduce point source nitrogen loads are available and periodic reissuance of permits under the Clean Water Act makes regulatory implementation of nitrogen discharges feasible. In contrast, nonpoint sources as covered by section 303 of the law rest on far less authoritative means. The new arrangements target nonpoint sources and are enormously complex. At the federal level those departments or agencies operating primarily on land (Agriculture, Interior), in coastal lands and waters (USEPA, and Departments of Defense, Homeland Security and Transportation) and primarily in coastal waters (Department of Commerce) need to operate in an environmentally coherent manner. Nitrogen matters were to receive increased attention by the Department of Agriculture and the USEPA. USDA was to reduce nutrient loads by concentrating on selected federal programs in priority counties (Executive Order 13508, section 401). This horizontal division of bay and watershed management obligations across agencies was further complicated by the vertical governance dimension, which extended management from federal to state, county, and local governments. The ability to act—even in a limited form—would require substantial political will and great skill on the part of the administrators.

Monitoring change has been particularly important with respect to oxygen levels in bottom waters of the Bay. To reduce hypoxia, action must be taken throughout the watershed and airshed. So this one measure becomes a way to evaluate both the ability of the Chesapeake Bay Commission to influence human behavior and the extent to which the low oxygen problem has been resolved. The results for the Chesapeake have not been good: agriculture continues to be a significant source of excess nitrogen. Ernst (2003) found that in the decades after the nutrient problem was originally noted and attributed to nonpoint sources agricultural interests have hamstrung efforts to address what he refers to as the Bay's primary environmental hazard. This finding amplifies other critiques that found limited restoration of dissolved oxygen levels, crabs, oysters, and other Bay systems.

Monitoring the natural environment for over 2 decades indicated little progress in restoring the Chesapeake (Williams et al. 2010). Chlorophyll-a, dissolved oxygen, and secchi depth, indicators of nitrogen enrichment have shown little improvement. Some have worsened. Fortunately, a reduction of nitrogen loads from sewage treatment plants in some tributaries have been linked to increasing submerged aquatic vegetation in those locations (Williams et al. 2010). Various reasons account for the lack of improvement. Evaluations have pointed to a lack of coordination, particularly between management of agriculture and suburban development (USEPA 2006, 2007; US Government Accountability Office 2008). Agriculture and urbanization of the watershed, targets of the tributaries approach in the 1992 agreement, have proven particularly difficult to address because of lack of political will. As a result the Bay remains in an apparently stable, yet degraded state (Boesch and Goldman 2009).

By 2010, a lack of significant water quality improvement and a redefinition of the problem forced EPA to return to TMDLs as the method for effective implementation. Originally established under the Federal Water Pollution Control Act amendments of 1972, the TMDL process called for states to identify waters where nonpoint sources were problematic and adopt means to control them. More specifically, the Act called for the determination of acceptable load by a contaminant and then the partitioning of that load between point and nonpoint sources with the ultimate goal of controlling both. In the Chesapeake, the nonpoint source load of nitrogen predominates. But as Houck (2002) and others have demonstrated, weak authority to control nonpoint sources has left success in doubt.

In late 2010, the USEPA completed a TMDL plan for the watershed and Bay (USEPA 2010). The goal of the current Chesapeake Bay TMDL is to reduce the annual flux of nitrogen to bay waters by 25%. USEPA requires that reductions identified for each state by the federal government will result in local area targets consistent with the state limits. To compel success the agency appears willing to expand coverage of and tighten requirements on point sources, increase federal enforcement actions, prohibit new discharges, limit grants, and revise federal water quality standards. Each of these measures may be viewed as a way to offset the limited authority for nonpoint sources available under the law.

To the extent that the biological sustainability trajectory of Bay waters is determined by nitrogen reductions, results are inadequate as viewed from almost every perspective. The Chesapeake is both a science-rich and a conflict-rich situation. Many of the additional nitrogen reductions will have to come from nonpoint sources. However, the identified TMDL approach has shown limited success when applied elsewhere. Furthermore to the extent that both the failed tributary strategy and the proposed TMDL strategy require voluntary action, the fate of the latter seems problematic. The importance of nonpoint sources in the Chesapeake coupled with weak institutional structure to deal with them makes progress toward more sustainable biological systems in the Bay uncertain.

Conclusions

Sustainability trajectories provide a useful perspective from which to view social change. The approach requires the consideration of values, choices, solutions, and most importantly institutions. Trajectories record the net effects, which can be positive or negative, of repeated choices related to environment and development. Coastal governance is an iterative process through which issues like nitrogen enrichment are repeatedly revisited. As a result, it is possible to examine the issues that arise as society grapples with the problem in different places and multiple times.

Sustainability ultimately relies on the ability to create, adopt, and implement new governance systems with new values and approaches. Information from the natural sciences about the state of the environment is but one dimension of sustainability, which goes a long way to explain why it is persuasive in some but not all situations. Since the goal requires new combinations of actions to advance sustainability and development, it ultimately rests on new sets of rules. As a result the ultimate test of a positive sustainability trajectory is the ability of a society to design, adopt, and implement new ways of governing human behavior that are consistent with the new goals. Positive sustainability trajectories are built on institutional changes that support new governance systems. Each of the cases is a test of a region's ability to create new governance systems consistent with sustainability of biological systems in coastal waters.

In the New York Bight, dumping of sewage sludge elicited growing debate through the 1970s and 1980s. The discussion revolved around concepts like unreasonable degradation and balancing that were elaborated through increasingly detailed natural science. However, the underlying values argument could not be resolved by more detailed understanding of biophysical systems and disputes frequently landed in the courts. As a result dumping practices continued albeit with modifications. Following the arrival of medical waste at ocean facing beaches in the New York area, an occurrence which was erroneously linked to sludge dumping in the public mind, the US Congress acted to ban ocean dumping and more effectively deal with medical waste. Passage of the Ocean Dumping Ban Act of 1988 clarified the values and suspended further dumping of sewage sludge. Since ocean disposal ceased to be an option, municipalities generating sludge converted more of it to a biosolid, which could be recycled. Together, the actions could ideally result in more sustainable coastal waters and agricultural systems as the nitrogen was transferred to where its impacts could be positive.

In upper Narragansett Bay, the means for change rested on a combination of new technology as well as the legal and administrative actions that required its application. A variety of sources deliver nitrogen to the upper Bay with the consequence of low oxygen events occurring usually late summer under specific physical conditions. Through various provisions of the Clean Water Act, federal and state environmental agencies demanded significant reductions of nitrogen in effluents from the major sewage treatment plants. A combination of administrative agency demands and state law made nitrogen removal a requirement in advance of convincingly completing a TMDL analysis for nitrogen in the upper Bay. By selecting BNR technologies and installing them, sewage treatment plants in the upper bay will soon be releasing far less nitrogen. This change should reduce episodic oxygen declines in the upper Bay and may reduce primary production and higher-level production in the lower Bay. In this setting, administratively mandated technological change supported by state law will result in more sustainable coastal waters along selected dimensions.

In the Chesapeake Bay persistent problems with low oxygen have been linked to nitrogen, most of which comes from atmospheric pollution, chemical fertilizer, and animal waste. The latter two are tied to the large amount of agricultural production in the watershed. Utilizing current management techniques for nonpoint sources is far more difficult, both technically and politically, than controlling nitrogen flows through point sources. Increasingly elaborate natural science conducted over decades at costs of hundreds of millions of dollars has refined understanding of the scientific and technical issues, but it has not resulted in new institutions that substantially limit nonpoint sources. Since values remain contested, the political will to create authoritative means to limit nonpoint sources have not been forthcoming. At present the initiatives to do so rest on a 2009 Executive Order and a more recent TMDL approach for the Bay. Federal agencies responding to the Executive Order have limited authority and, at times, the will to address the problem, and in other settings when the TMDL process has been contested, progress has been limited. The predominance of nonpoint sources in the Chesapeake makes creating more sustainable biological systems unlikely unless values are clarified and authoritative governance systems are created.

Viewing nitrogen flows to urban waters from the perspective of sustainability trajectories exposes several dimensions of the problem. First, the technological solutions invoked to date provide solutions that, while useful, are only partial. Effective control of nitrogen flows will require far more effective means to limit leakage from agricultural and other uses of the watershed and airshed. Both cost and equity concerns may limit further nitrogen reductions at point sources. Additional solutions may lie in changing human diets to be less nitrogen intensive so that watershed sources are reduced and by changing city metabolism to soften downstream nitrogen impacts. The latter may include some form of cap and trade. Second, the relationship between sustainability science and political will faces major challenges when powerful interests are opposed to change. This is most clearly illustrated in the conflict between agricultural and water quality interests in the Chesapeake Bay, but it also occurs elsewhere. Where values are contested and interests strong, additional information about the environment may not lead to meaningful action. Third, sustainability trajectories as described here may be optimized for any of several dimensions that range from the natural to the social and economic. In absence of accepted and predominant values, trajectories can become double-edged swords that advance one dimension of sustainability at the expense of others. In upper Narragansett Bay, the urban poor will pay higher wastewater treatment bills to improve water quality, which raises equity issues. Finally, in situations where values are clear and solutions are known, the creation of appropriate institutions becomes paramount. Ultimately selecting and implementing authoritative means to meet targets determines whether a sustainability trajectory will be positive or negative. For New York and Rhode Island, legislative and administrative means to create new governance systems were forthcoming. The circumstance in the Chesapeake remains uncertain. As matters of sustainability before the society become more urgent, institutional design in circumstances where values remain contested will rapidly grow in importance.

References

Alexander R, Smith R, Schwartz G et al (2001) Atmospheric nitrogen flux from the watersheds of major estuaries of the United States: an application of the SPARROW watershed model. In: Valigura R, Alexander R, Castro M et al (eds) Nitrogen loading in coastal water bodies: an atmospheric perspective. American Geophysical Union, Washington

Birkland T (2011) An introduction to the policy process. M. E. Sharpe, Armonk

Boesch D, Goldman E (2009) Chesapeake Bay, USA. In: McLeod K, Leslie H (eds) Ecosystem-based management for the oceans. Island Press, Washington

Boesch D, Burroughs R, Baker J et al (2001) Marine pollution in the United States. Pew Oceans Commission, Arlington

Boyer E, Goodale C, Jaworski N et al (2002) Anthropogenic nitrogen sources and relationships to riverine nitrogen export in the northeastern U.S.A. Biogeochemistry 57(58):137–169

Burroughs R (1988) Ocean dumping: information and policy development in the USA. Mar Policy 12:96–104

Burroughs R (2011) Coastal governance. Island Press, Washington

Canfield D, Glazer A, Falkowski P (2010) The evolution and future of Earth's nitrogen cycle. Science 330:192–196

Capper J, Power G, Shivers F (1983) Chesapeake waters: pollution, public health, and public opinion, 1607–1972. Tidewater Publishers, Centreville

Chesapeake Bay Agreement of 1987, www.chesapeakebay.net/content/publications/cbp_12510.pdf; last accessed on April 10, 2011

Chesapeake Bay Commission. History of the commission, http://www.chesbay.state.va.us/history.html; undated, last accessed on April 10, 2011

Chesapeake Bay Program, www.chesapeakebay.net; undated, last accessed April 10, 2011

Chesapeake Executive Council, http://www.chesapeakebay.net/exec.htm; last accessed on April 10, 2011

Clark T (2002) The policy process: a practical guide for natural resource professionals. Yale University Press, New Haven

Clark W (2010) Sustainable development and sustainability science. In: Levin S, Clark WC (eds) Toward a science of sustainability. Princeton University, Princeton

Costanza R, Greer J (1995) The Chesapeake Bay and its watershed: a model for sustainable ecosystem management? In: Holling C, Light S, Gunderson L (eds) Barriers and bridges to the renewal of ecosystems and institutions. Columbia University Press, New York

Council on Environmental Quality (1970) Ocean dumping: a national policy. U.S. Government Printing Office, Washington, DC

Deacutis C (2008) Evidence of ecological impacts from excess nutrients in upper Narragansett Bay. In: Desbonnet A, Costa-Pierce B (eds) Science for ecosystem-based management: Narragansett Bay in the 21st century. Springer, New York

Diaz R, Rosenberg R (2008) Spreading dead zones and consequences for marine ecosystems. Science 321:926–929

Ernst H (2003) Chesapeake Bay blues: science politics, and the struggle to save the Bay. Rowman & Littlefield, Lanham

Ernst H (2010) Fight for the Bay: why a dark green environmental awakening is needed to save the Chesapeake Bay. Rowman & Littlefield, Lanham

Fischer J, Manning A, Steffen W et al (2007) Mind the sustainability gap. Trends Ecol Evol 22:621–624

Fulweiler R, Nixon S, Buckley B (2010) Spatial and temporal variability of benthic oxygen demand and nutrient regeneration in an anthropogenically impacted New England estuary. Estuar Coast 33:1377–1390

Government Accountability Office (2008) Chesapeake Bay program: recent actions are positive steps toward more effectively guiding the restoration effort. GAO-08-1033T, Washington, DC

Han Y, Li X, Nan Z (2011) Net anthropogenic nitrogen accumulation in Beijing metropolitan region. Environ Sci Pollut Res 18:485–496

Hassett B, Palmer M, Bernhardt E et al (2005) Restoring watersheds project by project: trends in Chesapeake Bay tributary restoration. Front Ecol Environ 3:259–267

Houck O (2002) The Clean Water Act TMDL program: law, policy, and implementation. Environmental Law Institute, Washington, DC

Howarth R, Boyer E, Pabich W et al (2002) Nitrogen use in the United States from 1961-2000 and potential future trends. Ambio 31:88–96

Howarth R, Swaney D, Boyer E et al (2006) The influence of climate on average nitrogen export from large watersheds in the northeastern United States. Biogeochemistry 79:163–186

Jansson A, Colding J (2007) Tradeoffs between environmental goals and urban development: the case of nitrogen load from the Stockholm County to the Baltic Sea. Ambio 36:650–656

Jordan T, Weller D (1996) Human contributions to terrestrial nitrogen flux. Bioscience 46:655–664

Kates R, Clark WC, Corell C et al (2001) Environment and development: sustainability science. Science 292:641–642

Kennedy C (2012) A mathematical description of urban metabolism. In: Weinstein MP, Eugene Turner R (eds) Sustainability science: the emerging paradigm and the urban environment. Springer, New York

Kennedy C, Cuddihy J, Engel-Yan J (2007) The changing metabolism of cities. J Ind Ecol 11:43–59

Lasswell H (1971) A pre-view of the policy sciences. American Elsevier Publishing Company, New York

Laws EA (2000) Aquatic pollution: an introductory text, 3rd edn. Wiley, New York

Lyytimaki J, Hilden M (2007) Thresholds of sustainability: policy challenges of regime shifts in coastal areas. Sustain Sci Pract Policy 3:61–69

Moore SJ (1992) Troubles in the high seas: a new era in the regulation of U.S. ocean dumping. Environ Law 22:913–951

Mueller J, Anderson A (1983) Municipal sewage systems. In: Myers E, Harding E (eds) Ocean disposal of municipal wastewater: impacts on the coastal environment, vol 1. National Oceanic and Atmospheric Administration, Washington, DC

National Advisory Committee on Oceans and Atmosphere (NACOA) (1981) The role of the ocean in a waste management strategy. US Government Printing Office, Washington, DC
National Research Council (1978) Multi-medium management of municipal sludge. National Academy of Sciences, Washington
National Research Council (1999) Our common journey: a transition toward sustainability. National Academy Press, Washington
New York State Department of Environmental Conservation (2006) Biosolids management in New York State, www.dec.ny.gov/docs/materials_minerals_pdf/bioreprt.pdf; last accessed on April 2, 2011
Nixon S (1995) Coastal marine eutrophication: a definition, social causes, and future concerns. Ophelia 41:199–219
Nixon S, Buckley B, Granger S et al (2008) Nitrogen and phosphorous inputs to Narragansett Bay: past, present, and future. In: Desbonnet A, Costa-Pierce B (eds) Science for ecosystem-based management: Narragansett Bay in the 21st century. Springer, New York
Obama B (2009) Chesapeake Bay protection and restoration. Executive Order 13508
Rabalais N, Gilbert D (2009) Distribution and consequences of hypoxia. In: Urban E, Sundy B, Malanotte-Rizzoli P et al (eds) Watersheds, bays, and bounded seas: the science and management of semi-enclosed marine systems. Island Press, Washington, DC
Saarman E, Prell W, Murray D et al (2008) Summer bottom water dissolved oxygen in upper Narragansett Bay. In: Desbonnet A, Costa-Pierce B (eds) Science for ecosystem-based management: Narragansett Bay in the 21st century. Springer, New York
Scavia D, Bricker S (2006) Coastal eutrophication assessment in the United States. Biogeochemistry 79:187–208
Tchobanoglous G, Schroeder E (1987) Water quality: characteristics, modeling, modification. Addison-Wesley, Reading
US Environmental Protection Agency (1995) Monitoring, research, and surveillance of the 106-mile deepwater municipal dump site and environs. EPA, EPA 842-R-95-001, Washington, DC
US Environmental Protection Agency (1999) Biosolids generation, use, and disposal in the United States. EPA, EPA 530-R-99-09, Washington, DC
US Environmental Protection Agency (2006) Saving the Chesapeake Bay watershed requires better coordination of environmental and agricultural resources. EPA, EPA OIG Report No. 2007-P-00004, Washington, DC
US Environmental Protection Agency (2007) Development growth outpacing progress in watershed efforts to restore Chesapeake Bay. EPA, EPA OIG Report No. 2007-P-00031, Washington, DC
US Environmental Protection Agency (2010) Final Chesapeake Bay TMDL. http://www.epa.gov/reg3wapd/tmdl/ChesapeakeBay/tmdlexec.html; last accessed on February 4, 2011
Turner RE, Rabalais N (2003) Linking landscape and water quality in the Mississippi River basin for 200 years. Bioscience 53:563–572
Wackernagel, M, Monfreda C, Moran D et al. (2005) National footprint and biocapacity accounts 2005: the underlying calculation method. www.footprintnetwork.org; last accessed on December 7, 2011
Water Environment Federation and American Society of Civil Engineers (2006) Biological Nutrient Removal (BNR) operation in wastewater treatment plants. McGraw-Hill, New York
Weinstein MP (2010) Sustainability science: the emerging paradigm and the ecology of cities. Sustain Sci Pract Policy 6:1–5
Williams M, Filoso S, Longstaff B et al (2010) Long-term trends of water quality and biotic metrics in Chesapeake Bay: 1986 to 2008. Estuar Coast 33:1279–1299
Wolman A (1965) The metabolism of cities. Sci Amer 213:179–190

Part V
Restoring and Rehabilitating Ecosystems: Return from the Precipice

There are many criteria for restoring degraded ecosystems, but they can be collectively distributed across a spectrum ranging from science-based restoration ecology to society-based ecological restoration. The differences are not trivial and may lead to widely varying goals for the restoration effort. Because nature and ecosystems are historically and culturally contingent ideas, Higgs (1997) also suggests that there is no one single, fixed, correct restoration for any particular site, although structure, composition, and function criteria may provide tight guidelines for success of the project. By Higgs' standards, the definition of good ecological restoration is rooted in ecological fidelity, but will also benefit from an expanded context (especially in setting goals and outcomes) by including societal values (economic efficiency and social, historical, political, moral, and aesthetic). The cultural element is also critical, not only because incorporating societal values enhances public acceptance of restoration and improves its chances of success, but also because virtually all lands have been influenced by human presence. Thus, the fabric of restoration is at once driven by ecological criteria (restore ecosystem function), as well as the likelihood that restoration end points may be something less than pristine, but societally acceptable. The traditional view of restoration as activities carried out on a site-by-site basis should give way to one where restoration occurs on a landscape scale and is an important component of regional planning.

Reference

Higgs ES (1997) What good is ecological restoration? Conserv Biol 11:338–348

Reversing Two Centuries of Wetland Degradation: Can Science Better Inform Policy and Practice?

Michael P. Weinstein, Steven Y. Litvin, and Michael G. Frisk

Abstract Perhaps more than any other ecotone, the land–water interface has been "reclaimed" solely for human uses—living space, ports and harbors, and agriculture—essentially extirpating other goods and services that these ecosystems provide. Although the importance of ecosystem services associated with wetland transition zones has been increasingly recognized in the past 60 years, the approach to "restoration" and "rehabilitation" has largely lacked scientific rigor. The status of coastal wetland restoration science is discussed herein with specific attention to design criteria that attempt to restore wetland functions and ecological fidelity. Methods for better integration of restoration science and practice to inform policy, and the quantification of restored functions are described within the context of three case histories.

Keywords Restoration ecology • Linking structure with function • Essential fish habitat • Case histories

---

M.P. Weinstein (✉)
PSEG Institute for Sustainability Studies, Montclair State University, Mallory Hall 116E, Montclair, NJ 07043, USA
e-mail: weinsteinmi@mail.montclair.edu

S.Y. Litvin
Hopkins Marine Station, Stanford University, Oceanview Boulevard, Pacific Grove, CA 93950-3094, USA
e-mail: litvin@stanford.edu

M.G. Frisk
School of Marine and Atmospheric Sciences, Stony Brook University, Stony Brook, NY 11794, USA
e-mail: mfrisk@notes.cc.sunysb.edu

M.P. Weinstein and R.E. Turner (eds.), *Sustainability Science: The Emerging Paradigm and the Urban Environment*, DOI 10.1007/978-1-4614-3188-6_17,

Introduction

> It is an open question whether ecosystem management will become a passing fad, an expansion of rigid bureaucratic procedures, or a sustaining foundation for learning to deal with interactions between people, nature, and economic activities (Holling 1996).

Restoration ecology straddles the interface of sustainability science and the reconciliation of human use of natural resources with the planet's ability to provide them. As the debate continues over whether humans have "escaped" the domination of natural laws or are still subject to them, Cairns (2000) expressed the concern that no matter how robust the science and technology of restoration ecology becomes, the science must enjoy societal acceptance of its dependence on ecosystem services as part of society's life support system. Unless this acceptance comes about, Cairns warns that the data will not be collected at the scale necessary to advance the science of restoration, and its development in a sustainability science framework will be hampered.

Restoration ecology also manages for change, fosters biodiversity and emphasizes the return of system functions, and goods and services to degraded ecosystems. An ecocentric framework for restoration is, therefore, an essential component of a transformation to global sustainability (Jackson and Hobbs 2009). Because humans dominate virtually all landscapes, the practice partially focuses on restoring ecosystem functions (e.g., flood storage capacity or storm buffering), that are not necessarily a return to "naturalness" (Stanturf et al. 2001; Weinstein and Reed 2005).

The science of restoration ecology also includes a body of theory for repairing damaged ecosystems (Palmer et al. 1997; Falk et al. 2006) and as these authors comment "the time is ripe for basic researchers to ask if current ecological theory is adequate for establishing the principles of restoration ecology." Yet, as Hildebrand et al. (2005) note, "the incredible complexity of nature forces us to simplify the systems we study in order to develop theory and generalities by reducing them to understandable subsets." Because ecosystems are inherently dynamic and exhibit nonlinearities and behavioral surprises, the ability to predict and manage restoration trajectories have been particularly vexing (Mitsch et al. 1998; Anand and Desrochers 2004; Ruiz-Jaen and Aide 2005). In addition, Hildebrand et al. (2005) assert that realistic goals should include multiple scientifically defensible endpoints of functional equivalence. In a thoughtful treatise, Ehrenfeld (2000) offered the following: (a) explicit recognition that no one-size-fits-all, goals have to be developed appropriately and individually for each project, and (b) that ecologists establish "probabilistic laws" to define the conditions under which it is desirable to address landscape-scale ecosystem processes; i.e., to determine the sets of conditions that mandate particular methods or goals for individual projects. Because wetlands are hydrologically, chemically, and biologically linked to the landscapes in which they occur, the "templates" for wetland restoration that comprise the various combinations of climate and hydrogeologic settings in a given geographic region, and cumulative alteration of landscapes therein, are likely the greatest constraint on successful restoration design (Bedford 1999). In addressing this particular issue, Kentula

(2000) described an emerging approach that develops a statistical representation or "model" of reference sites as the standard for comparison.

There are also questions related to community stability, resilience and persistence; all central to understanding/predicting whether a restored system will be self-sustaining. A primary challenge for restoration ecologists is to develop tools for assessing acceptable levels of variability in restored systems, most appropriately in a regional or landscape context and within some "bound of expectation" (White and Walker 1997; Weinstein et al. 1997; SER 2004; French 2005).

Thus, the evaluation of restored functions include measures of *processes* such as primary or secondary production, but may also reflect considerations of biogeochemical cycling, food web structure, food quality, habitat connectivity, biological interactions, refuge from predators, keystone species, donor control (Polis and Strong 1996), microhabitat structure, and access to resources. Many species exhibit complex life histories that place them in different parts of the landscape at different times, but their overall success may depend on the quality of specific habitats as critical "bottlenecks" in their life-history sequence. For example, marine transient finfish at mid-latitudes are characterized by life-history traits that evoke a "coastal conveyor belt" with adults spawning offshore and near estuaries, and young spending their first year of life in various estuarine habitats including tidal wetlands (Weinstein et al. 2009a). Young-of-year complete the cycle by accompanying adults offshore during their autumn migration to overwintering grounds. It is likely that the "quality" of the estuarine habitats, especially tidal wetlands and seagrass meadows is reflected in the growth and survival of young-of-year marine transients and is a critical aspect of their successful recruitment to adult stages. Restoration ecology should embrace these considerations.

Linking Structure to Function: The Salt Marsh Paradigm and Secondary Production

Teal's (1962) mass balance model for a salt marsh near Sapelo Island, Georgia was soon followed by Odum's (1968) outwelling hypothesis, and as a result, coastal wetlands and their detrital production were soon being depicted as the "great engine" driving much of the secondary production of near shore coastal waters (see also Turner et al. 1979; Weinstein 1981). The fundamental view of a detritus-driven system was soon challenged, however, by Haines (1979) who recognized that finfish, as well as, other primary producers (phytoplankton and benthic microalgae) also contributed substantially to nutrient flux from the salt marsh to open waters. Haines (1979) commented that the "true" nursery-ground of the estuary "was perhaps not so much the large open waters and sounds as the salt marshes and narrow tidal creeks." She added that the major export of marsh plant production might occur "not as particulate detritus but as living organisms." At about the same time, Weigert and Pomeroy (1981) stated that "our present view of the food web of the marsh and estuary suggests that the preservation of fisheries depends as much upon the protection of the smaller tidal creeks as upon protection of the marsh and its *Spartina* production." A year after Haines published her "emerging paradigm," Nixon (1980) reviewed the concept of outwelling and concluded that the average passive export of organic matter (particulate and dissolved) was relatively small, amounting to between 100 and 200 gC m^{-2} $year^{-1}$ for tidal wetlands on the mid-Atlantic and Gulf coasts of the United States.

Haines and Nixon's views stimulated an era of intense research for refining our understanding of functional links between salt marshes and the estuary/coastal zone. Thus, the "outwelling" concept (Odum 1968) has become but a single component in an evolving view of marsh function and the links between primary and secondary production. Today, the Haines' view is still undergoing modifications, and we are slowly unraveling the complexities of nutrient exchange, and the links between primary producers and the marsh/estuary fauna. The notion of the marsh drainage, especially the interface between tidal creeks and the marsh plain, serving as ecological "hotspots" (*sensu* Simenstad et al. 2000), and as a potential refugium from predators gained popularity in the 1970s (reviewed by Boesch and Turner 1984). *Spartina* spp. and many other marsh plants decompose relatively quickly, and this in situ production may be available to consumers by the end of the first growing season (Fry et al. 1992; Newell 1993). Benthic microalgae and many phytoplankton with their high palatability are also readily and efficiently assimilated by many consumers (Currin et al. 1995; Sullivan and Moncreiff 1990). Although progress has been made in understanding how marshes "work," we have also learned that the story is far more complicated than originally thought (Turner 1977; Peterson et al. 1994; Peters and Schaaf 1991; Mallin et al. 1992; Polis et al. 1997; Deegan et al. 2000; Winemiller et al. 2007; Dame and Christian 2008).

Marsh Physiography

From a restoration standpoint, the physiography of the salt marsh is a critical link in the dynamics and transfer of primary production to consumers. The physiographic features of the marsh that contribute to primary and secondary production include: elevation, drainage characteristics and surface rugosity that expands "edge" and influences the hydroperiod (Kneib 1997; Zimmerman et al. 2000; Larkin et al. 2008); access to the intertidal marsh for fauna (Rozas et al. 1988); predation refugia (McIvor and Odum 1988; Beck et al. 2001, 2003); and interspersed standing water for foraging by resident fishes and wading birds, and resting areas for waterfowl (Rubino 1991).

"Donor Control" and Restoration Planning

Marine transients may also benefit from tidal salt marshes and their production *without directly occupying these habitats*. Many are highly mobile, and tend to cross habitat boundaries in their quest for food and shelter. They are generally not habitat specialists but are rather opportunistic in utilizing the resources of the estuarine landscape. Restoration planners should and must, therefore, view restoration goals within the context of the habitat mosaic and the exchange of materials and organisms between adjacent habitats (e.g., salt marshes and the open waters of the estuary). Stated simply, salt marshes do not function in isolation when supporting estuarine secondary production, but are integrated components of larger systems (Weinstein et al. 2005). Moreover, the open waters of the estuary may be donor-controlled, i.e., they are systems in which the rate of import, availability, or dynamics of allochthonous resources (such as products of the salt marsh), is controlled by external donor systems rather than by consumers. Indeed, consumers may be more abundant when supported by allochthonous resources than if supported solely by the in situ resources of open waters (Polis et al. 1995). The latter concept is critical in the context of restoration ecology, because failure to account for trophic subsidies in the open estuary may result in restoration designs that have negative feedback on the recruitment success of numerous marine transients.

Childers et al. (2000) captured these concepts in their description of the interaction among estuarine habitats supporting fisheries. Their conceptual model posits integrated subsystems linked by an overlying water column that mediates functional processes across subsystem boundaries. Nutrient and organic matter flux associated with the movements of animals, especially juvenile marine transients, were also recognized as important vectors transcending system boundaries. The question of whether specific habitats confer disproportionate survival advantage to young marine transients is still rigorously debated (Beck et al. 2001). In our view, trophic subsidies to donor-controlled systems may confer survival advantages on young nekton.

Essential Fish Habitat, Restoration Design, and Higher Order Metrics of Restoration Success

Restoration efforts can also be evaluated within the context of essential fish habitat (EFH) by integrating the factors affecting fish survival and well-being during their life cycle (Able 1999). The degree to which a natural or restored habitat is utilized is presumed dependent on its value. In restored sites, habitat value is maximized once it has reached its restoration asymptote (Weinstein et al. 1997). The application of EFH to fishery management and restoration design necessitates the analysis of habitat information in a hierarchical or matrix fashion. At the least informative level (Tier I), the presence or absence information may be used to infer the potential value of habitats, albeit with a high level of uncertainty. At increasingly complex levels, habitat value becomes a function of the relative abundance or density of individuals at different locations (Tier II). At the next level, growth, reproduction, and survival rates, if available, are used with the assumption that the habitats contributing most to productivity should be those supporting the highest levels of these parameters (Tier III). Finally, production rates can be used to directly relate species or life stages to types, quantity, quality, and location of essential habitats (Tier IV). There has been a relatively slow evolution of restoration success criteria to include the upper tiers of EFH (III and IV).

In the remainder of this chapter, we focus on the integration of life-history strategy and landscape scale considerations in restoration planning based on our previous research on marine transients and estuarine resident finfish in the Delaware Bay, and Hudson River estuaries, USA. We adopt, but go beyond the premise introduced by Simenstad and Cordell (2000), that "the fundamental approach we recommend is 'self-monitoring,' letting the fish test whether the occupation of a restored habitat provides residence time, foraging success, or growth equivalent to that achieved in a comparable reference habitat." Rather, we address secondary production and Tier III and IV EFH parameters as potential endpoints to measure the outcome and success of restoration practices. The three case histories we present do not make direct comparisons between reference and restored habitats (although we have done this), but from a restoration ecology perspective are intended to assist future wetland restoration designs, not only to consider specific processes, but also to *promote exchange* of materials and organisms between the habitat being restored and the adjacent estuary; i.e., the donor control function of wetlands. Case History I focuses on the growth and survival of a marsh resident finfish, the common mummichog, *Fundulus heteroclitus* and stresses the deposition of energy reserves for overwintering survival at the end of the first year (Tier III, EFH). Case History II uses bioenergetics modeling in a "whole estuary" approach to estimate the nursery value of estuarine regions comprised of marsh and open waters for young-of-year weakfish, *Cynoscion regalis* (Tier III, EFH), and Case History III addresses the response of macroscale tidal salt marsh restoration within the context of secondary production of species that depend on these habitats and/or their products (Tier IV, EFH). Together, the three case histories demonstrate advances in the science of restoration ecology that go far beyond structural characteristics of degraded and restored tidal marshes to address the components of functional equivalency of restored sites.

Case History I (EFH Tier III): Biochemical Condition of a Marsh Resident Finfish, *Fundulus heteroclitus*

Carnivorous fishes are reliable indicators of the condition of complex ecosystems because they are the tertiary link in the food web. Thus, the magnitude of protein and fat deposition and the level of fat reserves can be used to not only assess the "degree of well-being" of fishes but can also serve to integrate the overall value of habitats in their production. Using this premise, we examined the concept of habitat quality for a marsh resident, the mummichog, *F. heteroclitus*, in relatively undisturbed and *Phragmites australis*-dominated tidal salt marshes along the mid-Atlantic

Coast of the United States. In addition to our earlier work on the trophic spectrum of this species (Wainright et al. 2000; Currin et al. 2003), an interesting "natural experiment" was available to us because (1) the species spends its entire life cycle within the confines of the marsh and has an extremely small home range (Valiela et al. 1977; Meredith and Lotrich 1979; Teo and Able 2003), and (2) many tidal salt marshes, particularly those with brackish salinities, are dominated by virtual monocultures of the invasive variety of the *P. australis* that is perceived to reduce habitat quality for *F. heteroclitus*, and general access to the marsh plain by nekton (Weinstein and Balletto 1999; Saltonstall 2002; Hagan et al. 2007). By adopting a whole system approach, we essentially had a "captive audience," one in each of two isolated marsh complexes located on the Hudson River estuary (Weinstein et al. 2009b), a polyhaline system dominated by *Spartina alterniflora* and a meso-oligohaline system dominated by an invasive variety of *P. australis*. In addition to others, the following questions were addressed in our work: (1) were there any differences in biochemical condition, principally the deposition of energy reserves, in mummichogs captured seasonally in the *S. alterniflora*-dominated "natural" and the *P. australis*-invaded salt marshes (Tier III EFH analysis); (2) were any differences related to size distributions of individuals in the populations; and (3) could biochemical condition ultimately serve as a success criterion to evaluate the functional success of wetland restoration?

The Use of Biochemical Condition as a Metric of Restoration Success

Biochemical condition of individual *F. heteroclitus* was evaluated on the basis of triacylglycerol (TAG), free fatty acid (FFA), and phospholipid (PL) composition and concentration. Other lipid classes such as cholesterol, fatty alcohols, and wax esters were not examined in detail but were included in the calculation of total lipid mass. It is generally accepted that the size of lipid stores and their composition can be used to predict whether a fish is ready to migrate, preparing to overwinter, or is likely to have future reproductive success (Ackman 1980; Shulman and Love 1999). Previous results of lipid class dynamic studies in young teleosts suggest that TAG is the primary form of lipid used in energy storage; therefore, this lipid class was selected as an important indicator of biochemical condition (Ackman and Eaton 1976; Lochmann et al. 1995, 1996; Lochmann and Ludwig 2003; Heintz et al. 2004; Weinstein et al. 2010). FFAs and phospholipids, however, can also contribute to energy metabolism (Ross and Love 1979; Yuneva et al. 1991; Henderson and Tocher 1987) and may be important in the reproductive cycle (Ackman 1980). We examined these in individual fish. TAG, FFA, and PL, all expressed in milligrams per gram dry weight for whole fish was extrapolated from extracted subsamples and converted to total storage quantities by adjusting to the dry weight of each fish.

Findings and Conclusions

The tradeoffs between energy allocation for growth, reproduction, and the laying down of sufficient storage reserves for periods of resource scarcity as "competing demands" in prereproductive organisms living in seasonal environments have been described by numerous authors (Walters and Juanes 1993; Fullerton et al. 2000; Post and Parkinson 2001). This is especially important in north-temperate fishes because experimental and field data suggest that energy availability is often limiting, i.e., fish in their natural environments tend to grow at less than their physiological optimum at a given temperature (Post and Parkinson 2001).

Although our comparisons on a dry weight (morphometric; EFH tier II evaluation) basis alone *did not detect differences* in somatic condition of *F. heteroclitus* populations in the two marshes (Fig. 1a), the examination of energy reserves in these fish after removing the potential confounding influences of the reproductive cycle and parasitization clearly indicated that significant differences occurred in TAG and FFAs levels (Fig. 1b).

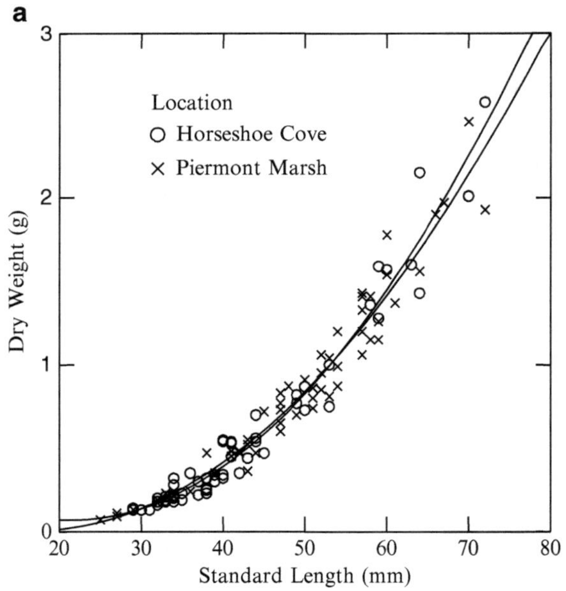

Fig. 1 (**a**) The relationship between length (mm) and somatic condition (dry weight in g) by location for mummichogs *Fundulus heteroclitus* captured in two tidal salt marshes, Horseshoe Cove and Piermont Marsh on the Hudson River estuary. (**b**) Total free fatty acids, triacylglycerol (TAG), and phospholipids vs. standard length (mm) in individual mummichogs (*F. heteroclitus*) captured at Horseshoe Cove (H) and Piermont (P) Marshes. All lipid values expressed in milligrams (mg)

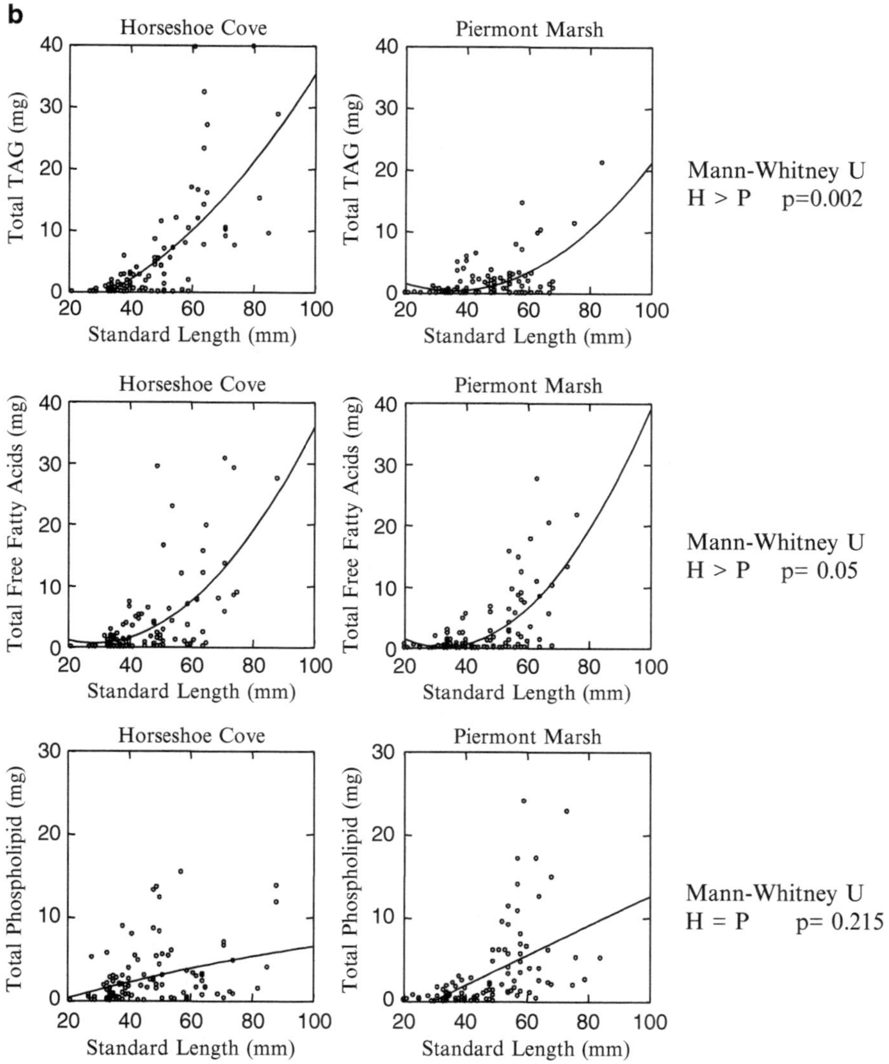

Fig. 1 (continued)

Supplementing tier II data with a tier III biochemical condition approach was, therefore, a more sensitive measure of the condition of individuals produced in these habitats. This conclusion is supported by Mommsen (1998) who suggested that a 100 g fish acquiring 1 g of lipid was unlikely to change in length, and although its weight gain was hardly detectable in the statistical noise, the fish had added a statistically significant amount of energy.

By focusing on energy reserves, principally TAG, we have been able to demonstrate that mummichogs residing in a polyhaline *S. alterniflora*-dominated tidal salt marsh were better able to acquire energy reserves for reproduction and overwintering survival than fish residing in a *Phragmites*-dominated marsh. Thus, *Phragmites* invasion and its consequent habitat impacts may be contributing to lower quality EFH for mummichogs (Weinstein and Balletto 1999; Hagan et al. 2007; Weinstein et al. 2009b).

Case History II (EFH Tier III): Use of Bioenergetics Models to Estimate the Nursery Value of Estuarine Habitats, Young-of-Year Weakfish (*C. regalis*)

Spatially explicit models of fish growth have been used to measure the quality of habitats for nekton production in a variety of species and aquatic systems by integrating variability in biotic and abiotic factors across habitats within a bioenergetics framework (Brandt et al. 1992; Brandt and Kirsh 1993; Mason et al. 1995; Demers et al. 2000; Luo et al. 2001). In this example, a mechanistic growth model, Fish Bioenergetics 3.0 (Hanson et al. 1997), was applied to a series of habitat "regions" within Delaware Bay (upper, middle and lower Bay; Fig. 2). Each region consisted of a marsh to open water gradient, and each had its own set of unique environmental conditions. While most models estimate growth from environmental conditions and the availability of prey, the approach adopted here was to estimate prey consumption from detailed growth estimates in juvenile weakfish that were recruited to the Bay in 1999 and 2001. The calculated rates of consumption ("realized" consumption) were compared to those expected if individuals were feeding at their maximum rate ("optimum" consumption), under ad libitum prey density. The goal was to understand how temporal and spatial variability in availability of food resources and temperature regime governed habitat use and value for juvenile weakfish.

Trawl survey data were used to estimate growth from the changes in the mean monthly weight of juvenile weakfish (Litvin 2005). A separate model was constructed for each cohort identified by length frequency analysis within each Delaware Bay region/year combination and analyzed for the duration that the cohort persisted. Fixed parameters of the model included prey energy density and the *initial* wet weight of individuals within cohorts (derived from the empirical data). The variable parameters included were: in situ temperature, energy density of juvenile weakfish and diet composition (shift from specialization on mysids in early recruits to >90% piscivory in larger individuals; Grecay and Targett 1996; Nemerson 2001).

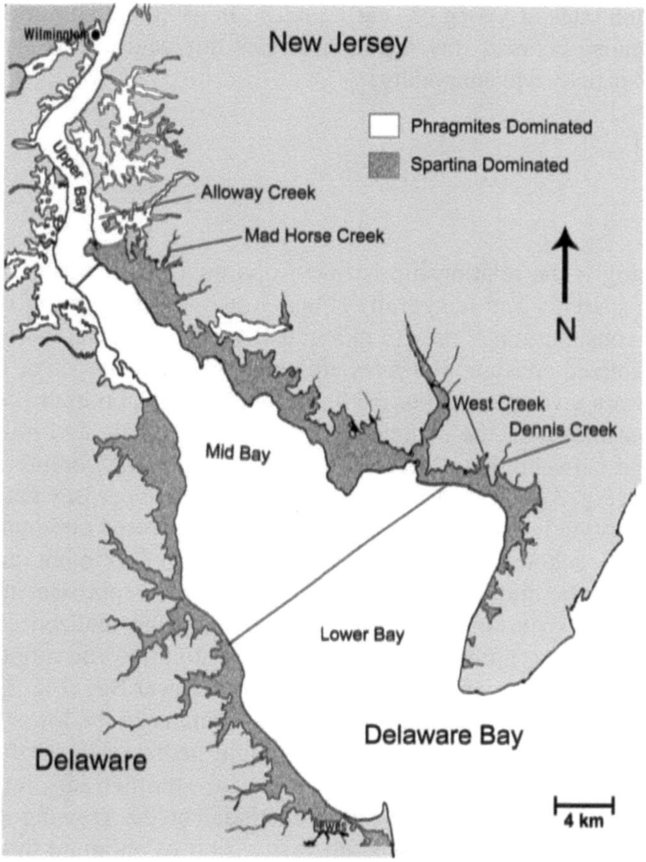

Fig. 2 Weakfish (*Cynoscion regalis*) sampling regions in Delaware Bay comprised of lower, middle ("mid"), and upper Bay and their associated tidal marshes

We modeled (1) realized consumption, which is estimated prey consumed under in situ temperature and empirical estimates of growth ("realized growth" derived from the trawl surveys), (2) optimal growth, which is the theoretical maximum growth under in situ temperature and ad libitum feeding conditions; and (3) optimal consumption, which is the prey consumed under optimal growth. From these results we calculated the excess demand, the proportional difference between the cumulative optimum, and the realized consumption over the period the cohort persisted (which is a relative measure of the suitability of a given habitat for fish production). The calculated realized and optimal growth and consumption and excess demand were compared to determine if food availability, temperature or other factors determined the production rates of young-of-year weakfish. In addition, these measures were compared among cohorts, Bay region and years to elucidate how habitat values varied both spatially and temporally.

Findings

Not surprisingly, the relationship between optimum and realized consumption varied both spatially and temporally (Figs. 3 and 4). Analyzed in terms of the cumulative consumption of prey (g individual^{-1}), it is important to note that optimum and realized consumption rates are equal when the slopes of their curves are equal over a given period. For the first cohort in 1999, optimum consumption increased steadily over the summer and fall in the middle and lower Bay but remained flat between days 200 and 240 in the upper Bay, before rising for a brief period (Fig. 3). The realized consumption in the upper Bay exceeded optimum consumption in the early summer (July) then mirrored optimum consumption until the cohort was no longer detected. The optimum and realized consumption in the middle and lower Bay regions were approximately equal in July and August with optimum consumption exceeding realized consumption during the remaining months. The optimum consumption for cohort 2 quickly outstripped the realized consumption, except in the lower Bay (Fig. 3). Optimum consumption in 2001 increased throughout the summer in the lower and middle Bay for cohort 1, and was followed by a slight decrease in the fall (Fig. 4). Optimum consumption in the upper Bay rose in June through July, fell in August, and then increased steadily during the remainder of the growing season. The realized consumption in all regions fell vastly short of optimum throughout the season. For the second cohort, the difference between optimum and realized consumption was substantial in the middle, but not the lower, Bay (Fig. 4). The excess demand (grams prey consumed per individual) for the first cohort varied substantially between regions and years (Fig. 5). The excess demand in 1999 ranged from −81% (the realized exceeded optimum consumption) through 78% in the middle Bay, and rose to 131% in the upper Bay, respectively. The surplus consumption in 2001 was markedly higher (407–505%), with peak values occurring in the upper Bay. This was driven both by changes in the estimated growth rates in the field (realized growth) and changes in optimum growth which rose from the upper through the lower bay in both years and was higher in 2001 in all regions (Fig. 5).

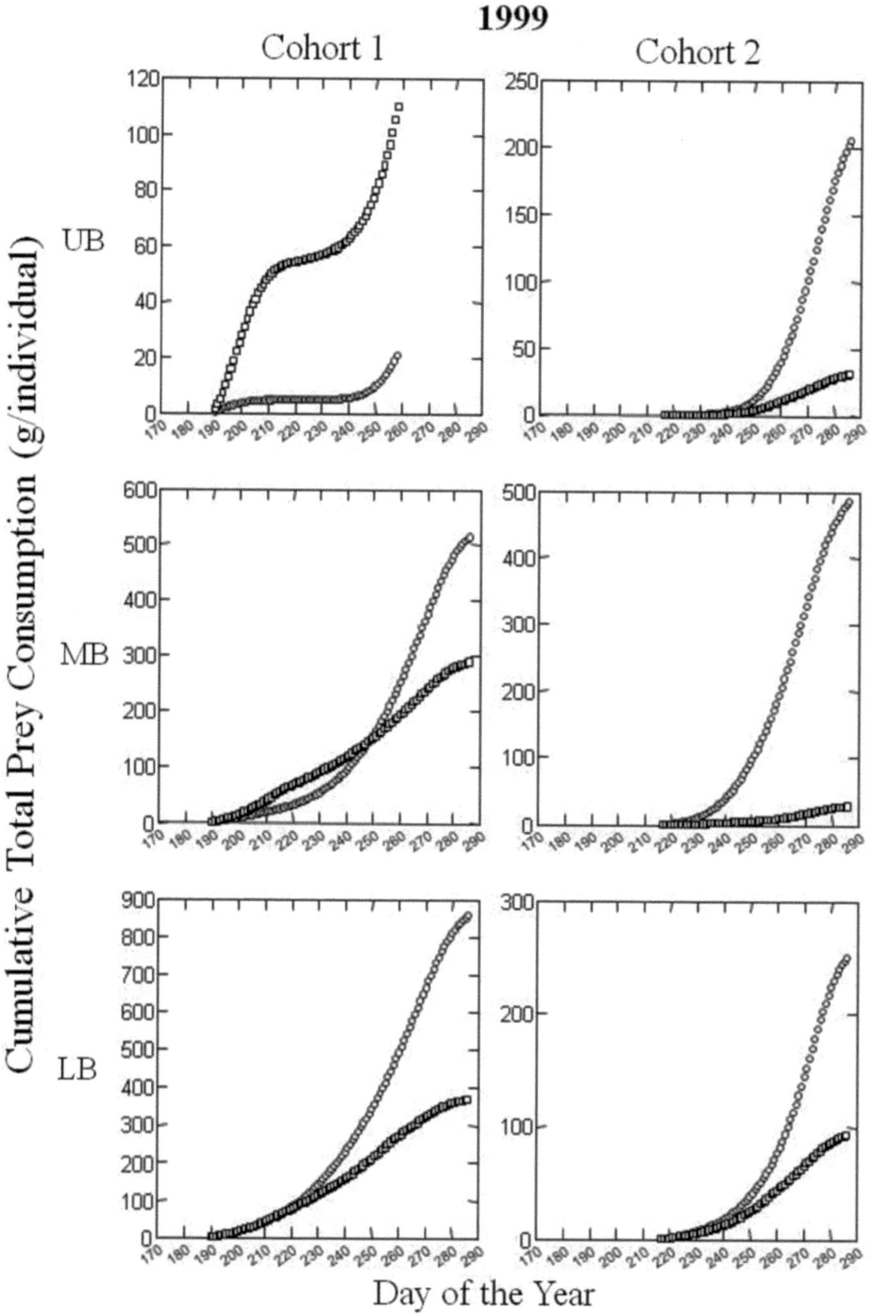

Fig. 3 The relationship between realized (*open squares*) and optimum (*open bullets*) consumption for juvenile weakfish captured in 1999, Delaware Bay, USA. LB, MB, and UB are lower, middle, and upper Bay, respectively

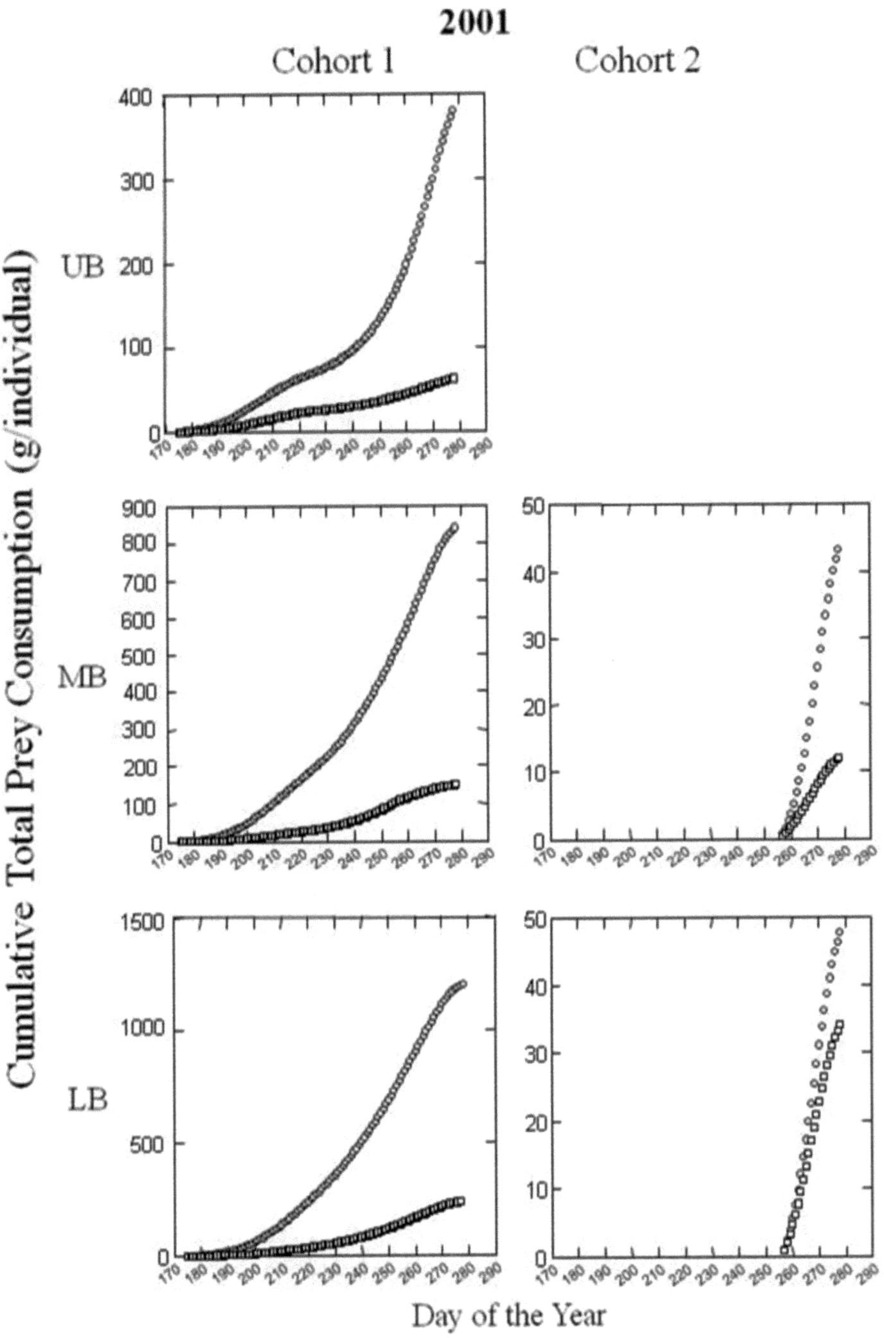

Fig. 4 The relationship between realized (*open squares*) and optimum (*open bullets*) consumption for juvenile weakfish capture in 2001, Delaware Bay, USA. LB, MB, and UB are lower, middle, and upper Bay, respectively

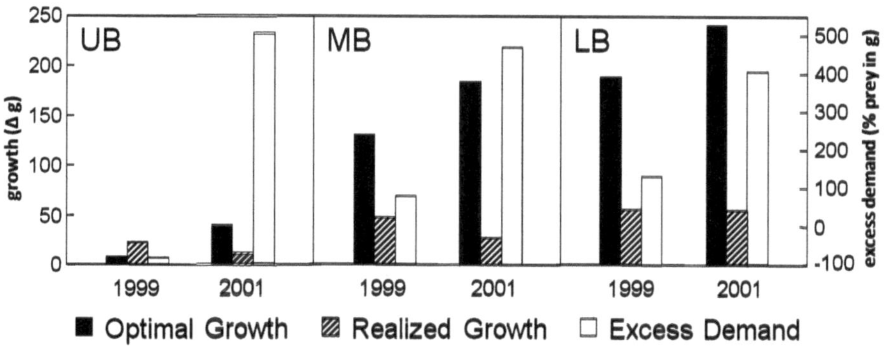

Fig. 5 Optimal growth, realized growth, and excess demand for cohort 1 from each Bay region (*lower* LB; *mid* MB; *upper Bay* UB) for 1999 and 2001

Summary and Conclusions

The trends in optimum consumption and excess demand provided insights into the relative suitability of the different segments of Delaware Bay for juvenile weakfish. In both years, estimates of optimum consumption suggested that the region with the physiological conditions most suitable for potential growth varied temporally. In July and August, optimum consumption estimates were higher in the lower Bay than those from the middle Bay, and both possessed higher values than the upper Bay region. This disparity in potential habitat quality dissipated as temperatures fell through September and October, with the optimal zone shifting to the upper Bay in the final days modeled. Between years, the differences in the mean estimates of excess demand (~100% and ~450% for 1999 and 2001, respectively) suggested that the value of Bay regions for the production of young weakfish was considerably lower in 2001. While it is difficult to determine if the variability in estimated habitat suitability between habitats and among years is driven by differences in prey supply, the expected reduction in juvenile weakfish foraging success due to high turbidity encountered in oligohaline habitats, or physiochemical considerations not accounted for in the model (see below), these results parallel the spatial patterns in juvenile weakfish condition and growth previously reported for this estuary and its marshes (Grecay and Targett 1996; Paperno et al. 2000; Litvin and Weinstein 2003; Litvin 2005).

Two factors, both attributable to the high river discharge rates in June and July of 2001, likely drove inter-annual differences in habitat suitability (USGS[1]). Juveniles recruited to the upper Bay in June through the early fall of 2001 faced increased physiological stress and lower potential growth rates than those from other regions

[1] USGS New Jersey Monthly Streamflow Statistics for Trenton, NJ (Site # 01463500); http://waterdata.usgs.gov/nl/nwis.

due to the interaction of high temperatures and low salinity (Lankford and Targett 1994; Grecay and Targett 1996; Paperno et al. 2000). In addition, an increase in the flow rates might have lead to "compression" of the meso and polyhaline regions and resulted in the increased density of piscivorous marine predators that might normally be restricted from oligohaline waters (Weinstein et al. 1980; Taylor 1987; Martino and Able 2003). Therefore, juvenile weakfish moving down Bay in 2001 into meso and polyhaline waters with superior physiochemical conditions likely faced increased predation risk, relative to low flow years, which in turn might affect acquisition of prey and growth (Walters and Juanes 1993; Sogard 1997).

A significant purpose of any modeling effort is to identify data needs and suggest future directions for research. We developed several recommendations base on this work. It is important to examine the interactive effects of temperature and salinity on the scope of growth to parameterize future bioenergetics models intended to gauge habitat value for estuarine fish. Knowing the choices in prey, aside from their energy density may also be important. The variability of postconsumptive constraints based on prey type, known to occur in young weakfish (Lankford and Targett 1997), were not considered although they may lead to consumption-dependent systematic errors in estimates of growth (Bajer et al. 2004). Although labor-intensive, obtaining estimates of spatial and temporal variability of prey from stomach content analysis, and the incorporation of this information into bioenergetic variables (that represent postconsumptive processes), would further increase model accuracy. Also, the methods to account for seasonal changes in habitat utilization patterns, e.g., estimating movements out of marsh habitats and downstream migration rates as individuals grew (Litvin and Weinstein 2004), should be accounted for when determining spatially explicit growth. Stable isotope analyses have demonstrated great utility as "biomarkers" for discerning habitat utilization patterns in juvenile weakfish and other species from Delaware Bay and the employment of this technique in conjunction with field measures of length and weight will improve the accuracy of growth estimates (Weinstein et al. 2000; Litvin and Weinstein 2003, 2004). These three considerations are also particularly important when using bioenergetics models to move beyond the "regional" approach utilized here to determine the value of specific marsh habitats for juvenile nekton in the context of the greater estuary. Estuarine and marsh habitats, even those separated by small distances, should be expected to have different environmental regimes as well as abundance and diversity of food resources. For restored marsh habitats, their position in the restoration trajectory may heavily influence these factors (Weinstein et al. 2005). In addition, understanding the habitat utilization patterns of species, like juvenile weakfish, potentially using a wide variety of estuarine habitats is critical to understanding the relative value of marsh habitats for fish production. For example, the physiochemical conditions in open waters of the Bay region in 2001 were sub-par, and so the biotic conditions in marsh habitats may have resulted in higher optimum and realized growth for juvenile weakfish. This may both increase the value of marsh habits for juvenile fish production in a given year, and buffer individuals leaving marshes to migrate toward the bay mouth from the depressed conditions in the open estuary. Together, these two situations would ultimately increase the relative contribution (per unit area) from marsh habitats to overall estuarine production. While the exclusion of these considerations does not preclude the use of the model to help understand the variables governing the demonstrated spatial and temporal stochastisity in the nursery value of estuarine habitats for young weakfish and other species, their incorporation into models would help to improve both their accuracy and utility as a tool for both ecologists and natural resource managers.

Case History III (EFH Tier IV): Estimating the Response of the Delaware Bay Ecosystem to Tidal Marsh Restoration

During the 1990s, 45.5 km^2 of wetland habitat were restored in the Delaware Bay ecosystem to offset mortality caused by power plant cooling water intakes (Teal and Weinstein 2002). The restoration effort resulted in a 3% increase in marsh habitat (Balletto et al. 2005; Hinkle and Mitsch 2005) and provided scientists with the opportunity to assess changes to system productivity and structural changes before and after restoration. A series of baseline and monitoring studies were conducted to quantify nekton assembly composition and usage of restored and reference marshes (Kimball and Able 2007; Nemerson and Able 2005; Jivoff and Able 2003; Able et al. 2008). These studies documented the impact of restoring habitat and demonstrated that the nektonic assemblage responded favorably to restoration. These studies, however, did not address overall system productivity and the structural changes resulting from restoration efforts.

In the following section, we summarize the approach and results reported in Frisk et al. (2011) who estimated the increment of new secondary production that resulted in the entire Delaware Bay ecosystem following restoration. Estimating system-wide impacts required distinguishing between the impacts of restoration and background variability in spatiotemporal patterns of productivity and ecosystem structure. The latter effort entailed estimating system productivity after restoration and simulating the proportion of biomass that would have been lost if restoration efforts had not taken place. To achieve this result required the development and parameterization of a mass-balanced time-dynamic ecosystem model fitted to observed time series of key species and then simulating a nonrestored system.

Assessing Restoration Using Ecopath with Ecosim

Details regarding the model structure of Ecopath with Ecosim (EwE) can be found in Christensen and Pauly (1992), Walters et al. (1997), and Pauly et al. (2000), and, for the model presented here, in Frisk et al. (2011). Ecopath was used to develop a mass-balanced network of trophically-linked biomass pools representing a static description of the ecosystem from detritus to upper-trophic level species. The Ecopath model provided the initial parameters used to fit the dynamic Ecosim model to time series of data for the Delaware Bay ecosystem. Ecosim uses a series of coupled delay-difference age/size-structured equations to model all species in the system. The flows between species are linked by both predator and prey consumption rates.

Parameter Inputs

The times series data are derived from long-term biological studies and harvest records used to estimate biomass and landings and for the development of stock assessments for key species in Delaware Bay (DNREC 1966–2003; NOAA Fisheries 2011). The demographic and diet data were derived from the literature or inferred from adjacent systems when necessary.

Measuring Habitat Restoration

Changes to ecosystem structure and productivity following restoration were reflected in the biological time series conducted in the Bay during 1996–2003. Therefore, the biomass gains resulting from restoration were captured in a model fitted to these data. To estimate the lost productivity had restoration not occurred, a second model was run that assumed a 3% decrease in the available marsh habitat. This was achieved by fitting the Ecosim model for 1966–2003 and applying forcing functions to decrease the production rates for marsh meiofauna and macrofauna for the years following restoration, 1996–2003. The difference between the two models' total system biomass was used to estimate the gains associated with restoration.

Model development. The modeling approach used 47 functional groups including: 27 fish species, 5 invertebrate groups, 4 multi-species benthic groups, 6 multi-species fish groups, 3 plankton groups, 1 shorebird group, and 1 marine mammal group (Frisk et al. 2011). The static mass-balanced Ecopath food-web model was developed for 1966 and served as the initial parameter estimates for the time-dynamic Ecosim model for the period 1966–2003. The Ecopath model required estimates of biomass (B), the ratio of production to biomass (P:B), consumption to biomass (Q:B), ecotrophic efficiency (EE), and diet data for all model groups. The Ecosim models were parameterized using catch, biological survey, diet, and demographic data. Time series of biomass (catch per unit effort) for eight species, catch time series for six species and fishing mortality for five species were fitted in Ecosim (Table 1).

Table 1 Data used for fitting the Ecosim model included time series of catch (*C*), biomass (*B*), and fishing effort (*F*) where numbers represent the length of time series in years

Species	C_{fitted}	$C_{subtracted}$	*B*	*F*
American eel		38		
Atlantic croaker	38		38	38
Atlantic menhaden	38		38	38
Blue crab	38			38
Bluefish	38		38	38
Clearnose skate			27	
Dogfish			27	
Horseshoe crab		30		
Oyster		38		
Spot		38		
Striped bass	38		23	38
Summer flounder	38			
Weakfish		38	27	
White perch		38		

Catch was statistically fitted (C_{fit}) or subtracted from model estimated biomass (C_{sub}). Stock assessments were conducted on Atlantic croaker, Atlantic menhaden, and bluefish providing biomass (CPUE) and fishing effort (fishing exploitation rate). Other biomass estimates were derived from the DNREC surveys except striped bass which came from ASMFC (2004)

The catches were subtracted from the estimated biomass for eight key species that did not have prior estimates of fishing mortality to ensure that the model produced stock sizes large enough to support the historical fisheries.

Findings

The results of the model runs indicated that restoration resulted in a net gain of 47.7 tons km^{-2} year^{-1} in system biomass. The biomass change was seen across a wide range of species and biomass groups, and had an average percent change of 1.2%, ranging from 4.3% for macrozooplankton to 1.3% decrease in blue crab (Fig. 6).

Restoration also resulted in 41 species increasing in biomass and four species showing slight decreases. The decreases likely resulted from food web interactions with groups that increased. Proportionally, consumer trophic species such as Atlantic menhaden, resident striped bass, macrozooplankton, and summer flounder showed the greatest gains.

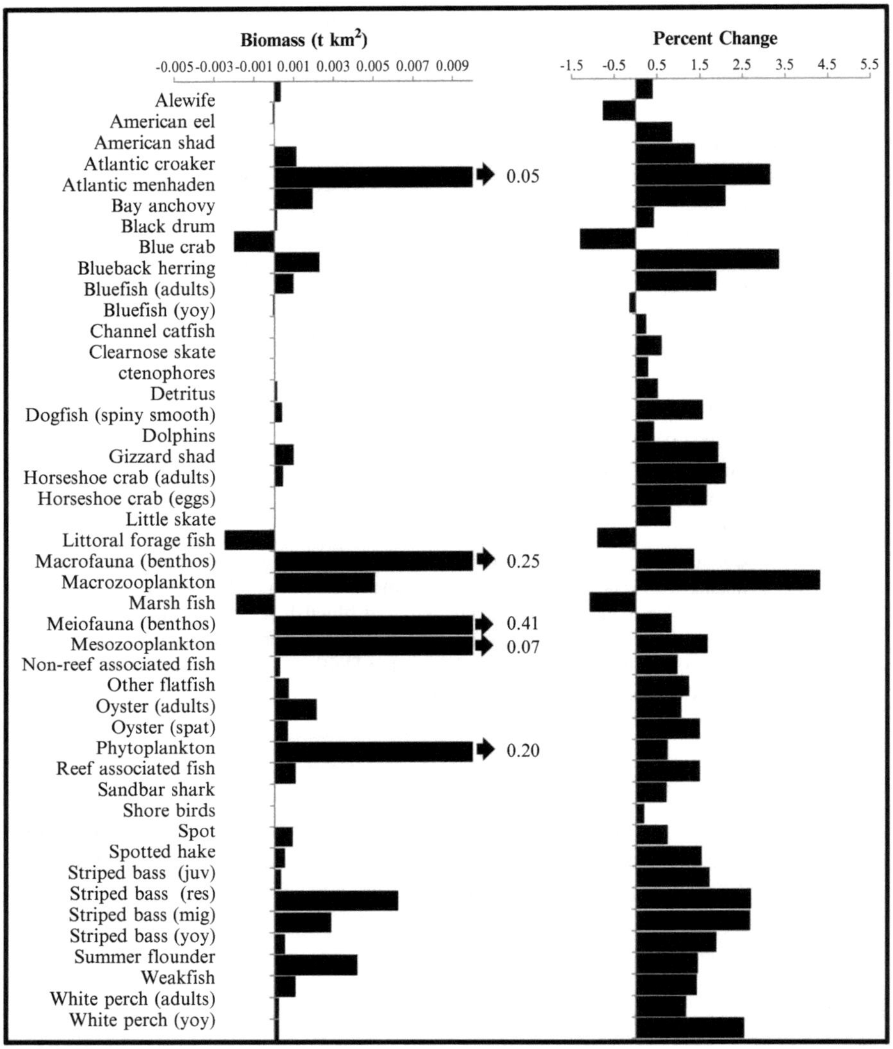

Fig. 6 Change in species' annual abundance resulting from restoration for the period 1996–2003 reported in biomass and percent change, where "yoy" indicates young of the year staged fish, "juv" is juvenile staged fish, "res" is a resident contingent, and "mig" is a migratory contingent

Summary and Conclusions

As with any modeling attempt, ecosystem complexities are not completely captured in the model structure. Ecosystem models require large amounts of information, and the reliability and availability of data varies by species and biological group. The Delaware Bay model is a compromise among realism, parameterization, and data limitations. Estimating restoration impacts utilizing data collected continually before and after restoration does not allow for direct comparisons of an ecosystem with and without restoration. Instead, the statistically fitted model was altered to reflect a simulated no-restoration system to estimate the loss of biomass had the marsh habitat not been created. The approach allows for a total system evaluation of restoration; however, the simulation approach may add uncertainty to the results.

Large-scaled restoration efforts require an enormous investment of time and money. The success and justification of these efforts should be judged by whether or not the degraded ecosystems can ultimately be rehabilitated to provide basic ecological goods and services. In Delaware Bay, researchers conducted numerous field studies documenting the success of newly created marsh habitat for use by native species and general production of finfish and shellfish (Kimball and Able 2007; Jivoff and Able 2003; Able et al. 2008). More broadly, Frisk et al. (2011) were able to show that the overall ecosystem biomass increased including several important commercial finfish. The results supported previous field estimates and demonstrated increases throughout the entire food web that were dependent on the increased marsh habitat. The results from the Frisk et al. (2011) mass-balanced time-dynamic ecosystem model of Delaware Bay were used to address the following questions: (1) Has restored habitat resulted in changes to the productivity of the ecosystem?; (2) Is there evidence that restoration can impact the structure and health of the Delaware Bay ecosystem?

The model results indicated that many species increased in biomass including ecologically important lower trophic groups such as macrofauna, macro-meso zooplankton, and higher trophic groups consisting of striped bass, Atlantic croaker, and summer flounder. The model estimates indicated that restoration of marsh habitat resulted in increased biomass throughout the food web. These results, when combined with previous field studies indicated that the restoration of 45.5 km^2 of marsh habitat increased productivity and restored ecologically meaningful amounts of goods and services to the Bay.

In addition to biomass gains, the model results indicated that restoration has the potential to alter the structural composition of Delaware Bay, and that there were slight changes in several ecosystem properties including productivity and system maturity metrics (see Frisk et al. 2011, for details). This has important implications for stemming over a century's loss of wetland habitat. Network analyses indicated that the Delaware Bay is in an immature state and suffers from decades of nutrient enrichment and pollution (Sharp 2010). It is encouraging that these model results indicate that restoration can reduce the impact of long-term alteration of estuarine ecosystems and potentially increase system maturity. The changes, however, were slight and even larger-scaled restoration efforts integrated with management strategies covering the entire drainage basin may be needed to further restore ecosystem function.

Synthesis

The need for linking restoration ecology, ecological restoration, and sustainability science are a *sin qua non* of the global sustainability transition. In this Age of Humans (Crutzen 2002), they form what Aronson and Vallejo (2006) term our "survival strategy" where "nonscientists and scientist work together in transdisciplinary efforts to imagine, develop, test and apply new methods, tools and approaches to the enormous [sustainability] challenges ahead." Moreover, managing our life support systems will require stewardship "from the inside" in ways that recognize our dependence on, and responsibility to, sustainably manage the systems that we are an integral part of. Aronson and Vallejo (2006) suggest that restoration projects should adopt broad suites of relevant, reliable and complementary traits or ecosystem attributes that when combined, reflect the structural and functional dynamics of the system.

The discipline of restoration ecology has now matured to where knowledge of natural systems can define a better pathway to the desired restoration outcomes (Larkin et al. 2008). Whether it is considering individual habitats in a system-wide integrated framework (Childers et al. 2000; Weinstein et al. 2005), reestablishing the physiographic heterogeneity (and concomitant physical access) of a salt marsh that we have set forth above, or other "landscape" features—hydrology/hydroperiod (Rozas et al. 1988), edge or "critical transition zones" (Baltz et al. 1993; Kneib 2003), drainage density (Kneib 1994, 1997), area of vegetated marsh (and its relationship to secondary production; Turner 1977; Zimmerman et al. 2000)—and organism interactions; trophic access (sensu Kneib 2003), predator–prey interactions (Boesch and Turner 1984; Deegan et al. 2000), ecological and engineering criteria can be combined into a much more quantitative approach to achieving "success" (or whatever we choose to call it; Zedler 2007). Despite the current debate over the issue of attaining habitat equivalency of marsh *functions*, we agree with Kneib (2003) that the rubic "build it and they will come" has a degree of validity with respect to organisms utilizing the restored habitat. Kneib notes that "early indications suggest that fishes do not discriminate between natural and excavated wetland channels," and that "there should be every expectation that nekton production from the restored created site has potential to rapidly match that of natural systems." But Kneib (2003) also noted that "site-specific bioenergetic and landscape constraints [while setting upper limits to production] should guide the development of realistic expectations and success criteria for marsh restoration designed to enhance estuarine nekton production." The simple underlying question then is: yes, we have restored a marsh, but what kind of marsh have we restored and will it fulfill our expectations with respect to secondary production goals for targeted species?

What seems clearly lacking in today's purview is a policy and practice that goes beyond developing the status and trend inventories for coastal wetlands (Stedman and Dahl 2008), or the promulgated guidelines for conservation, restoration and management that lack rigorous ecological criteria for meeting design goals. Clearly, the importance of coastal wetlands as EFH is recognized by resource agencies: "coastal wetlands provide valuable habitat for the vast majority of commercially and recreational marine species" (Stedman and Dahl 2008). Although more than $28 million has been allocated by the National Oceanographic and Atmospheric Administration's (NOAA) for habitat restoration (J. Rapp, personal communication) there seems to be a great paucity of process and function criteria, in general, and especially at higher levels of EFH analysis. In NOAA's "Science-Based Restoration Monitoring of Coastal Habitats" guidelines for developing a monitoring plan (Thayer et al. 2003), explicit recognition is given to the need to develop testable hypotheses to "determine progress toward restoration goals," yet the examples given of postimplementation monitoring are large structural criteria, and no functional criteria are proposed above the level of Tier II, EFH; i.e., density and composition of organisms. A cursory review of Progress Reports submitted under the NOAA Restoration Center's Community-based Restoration Program (CRP), Progress Report Narrative Formats includes only biological inventories (restricted at or below EFH, Tier II) and/or topographical/structural parameters.

We can and must do better. Simenstad et al. (2006) note that "while desirable functions may result from the structure of ecosystems, it is typically the dynamics of ecosystem processes that sustain the structure at the landscape scale or in some cases may even be the underlying mechanism behind the function." Further, the relationship between salt marsh restoration and attributed functions is highly scale-dependent, nonlinear, and dictated by thresholds (Simenstad et al. 2006). We agree that the long-term performance of a restored site depends on reintroducing natural dynamics and disturbances into the system (Middleton 1999), also within the context of life-history requirement of extant flora and fauna. Perhaps these restoration efforts should be approached as "natural experiments" to be quantified for their return of desired ecosystem goods and services, and, as Simenstad et al. (2006) note, be treated in a manner that enables learning from the results. Performance criteria should include more process-based metrics to help ensure success.

Virtually all restoration scientists agree that tidal marshes are complex systems that require the best applications of science and engineering principles (Weinstein et al. 1997) to achieve the goals of a particular restoration design; especially when these are systems whose physiographic province affecting virtually every component of secondary production support is measured in centimeters (Vivian-Smith 1997). As noted above, the progress we have made in understanding how marshes "work" is far from complete, and research should not become subordinate to practice in advancing the goals of restoration. Ongoing research can provide important perspectives on the results achieved, and will allow us to correct mistakes, and suggest new approaches for defining success (Kentula 2000).

It is beyond the scope of this chapter to attempt a comprehensive review of the application of restoration science to the design and success of tidal wetland restorations in supporting secondary production except to say that quantitative studies are far and in-between in the published literature. A relatively recent review by Borde et al. (2004) helps make the point. Beginning in 1998, more than 550 citations from scientific journals, books, technical reports and proceedings were surveyed for "innovations" in coastal restoration. This effort supported NOAA's attempts "to advance the science of restoration ecology" including research on coastal ecosystem structure and function. Although the review of the literature suggested that restored salt marshes were functioning to increase the growth, production, and resilience of fish populations, there were no specific recommendations to incorporate these functional criteria into goal setting and success criteria, nor have we seen inclusion of parameters like EFH Tier III and IV criteria built into restoration planning. Rather, the authors' summary of "innovative methods and techniques" to our knowledge have yet to find their way into NOAA Restoration Center design protocols, monitoring techniques, nor project success. We are not saying it will be easy, but simply recognizing that is should and must be done.

Finally, we leave the reader with a research approach proposed by Choi (2004). It serves both as a useful take home message and template for future progress. In synthesizing the need for a "futuristic approach" to restoration, Choi proposed that we:

1. Set realistic and dynamic (rather than static) goals for future, instead of past, environments
2. Assume multiple trajectories acknowledging the unpredictable nature of ecological communities and ecosystems
3. Take an ecosystem or landscape approach, instead of ad hoc gardening, for both structure and function
4. Evaluate restoration progress with explicit criteria
5. Maintain long-term monitoring of restoration outcomes

Of course, we recommend further that these efforts be supported by strong ongoing and fully funded restoration science research!

Acknowledgments Much of our research has been supported by NOAA (Saltonstall-Kennedy, Aquatic Nuisance Species Program, and Sea Grant), EPA, NJWRRI, the PSEG Corporation (Marsh Ecology Research Program, and Estuary Enhancement Program), and USGS (State Partnership Program). This is contribution number ISS-2012-0126 from the PSEG Institute for Sustainability Studies.

References

Able KW (1999) Measures of juvenile fish habitat quality: examples from a national estuarine research reserve. Am Fish Soc Symp 22:134–147

Able KW, Grothues TM, Hagan SM et al (2008) Long-term response of fishes and other fauna to restoration of former salt hay farms: multiple measures of restoration success. Rev Fish Biol Fish 18:65–97

Ackman RG (1980) Fish lipids, part I. In: Connell JJ (ed) Advances in fish science and technology. Oxford, London

Ackman RG, Eaton CA (1976) Variations in fillet lipid content and some lipid-iodine value relationships for large winter Atlantic herring from southeastern Newfoundland. J Fish Res Board Can 33:1634–1638

Anand M, Desrochers RE (2004) Quantification of restoration success using complex systems concepts and models. Rest Ecol 12:117–123

Aronson J, Vallejo R (eds) (2006) Restoration ecology. Blackwell Publishing, Malden

ASMFC (2004) Stock assessment report for Atlantic striped bass: catch-at-length-based VPS, and tag release/recovery based survival estimation. SBTC Report #2004–4:93

Bajer PG, Whitledge GW, Hayward RS (2004) Wide spread consumption dependent systematic error in fish bioenergetics models and its implications. Can J Fish Aquat Sci 61:2158–2167

Balletto JH, Heimbuch MV, Mahoney HJ (2005) Delaware Bay salt marsh restoration: mitigation for a power plant cooling water system in New Jersey, USA. Ecol Eng 25:204–213

Baltz DM, Rakocinski C, Fleeger JW (1993) Microhabitat use by marsh edge fishes in a Louisiana estuary. Environ Biol Fish 36:109–126

Beck MW, Heck KL Jr et al (2001) The identification, conservation, and management of estuarine and marine nurseries for fish and invertebrates. Bioscience 51:633–641

Beck MW, Heck KL Jr et al (2003) The role of nearshore ecosystems as fish and shellfish nurseries. Issues Ecol 11:1–12

Bedford BL (1999) Cumulative effects on wetland landscapes: links to wetland restoration in the United States and Southern Canada. Wetlands 19:775–788

Boesch DF, Turner RE (1984) Dependence of fishery species on salt marshes: the role of food and refuge. Estuaries 7:460–468

Borde AB, O'Rourke LK, Thom RM et al (2004) National review of innovative and successful coastal habitat restoration. Prepared for the National Oceanic and Atmospheric Administration, Coastal Services Center. Battelle Memorial Institute, Sequim, WA

Brandt SB, Kirsh J (1993) Spatially explicit models of striped bass growth rate potential in Chesapeake Bay. Trans Am Fish Soc 122:845–869

Brandt SB, Manson DM, Patrick EV (1992) Spatially-explicit models of fish growth rate. Fisheries 17:23–35

Cairns J Jr (2000) Setting ecological restoration goals for technical feasibility and scientific validity. Ecol Eng 15:171–180

Childers DL, Day JW Jr, McKellar HN Jr (2000) Twenty more years of marsh and estuarine flux studies: revisiting Nixon. In: Weinstein MP, Kreeger DA (eds) Concepts and controversies in tidal marsh ecology. Kluwer Academic, Dordrecht

Choi YD (2004) Theories for ecological restoration in changing environments: toward 'futuristic' restoration. Restor Ecol 19:75–81

Christensen V, Pauly D (1992) Ecopath-II—a software for balancing steady-state ecosystem models and calculating network characteristics. Ecol Model 61:169–185

Crutzen PJ (2002) The Anthropocene: geology of mankind. Nature 415:23

Currin CA, Newell SY, Paerl HW (1995) The role of standing dead *Spartina alterniflora* and benthic microalgae in salt marsh food webs: considerations based on multiple stable isotope analysis. Mar Ecol Prog Ser 121:99–116

Currin CA, Wainright SC, Able KW, Weinstein MP, Fuller CM (2003) Determination of food web support and trophic position of the mummichog, *Fundulus heteroclitus*, in New Jersey smooth cordgrass (*Spartina alterniflora*), common reed (*Phragmites australis*), and restored salt marshes. Estuaries 26(2B):495–510

Dame JK, Christian RR (2008) Evaluation of ecological network analysis validation of output. Ecol Model 210:327–338

Deegan LA, Hughes JE, Rountree RA (2000) Salt marsh ecosystem support of marine transient species. In: Weinstein MP, Kreeger DA (eds) Concepts and controversies in tidal marsh ecology. Kluwer Academic, Dordrecht

Demers E, Brandt SB, Barry KL et al (2000) Spatially explicit models of growth rate potential. In: Hobbie JE (ed) Estuarine science: a synthetic approach to research and practice. Island Press, Washington

DNREC (Division of Natural Resources and Environmental Control) (1996–2003) Coastal Finfish assessment survey, annual reports, project F-42-R-8, Dover, DE

Ehrenfeld JG (2000) Defining the limits of restoration: the need for realistic goals. Rest Ecol 8:2–9

Falk DA, Palmer MA, Zedler JB (2006) Foundations of restoration ecology. Island Press, Washington

French PW (2005) Managed realignment—the developing story of a comparatively new approach to soft engineering. Estuar Coast Shelf Sci 67:409–423

Frisk MG, Miller TJ, Latour RJ et al (2011) Assessing biomass gains from marsh restoration in Delaware Bay using Ecopath with ecosim. Ecol Model 222:190–200

Fry B, Hullar M, Peterson BJ et al (1992) DOC production in a salt marsh estuary. Arch Hydrobiol 37:1–8

Fullerton AH, Garvey JE, Wright RA, Stein RA (2000) Overwinter growth and survival of large-mouth bass: interactions among size, food, origin, and winter severity. Trans Am Fish Soc 129:1–12

Grecay PA, Targett TE (1996) Spatial patterns in condition and feeding of juvenile weakfish in Delaware Bay. Trans Am Fish Soc 125:803–808

Hagan SM, Brown SA, Able KW (2007) Production of mummichog (*Fundulus heteroclitus*): response in marshes treated for common reed (*Phragmites australis*) removal. Wetlands 27:54–67

Haines EB (1979) Interactions between Georgia salt marshes and coastal waters: a changing paradigm. In: Livingston RJ (ed) Ecological processes in coastal and marine systems. Plenum, New York

Hanson PC, Johnson TB, Schindler DE et al (1997) Fish bioenergetics 3.0 for windows. University of Wisconsin Sea Grant Institute, Madison

Heintz RA, Bonita DN, Hudson JH et al (2004) Marine subsidies in freshwater: effects of salmon carcasses on lipid class and fatty acid composition of juvenile coho salmon. Trans Am Fish Soc 133:559–567

Henderson RJ, Tocher DR (1987) The lipid composition and biochemistry of freshwater fish. Prog Lipid Res 26:281–347

Hildebrand RH, Watts AC, Randle AM (2005) The myths of restoration ecology. Ecol Soc 10:19–29

Hinkle RL, Mitsch WJ (2005) Salt marsh vegetation recovery at salt hay farm wetland restoration sites on Delaware Bay. Ecol Eng 25:240–251

Holling CS (1996) Surprise for science, resilience of ecosystems and incentives for people. Ecol Appl 6:733–736

Jackson ST, Hobbs RJ (2009) Ecological restoration in the light of ecological history. Science 325:567–569

Jivoff PR, Able KW (2003) Evaluating salt marsh restoration in Delaware Bay: the response of blue crabs, *Callinectes sapidus*, at former salt hay farms. Estuaries 26:709–719

Kentula ME (2000) Perspectives on setting success criteria for wetland restoration. Ecol Eng 15:199–209

Kimball ME, Able KW (2007) Tidal utilization of nekton in Delaware Bay restored and reference intertidal salt marsh creeks. Estuar Coast 30:1075–1087

Kneib RT (1994) Spatial patterns, spatial scale and feeding in fishes. In: Stouder DT, Fresh KL, Feller RJ (eds) Theory and application in fish feeding ecology. University of South Carolina Press, Columbia

Kneib RT (1997) The role of tidal marshes in the ecology of estuarine nekton. Oceanogr Mar Biol Annu Rev 35:163–220

Kneib RT (2003) Bioenergetic and landscape considerations for scaling expectations of nekton production from intertidal marshes. Mar Ecol Prog Ser 264:279–296

Lankford TE, Targett TE (1994) Suitability of estuarine nursery zones for juvenile weakfish (*Cynoscion regalis*): effects of temperature and salinity on feeding, growth and survival. Mar Biol 119:611–620

Lankford TE, Targett TE (1997) Selective predation by juvenile weakfish: post-consumptive constraints on energy maximization and growth. Ecology 1997:1049–1061

Larkin DJ, Madon SP, West JM et al (2008) Topographic heterogeneity influences habitat use of experimentally restored tidal marsh. Ecol Appl 18:483–496

Litvin SY (2005) Trophic linkages, movements, condition and energetic of a juvenile marine transient in the Delaware Bay Estuary: implications for the role of habitat in recruitment success. PhD Dissertation, Rutgers University, New Brunswick, NJ

Litvin SY, Weinstein MP (2003) Life history strategies of estuarine nekton: the role of marsh macrophytes, microphytobenthos and phytoplankton in the trophic spectrum. Estuaries 26(B): 553–653

Litvin SY, Weinstein MP (2004) Multivariate analysis of stable isotope ratios to infer movements and utilization of estuarine organic matter by juvenile weakfish (*Cynoscion regalis*). Can J Fish Aquat Sci 61:1851–1861

Lochmann SE, Ludwig GM (2003) Relative triacylglycerol and morphometric measures of condition in sunshine bass fry. N Am J Aquacult 65:191–2003

Lochmann SE, Maillet GL, Frank KT et al (1995) Lipid class composition as a measure of nutritional condition in individual larval Atlantic cod (*Gadus morhua*). Can J Fish Aquat Sci 52:1294–1306

Lochman SE, Maillet GL, Taggart CT et al (1996) Effects of gut contents and lipid degradation on condition measures in larval fish. Mar Ecol Prog Ser 134:27–35

Luo JG, Hartman KJ, Brandt SB et al (2001) A spatially-explicit approach for estimating carrying capacity: an application for the Atlantic menhaden (*Brevoortia tyrannus*) in Chesapeake Bay. Estuaries 24:545–556

Mallin MA, Burkholder JM, Sullivan MJ (1992) Contributions of benthic microalgae to coastal fishery yield (comment). Trans Am Fish Soc 121:691–693

Martino EJ, Able KW (2003) Fish assemblages across the marine to low salinity transition zone of a temperate estuary. Estuar Coast Shelf Sci 56:969–987

Mason DM, Goyke AP, Brandt SB (1995) A spatially explicit bioenergetics measure of habitat quality for adult salmonids: comparison between Lakes Michigan and Ontario. Can J Fish Aquat Sci 52:1527–1583

McIvor CC, Odum WE (1988) Food, predation risk, and microhabitat selection in a marsh fish assemblage. Ecology 69:1341–1353

Meredith WH, Lotrich VA (1979) Production dynamics of a tidal creek population of *Fundulus heteroclitus* (Linnaeuas). Estuar Coast Mar Sci 8:99–118

Middleton BA (1999) Wetland restoration, flood pulsing and disturbance dynamics. Wiley, New York

Mitsch WJ, Wu X, Nairn RW et al (1998) Creating and restoring wetlands. Bioscience 48:1019–1030

Mommsen TP (1998) Growth and metabolism. In: Evans DH (ed) The physiology of fishes, 2nd edn. Academic, New York

Nemerson DA (2001) Geographic, seasonal and ontogenetic patters in the trophic dynamics of five fish species of Delaware Bay, USA. PhD dissertation, Rutgers University, New Brunswick, NJ

Nemerson DM, Able KW (2005) Juvenile sciaenid fishes respond favorably to Delaware Bay marsh restoration. Ecol Eng 25:260–274

Newell SY (1993) Decomposition of shoots of a salt-marsh grass: methodology and dynamics of microbial assemblages. Adv Microbiol Ecol 13:301–326

Nixon SW (1980) Between coastal marshes and coastal waters—a review of twenty years of speculation and research on the role of salt marshes in estuarine productivity and water chemistry. In: Hamilton R, MacDonald KB (eds) Estuarine and wetland processes. Plenum Press, New York

NOAA Fisheries (2011) http://www.st.nmfs.noaa.gov/st1/commercial/index.html and http://www.st.nmfs.noaa.gov/st1/recreational/index.html June 1, 2011

Odum EP (1968) A research challenge: evaluating the productivity of coastal and estuarine water. In: Proceedings of the 2nd Sea Grant conference, Graduate School of Oceanography, University of Rhode Island, Kingston, RI

Palmer MA, Ambrose RF, Poff NL (1997) Ecological theory and community restoration ecology. Rest Ecol 5:291–300

Paperno R, Targett TE, Grecay PA (2000) Spatial and temporal variation in recent growth, overall growth and mortality of juvenile weakfish (*Cynoscion regalis*) in Delaware Bay. Estuaries 23:10–20

Pauly D, Christensen V, Walters C (2000) Ecopath, ecosim, and ecospace as tools for evaluating ecosystem impact of fisheries. ICES J Mar Sci 57:697–706

Peters DS, Schaaf WE (1991) Empirical models of the trophic basis for fishery yield in coastal water of the eastern USA. Trans Am Fish Soc 120:459–473

Peterson BJ, Fry B, Hullar M et al (1994) The distribution and stable carbon isotopic composition of dissolved organic carbon in estuaries. Estuaries 17:111–121

Polis GA, Strong DR (1996) Food web complexity and community dynamics. Am Nat 147:813–846

Polis GA, Holt RD, Menge RA et al (1995) Time, space, and life history: influences on food webs. In: Polis GA, Winemiller K (eds) Food webs: integration of patterns and dynamics. Chapman & Hall, New York

Polis GA, Anderson WB, Holt RD (1997) Toward an integration of landscape and food web ecology: the dynamics of spatially subsidized food webs. Annu Rev Ecol Syst 28:289–316

Post JR, Parkinson EA (2001) Energy allocation strategy in young fish: allometry and survival. Ecology 82:1040–1051

Ross DA, Love RM (1979) Decrease in the cold-store flavor developed by frozen fillets of starved cod (*Gadus morhua* L.). J Food Technol 14:115–122

Rozas LP, McIvor CC, Odum WE (1988) Intertidal rivulets and creek banks: corridors between tidal creeks and marshes. Mar Ecol Prog Ser 47:303–307

Rubino GD (1991) Chronicling geologic processes on a tidal marsh from aerial photography. Master's thesis, University of Delaware, Newark

Ruiz-Jaen MC, Aide TM (2005) Restoration success: how is it being measured? Rest Ecol 13:569–577

Saltonstall K (2002) Cryptic invasion by a non-native genotype of the common reed, *Phragmites australis,* into North America. Proc Natl Acad Sci 99:2445–2449

SER (Society of Ecological Restoration, International Science & Policy Working Group) (2004) The SER international primer on ecological restoration. Society for Ecological Restoration International, Tucson, AZ. http//www.ser.org

Sharp JH (2010) Estuarine oxygen dynamics: what can we learn about hypoxia from long-time records in the Delaware Estuary? Limnol Oceanogr 55:535–548

Shulman GE, Love RM (1999) The biochemical ecology of marine fishes. In: Southward AJ, Tyler PA, Young CM (eds) Advances in marine biology, vol 36. Academic, London

Simenstad CA, Cordell JR (2000) Ecological assessment criteria for restoring anadromous salmonid habitat in Pacific Northwest estuaries. Ecol Eng 15:283–302

Simenstad CA, Hood WG, Thom RM et al (2000) Landscape structure and scale constraints on restoring estuarine wetlands for Pacific coast juvenile fishes. In: Weinstein MP, Kreeger DA (eds) Concepts and controversies in tidal marsh ecology. Kluwer Academic, Dordrecht

Simenstad CA, Reed DJ, Ford M (2006) When is restoration not? Incorporating landscape-scale process to restore self-sustaining ecosystems in coastal wetland restoration. Ecol Eng 26:27–39

Sogard SM (1997) Size selective mortality in the juvenile stage of teleost fishes: a review. Bull Mar Sci 60:1129–1157

Stanturf JA, Schenholtz SH, Schweitzer CJ et al (2001) Achieving restoration success: myths in bottomland hardwood forests. Rest Ecol 9:189–200

Stedman S, Dahl TE (2008) Status and trends of wetlands in the coastal watersheds of the Eastern United States 1998 to 2004. National Oceanic and Atmospheric Administration, National Marine Fisheries Service and U.S. Department of the Interior, Fish and Wildlife Service, Washington, DC

Sullivan MJ, Moncreiff CA (1990) Edaphic algae are an important component of salt marsh foodwebs; evidence from multiple stable isotope analyses. Mar Ecol Prog Ser 62:149–159

Taylor ET (1987) Food habits of dominate piscivorous fishes in Delaware Bay, with special reference to predation on juvenile weakfish. Master's thesis, University of Delaware, Newark, DE
Teal JM (1962) Energy flow in the salt marsh ecosystem of Georgia. Ecology 43:614–624
Teal JM, Weinstein MP (2002) Ecological engineering, design, and construction considerations for marsh restorations in Delaware Bay, USA. Ecol Eng 18:607–618
Teo SLH, Able KW (2003) Habitat use and movement of the mummichog (*Fundulus heteroclitus*) in a restored salt marsh. Estuaries 26:720–730
Thayer GW, McTigue TA, Bellmer RJ et al (2003) Science-based restoration monitoring of coastal habitats, Volume One: a framework for monitoring plans under the Estuaries and Clean Water Act of 2000 (Public Law 160-457). NOAA Coastal Ocean, Program decision analysis series no. 23, vol 1. NOAA National Centers for Coastal Ocean Science, Silver Spring, MD
Turner RE (1977) Intertidal vegetation and commercial yields of penaeid shrimp. Trans Am Fish Soc 106:411–416
Turner RE, Woo SW, Jitts HR (1979) Estuarine influences on a continental shelf plankton community. Science 206:218–220
Valiela I, Wright JE, Teal JM et al (1977) Growth, production and energy transformations in the salt marsh killifish, *Fundulus heteroclitus*. Mar Biol 40:135–144
Vivian-Smith G (1997) Microtopographic heterogeneity and floristic diversity in experimental wetland communities. J Ecol 85:71–82
Wainright SA, Weinstein MP, Able KW et al (2000) Relative importance of benthic microalgae, phytoplankton and detritus of smooth cordgrass (*Spartina*) and the common reed (*Phragmites*) to brackish marsh food webs. Mar Ecol Prog Ser 200:77–91
Walters CJ, Juanes F (1993) Recruitment limitation as a consequence of natural selection for use of restricted feeding habitats and predation risk taking by juvenile fishes. Can J Fish Aquat Sci 50:2058–2070
Walters C, Christensen V, Pauly D (1997) Structuring dynamic models of exploited ecosystems from trophic mass-balance assessments. Rev Fish Biol Fish 7:139–172
Weigert RG, Pomeroy LR (1981) The salt-marsh ecosystem: a synthesis. In: Pomeroy LR, Weigert RG (eds) The ecology of a salt marsh. Springer, New York
Weinstein MP (1981) Plankton productivity and the distribution of fishes on the southeastern U.S. continental shelf. Science 214:351–352
Weinstein MP, Balletto JH (1999) Does the common reed, *Phragmites australis* reduce essential habitat for fishes? Estuaries 22(3B):793–802
Weinstein MP, Reed DJ (2005) Sustainable coastal development: the dual mandate and a recommendation for "commerce managed areas". Rest Ecol 13:174–182
Weinstein MP, Weiss SL, Walters MF (1980) Multiple determinants of community structure in shallow marsh habitats, Cape Fear River estuary, North Carolina, USA. Mar Biol 58:227–243
Weinstein MP, Balletto JH, Teal JM et al (1997) Success criteria and adaptive management for a large-scale wetland restoration project. Wetlands Ecol Manage 4:111–127
Weinstein MP, Philipp KR, Goodwin P (2000) Catastrophes, near-catastrophes and the bounds of expectation: wetland restoration on a macroscale. In: Weinstein MP, Kreeger DA (eds) Concepts and controversies in tidal marsh ecology. Kluwer Academic, Dordrecht
Weinstein MP, Litvin SY, Guida VG (2005) Consideration of habitat linkages, estuarine landscapes and the trophic spectrum in wetland restoration design. J Coast Res 40:51–63
Weinstein MP, Litvin SY, Guida VG (2009a) Is global climate change influencing the overwintering distribution of weakfish (*Cynoscion regalis*, Bloch & Schneider, 1801)? J Fish Biol 75: 693–698
Weinstein MP, Litvin SY, Guida VG (2009b) Essential fish habitat and wetland restoration success: a Tier III approach to the biochemical condition of the common mummichog, *Fundulus heteroclitus* in common reed, *Phragmites australis* and smooth cordgrass, *Spartina alterniflora* dominated salt marshes. Estuar Coast 32:1011–1022
Weinstein MP, Litvin SY, Guida VG (2010) Stable isotope and biochemical composition of white perch in a *Phragmites* dominated salt marsh and adjacent waters. Wetlands 30:1181–1191

Winemiller KO, Akin S, Zeug SC (2007) Production sources and food web structure of a temperate tidal estuary: integration of dietary and stable isotope data. Mar Ecol Prog Ser 343:63–76
White PS, Walker JL (1997) Approximating nature's variation: selecting and using reference sites and reference information in restoration ecology. Restor Ecol 5:338–349
Yuneva TV, Shulman GE, Shcepkina AM (1991) The dynamics of lipid characteristics in horse-mackerel during swimming. Zhur Evol Biochim Physiol 27:730–735
Zedler JB (2007) Success: an unclear, subjective descriptor of restoration outcomes. Restor Ecol 25:162–168
Zimmerman RJ, Minello TM, Rozas LP (2000) Salt marsh linkages to productivity of penaeid shrimps and blue crabs in the northern Gulf of Mexico. In: Weinstein MP, Kreeger DA (eds) Concepts and controversies in tidal marsh ecology. Kluwer Academic, Dordrecht

Changing Nature: Novel Ecosystems, Intervention, and Knowing When to Step Back

Eric Higgs

Abstract Climate change, species invasions, and changes in social practices and cultural beliefs about nature are creating new ecosystems, some of which have no apparent roots in the past. The emergence of hybrid (familiar ecosystems with new combinations) and novel (unfamiliar) ecosystems challenges conventional ecological restoration practices, which places reliance on robust notions of historical fidelity. There is an extent to which the science and practice of restoration can be adapted to cope with significant change and discontinuities, but beyond a certain point, yet unknowable, it may be necessary to look ahead to emerging practices that blend the important qualities of restoration with wild or regenerative design.

Keywords Ecological restoration • Novel and hybrid ecosystems • Historical fidelity • Wild design • Ecological intervention

Intervening in Ecosystems

How do we intervene respectfully in ecosystems under conditions of rapid change? This is emerging as a central question in sustainability science. The pace of environmental (e.g., climate), ecological (e.g., species invasions) and cultural (e.g., treating nature and natural processes as providing services that can be optimized) change is accelerating, and there is the prospect that management based on conventional historical references, notably ecological restoration, will also need to evolve. The challenge is knowing when change has exceeded ecological

E. Higgs (✉)
School of Environmental Studies, University of Victoria, Social Sciences and Math Building B243, 3800 Finnerty Road (Ring Road), Victoria, BC V8P 5C2, Canada
e-mail: ehiggs@uvic.ca

M.P. Weinstein and R.E. Turner (eds.), *Sustainability Science: The Emerging Paradigm and the Urban Environment*, DOI 10.1007/978-1-4614-3188-6_18,

thresholds that frustrate conventional approaches, decision points for which at present we have no solid basis for action. At the same time, it will be important to know when interventions are critical for the well-being of ecosystems, and when they are merely meddling. In the words of Kenny Rogers' immortal lyrics for The Gambler: "You got to know when to hold 'em, know when to fold 'em, know when to walk away and know when to run." The task ahead is not in any sense straightforward.

In this chapter, I provide background on recent work in ecological restoration theory and practice with the aim of (1) emphasizing the importance of historical knowledge, and (2) describing the challenges that rapid environmental, ecological and cultural changes bring to classical restoration approaches and theories. Over the last decade I have focused on advancing ecological restoration, and in particular in exploring how our impulse to restore is both an artifact of, and antidote to, life in an advanced technological society. Ecological restoration is powerful because it adds value to ecosystems through community and personal engagement: people are enlivened through their efforts at restoring ecosystems. All of this made perfect sense in a world that had not yet grasped the implications of rapid environmental and ecological change, not to mention fast changing perspectives on technology, equity, and financial systems. In combination, these ecological, environmental and cultural drivers will push some ecosystems outside the long-term range of historical variation, and thereby pull out a primary basis upon which ecological restoration has depended. For other ecosystems, there is likely to be hybridization between nonanalogue or novel ecosystems and ones that still rest in historical precedent. Thus, it is a time for assessing how a world of novel ecosystems will challenge the science and practice of ecological restoration, and what new approaches are needed. In this respect I argue for a virtues approach—based on virtue ethics, in which the character of the moral agent(s) rather than the outcome of an action is the driving force behind moral behavior—that will point us toward how to think about respectful intervention in ecosystems, and also to considering what might lie beyond a traditional notion of restoration.

The prospect of a world of novel ecosystems disturbs traditional views on restoration, and threatens to unsettle how people and natural processes interrelate (Hobbs et al. 2009). At bottom is the threat that, despite generosity by restoration practitioners and reciprocity between restorationists and ecosystems, restoration will give way to interventions that magnify human interests at the expense of ecosystems: in the future will ecosystem services not just be restored, but amplified?; will *smart* ecosystems deliver on particular functional and aesthetic goals even more than they do presently? (Higgs 2003). Even the very term, *novel ecosystem*, suggests innovation and opportunity. It is difficult to know the extent to which in our lifetimes or our children's lifetimes there will be a need to address these nonanalogue or novel ecosystems. Acting now and understanding the potential consequences of change is the mark of precaution and diligence, and is at the heart of good sustainability science.

Ecological Restoration and History

Ecological restoration, "the practice of assisting the recovery of ecosystems that have been damaged, degraded, or destroyed" (Society for Ecological Restoration 2004), has emerged over the last 20 years as a defining environmental management practice. The confluence of better ecological science and understanding, and a pressing realization of the magnitude of degradation and the limits of preservation of remnant intact ecosystems, has pushed restoration forward (National Parks Directorate 2007). The scale of projects range enormously. In Victoria, British Columbia, where I live, there are dozens of small projects focused on recovery of nationally significant threatened ecosystems defined by the charismatic Garry oak (*Quercus garryana*); a combination of backyard homeowner initiatives, volunteer invasive species removal, nongovernment projects, and government sponsored networks and initiatives. The work is hands-on, and people from all walks of life are involved. There is strong technical and scientific support, but such expertise is not predominant (Fuchs 2001). At a much larger scale, the restoration of the Florida Everglades involves teams of specialists organized around a large vision, dozens of individual projects, and funding at a level not often associated with community-based projects (Sklar et al. 2005).

The bases of restoration have been formulated in many ways (Hobbs and Norton 1996; Falk et al. 2006). In almost all cases, successful restoration depends on two keystone concepts (Higgs 2003): ecological integrity and historical fidelity. There are many ways of describing the integrity (or health) of an ecosystem, but the essence is that we can determine either directly or through proxy data what a well-functioning ecosystem should be (White and Walker 1997; Egan and Howell 2001). Restoration, in the strictest sense, depends on historical knowledge to ascertain ecological integrity. The premise is that ecosystems functioned better at some time prior to a defined or ongoing disturbance. The goal is, therefore, to return the ecosystem to conditions that existed in the past. Of course, this is a challenging concept in practice. Historical referents may be difficult or nearly impossible to find. Site or surrounding conditions may have changed significantly such that reflecting historical conditions may no longer make sense. Or, it may make more sense to reconfigure historical conditions, but then allow the ecosystem to shift within a defined range of historic variation (Duarte et al. 2009). These, and other options, fuse the need for ecological integrity and historical fidelity.

There are two primary problems with history, or more precisely, *historicity* (the quality of being historical), for ecological restoration. First, ascertaining historical references is painstaking work. Second, environmental and ecological change is pushing ecosystems outside of historical ranges. The Mountain Legacy Project, an interdisciplinary applied research project focused on using repeat photography based on a vast collection of systematic historical topographic survey photographs from Western Canadian mountains to chart historical conditions and how ecosystems have changed, exemplifies the challenge of using historical knowledge (MacLaren 2006; Higgs 2008). As part of the Project, Will Roush embarked on a study of rising treeline ecotone in Kootenay National Park. Using historical and repeat photographs he ascertained an apparently dramatic elevation in treeline and change in the density of upper subalpine forests. At the landscape level, and using a latitudinal gradient of photographs, he showed the Kootenay ecosystem changes were consistent with other regions in the Canadian Rockies. Intensive fieldwork focused on disentangling components-level drivers for the change (e.g., microtopographic conditions; aspect) from constraints-level drivers (e.g., climate). His study showed conclusively that the dramatic change in forest composition is driven by climate, a finding that demonstrates potential diminishment of alpine ecosystems and the difficulty of setting recovery targets given the rates of change (Roush 2009; Higgs and Roush 2011).

Historical knowledge exacts specific obligations by compelling detailed research about past conditions that can be used to set goals for ecological restoration. With changing climate (IPCC 2007), biogeoclimatic envelopes shift beyond particular limits, species, assemblies and entire ecosystems will shift (Harris et al. 2006). New niches will open, but some historical niches will fall away. Historical knowledge will come into play by providing intelligence on long-term ecological characteristics and trajectories, but it will not serve in the same way as a specific guide for intervention. Indeed, history may play a strong role in the future determination of ecological interventions because reliable information about continuity becomes more, not less, critical (Higgs 2012).

Novelty

History has served as the anchor in our understanding of how to respectfully intervene in ecosystems. Peering into the past ecologists could discern the condition of predisturbance ecosystems and study the sources and effects of degradation. While the details of the restoration goals were complicated by historical and social contingency, the overall mission remained the same. Now, the development of ecosystems that differ in pattern and function from those in the past and present is increasingly understood as a consequence of anthropogenic climate change, globalization of species distributions, and changing human activities. Such ecosystems are variously described as *novel*, *no-analogue*, or *emerging* (Hobbs et al. 2009).[1]

A novel ecosystem assembles in response to environmental (climate), ecological (species invasions) and cultural (new attitudes toward nature) drivers, and produce a distinctive and previously undescribed metastable assembly. All ecosystems are novel in the trivial sense that there are myriad species assemblies and that these change constantly. It is also the case that highly contrived ecosystems, urban parks for instance, are in significant respects novel when they bring together an admixture of species that have not existed before. An ecosystem is distinctive in these respects,

[1] I prefer the term, "novel," since the implication is that not only are ecosystem components and processes new but the challenges for ecological restoration are also new.

but this is not the essence of a novel ecosystem. Contrived and heavily managed ecosystems such as parks and gardens, while distinctive, are not novel in the sense of an emerging and previously undocumented assemblage (Hobbs et al. 2009).

Relatively stable formations give us the ecosystems that are familiar to human experience and human time scales. For instance, local Garry oak woodlands arose in their present configuration several thousand years after the last withdrawal of glacial ice from southern Vancouver Island. The savannah-like ecosystems that are familiar in the present arose in response to indigenous land management practices: tilling, selective harvesting of nutritional plant materials, and prescribed fire (MacDougall et al. 2004). Thus, what greets the modern eye in greater Victoria (gnarly oaks with an open understory of shrubs, forbs, and grasses) has adapted over several thousand years to a wide variety of temperature and precipitation changes. Indeed, change is a normal quality of ecosystems that are subject to direct and indirect disturbance. What is distinctive about the present era is the *rate and extent* of disturbance by climate-induced effects (e.g., intensifying storm events, elevated temperatures, changing temperature and moisture patterns) and globalized movement of species that reshape their new host ecosystems (e.g., Tamarisk in the United States; Kerns et al. 2009). One can simply turn back the clock to the pre-Holocene or earlier and find environmental conditions that describe an analog for today's shifting climate. However, this stretches beyond the traditional Holocene reference period that guides much of ecological restoration, and misses the significance of novel ecosystems in the present Anthropocene Epoch (Crutzen 2002). The pace of change along with the erosion of historical biogeographic barriers to species movement and intensifying human activity collide to create ecosystems that are either entirely novel or are relatively stable hybrid formations of historical and novel patterns and functions.

Sitting at the cusp of major changes in climate and intensifying species invasions, not to mention massive change in the way we think about ourselves and technology and act on these beliefs (Mau and Institute Without Boundaries 2004; Homer-Dixon 2006), it is difficult to comprehend how ecosystems will unfold in the future, and how policies and practices will adapt to changing conditions. For example, Stone et al. (2008) reported on apparently novel interactions between the Turkey oak (*Quercus cerris*), native and invasive gall wasps, and two songbird species. The oak was native to the United Kingdom prior to the last glaciation, but has not been present until it was planted ornamentally in the last 300 years; it did not disperse naturally to its former range. The oak has become a reservoir food source for blue and great tits (*Cyanistes caeruleus* and *Parus major*), which are laying eggs earlier in the season in response to climate signals. The oak also acts as a host for gall wasps (*Hymenoptera*: *Cynipidae*, *Cynipini*, many of which are also nonnative but migrating northwards), which provide critical food for the tit population prior to the emergence of traditional food sources. The native and nonnative gall wasps are interacting in novel ways. There are significant conservation and restoration questions, and certainly it is not resolved whether the interactions are positive (e.g., securing the tit population) or negative (e.g., supporting increased invasive insect populations). At what point does a species become native? Is evidence of analogous multitrophic interactions detected in the fossil record sufficient to consider these novel interactions historically rooted? Is intervention necessary, and if so, of what kind?

Interactions of this kind will become more common as the lag in ecological response catches up to environmental change, and more study is done to increase awareness of novel ecosystems. The case of the Turkey oak is illustrative of the complexity of these issues, and points to an ambivalence at the heart of emerging ecosystems. As conditions change significantly, historical ecosystems shift eventually to novel ecosystems. Along the way, there are conditions that blend novelty and historicity, and are hybrid. As with most continuous change, the extreme values are relatively easier to understand; what lies in the middle will cause greater difficulty in knowing how to respond. At the boundary layer of hybrid and novel ecosystems is a threshold that limits the degree of practical ecological restoration. Beyond this point, and the threshold will vary from ecosystem to ecosystem, the conditions are such that restoration is all but impossible (Hobbs et al. 2009). Recent examples of salinized former agricultural lands that cannot be practically restored to historic Jarrah forests in Western Australia point to the difficulties that lie ahead for classical restoration (Yates et al. 2000). Extremely rare ecosystems, or those with very high cultural value, may be restored using heroic means (i.e., means beyond which conventional norms would suggest is appropriate effort or expenditure of resources), but for most practical purposes the restoration to historic conditions would become more difficult.

Inevitably our cultural values about nature, conservation, and restoration will change alongside shifting ecosystems. The guidance of history will shift from providing ecosystems of reference to offering insights about disturbance events and degradation patterns (Higgs 2012). The focus is shifting also to understanding ecosystems as providers of services that have direct and indirect human benefits as well as intrinsic value (Aronson et al. 2007). Ecological intervention, and especially restoration, will increasingly be configured as an economic and development imperative. This is apparent already, for example, in the development of ecological services as an organizing framework for understanding the value of ecosystems (Millennium Ecosystems Assessment 2005). A greater emphasis on services, for example in the revised regulation of the Reducing Emissions from Deforestation and Forest Degradation (REDD+), demonstrates a focus on processes over composition in the conservation and restoration of ecosystems aimed at carbon sequestration (Chazdon 2008; UN-REDD Programme 2009; Alexander et al. 2011). Hobbs et al. (2009) propose that novel ecosystems require different consideration for conservation and restoration: Is the system capable of maturing along a stable trajectory? Is the system resistant and resilient? Is the system thermodynamically efficient? Is the system providing ecosystem goods and services? Is it providing opportunities for individual or community engagement?

The social, moral, and ecological challenges thrown up by the arrival of hybrid and novel ecosystems are difficult to overstate. We have a limited track record of understanding present and historical landscape (including human systems) and interpreting species dynamics (Liu et al. 2007). What of the challenges in sorting out rapidly changing systems? The present values attached to nature are rooted primarily in a place-based and species-based management. A transition to process- and services-based models of nature will rub uncomfortably against deeply-embedded models of nature (Cole and Yung 2010). If accepted climate predictions by 2100 occupy even the lower range, there will be places where it is unlikely to prove practical to hold on to certain species and in some cases to entire assemblies and ecosystems (Hamann and Wang 2006; Wiens et al. 2009). A significant challenge is knowing how to ascertain when an ecosystem has shifted away from historical antecedents and is functioning as a hybrid or novel ecosystem. The question of when to intervene will become more difficult, not only because present predictive capabilities will be tested but it is also likely more surprises will emerge. It may also be the case, although not strictly necessary, that hybrid and novel ecosystems will be difficult to restore as a consequence of new environmental conditions (e.g., salinized soils) and species (e.g., trenchant invaders).

Accepting and Resisting Change

The challenges for classical restoration in the face of largely unknown climate impacts and inexorable and intractable species invasions are enormous. However, Erika Zaveleta, an ecologist at the University of California, Santa Cruz, commented in a 2009 presentation on resilience and ecosystems, "the fact of change need not mean that we accept any change." Understanding human agency in both creating and *resolving* problems is vital. The long lag effects of elevated CO_2 in the atmosphere make changing climate both inevitable and long in duration. Actions globally in reducing carbon emissions will help limit long-term consequences, but there is no scenario that will easily return the background condition to any recent baseline (IPCC 2007). To make matters more complicated, the response of ecosystems will vary at all scales working downwards from dynamic global weather patterns and upwards from site- and micro-site characteristics. We need global and regional models of climate impacts, but spatial variation will make it difficult to predict with precision exactly what will result. On east–west ridges in the Rocky Mountains of Canada, for example, vegetation response on southerly slopes is almost always different than what is found sometimes just a few meters away on the northerly aspect (Higgs and Roush 2011). Such complications are generally more than can easily be assimilated with existing models.

The classical view of ecological restoration with its dependence on historical continuity will be adapted to meet new challenges. The impetus to intervene in ecosystems to meet defined ecological and cultural goals will remain strong. There are circumstances where intervention is desirable, either because the ecological condition supports restoration (or a modified version of it), or social and cultural values determine that greater investment is worthwhile to achieve desirable objectives. In Victoria, British Columbia, for example, present climate models suggest by 2050 warmer (3–4°C in winter, 2–3°C in summer) and slightly wetter conditions (20–30% in winter; unchanged or very little in summer), which will likely continue to favor the Garry oak dominated ecosystems (Pacific Climate Impacts Consortium 2011). There may well be local conditions that prove too much for the oak, and so a sensible adaptive approach may be to propagate Garry oaks in locations that favor appropriate new conditions and shift provenance to sources from further south along a climatic gradient. Thus, classical restoration will persist as long as climate impacts remain locally modest and cultural support for legacy ecosystems remains strong.

Intervention of any kind imposes responsibilities best met by a framework that embeds ethical practices. Such obligations have been met in ecological restoration by appeal to ecological integrity, historical fidelity, and other important factors such as public engagement (Society for Ecological Restoration 2004; National Parks Directorate 2007). With novel ecosystems, the moral landmarks are more difficult to ascertain, especially under conditions of rapid and disorientating change. The problem is complicated by the fact that ecological changes are increasingly viewed through a window of economic and social benefit. Poverty alleviation and development goals are vital not only in improving human welfare but also in providing supportive conditions for ecosystem protection, conservation, and restoration (Millennium Assessment 2005; Sachs 2005).

An ethical framework that is gaining traction focuses on the virtues required to achieve a world in which ecosystems flourish. Thompson and Bendik-Keymer (2012) argue that we need virtues that begin with ecological restoration and at the same time account for the changes that are anticipated with environmental change. Throop (2010) for example, proposes the virtues of humility, sensitivity, self-restraint, and respect for the other; as intervenors in ecosystems we seek to uphold our awareness that ecosystems are always more complex than we can imagine, they require careful adaptive interventions, they benefit from always asking what is the least not the most intervention, and they demand a profound respect as something other than ourselves. It is not clear as yet whether novel ecosystems will require novel virtues or will simply reinterpret existing virtues of the kind Throop has proposed (Thompson and Bendik-Keymer, 2012). Perhaps historicity (defined earlier in this chapter) will emerge as a new virtue that compels awareness of historical continuities and discontinuities in shaping ecological intervention. There are few anchors in a rapidly changing world, but the very act of seeking historical knowledge provides an important prudential consideration and limits on our ambitions.

Wild Design

How best to manifest existing and new virtues in hybrid and novel ecosystems? One approach is a variant of design practice; *wild design* (Higgs and Hobbs 2010). Virtually every form of restoration is an explicit intervention with practical and ethical consequences. Developing basic approaches to restoration practice (SER 2004) and detailed principles and guidelines (National Parks Directorate 2007) provide the outlines of good practice. However, a more explicit overarching model is needed to situate effective, efficient and engaging restoration practice in a rapidly changing world. Design, with its deliberative underpinnings and openness to creativity, provides a starting point (Borgmann 1995; Higgs 2003).

A signal problem with design as a metaphor for restoration is that it gives too much attention to human interests (see below; section "Problems with Design and Intervention"). Design arose in the twentieth century in response to needs for better integrated and higher functioning products, placements, and systems (Buchanan 1992). Extending to landscape design, it became a powerful way of encoding both aesthetic and practical intentions (Higgs 2003). However, these intentions are primarily human; can they be otherwise? The essence of wild design extends the evolution of design to whole systems, and acknowledges that design of ecosystems is always about both human and natural processes. Thus, a practitioner of wild design is mindful of ecosystem form and function, and at the same time recognizes hubris that accompanies too much confidence about deliberate ecosystem intervention (Borgmann 1995).

Wild design fits the practice of ecological restoration well through immediate acknowledgement of the interplay of human and ecological processes. This reflects the inclination of restorationists to "listen" well to the ecosystem, and to impose human intentions to the extent of regenerating ecological integrity in relation to historical qualities. Wild design is situated at the center of learning and intervention. We build understanding about places, and we intervene based on this knowledge; wild design provides a formal conduit for this shift from theory to practice. There are seven underlying principles: clarity, fidelity, resilience, restraint, respect, responsibility, and engagement (Higgs and Hobbs 2010). The process of wild design moves from ecological understanding, in which the problem is informed by ecological and cultural knowledge, to an understanding of how the general wild design principles are to be interpreted for a specific place (local characteristics, implementation, engagement). There is feedback among the three levels (ecological understanding, wild design principles, place-specific wild design) to ensure continuous assessment of evolving values, principles, knowledge, and action.

The seven principles of wild design can also be considered as virtues of ecological restoration (Table 1). Indeed, they can also be extended to address ecological intervention in novel ecosystems. The shift in importance of historical qualities in the determination of intervention goals places greater weight on both the individual principles/virtues, and the very idea of a virtues approach to ecological intervention. A virtues approach to ecological restoration and intervention has a number of advantages. First, virtues hold particularly human practices to knowable standards that operate across ecosystems and cultures without denying the importance of local conditions and viewpoints. Second, the exercise of a virtue is always in response to specific conditions and constraints. For example, the virtue of *fidelity* demands that historical qualities be acknowledged and understood in a restoration or intervention project. In heavily impacted systems subject to rapid change, precise historical references may be figurative rather than literal. In a heavily urbanized contaminated site subject to recovery efforts, the prescription for ecological intervention will borrow heavily from historical site qualities for the design even if historical species assemblies are either unrealistic or very difficult (e.g., a combination of altered soils and temperature change). The act of understanding history gives us a deeper appreciation of the challenges of intervention, and tends to expand our potential for

Table 1 Adapted from "wild design" (Higgs and Hobbs 2010)

Principle/virtue	Guideline	Wild design questions
Clarity	Clear goals and objectives to ensure the transparency of values, and allow for careful negotiation of differences	Have goals and objectives been established? Have value claims been made explicit? Are goals and objectives consistent with the interests of those concerned with the intervention?
Fidelity	Entails careful historical research to understand past conditions of the system, and to assess these past conditions against present functions, structures, and patterns	What is known historically about the ecosystem? Have all sources of knowledge been explored? What signals in the contemporary ecosystems can be inferred from historical information?
Resilience	Ensure autogenic functioning is restored to an ecosystem and that an ecosystem has appropriate resources to cope with external perturbations	What are the functional requirements of the ecosystem? At what point can the ecosystem be autogenic? How much continued intervention or management will be required?
Restraint	Less intervention is better than more	Are the means in place to assess the impact of an intervention? Is it well understood where the line is drawn between too little and too much intervention? Is precaution central to intervention strategies?
Respect	Interventions are always proxies for what we believe appropriate to a particular ecosystem	Are scientists, managers, and concerned members of the public aware that interventions are simply the best present-day approximation of what is best for the ecosystem, and that the values underlying these approximations may shift?
Responsibility	Responsibility includes wide knowledge of techniques and projects, operating according to high ethical standards, and striving to allow ecosystems to flourish	Are intervention practitioners properly trained? Is there a code of practice in place that guides professional conduct? Is it clear that the qualities of the ecosystem come before human interests (although there is room for human values, too)?
Engagement	Strong reciprocal ties are needed between people and ecosystems to ensure successful and durable interventions	Have people been brought into the process of designing the intervention early? Is the role of the concerned public substantial? Who is accountable for the outcome of the intervention? Is community support strong and growing?

responsibility, respect and engagement. A virtues approach, including the grouping of traditional virtues along with the development of new ones, ensures our moral understanding keeps pace with our intervention in ecosystems.

Problems with Novelty, Design, and Intervention

The apparent diminishment of historical knowledge at the center of ecological restoration under novel future conditions gives rise to the science of intervention ecology for addressing novel ecosystems (Hobbs et al. 2011), and further underscores the value of a flexible, adaptable and open process of wild design (Higgs and Hobbs 2010). These three notions—novelty, design, and intervention—should cause us unease, and there are several reasons for this.

Novelty. For some, the idea of novelty may imply a positive value, and especially in societies given over to valuing novelty, constant change, innovation, and planned obsolescence (Frank 1997). The sense that a novel ecosystem is somehow better because it is new is an anathema to arguments put forward by Hobbs et al.: "Decisions about how much conservation and restoration investment is appropriate will depend on shifting cultural values about historic fidelity and ecological integrity, sentimentality about ecosystems of the past, local species diversity, priorities for livelihood and sustainability (i.e. historically faithful restorations vs. ecosystem services-oriented projects), and designs for resilience" (2009, p. 604). The term *novel ecosystems* is invoked to provoke awareness of an impending cluster of ecosystem management challenges, not to promote novelty as a positive value.

Underneath a concern about novelty is pervasive pattern characteristic of advanced technological cultures: the conversion of things that matter into increasingly efficient and effective commodities, including nature and ecosystems (Borgmann 1984; Strong 1995; Higgs 2003). Of greater concern is the tendency to separate commodities from their attendant machinery (e.g., the music we hear and the equipment used to produce it; ecosystem services and the complex processes that produce them) and thereby increasing the distance between actions and their consequences. This separation attenuates the moral resolve to do what one might otherwise do outside of the influence of such devices, and at the same time makes it more difficult to focus on things that matter (Borgmann 1992). This view also amplifies Langdon Winner's earlier observation of a tendency in technological society to invert means and ends (e.g., allowing a new computer to reshape one's workflow) and thereby producing a condition of "reverse adaptation" (1977).

There is a greater likelihood that once stripped of considerations such as historicity, and fueled by innovations such as ecosystem services, novel ecosystems will shift increasingly from focal subject to commodity. The revised intent will be on the use of ecosystems to meet particular ends, whether these are ecological (in the sense of integrity and history), service delivery, or cultural. There may be greater flexibility in how we see appropriate end states for ecosystems of the future. Attendant with these changes is a likelihood that the greater extent that novel ecosystems form part of an intervention toolkit and become lodged in people's minds, the more that appropriate interventions will support novelty. This is consistent with the adoption of many new technological innovations. Thinking of technology as a pattern—commodification, comfort with rapid change, and the inversion of means and ends—can be extended from the paradigmatic instances of technology to less traditional milieux such as ecosystem intervention. Thus, the development of a refined view of ecological intervention much be understood against the backdrop of technological culture.

Design. Wild design is intended at least in part to address these concerns. Based on seven principles/virtues (Table 1), the explicit recognition of human responsibility for human problems allows for a thoughtful approach to ecosystems under stress and rapid change. Ecological integrity and historical fidelity still matter (the latter in different ways), but they are joined with other considerations that invoke a relational understanding of ecosystems. The deliberate recognition of restoration as a design process invokes responsibility for intervention, and establishes the possibility of a growing professional commitment to responsible design. However, there are aspects of design in general that cause concern. Not the least is a recognition that design as a professional activity grew in lockstep with the evolution of a technological society (Buchanan 1992). Design, therefore, may be an artifact of technological society and its existence predicates and furthers technological patterns of belief and action. This viewpoint is held by Throop and Purdom, who argue for the metaphor of healing instead of design for ecological restoration (Throop and Purdom 2006).

Design alone, with its technological connotations, falls short of what is needed (Higgs 2006). However, *wild* design acknowledges human responsibility and calls upon ingenuity, creativity, and the seven virtues described earlier in the chapter to rise to the challenge of novel ecosystems. Healing works as a metaphor, but begs the question of what is being healed under conditions of rapid change. There is also the risk of ignoring human agency, which serves to submerge the importance of deliberate action in some cases, and may also tend to naturalize ecosystems that are significantly co-evolved with human practices (e.g., in the Global South, where the restoration of ecological approaches to agriculture is more consistent with long-term ecological patterns).

Intervention. The principle of *restraint* becomes a key virtue of the future. Knowing when to leave well enough alone makes it possible to intervene modestly. Of course, understanding the divide between appropriate and inappropriate intervention is not either mechanical or straightforward. Cultural variations in belief about intervention, and refining these beliefs through broad conversations and social negotiation of shared values, will enlarge our understanding. Good intervention, like good ecological restoration, depends on being technically and scientifically proficient as well as being politically, economically, culturally, and morally aware. Restraint is perhaps the obverse of courage. Both virtues are needed in a world of novel ecosystems: courage to know when intervention is appropriate; restraint to know when to step back.

The fact of rapid environmental, ecological, and cultural change compels us to new ways of engaging or intervening in ecosystems. Recognizing that historically-centered ecosystems are giving way to hybrid and novel ones suggests that our practices and ethics must shift, too. The anticipation of change and the development of robust standards of practice is surely a better approach than a piecemeal post hoc response. Or, is it? Is it better to hold tight to traditional notions of ecological restoration in the hope of flexibly adapting these practices to meet oncoming challenges? Will admission of novel ecosystems lead to acceptance, and undermine our concerns about the source of the change?

These questions will only be answered accurately in time. My inclination is to retain the practice of ecological restoration, with its reliance on historical outlook, and to adapt flexibly as new challenges are introduced. The emerging science of intervention ecology will offer bold insights into how best to anticipate and respond to novel ecosystems. However, a problem with intervention is that it signals a moral neutrality as regards appropriate outcomes: it is not clear what of many potential aims are the proper ones for intervention beyond the mere act of intervening. This stands in distinction to restoration ecology and ecological restoration, both of which as science and practice function according to the specific moral impetus of getting a system back to what it resembled (however flexibly this is interpreted). Stripped of such obvious goals, intervention may fall prey to moral ambiguity. If intervention is alright, what it wrong with more intervention? And, why not intervene in ways that maximize ecological productivity and ultimately human goods and services? Making ecosystems in our own image is something that societies around the globe have done effectively for thousands of years. The emerging challenge is how *not* to do this, but instead to give priority to the wildness of natural processes and at the same time acknowledge the importance of respectful and restrained human activities, including restoration and intervention, alongside these ecosystems.

Regeneration Ecology

There may be another way out of this moral ambiguity by conceiving a new science and practice that invests the insights of ecological restoration but allows for greater latitude in dealing with rapid change. The science of *regeneration ecology* and the broader practice of *ecological regeneration* may be a more flexible alternative to restoration (Higgs 2003, p. 129). More than two decades ago, Jordan et al. wrote about the terminological confusion over restoration ecology, going so far as to suggest an alternative concept, *synthetic ecology* (Jordan et al. 1987). At that time, the momentum of restoration was sufficiently strong that it made little sense to invoke a new term. As the context in which restoration is practiced changes, and quickly, and especially now that alternative covering terms such as intervention ecology are proposed, *regeneration* deserves attention (Hindle 2006).

At the heart of regeneration is the idea of creating something again. *Regeneration ecology* can takes parts of its cue from regenerative biology, which uses techniques in cell and development biology to develop therapies for debilitating human conditions (this analogy is problematic in some respects because regenerative biology depends on an intensely technological approach). Regeneration *ecology* would focus new techniques and insights on ecosystems whether historical, hybrid, or novel. *Ecological regeneration* is already used as a term to describe an integrated practice of natural or ecological design. For example, the Natural Building Network, a US-based nongovernment organization, supports: "ecological regeneration, social justice, the building of community and economic opportunity, and the recognition of indigenous wisdom as essential in creating healthy, beautiful, and spiritually-uplifting habitation for everyone" (Natural Building Network 2011). Regenerative design is a process-oriented approach to systems design (Lyle 1994), and made popular by William McDonough and Michael Braungart in the form of *cradle-to-cradle* design (2002). Several design firms, such as Biohabitats and Regenesis, have picked regeneration as a central organizing principle.

Nature is changing faster than traditional approaches such as ecological restoration can cope, and thus begins a search for new ways of engaging with hybrid ecosystems and novel ecosystems. A synthesis of ecological restoration, with its attention to ecological integrity and historical fidelity, and wild or regenerative design, could result in an emerging science of intervention or regeneration ecology (and practice of ecological intervention or regeneration), which would join the pantheon of sustainability science. Working toward responsible ways of intervening in ecosystems undergoing rapid change is one of the most difficult challenges that lies ahead for ecologists and all those engaged with protecting, conserving, restoring, and managing ecosystems.

References

Alexander, SCR et al. (2011) Opportunities and challenges for ecological restoration within REDD+. Rest Ecol. 19:683–689

Aronson J, Milton S, Blignaut J (2007) Restoration natural capital: science, business, and practice. Island Press, Washington

Borgmann A (1984) Technology and the character of contemporary life. University of Chicago Press, Chicago

Borgmann A (1992) Crossing the postmodern divide. University of Chicago Press, Chicago

Borgmann A (1995) The depth of design. In: Buchanan R, Margolin V (eds) Discovering design: explorations in design studies. University of Chicago Press, Chicago

Buchanan R (1992) Wicked problems in design thinking. Design Issues 8:5–21

Chazdon R (2008) Beyond deforestation: restoring forests and ecosystem services on degraded lands. Science 320:1458–1460

Cole D, Yung L (eds) (2010) Beyond naturnalness: rethinking park and wilderness stewardship in an era of rapid change. Island Press, Washington

Crutzen P (2002) Geology of mankind. Nature 415:23

Duarte CM et al (2009) Return to Neverland: shifting baselines affect eutrophication restoration targets. Estuar Coast 32:29–36

Egan D, Howell E (2001) The historical ecology handbook: a restorationist's guide to reference ecosystems. Island Press, Washington

Falk D, Palmer M, Zedler J (2006) Foundations of restoration ecology. Island Press, Washington

Frank T (1997) The conquest of cool: business culture, counterculture, and the rise of hip consumerism. The University of Chicago Press, Chicago

Fuchs M (2001) Towards a recovery strategy for Garry oak and associated ecosystems in Canada: ecological assessment and literature review. Technical report GBEI/EC-00-030. Environment Canada, Canadian Wildlife Service, Pacific and Yukon Region

Hamann A, Wang T (2006) Potential effects of climate change on ecosystem and tree species distribution in British Columbia. Ecology 87:2773–2786

Harris J, Hobbs R, Higgs E et al (2006) Ecological restoration and global climate change. Rest Ecol 14:170–176

Higgs E (2003) Nature by design: people, natural process and ecological restoration. MIT Press, Cambridge

Higgs E (2006) Restoration goes wild: a reply to Throop and Purdom. Rest Ecol 14:500–503

Higgs E (2008) Twinning reality, or how taking history seriously changes how we understand ecological restoration in Jasper National Park. In: MacLaren I (ed) Culturing wilderness in Jasper National Park: studies in two centuries of human history in the upper Athabasca river watershed. University of Alberta Press, Edmonton, Canada

Higgs E (2012) History, novelty and virtue in ecological restoration. In: Thompson A, Bendik-Keymer J (eds) Ethical adaptation to climate change: human virtues of the future. MIT Press, Cambridge

Higgs E, Hobbs R (2010) Wild design. In: Cole D, Yung L (eds) Beyond naturnalness: rethinking park and wilderness stewardship in an era of rapid change. Island Press, Washington

Higgs E, Roush W (2011) Restoring remote ecosystems. Rest Ecol 19:553–558

Hindle T (2006) The regeneration of nature: an alternative to ecological restoration. PhD dissertation, Florida Atlantic University, Boca Raton

Hobbs R, Norton D (1996) Towards a conceptual framework for restoration ecology. Restor Ecol 4:93–110

Hobbs R, Higgs E, Harris J (2009) Novel ecosystems: implications for conservation and restoration. Trends Ecol Evol 24:599–605

Hobbs R, Hallett L, Ehrlich P, Mooney H (2011) Intervention ecology: applying ecological science in the 21st century. Bioscience 61:442–450

Homer-Dixon T (2006) The upside of down: catastrophe, creativity, and the renewal of civilization. Knopf, Toronto, Canada

IPCC (2007) Climate change 2007: synthesis report. Contribution of working groups I, II and III to the fourth assessment report of the intergovernmental panel on climate change. IPCC, Geneva, Switzerland

Jordan W III, Gilpin M, Aber J (1987) Restoration ecology: a synthetic approach to ecological research. Cambridge University Press, New York

Kerns B, Naylor B, Buonopane M et al (2009) Modeling Tamarisk (Tamarix spp.) habitat and climate change effects in the Northwestern United States. Invasive Plant Sci Manage 2:200–215

Liu J, Dietz T, Carpenter S et al (2007) Complexity of coupled human and natural systems. Science 317:1513–1516

Lyle J (1994) Regenerative design for sustainable development and design for human ecosystems. Wiley, New York

MacDougall A, Beckwith B, Maslovat C (2004) Defining conservation strategies with historical perspectives: a case study from a degraded oak grassland ecosystems. Conserv Biol 18:455–465

MacLaren I [with Higgs E, Zezulka-Mailloux G] (2006) Mapper of mountains: M.P. Bridgland in the Canadian rockies, 1902–1930. University of Alberta Press, Edmonton, Canada

Mau B, Institute Without Boundaries (2004) Massive change. Phaidon, London

McDonough W, Braungart M (2002) Cradle to cradle: remaking the way we make things. North Point, San Francisco

Millennium Ecosystem Assessment (2005) Ecosystems and human well-being: general synthesis. Island Press, Washington

National Parks Directorate (2007) Principles and guidelines for ecological restoration in Canada's protected natural areas. Parks Canada and the Canadian Parks Council, Ottawa

Natural Building Network (2011) http://nbnetwork.org/about. Accessed 30 July 2011

Pacific Climate Impacts Consortium (2011) BC coast climate impacts. http://pacificclimate.org/resources/climateimpacts/bccoast. Accessed 31 Jan 2011

Roush W (2009) A substantial upward shift of the alpine treeline ecotone in the southern Canadian Rocky Mountains. MA thesis, University of Victoria, Victoria

Sachs J (2005) The end of poverty: economic possibilities for our time. Penguin Books, New York

Sklar F, Chimney M, Newman S et al (2005) The ecological–societal underpinnings of Everglades restoration. Front Ecol Environ 3:161–169

Society for Ecological Restoration International Science & Policy Working Group (2004) The SER international primer on ecological restoration. Accessed from http://www.ser.org

Stone G, Van der Ham R, Brewer J (2008) Fossil oak galls preserve ancient multitrophic interactions. Proc R Soc Lond Ser B 75:2213–2219

Strong D (1995) Crazy mountains: learning from wilderness to weigh technology. SUNY Press, Albany

Thompson A, Bendik-Keymer J (eds) (2012) Ethical adaptation to climate change: human virtues of the future. MIT Press, Cambridge

Throop W (2010) Environmental virtues and the aims of restoration. In: Thompson A, Bendik-Keymer J (eds) Ethical adaptation to climate change: human virtues of the future. MIT Press, Cambridge

Throop W, Purdom R (2006) Wilderness restoration: the paradox of public participation. Rest Ecol 14:493–499

UN-REDD Programme (2009) Multiple benefits—issues and options for REDD. UN-REDD Program

White P, Walker J (1997) Approximating nature's variation: selecting and using reference information in restoration. Rest Ecol 5:338–349

Wiens J, Stralberg D, Jongsomjit D et al (2009) Niches, models, and climate change: assessing the assumptions and uncertainties. Proc Natl Acad Sci 106:19729–19736

Winner L (1977) Autonomous technology: technics-out-of-control as a theme in political thought. MIT Press, Cambridge

Yates C, Hobbs R, Atkins L (2000) Establishment of perennial shrub and tree species in degraded *Eucalyptus salmonophloia* (Salmon Gum) remnant woodlands: effects of restoration treatments. Rest Ecol 8:135–143

Knocking on Doors: Boundary Objects in Ecological Conservation and Restoration

Jac. A.A. Swart and Henny J. van der Windt

Abstract It is often argued that in the fields of conservation and restoration, research, practice, public perceptions, and societal interests should not only engage one another but also be integrated in order to guarantee success in the long term. Moreover, there is need for concepts and practices that are flexible enough to be acceptable to different parties and still have a common meaning. Such concepts and practices have been labeled "boundary objects." Here, we describe the concept of "natural limits" and the practice known as the "hands on the tap approach" as successful examples of boundary objects introduced into the discussion of gas exploitation in the Dutch Wadden Sea area. While the concept of natural limits focuses primarily on natural issues, in many restoration projects, societal issues—for example, protection against flooding—are often of at least comparable importance, especially in highly populated areas where many stakeholders are involved. The concept of social limits, on the other hand, refers to widely accepted "limit" values for important societal parameters, for example, safety, agriculture, and recreation. How these "social limits" can be taken into account is discussed in relation to a number of Dutch projects, including dune management, the protection of meadow birds, brook valley restoration, and the introduction of ungulates. Links between social and natural limits in environmental standard setting are addressed along with the issue of communication.

Keywords Ecological restoration • Nature conservation • Boundary objects • Natural and social limits

J.A.A. Swart (✉) • H.J. van der Windt
Science & Society Group, Energy and Sustainability Research Institute Groningen, PO Box 221 9700 AE, Groningen, Netherlands
e-mail: j.a.a.swart@rug.nl

M.P. Weinstein and R.E. Turner (eds.), *Sustainability Science: The Emerging Paradigm and the Urban Environment*, DOI 10.1007/978-1-4614-3188-6_19,

Introduction

Conservation and restoration are outstanding examples of discourses that bring nature and society together. They apply scientific knowledge to a socially sanctioned undertaking, i.e., conserving or restoring natural areas. However, the border region between nature and society is marked by contrasting views, not only because different scientific disciplines may be involved but also because various visions of nature and the relationships between nature and humans play an important role. Sometimes these views converge, while at other times they diverge and lead to conflict. In this chapter, we aim to review the border region between nature and society using the concept of boundary objects.

This concept was introduced by the sociologist Susan Leigh Star and her colleague James Griesemer to understand how people are able to establish shared meanings on pivotal issues despite coming from different social worlds (Star and Griesemer 1989). The authors describe, as an example, the establishment of the Museum of Vertebrate Zoology in Berkeley, California. Originally, this museum arose from private collections and was therefore based on popular and amateur culture. The successful establishment of the museum with a scientific and professional mission may therefore be seen as an example of the integration of different knowledge traditions (professional, amateur, and administrative). How did this occur? According to Star and Griesemer, a central factor was the establishment of so-called boundary objects which served as an interface between the different communities of practice. Such objects can function in this way because they "are both plastic enough to adapt to local needs and constraints of the several parties employing them, yet robust enough to maintain a common identity across sites" (Star and Griesemer 1989).

Boundary objects may have concrete or more abstract forms. For example, a map may represent both an administrative unit for administrators and a nature reserve for conservationists. In spite of these different meanings, it may nevertheless function as a communicative device enabling cooperation and thus as a concrete boundary object. An example of a more abstract boundary object is the species concept, which functions as a pivotal unit in biological theories and as a basis for collector's items for amateur biologists. As long as the different interpretations do not blockade each other, boundary objects may facilitate undertakings that require the involvement of different parties.

Ecological Restoration

Ecological restoration—probably even more so than conservation and preservation—also faces the challenge of reconciling different perspectives, knowledge traditions, and interests. Many restoration projects are planned in areas with a history of human utilization that—from a nature conservation point of view—have been negatively affected by such activities as agriculture, mining, and water management. Restoration ecologists must therefore often deal with points of view about natural areas that are related to former practices. However, here we immediately confront a potentially important source of conflict with respect to what counts as a negative effect of human utilization. The answer may differ for different social groups involved in the practice of restoration: professional ecologists, administrators and officials, amateur naturalists, former and current users of the area, and, last but not least, laypeople who aim to use or enjoy nature. These groups may have different perceptions about what nature is or should be and whether a natural area has been negatively affected or not by former practices (Swart et al. 2001; Van der Windt et al. 2007).

A recent example of a powerful clash of perceptions is the conflict over the Dutch government's plan to create new salt marshes by flooding formerly reclaimed land in the Hedwigepolder region of the southwest Netherlands (Nijpels et al. 2008). The plan was meant as a compensation measure for the loss of natural areas due to the deepening of the Western Scheldt to guarantee access to the harbor at Antwerp. In this conflict, local people expressed their admiration for the natural quality of the current polder landscape and protested strongly against the proposal to flood the area, which they argued showed a lack of respect for the efforts of their ancestors who had reclaimed the land from the sea. They even accused conservationists and restoration ecologists of destroying nature. Ironically, the whole project had been initiated to increase the natural quality of the area, but clearly only according to the thinking of restoration ecology professionals and policymakers (Floor 2009).

Other conflicts concerning conservation or restoration that have arisen between various stakeholders such as farmers, fishermen, and governments were not only due to the various interests involved but also due to opposing visions of nature (Worster 1977; Kwa 1987; Swart et al. 2001; Higgs 2003; Swart and Van der Windt 2005; Buijs 2009; Bauer et al. 2009; Drenthen et al. 2009; Runhaar and Van Nieuwaal 2010). Moreover, it was not only professional ecologists, laypeople, or stakeholders who differed in their perceptions or attitudes, and who collided on the meaning of nature. Within the professional community itself individuals also disagreed, as the field of ecology is characterized by different schools of thought, traditions or paradigms, and policy aims, resulting in different visions of what nature is or what it should be (Kwa 1987; Swart and Van Andel 2008; Turnhout et al. 2008). For example, in the Netherlands, the area of Oostvaardersplassen is considered by ecologists who focus on natural processes as being a valuable, almost completely natural environment, whereas ecologists who stress biodiversity consider it to be a rather poor natural area (Kuiters 2005).

Gas Exploitation in the Wadden Sea

It is true that the concept of boundary objects may be considered rather vague, not to mention that the concept itself might be considered a boundary object! Nevertheless, the recognition and fostering of such objects, whether they are concrete or abstract, may contribute to developing and maintaining coherence across intersecting or colliding practices, as occurs in many restoration and conservation projects.

An example of what appears to be a successful application of a boundary object is found in the recent history of the management of the Dutch Wadden Sea (Swart and Van der Windt 2007). This is a shallow sea located on the rim of the northwest European continent, consisting of tidal flats, salt marshes, and small islands. The Wadden Sea area is characterized by a fairly high level of primary production and functions as a nursery for many species of fish. The nearby islands, salt marshes, and intertidal areas are important resting and wintering sites for migratory birds.

However, because of its location near such a densely populated country as the Netherlands, it is highly vulnerable to human exploitation in relation to embankment, nutrient loading, gas exploitation, and fisheries (Weinstein et al. 2007). Since the late 1960s, in particular, many debates have taken place regarding several types of exploitation (Wolff 1992, 2000a, b; De Jong 2006; Swart and Van Andel 2008). In 2004, in response to the continuing debate on the use of the Wadden Sea for gas exploitation, a committee of politicians was established to develop recommendations for the sustainable use of the Wadden Sea. The committee argued that humans had been active in this area for many centuries, and that human use need not and should not hamper nature protection and nature restoration. According to the committee, both the impact of human activities on pivotal natural processes and the acceptable limits of these activities can be determined. In its final report, the committee proposed the concept of "natural limits" for the sustainable protection and development of the "Wadden Sea and the preservation of the unique open landscape" (Meijer et al. 2004).

According to the committee, a natural limit is *a framework of limit values for the most important natural parameters on which there is scientific consensus, such that if these limits are not exceeded, natural processes will continue undisturbed* (Meijer et al. 2004; translated). Thus, the concept of natural limits implied the recognition that beyond a certain level of human impact an ecosystem will significantly change. The committee applied this concept to the case of gas exploitation in the Wadden Sea. The report recognized that gas exploitation in this area may have negative effects, especially due to soil subsidence as a consequence of decreasing gas pressure deep underground, and that exploitation might also lead to disturbing effects on ecosystem functions through, for example, plant and transport facilities. Of these issues, the committee considered soil subsidence to be the most potentially negative effect, but argued that if the natural process of sedimentation could compensate for soil subsidence, no real adverse impacts would occur. The committee concluded further that, by taking into account the expected rise in sea level, subsidence could indeed be compensated for by sedimentation. As a matter of fact, the vision of the committee implied that during the period of gas exploitation, it was permissible to counteract the process of sea bed rise that would otherwise occur through natural processes such as sedimentation. Thus, the proposed natural limit of gas exploitation was "no actual soil subsidence," which would be determined by the combination of gas exploitation, sedimentation, and rising sea level.

It also appeared to be possible to monitor the subsidence effect and control the level of gas exploitation such that it would not exceed the proposed natural limit. The possibility of controlling gas exploitation levels in this manner was called "exploitation with hands on the tap." The result of the committee's advice was the acceptance of gas exploitation in the Wadden Sea by the Dutch government as well as by most environmental and nature protection groups, under the condition that part of the expected profit from the gas would be used to fund research and the protection of nature in the Wadden Sea. This effort concluded a societal debate that had arisen several times during the last decades. Although the acceptance of the concrete natural limit of "no soil subsidence" together with the "hands on the tap" approach was also facilitated by the funding conditions and aided by the economic power of the gas company, they can also be considered to be successful boundary objects as they reconciled contrasting views on the acceptability of gas exploration in the Wadden Sea.

It is important to note, however, that this approach including the concept of natural limits was not put forward by ecologists but by politicians and policymakers who were informed by ecologists and other experts on scientific and technical issues such as sea level rise metrics, sedimentation rates, and subsidence levels through gas mining activities. Although there is some overlap, we see different roles for the stakeholders involved: politicians and policymakers focusing on basic conditions and rules, and scientific experts providing substantiated information on ecosystems, physical systems, and their limits. Thus, the concept of natural limits and the practice of the hands on the tap approach were boundary objects for diverse groups. For the politicians, they were a means of resolving a severe political and societal conflict about what should be considered protectable nature in the Wadden Sea, functioning as policy concepts related to the notion of feasibility. For the ecologists and conservationists involved, the concept of natural limits was considered a means to place limits on what was acceptable from a "scientific viewpoint" framed in relation to concepts such as resilience and carrying capacity (Dankers et al. 2008). For the gas company and ministerial policymakers, they were a means of making exploitation acceptable, as they provided certainty about what was actually possible. For these stakeholders, the concept of natural limits functioned as a so-called "environmental utilization space," which refers to the notion of a sustainable use of natural resources (Opschoor and Weterings 1994). The natural limit can be considered as the utmost limit of utilization, taking into account a safety margin, which may still be considered as sustainable (Fig. 1).

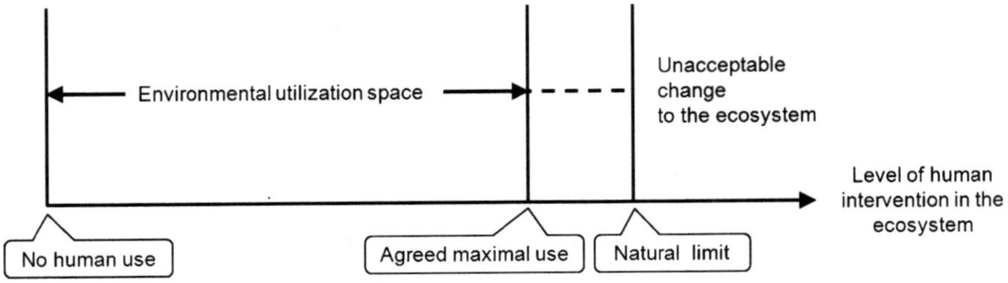

Fig. 1 Schematic representation of the concept of natural limits in relation to the concept of sustainable environmental use. The natural limit represents the utmost level of resource utilization which would not lead to an unacceptable change in the functioning of the ecosystem. The *dotted line* after the environmental utilization space represents a safety margin (After: Swart and Van der Windt 2007)

That the concept of natural limits and the hands on the tap approach could successfully function as a boundary object in the case of gas exploitation in the Wadden Sea was facilitated not only by its flexible meaning to different parties and because of favorable societal conditions (political and economic power and funding) but also by the already established conservation aims for the Wadden Sea, since the Dutch government had decided as early as the 1980s to take "naturalness" as a starting point for the management of the region (Ministry of VROM et al. 2007). The concept of natural limits has recently been elaborated in more detail and has been applied to food availability and bird population dynamics and to the effects of soil subsidence on the quality of salt marshes, sand flats, mudflats, and intertidal mussel beds (Dankers et al. 2008; Baptist et al. 2010).

Knocking on Nature's and Society's Doors

The negative impact of society on nature has been recognized worldwide and can be summed up in the saying that society is increasingly "knocking on nature's door." However, in recent decades, a new strategy has developed in relation to this encroachment in the form of ecological restoration (Aronson and Van Andel 2005) that aims to improve natural quality and extend and create natural areas and reserves. According to the Society for Ecological Restoration (SER), such restoration "initiates or accelerates the recovery of an ecosystem with respect to its health, integrity and sustainability" (SER 2004).

Because it has an offensive component, restoration may meet more societal resistance than conservation and preservation, which may be considered more defensive in nature. Ecological restoration attempts to "knock on society's door," but in order to be successful, restoration ecologists must pay serious attention to how society reacts and find ways to respond. In other words, rather than "natural limits," "societal limits" may be the starting point for developing boundary objects that facilitate the interactions among actors from different social worlds. Inspired by the definition of the concept of natural limits, we define such social limits as: *a framework of limit values for the most important societal parameters on which there is social and scientific consensus, such that if these limits are not exceeded these essential societal processes or conditions will go on undisturbed.* We may think of societal limits or barriers as resulting from considerations about food production, safety, security, and health, but also from deeply rooted cultural, ethical, and aesthetic considerations. Connecting different actors also means looking for social consensus. The challenge is to determine the societal limits and to let them function as boundary objects which do not appear as barriers but as drivers of ecological restoration and conservation.

An example of a societal barrier is the public fear of flooding by rivers and the sea, which is especially relevant in a country that largely resides below sea level. Protection against flooding by rivers or the sea is therefore important and often an essential condition in restoration and conservation projects. The classic approach in the Netherlands is dyke and dune fortification. However, in recent decades, the concept of dynamic coastal management has been successfully applied as an alternative in several areas. This approach makes use of sand supplementation and natural sea currents to move sand to places where it can contribute to flood protection. As such it may be considered a more natural form of coastal fortification. Currently, there is scientific evidence demonstrating that this technique ensures very high levels of security against flooding even in the case of the rising sea levels, and that it is even better than the traditional method of immobilizing sand volumes in risky locations (Arens and Overdiep 2008).

Dynamic coastal management also provides opportunities for ecological dune management and ecological restoration measures such as dismantling formerly constructed sand dikes in order to create new habitats within the dune landscape, leading to greater biodiversity and the occurrence of natural processes. Such measurements may even contribute to safety, as controlled flooding may result in a rise of the low lands behind the dunes and dikes through the process of sedimentation. Ensuring security against flooding can be considered a social limit, and dynamic coastal management may function as a boundary object as it provides society with better flood prevention barriers and restoration ecologists with opportunities for ecological restoration. The method of dynamic coastal management is accepted by a number of regional authorities, by most environmental groups, and by the Dutch Ministry of Infrastructure and the Environment responsible for protecting the land from encroachment by the sea. In contrast, a number of local communities still have their doubts (Arens and Overdiep 2008).

Societal barriers may also be rooted in culture and history. The failure to realize nature compensation in the Western Scheldt, described earlier in this chapter, was attributed to a cultural mismatch between the national government and restoration ecologists on the one hand and local stakeholders on the other with respect to what counts as nature. But there are also cases where such different points of view were ultimately reconciled; e.g., in the restoration plan for Gaasterland in the Dutch province of Friesland, the national government aimed to transform current rural areas into a more wilderness type of landscape (Kuindersma et al. 2006). Because the original plan did not account for the significance of the landscape in the region's identity, and the local community's need to decide its own future, it met huge resistance from local authorities, citizens, farmers, and other local and provincial stakeholders. In response, the government devised a new plan in which only a small part of the landscape originally envisaged as wilderness was retained, and in the remaining areas, nature-friendly agricultural management was introduced and farmers were funded to take responsibility for the management of birdlife and vegetation to ensure an increase in biodiversity.

Implementation of the new plan also entailed the use of a so-called nature credit points scheme for measuring nature quality: all relevant organisms, populations, and landscape types were rated according to this scheme. Based on an agreement between the parties involved, restoration aims were formulated in terms of these natural credit points, and it was expected that after a period of two decades, this adapted plan would produce results similar to the original plan (Kuindersma et al. 2006). Agricultural nature management using the credit points system thus functioned as a boundary object. It emphasized the importance of the original landscape and culture as a social limit, it provided society with an attractive landscape and ecologists and nature organizations with opportunities to achieve their ecological aims, and it encompassed the rights of the locals to determine the design of their region's landscape. The plan could only be successful if all of the diverse stakeholders were involved at all stages of the decision-making and management process, an essential ingredient for the successful establishment of such a boundary object.

An appreciation of the landscape is not the only element that may function as a boundary object—an appreciation of certain animals may also function in this way. An example is the lapwing (*Vanellus vanellus*) in the Dutch province of Friesland. There is a strong tradition among farmers and other locals of collecting lapwing eggs for human consumption (Steenkamp and Rip 1984; Van der Windt 1995). This is an integral part of the sociocultural tradition of the Frisian countryside and every year the first lapwing egg found is offered to the highest authorities of the province (and in former times to the queen of the Netherlands). For most Frisians, the lapwing is a symbol of spring, country life, the beauty of the countryside, and a harmonious relationship between humans and nature.

However, some decades ago, it became clear that the lapwing population had decreased. Conservationists, especially those from other provinces, called for an end to the egg-collecting tradition and for the creation of lapwing reserves. This request resulted in a lengthy and heated debate. The non-Frisian conservationists faced a societal barrier in the practices and attitudes of the Frisian people. Dutch bird protection legislation and the nature reserve approach were heavily criticized by the egg collectors. There were uncertainties about the cause of the decline in the lapwing population. Farmers and hunters, for example, were convinced that it was not egg collecting but predation by crows and foxes that was the main cause of the decrease. The solution was also unlikely to be simple. Strict reserves were impossible for several reasons, including their possible ineffectiveness due to the mobility of the birds. After some time, each of the stakeholders involved took initiatives in order to reach agreement. One initiative was a study to determine the cause of the decline of the species; another included an intensive protection program, consisting of a system of nest protection by the egg collectors themselves, in conjunction with a restriction of egg collection to specific periods and a flexible system of agricultural nature management by farmers. Thus, rather than becoming a divisive factor, the lapwing became a boundary object. All of the social groups agreed that the lapwing should survive, that nature reserves need not be the only solution, and that the lapwing would remain a symbol of the Frisian countryside.

Language and World Views

To overcome societal barriers, language is also important, with particular terms functioning as boundary objects that have different meanings for stakeholders from various social worlds. As an example, in a large restoration plan for a number of brook valleys in the Dutch provinces of Groningen and Drenthe (close to the Wadden Sea), the term "Green River" was adopted as a notion that appealed to all parties (Van der Windt and Swart 2008). Due to reclamation and canalization in the past, the brooks were not functioning very well ecologically. Therefore, conservationists, together with ecologists, recreation organizations, and fishermen created an ambitious restoration plan. Subsequently, the local authorities and farmers also became involved in the planning process. Eventually, despite the different interests and perspectives of the parties, they came to an agreement. For conservationists and restoration ecologists, the term Green River expressed the necessity of creating possibilities for the migration of fish as well as mammals such as the otter. The term "green"—and thus not "natural"—was attractive to stakeholders with a recreational or farming background. In addition, the term "river"—rather than "canal"—opened discussion on new approaches to the water management of large river beds, which involved the use of nature rather than fighting against it. Even the local water authority, which had the greatest reservations because of their traditional focus on safety, agreed to the plan, accepting that safety was guaranteed by its provisions.

Taking world views and language into account may also mean addressing ethical issues, especially when animals are involved. For example, the introduction of large herbivores for restoration and conservation goals in the Dutch nature reserve of Oostvaardersplassen led to an intense debate concerning their fate. The Dutch State Forest Service had introduced Heck cattle and Konik horses in the early 1990s (Vera 2009) and under minimal management intended that they would "de-domesticate" and in succeeding generations return to a natural way of life. However, the winters proved very difficult because of the limited availability of food and the harsh weather conditions. On several occasions, there was starvation among the animals, leading to political and public disquiet (ICMO 2006; ICMO2 2010), especially after images of starving deer and cows were shown on Dutch television.

The public resistance to this form of wild-life management had different sources, but in all cases the concern involved strong feelings with respect to animal welfare. Thus, the attitude to animal welfare became an important social barrier to the use of introduced species in the nature management process. According to the critics, humans had bred these animals and introduced them into the reserve. On this basis, they were seen as primarily domesticated animals and to let them starve and die clearly did not correspond with this status. Moreover, it was argued that their natural habitat in Oostvaardersplassen had been restricted by fencing and that natural forces such as predation, for example, did not affect the populations. From this standpoint, these animals were primarily considered to be under the control of human beings and thus their responsibility (RDA 2005). The supporters of de-domestication management meanwhile argued that intervention to prevent starvation would ultimately undermine the self-sufficiency of the population and would inhibit its capacity to adapt to the natural circumstances. They argued that high mass starvations also occur in other natural reserves, for example, the Serengeti reserve in Africa (Vera 2009).

The foregoing conflict can be considered as a clash between zoo-ethical and eco-ethical approaches. In order to prevent a stalemate, a "care for the wild" approach was suggested (Swart 2005; Swart and Keulartz 2011), where both wild and domesticated animals were seen as dependent on their respective environments, whether natural, human influenced, or somewhere in between. The recognition of this dependence made it possible to reconcile the different positions and to implement a form of management that involved monitoring and preventative culling so that mass starvation could be prevented (ICMO2 2010). The "care for the wild" approach may therefore be considered as a boundary object in the conflict over the management of ungulates in the Oostvaardersplassen.

Prying the Doors Open

Whether one starts out from one side or the other of the border between society and nature, dealing with uncertainty by using flexible and transparent procedures is essential (Gross 2010). Just as environmental impact assessment procedures are applied to evaluate the environmental and natural consequences of a policy plan or program, societal impact assessments might be used to assess the effects of restoration projects on society (Petts 1999). Such an assessment could clarify the impact of a restoration project on agriculture, landscape, health, safety, the rural economy, recreation opportunities, and so on, suggesting ways to alter plans in order to minimize the societal impact or to combine the plans with measures that may compensate the adverse effects of ecological restoration. It may also reveal approaches and solutions that were not considered before.

Flexible socioecological standard-setting procedures are needed to take into account the uncertainties and the various value bandwidths associated with both natural and societal parameters. Moreover, it is important that insights into and information on standards are revealed, communicated, and discussed by all of the participants involved. In conservation and restoration practices in which so many different parameters, circumstances, visions, and uncertainties may figure, consensus building is difficult and requires participative events such as public meetings and debates, and the understandable representation of benchmarks, conditions, and limits. The credit points scheme discussed above may function as one such representative tool but remains rather abstract.

An example of a rather communicable tool is the so-called AMOEBA approach used in Dutch policymaking (Ministry of V&W et al. 1989; Ministry of V&W 1990; Rutgers et al. 2005). AMOEBA is an acronym for "A general Method of Ecological and Biological Assessment," and was developed by ecologists in cooperation with water managers (Ten Brink and Colijn 1990; Ten Brink et al. 1991). The approach consists of a set of parameters enabling the assessment of the required or desired ecological quality of ecosystems. Most parameters represent species abundance (e.g. population sizes). A reference value and a current value are estimated for each parameter and put into a radar diagram such that the reference values are set in a circle with a value of 100% and the current values are shown as a percentage of the reference value for each parameter. Accordingly, all points are connected by a line for a better visualization. The reference values may be chosen from a particular year or from another area that functions as an ideal example, or they may be based on theoretical grounds. The resulting figure can present the current ecological condition of an area in one glance (Fig. 2).

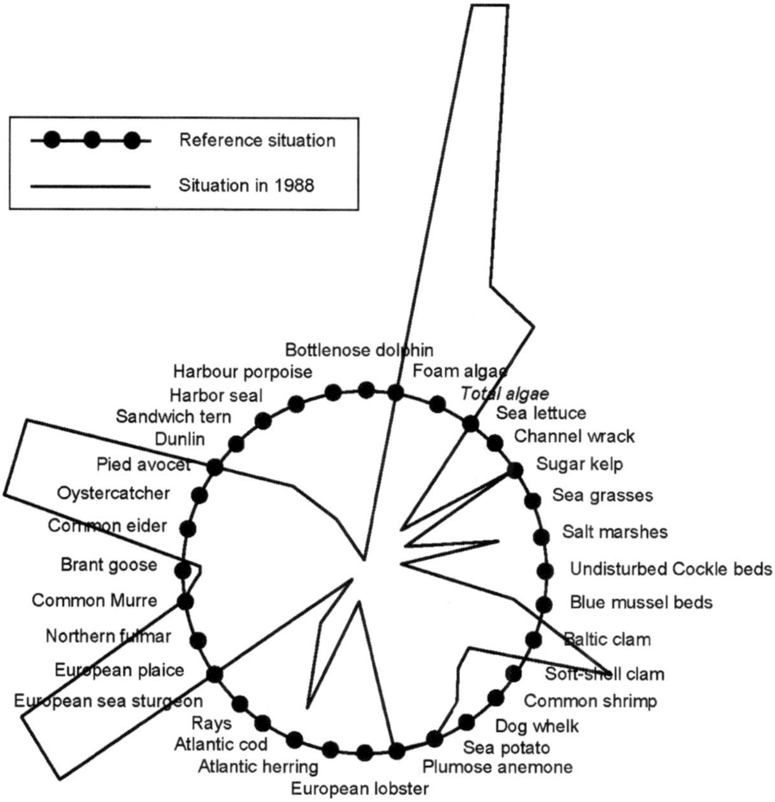

Fig 2 Example of the Amoeba figure for the North Sea for 1988 (After: Ministry of V&W et al 1989). The parameters represent main elements of the North Sea ecosystem: plants, invertebrates, birds, fish, mammals, and typical niches. The values of these parameters represent, e.g., population size, number of breeding pairs, hectares, etc (see Ten Brink et al. 1991 for a more extensive description). In most cases, these values are much too low or high as compared to the reference situation, which represents the situation around 1930

We believe that the AMOEBA approach can also be used to make the concepts of natural and social limits more communicable. Using reference values, ecological science, and practical experience, natural limits can be estimated, distinguishing the upper and lower values, because the value of a parameter can be either too low or too high from an ecological point of view (e.g. underpopulation or overpopulation of a species). Natural limits can thus—taking into consideration the definition cited earlier in this chapter—be seen as hard bandwidths around reference values for chosen parameters of an ecosystem. Similarly, we may construct an AMOEBA-like graphics of the social limits for an area of interest using parameters such as flood risk, emission levels in industry or agriculture, recreation intensity, or transport intensity. Accordingly, we can compare the actual values and the natural limits with social limits. This gives us information on potential conflicts, opportunities, benchmarks, and trade-offs for the chosen set of parameters.

Of course we must recognize that representation and assessment tools such as the AMOEBA approach and the credit points system do have shortcomings, for example, they do not take into account the possible mutual dependence of parameters. Moreover, the set of parameters chosen also depends on scientific paradigms, visions of nature, and specific interests. Thus, every set chosen implies another set of parameters that were not chosen. Consequently, the process of determining the assessment methodology should also be part of the decision-making procedure.

Conclusion

While acknowledging that ecological restoration involves an offensive approach, the social acceptance of restoration measures cannot be reached simply by stressing the importance of nature and natural processes alone as ecologists and ecological restoration projects sometimes seem inclined to do. Rather, they should emphasize that ecological restoration can also contribute to societal goals. In such cases, restoration goals may function as boundary objects that allow different actors from different social worlds to reach agreement. We have discussed a number of examples of such boundary objects in this chapter and many more examples can be found in the current practice of ecological restoration and conservation. For example, concepts of natural capital (Aronson et al. 2007) or paying for environmental services (Pascual and Perrings 2007) may be considered as boundary objects that are able to reconcile different social worlds.

The role of these types of boundary objects—linking societal and restoration goals aims—is especially important in densely populated areas such as Western Europe and many coastal areas throughout the world (Weinstein et al. 2007). In these areas, natural limits have to be taken into account from ecological or conservation points of view, but at the same time the social limits must also be recognized. It is therefore important to develop representation and assessment tools that may assist in finding and determining such boundary objects.

Based on our overview, we conclude that boundary objects may vary, from being more society-based to more science-based, from existing procedures and management styles to new terms, from landscapes to species, and from practices to concepts. The establishment of boundary objects is not easy. Sometimes they develop from conflicts, as in the cases of gas exploitation in the Wadden Sea, dynamic dune management, and agricultural nature management. Sometimes they simply emerge, as in the case of the lapwing, while in others they are consciously constructed, as in the case of the Green River project or the AMOEBA approach. The devising or use of these tools should not be seen as purely strategic and rhetorical, but must be seriously considered and substantiated by scientific and societal support. This also requires openness: the willingness to include different types of knowledge and different views, as well as a fair system of checks and balances and transparent planning procedures. Ecological restoration requires that we take into account societal conditions and perceptions of nature right from the beginning and not only after restoration aims have already been determined.

References

Arens B, Overdiep T (2008) Vijftien jaar dynamisch zeereepbeheer op de waddeneilanden [Fifteen years of dynamic coastal management on the Frisian Islands]. Duin 31:3–5

Aronson J, Van Andel J (eds) (2005) Restoration ecology: the new frontier. Blackwell, Oxford, UK

Aronson J, Milton SJ, Blignaut JH (2007) Restoring natural capital. science, business, and practice. Island Press, Washington, DC

Baptist MJ, Dankers N, Bastmeijer CJ et al (2010) User limits or natural limits: can we set limits to human use, based on a natural functioning of the Wadden Sea? In: Marencic H, Eskildsen K, Farke H et al (eds) Science for nature conservation and management: the Wadden Sea ecosystem and EU Directives. Proceedings of the 12th International Scientific Wadden Sea Symposium in Wilhelmshaven, Germany, 30 March–3 April 2009, pp. 193–198

Bauer N, Wallner A, Hunziker M (2009) The change of European landscapes: human-nature relationships, public attitudes towards rewilding, and the implications for landscape management in Switzerland. J Environ Manage 90:2910–2920

Buijs AE (2009) Lay people's images of nature: comprehensive frameworks of values, beliefs, and value orientations. Society and Natural Resources 22:417–432

Dankers N, Baptist MJ, Bastmeijer CJ et al (2008) Natuurgrenzen in de Waddenzee: een verkenning voor beleid en beheer [Natural limits in the Wadden Sea: an exploration for policymaking and management] Report C067/08 Wageningen Imares. Texel, The Netherlands

De Jong F (2006) Marine eutrophication in perspective: on the relevance of ecology for environmental policy. Springer, Berlin

Drenthen M, Keulartz J, Proctor J (2009) New visions of nature. Complexity and athenticity, Springer, Dordrecht

Floor J (2009) Ontpolderen langs de Westerschelde. Invloed van lokale en regionale actoren en legitimiteitsproblemen in het beleidsproces (2005–2009) [De-reclamation along the Western Scheldt. The influence of local and regional actors and legitimization factors in the policy process (2005-2009)]. Master's thesis Environmental Policy Group (ENP) Wageningen University. http://edepot.wur.nl/13732

Gross M (2010) Ignorance and surprise. Science, society, and ecological design. MIT, Cambridge

Higgs E (2003) Nature by design. People, natural process, and ecological restoration. MIT, Cambridge
ICMO (2006) Reconciling nature and human interests. Report of the International Committee on the management of large herbivores in the Oostvaardersplassen. Wing rapport 018, The Hague, Wageningen, The Netherlands
ICMO2 (2010) Natural processes, animal welfare, moral aspects and management of the Oostvaardersplassen. Report of the Second International Commission on management of the Oostvaardersplassen (ICMO2). Wing rapport 039, The Hague, Wageningen, The Netherlands
Kuiters L (2005) Grote grazers en behoud van biodiversiteit: Na 30 jaar beheerpraktijk nog veel vragen onbeantwoord [Large herbivores and conservation of biodiversity: many questions still after 30 years of management]. Vakblad Natuur, Bos en Landschap 2:6–10
Kuindersma W, Molenaar JG, Kolkman G et al (2006) Het experiment Gaasterland [The Gaasterland experiment]. Wageningen, Alterra
Kwa CL (1987) Representations of nature mediating between ecology and science policy: the case of the International Biological Programme. Soc Stud Sci 77:413–442
Meijer W, Lodders-Elfferich PC, Hermans LMLHA (2004) Ruimte voor de wadden [Space for the mud flats]. The Hague, Ministry of LNV
Ministry of V&W (1990) Wadden Aktieplan, discussienota [Wadden Sea Action Plan, discussion paper], The Hague/Haren, The Netherlands
Ministry of V&W, Ministry of VROM, Ministry of LNV (1989) Derde Nota Waterhuishouding [Third Policy Document on Water Management]. SDU: The Hague, The Netherlands
Ministry of VROM, Ministry of LNV, Ministry of V&W, Ministry of EZ (2007) Ontwikkeling van de wadden voor natuur en mens [Development of the mud flats for nature and humans]. The Hague, The Netherlands
Nijpels C, Heip CHR, Hulscher S et al (2008). Wennen aan de Westerschelde. Advies Commissie Natuurherstel Westerschelde: Alternatieven voor ontpoldering Hertogin Hedwigepolder. [Getting used to the Western Scheldt. Advice of the Committee on Nature Recovery, Western Scheldt: Alternatives for the reclamation of the Duchess Hedwige polder] http://www.rijksoverheid.nl/documenten-en-publicaties/rapporten/2008/11/21/wennen-aan-de-westerschelde.html
Opschoor JB, Weterings R (1994) Environmental utilisation space: an introduction. Milieu 9: 198–205
Pascual U, Perrings C (2007) Developing incentives and economic mechanisms for in situ biodiversity conservation in agricultural landscapes. Agr Ecosys Environ 121:256–268
Petts J (ed) (1999) Handbook of environmental impact assessment, vol 1, Environmental impact assessment: process, methods and potential. Blackwell Science, Oxford
RDA (2005) Advies over de wintersterfte 2004-2005 van grote grazers in de Oostvaardersplassen [Advice about the deaths of large herbivores during winter 2004-2005 in the Oostvaardersplassen]. RDA 05/8. Raad voor Dieraangelegenheden: The Hague, The Netherlands
Runhaar R, Van Nieuwaal K (2010) Understanding the use of science in decision-making on cockle fisheries and gas mining in the Dutch Wadden Sea: Putting the science–policy interface in a wider perspective. Environ Sci Pol 13:239–248
Rutgers M, Mulder C, Schouten AJ et al (2005) Typeringen van bodemecosystemen Duurzaam bodemgebruik met referenties voor biologische bodemkwaliteit [Typology of soil ecosystems. Sustainable soil use with references for soil quality]. RIVM, Bilthoven, The Netherlands
SER (2004) The SER International Primer on Ecological Restoration. http://www.ser.org/content/ecological_restoration_primer.asp
Star SL, Griesemer JR (1989) Institutional ecology. Translations and boundary objects: Amateurs and professionals in Berkeley's Museum of Vertebrate Zoology, 1907-39. Soc Stud Sci 19: 387–420
Steenkamp F, Rip A (1984) The meadow birds organize a symposium. In: George Sarton Centennial, communication and cognition. Ghent, Belgium
Swart JAA (2005) Care for the wild. Dealing with a pluralistic practice Environl Values 14:251–263
Swart JAA, Keulartz J (2011) The wild animal in our society. A contextual approach to the intrinsic value of animals Acta Biotheoret 59:185–200

Swart JAA, Van Andel J (2008) Rethinking the interface between ecology and society. The case of the cockle controversy in the Dutch Wadden Sea J Appl Ecol 45:82–90
Swart JAA, Van der Windt HJ (2005) Visions of nature and environmental sustainability: Shellfish fishing in the Dutch Wadden Sea. Rest Ecol 13:83–192
Swart JAA, Van der Windt HJ (2007) Hoe hard zijn natuurgrenzen? [How solid are natural limits?] In: Raad voor Wadden (ed) Natuurgrenzen voor dagelijks gebruik. Advies over de toepassingsmogelijkheden voor natuurgrenzen. [Natural limits for daily use. Advice on possibilities for the application natural limits] 2007/03. Raad voor de Wadden, Leeuwarden
Swart JAA, Van der Windt HJ, Keulartz L (2001) Valuation of nature in conservation and restoration. Rest Ecol 9:230–238
Ten Brink BJE, Hosper SH, Colijn F (1991) A quantitative method for description and assessment of ecosystems, the AMOEBA approach. Mar Poll Bull 23:265–270
Ten Brink BJE, Colijn F (eds) (1990) Ecologische ontwikkelingsrichtingen zoute wateren [Ecological development directions for salt water]. Ministry of Transport, Public Works and Water Management, The Hague
Turnhout E, Hisschemöller M, Eijsackers H (2008) Science in Wadden Sea policy: from accommodation to advocacy. Environ Sci Pol 11:227–239
Van der Windt HJ (1995) En dan: wat is natuur nog in dit land? [After all: what is still nature in this country?]. Boom, Meppel
Van der Windt HJ, Swart JAA (2008) Ecological corridors, connecting science and politics: the case of the Green River in the Netherlands. J Appl Ecol 45:124–132
Van der Windt HJ, Swart JAA, Keulartz J (2007) Nature and landscape planning: exploring the dynamics of valuation, the case of the Netherlands. Lands Urban Plan 79:218–228
Vera FWM (2009) Land-scale nature development—the Oostvaardersplassen. Brit Wildl 20:28–36
Weinstein MP, Baird RC, Conover DO et al (2007) Managing coastal resources in the 21st century. Front Ecol Environ 5:43–48
Wolff WJ (1992) The end of a tradition: 1000 years of embankment and reclamation of wetlands in the Netherlands. Ambio 21:287–291
Wolff WJ (2000a) The south-eastern North Sea: losses of vertebrate fauna during the past 2000 years. Biol Conserv 95:209–217
Wolff WJ (2000b) Causes of extirpations in the Wadden Sea an estuarine area in the Netherlands. Conserv Biol 14:876–885
Worster D (1977) Nature's economy. Cambridge University Press, Cambridge

Sustainability: More About the Toolmaker than the Tools

R. Eugene Turner

Abstract A sustainable system is not necessarily a high-quality one, but it could be. We could, for example, "survive" on the desperate edge, as the remnants in a self-fouled and deteriorating environment. Why won't a future sustainable system be just another industrial model of mass efficiency and throughput? Perhaps the incompatible outcomes are a choice between the sometimes nearly invisible civilizing aspects of culture nurturing respect, equality, and cooperation on one hand, and the greed and self-indulgences undermining social tolerance, empathy, and cooperation that ends up promoting violence and dehumanization. The human heritage is subtle, indestructible, and worth nurturing if we want that hospitable sustainable system. But, assuming that a kind of social osmosis will be sufficient to sustain justice and fairness is wrongheaded and dismisses the historical examples. A new cultural narrative is needed to override the maladaptive dissonance preventing formation of sustainable systems. This narrative will be anchored in personal initiatives, incorporates an appreciation of our evolved heritage, and is informed by intentional social learning within groups and occasional social punishment.

Keywords Humane sustainable systems • Cooperation • Altruism • Social learning • Governance • Societal behavior

Our Heritage

> A Talmudic saying: "No one is the owner of his instincts; but controlling them, that is civilization" (Elie Wiesel, New York Times, 21 May 2011).

R.E. Turner (✉)
Department of Oceanography and Coastal Sciences, Louisiana State University, Baton Rouge, LA 70803, USA
e-mail: euturne@lsu.edu

M.P. Weinstein and R.E. Turner (eds.), *Sustainability Science: The Emerging Paradigm and the Urban Environment*, DOI 10.1007/978-1-4614-3188-6_20,

> When human beings lose their connection to nature, … then they do not know how to nurture their environment healing our society goes hand in hand with healing our personal, elemental connection with the phenomenal world (Chögyam Trungpa, Founder, Naropa University).

Imagine looking through the windows of your building and seeing something move on the prairie—yes, just over there. Maybe you can see the First Nation hunter put on a wolf or buffalo skin? He crawls for hours through the Blue Stem prairie under a hot sun collecting cat burrs in his hair, breathing dust, and accumulating small nicks to his skin as he follows the ambling bison herd (Fig. 1). The hunter

Fig. 1 "Buffalo Hunt in Wolf Masks" by George Catlin (1832) in the upper Missouri River (Original Painting held in Smithsonian American Art Museum, Washington, DC)

wants to be near a skittish 2,000-lb bull that could easily kick the femur out of his skin, or rumble away with 5,000 other buffalo—all because a hawk screeches. The hunter might use a bow and arrow at 3 m, or perhaps he puts a short spear through the animal's tough underbelly—it is a strong thrust under the ribs and into the soft heart. He twists and rams it in and out rapidly using both hands, draining the cardiac pump of its life force. The animal bellows, falls to its knees and then on its side, heaves a last breath, and stares at the sky. The hunter makes sure that the bull is dead then gives thanks, and washes in a creek, being mindful of what has just happened. There is a grassy-dry-matted-furred buffalo-coat smell to the hide as flies land on the dried blood. The carcass is processed on site. There is food for a week and raw materials for clothing and tools.

This day was once part of living closely, and deeply, within the natural world. There are no buffalo herds roaming the Plains today and the trails are straight and paved. Brick and lumber frame the windows you are looking through. There are electrical wires behind the walls. Was this a mirage? Did it really happen? If it did happen, then what remains?

Our hominoid ancestors may have been forest dwellers 6–8 million years ago, and *Homo sapiens* as early as 200,000 years ago, and most likely originating in Africa.[1] They came to North America 10,000 years ago—that is, about 600–700 generations ago. The senses were used and appreciated to survive, and not *just* survive, but to form communities with relationships that could be quite touching. The Neanderthals, for example, buried their dead in a grave of flowers. People thanked their animated world for blessings, painted cave walls with graceful imagery (Fig. 2), and had a sense of proportion and color and design in their clothing and cook ware. Cloth reminders were hung on trees to recognize spirits in ways that seem strange to modern people. Yes, survival required hard work and risk and there were mortalities. Survival happened within the context of the whole set of experiences that were totally dependent on the physical and mental senses. It was fully a

Fig. 2 Three horses facing each other in a Chauvet Pont d' Arc cave painting. The sensuous lines, shading, and posture were drawn 30,000–32,000 years BCE. *Source*: http://www.metmuseum.org/toah/hd/chav/hd_chav.htm

[1] Fossil records suggest modern humans emerged in sub-Saharan Africa about 200,000 years ago, but their dispersal is thought to have begun about 70,000 years ago and resulted in very little inbreeding with the Neanderthals. The earliest evidence of modern humans appears in Australia, dating to about 50,000 years ago.

world of relationships that evolved from pre-existing relationships with animals and plants, and of the environment and its inhabitants. We have that legacy in our genes, our physical potential, and our neural networks. It is a set of pre-requisites inherited at conception and given potential at birth. It is unavoidably a part of us.

This inheritance should be considered the all of "you," because we don't acquire genetic traits rapidly. At 20 years per generation, our oil-based society of 100 years is five generations distant from the 5,000 to 10,000 generations since *H. sapiens* evolved and a few hundred thousand generations since *Homo* sp. showed up. Yet we have such little knowledge of this tethering to our past because of a constant set of diversions emphasizing a mere part of this inheritance. Our daily connections to the natural world are covered over so much that most of us don't know local birds, plant seasons, where milk comes from, the smell of different kinds of snow, of dirt, or of silence. We are, it seems, several trophic levels apart from the reality of food source, food preparation, and how to consume it while being fully present. The separations may seem to be about the external or physical world, but are carried within the mind of individuals and groups.

Examples for Individuals

It's not that we must return to some idyllic or imaginary past, but that we must appreciate that this heritage is within us, and that it has qualities to respect and nurture. But the nuances of those qualities may seem so invisible in modern life that they seem to not exist. Here are some examples of how that sometimes-subterranean legacy is exposed.

Recovery from Gallbladder Surgery

Ulrich (1984) examined the records of different groups of patients who had undergone gall bladder removal. The first group had a window view of a natural scene (trees) and the second group looked at a brick wall (Fig. 3). The two groups received the same post-operative medical attention and lived on the second and third floors of the same hospital wing. Their rooms were of similar dimensions and had one window. The first group recovered faster than the second group, with an average stay in the hospital of 8.0 days, compared to 8.7 days for the group looking at the brick wall. They received fewer negative evaluations by staff (1.1 vs. 4.0 per patient), and took fewer painkillers.

Greenness and Psychological Health

Other studies reveal how exposure to natural systems has a restorative effect on indicators of stress or well-being (Kaplan 1995; Ulrich 2007), including attention-deficit

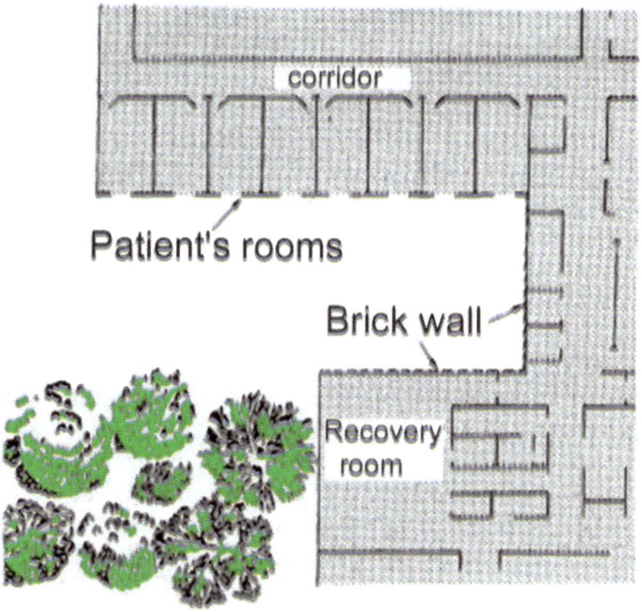

Fig. 3 The two views of patients undergoing gall bladder removal (Redrawn from Ulrich 1984)

activity (Kuo and Taylor 2004). Ulrich et al. (1991), for example, had 120 subjects view a stressful movie, and then exposed them to six different natural and urban settings while monitoring changes in various indices of stress (e.g., blood pressure, heart rate, muscle tension, skin conductance). Recovery was faster and more complete when subjects were exposed to natural rather than urban environments. The natural setting was accompanied by sustained attention. Taylor et al. (2001) found that inner-city girls (and interestingly, not boys) were positively affected by the naturalness of the view from their high-rise urban homes. For girls, the variance in test concentration, impulse inhibition, and delay of gratification was all directly related to their self-discipline, and 20% of that was related to the naturalness of the small natural setting near their home.

Examples for Societies

It was only a few hundred generations ago that our ancestors lived without written communication, having evolved to that state in intimate contact with the natural world by using all of their senses (Abram 1996) and in intimate social gatherings. They had to cooperate to survive and so it is easy to imagine that evolutionary pressures are involved to support cooperation and to dissuade maladaptive tendencies. Children, for example, spontaneously begin to demand fairness/equal treatment for others ("pro-sociality") between 4 and 7 years old, at the same time that their interest in fairness begins to favor their peers (Fehr et al. 2008). These and other inheritances are expressed in individual and group preferences. Their influence may be nuanced, for example, as in the innate aversion to unequal rewards for equal work by monkeys and dogs (Brosnan and de Waal 2003; Range et al. 2009). They may be complex, like the decisions affecting cooperation or defectors in group dynamics (Semmann et al. 2003); and they may be subtly hidden in the social matrix like the changes in altruism as population density rises (Levine 2003). The political behavior of chimpanzees and early human societies suggests that *we have evolved to maximize our interactions at the small-scale, at the personal scale* (de Waal 2007; Boehm 1999). Gandhi (1909) emphasized how the basic goodness and cooperative nature of people at this personal scale *was* the history of the world, and that the written formal history was a thin veneer of exceptions covering over these mostly invisible daily interactions.

This desire for fairness and equal access is one of the fundamental requirements for peaceful cooperation at the personal level and the group level. Wilkerson and Pickett (2009a, b) have illuminated the dozens of strong relationships between income inequality and social problems, including mental illness, incarceration rates, teenage pregnancy, illiteracy, obesity, drug abuse, and education performance (Fig. 4). The social stratification in income distribution reflects strong societal differences in material inequities permeating developed and underdeveloped countries. The lessons of history are that strong social inequalities cannot be sustained.

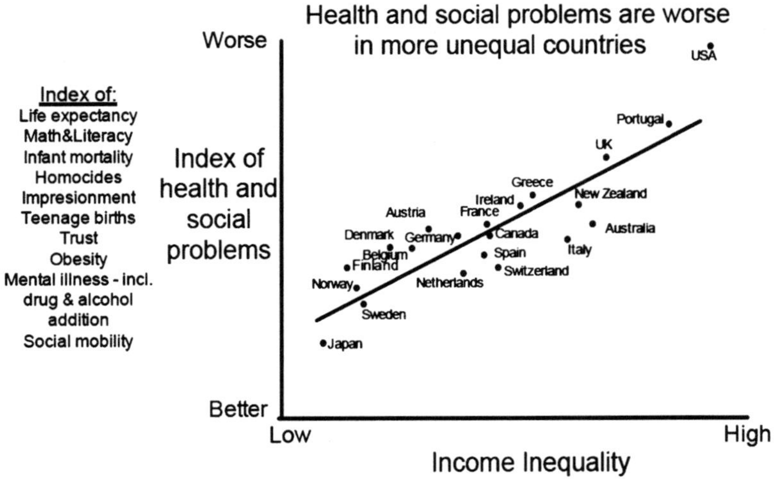

Fig. 4 There are many direct relationships between the scale of income disparity and negative social attributes that are discussed in Wilkerson and Pickett (2009a, b). This is one example (Redrawn from http://www.slideshare.net/equalitytrust/the-spirit-level-slides-from-the-equality-trust)

Contrasting Options

Sustainable systems, or nearly sustainable ones, existed in the New World before Europeans arrived. They were not static, and obviously did not successfully resist the more aggressive invaders intent on a growth economy whose wealth was distributed in a highly stratified society. Their social structure is an example that we might learn from as we re-invent our governance for the new conditions.

Balinese Rice Farmers

The Indonesian island of Bali has a water distribution network dependent on the rains whose amount is seasonal and dependent on elevation. J. Stephen Lansing studied this system for decades and wrote an outstanding summary of the linked geology, history, culture, and ecology of this resilient landscape as it was challenged in the modern era (Lansing 2007; Fig. 5). Water is distributed by a system of canals, tunnels, streams, and weirs. The availability of water determines the number and kinds of crops that are grown and pests. The pests come from adjacent fields and can be controlled by large-scale withdrawal of their habitat via drying fields and synchronous planting/harvest rhythms. So there is a kind of

Fig. 5 The cover from Lansing's book "Priests and Programmers" illustrating a temple up the mountain that regulates water flow

benefit–cost calculation made to control pests by timely and large-scale water withdrawal, and having water delivered for the crops. And this balancing was orchestrated by the temple priests who attentively dealt with the strictly agricultural, as well as the related social issues. To the Balinese, the relationship between the natural environment and the social environment were interdependent, and not linear. Effective management required cooperation across all watersheds, and even beyond. Lansing says that "essentially, water temples establish symbolic connections between productive groups and the components of the natural landscape that they seek to control. The natural world surrounding each village is not a wilderness but an engineered landscape of rice terraces, gardens, and aqueducts created by the coordinated labor of generations of predecessors. Anthropomorphic deities evoke this residual human presence in an engineered landscape."

The very success of the system kept it invisible to the colonial Dutch and then the post-World War II administrations. They were illuminated to the non-Balinese only in the early 1970s. It was then that the Green Revolution undercut the traditional cropping system in a way that resulted in competition for water and unresolved dissatisfactions, a patchwork of habitats harboring many pests that were never constrained, and poor harvests. The role the intricate and self-sustaining social network was eventually recognized and re-fortified, and the formerly invisible threads of cooperation was sanctified by bureaucrats who recognized a system that they could not do without. The top–down approach control structure was minimized and the bottom–up and integrated social network it harmed was being restored.

Columbus

The historian Howard Zinn (2003) describes Columbus as an excellent navigator and a brutal colonialist. He quotes the journals of a Christian priest (Las Casas) who said the Spaniards "thought nothing of knifing Indians by tens and twenties and of cutting slices off them to test the sharpness of their blades." Half of the 250,000 Indians on Haiti died within 2 years from murder, mutilation, or suicide. Las Casas estimated that "from 1494 to 1508, over three million people had perished from war, slavery, and the mines. Who in future generations will believe this? I myself writing it as a knowledgeable eyewitness can hardly believe it." Columbus loaded 500 Awarak slaves on the return voyage from the Caribbean, and 200 died on the way.

These cultural atrocities attributed to Columbus were not the exception, of course. The English spread smallpox in woolen goods given to the Mi'kmaq of Nova Scotia (circa 1713) and the 24 Nov 1752 reward for their scalps was in effect for more than 200 years (Paul 2000). What is instructional, however, were the Native American societies existing in Columbus's time. The Awaraks had greeted him with a hospitality not found in Europe. They shared all they had with him, as they did with themselves. Zinn (2003) writes that: "Columbus and his successors were not coming into an empty wilderness, but into a world which in some places was as densely populated as Europe itself, where the culture was complex, where human relations were more egalitarian than in Europe, and where the relations among men, women, children and culture were more beautifully worked out than perhaps any place in the world." And that "they were people without a written language, but with their own laws, their poetry, their history kept in memory and passed on, in an oral vocabulary more complex than Europe's, accompanied by song, dance, and ceremonial drama. They paid careful attention to the development of personality, intensity of will, independence and flexibility, passion and potency, to their partnership with one another and with nature." Zinn (2003) says that that 2% of the population owned 95% of the land in Europe, at the very time that there were sustainable societies of fairness and harmony in the New World. This is quite a contrast.

Deganawidah

Not all societies in the New World were in harmony, of course. The transformative agents that arose demonstrate that fundamental changes could occur within a generation and illustrate some of the elements that sustain their influence. Deganawidah was one of these influences. He was born to a Huron woman in what is now New York circa five centuries before Columbus arrived. The Huron were a warring tribe then and rejected his message of peace. Deganawidah left them to begin a pilgrimage that eventually united a confederation of Haudenosaunee tribes (the Great Iroquois Confederation), which the Huron eventually did join (Johansen and Mann 2000). Deganawidah was accompanied by Ayowenta[2] who he rescued from both bitterness and cannibalism.[3] Red Elk's description of Deganawidah's teachings (http://www.manataka.org/page1639.html) speaks of the Way of Great Peace, which taught the need for balance within society that began within the individual. When there was a balance of male/female within the individual, then peace expands out to the other gender, then to family, and eventually to other tribes. The tribe was governed as a living system of human consciousness for all, and was not focused on the individual. The decision matrix for the tribe was complex and inclusive. No individual, gender, or age group made a decision for the tribe. The males on the war councils could be removed by the Council of Grandmothers who may have nominated them. Further, the women made moccasins and so no war would last long without their support. Their governance structure and attitudes formed the basis for much of the Constitution of the United States and the US Declaration of Independence.

[2] Ayowenta is the Onondaga spelling of Hiawatha, who is *not* the imaginary person of H.W. Longfellow's poem.

[3] Ayowenta is sometimes also known as the "Translator" because of Deganawidah's presumed speech impediment.

Modern Analogues to Work with?

An explicit societal effort is required to overcome those parts of our genetic inheritance that produces undesirable outcomes. The good news is that our inheritance also includes the indestructibly decent and desirable possibilities to do that. It is not that there are "bad behaviors" to be pushed away, but that we need to be inquisitively aware of them, lest they become manipulated by omission or active external influences. Here are two simple examples: (1) my experience in England, at times, is that people are expected to stand in a line when waiting on a bank teller, at a post office, or for a ticket. If someone ignores the line to squeeze ahead of others, then more than one person will tell them quite directly to "queue up," "take your turn, mate" or "we all want to get this done, just like you; now get to the back of the line." In other countries, there are sharp elbows and no line. It is a stronger social instruction if more than one individual responds; (2) there is a "social trap" that I use in class to discuss the social contract. It involves a mattress falling of the roof of a car blocking the road to the cars behind; the driver carrying the mattress does not realize that the mattress fell off, and continues on, but the mattress is now blocking the road. One option for the first driver behind the mattress now lying on the road is to wait for an opening in the oncoming traffic. This takes time and the traffic logjam in that lane grows as each driver sequentially finds the mattress blocking the path forward. It only takes one person to move the mattress to the side of the road so that everyone can continue without further delays.

Viticulture

Vineyards are ecosystems managed for decades. The producers are usually skilled, but I doubt that most think they know everything about making and marketing wine, or that there are no more surprises. Many vineyards are more than 100 years old, and are the result of transplanted expertise whose viticultural roots are perhaps thousands of years old. These vineyards and others in the world have formal and informal means to improve the harvest—governmental policies, extension agents, scientists, accountants, familial apprenticeships, and export–import industries. Developing and maintaining sustainable systems is hardly so simple or well evolved as vineyard management. But there are similarities. The interactions must evolve, there are external influences, and we tend to think of ecosystems as producers of something. Ecosystem management views are often subsumed by an economic view of the world, so a product–service relationship may be espoused. I think this is shallow view, but it is an influential view. And in the process of selling a product, society has often hypocritically promoted its abuse. There is often a great deal of hypocrisy, for example, with governmental and personal promotion/acceptance of the well-documented and unhealthy consequences from drinking and smoking for the consumer and those around them.

Smoking Laws

The present restrictions on smoking in public spaces were difficult and took years to implement (and are not completely in place) despite an incredible array of research results, support from health professionals, and angry non-smokers. The tobacco industry executives did not go to the funerals of our relatives dying from emphysema. Those that went through the social change might recall the struggle to minimize the discomfort of telling our smoking friends that they could not smoke in our homes, or that we would not be able to socialize in "smoky" joints with them. Parents had to persuade children that it was not a "cool" behavior within the context of a torrent of advertising saying otherwise. And this was just a change in smoking laws, which seems like light work compared to what it will take to make the transition into sustainable systems!

Wetland Protection

Are there good predictors of success (and failure)? For example, what makes for a successful restoration program when viewed at a societal level? LaPeyre et al. (2001) addressed both of these questions in an interesting and straightforward way using a traditional scientific analysis, and with results that non-scientists might appreciate (Fig. 6). They conducted a statistical analysis of factors

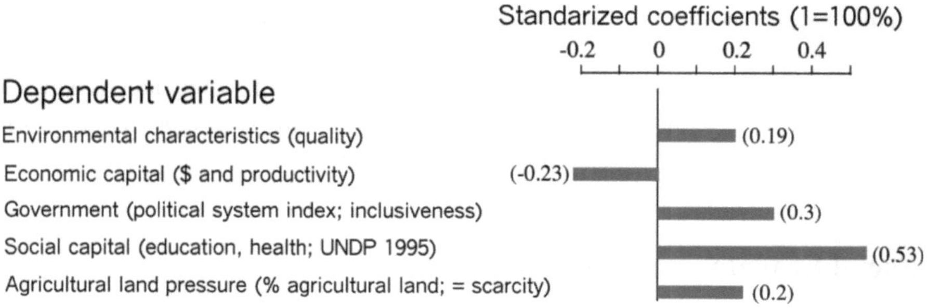

Fig. 6 A statistical analysis of factors influencing wetland protection and management in 90 nations (conventions, area, participation). The stronger the response, the more influence that factor has on environmental quality (Adapted from LaPeyre et al. 2001)

influencing wetland protection and management in 90 nations. The measure of success was an amalgam of several indicators of wetland protection, including the number of RAMSAR sites and when they received designation, the total wetland area protected, and the percentage of wetland area protected. Various social and economic factors were examined to see how much they influenced the dependent variable (wetland protection). The independent variables included national indices developed by others on: (1) social capital (e.g., health and education), (2) economic indicators (economic growth, trade, per capita earnings, investment), (3) governmental characteristics (various indicators of shared governance and openness), (4) environmental characteristics (air and water quality, government actions in protection treaties, and citizen participation), and (5) land pressure (primarily agricultural development).

The essential points of what they found are summarized in Fig. 6. The factors that had a positive influence on the quality of wetland protection and management were the quality of the environment, the openness and inclusiveness of government, the quality of the nation's social development, and, lastly, the expansion of agriculture. A negative influence was the degree of economic capitalization. I interpret these results as support for the following conclusions. First, the scarcity of wetlands brings appreciation for their losses and value, so that when agricultural expansion results in a regression of wetland area, then people and governments tend to respond with a heightened sense of the need for wetland protection/management. Second, this response is more likely to occur when citizens appreciate environmental quality and also have the means of responding. But this favorable response is more likely to happen when the governmental structure is receptive to these reactions through all types of interactions, including through local, regional, national, and non-governmental organizations. The negative relationship between wetland protection and indices of economic capital is due to the pressures for using the wetland area or capturing the social services of wetlands (a public benefit) for private gain. Greed, in other words, is not known as being an altruistic social attribute, or, to put it more benignly, economic activity is an imperfect measure of the general welfare.

My argument is that there are reciprocal interactions between environmental quality (hence sustainable systems), and the society's interests or appreciation for them. If social capitalization is necessary to initiate and sustain environmental quality, then isn't there a positive feedback between the health of society and its ecosystem (and I will leave the definition of "health" for the reader to interpret in their own situation—I am asking only for one to think about it)?

Social Learning and Punishment

Experiments and comparative analyses within and among cultures have unequivocally demonstrated that many individuals will be "fair" in situations when they have no recognizable benefit in terms of reproduction, food, economics, or status, and that they will punish unfairness. Perhaps paradoxically, the role of punishment is a positive influence on cooperation (punishment must not generate significant permanent counter-acting resentment), promotes social learning and supports institutions governing resources held in common (Sigmund et al. 2010). The drain of free riders (hitch hikers) is held in check, and the costs of punishment are outweighed by the increased gains from cooperation (Gächter et al. 2008), but only "if complemented by strong social norms of cooperation" (Herrmann et al. 2008). Social learning is, therefore, not entirely dependent on hereditary influences (Henrich et al. 2010).

The Edge of the Precipice

> It is not necessary to "go back" in time to be the kind of creature you are. The genes from the past have come forward to us. I am asking that people change not their genes but their society, in order to harmonize with the inheritance they already have (Shepard 1996).

This is already a crowded planet and getting warmer. There are 107 million km^2 of habitable land on earth and seven billion people. That means that there are 1.53 ha (3.76 ac) per person. From that 1.5 ha per person come the materials for cell phones, cars and computers, trucks, cement, etc. The land is where people labor to provide the beans for a double-latte grande, wheat, flowers, fruits, and fructose. Energy reserves are dwindling and water is already re-cycled multiple times. If we are going to build sustainable systems that are desirable, then, based on the examples discussed, it will not be helpful to stratify the remaining resources further between haves and have-nots. Doing so is incompatible with our deeply social nature that is cooperative and that values fairness and equitability. This book is about building sustainable systems, which is to say that it is about social contracts.

A sustainable system does not have to be a pleasant one, *but it could be*. The science of sustainability brings clarity, and is absolutely essential to understand the qualities of managing coupled human–natural systems. But why will a sustainable system not be just another industrial model of efficiency, of mass throughput, and based on mass survival, but not quality? Can't a dictatorial social system with a stratified social structure survive for quite awhile? Or will it eventually fail because the social contract between individuals is unenforceable? We expect quality in our sustainable systems, but that will not happen, I think, without a pervasively ubiquitous and conscious effort at the individual level, which is what the Great Iroquois Confederation—just one example of many—was all about.

The dissociation of the individual from the natural world is continuing (Pergams and Zaradic 2008) and it has consequences. We can and must acknowledge and work with that disconnection at the personal level to be helpful working within society. I am not arguing, however, to recreate harsh environments, but to respect the human aspects that supported the genetics, or that the genetics gave life to. I think it a reasonable hypothesis, therefore, that the restoration of a few hectares of habitat is, potentially at least, also about societal renewal and health. In the process of restoring ecosystems, individuals and society are re-establishing their relationships with the larger environments they live in, depend upon, learn from, and which many believe must be nurtured if the much-used term "ecosystem sustainability" has meaning.

This is not going to be easy and we are working against our evolutionary nature in some ways. We spent our evolutionary path, for example, trying to find satisfactory food that tastes good. We now have it at every corner, and many of us are obese. I am not saying to stop eating. We ambled along our evolutionary walk gaining some predictability over food supplies, and now we have the "Dead Zone" at the end of the Mississippi River created as a result of land use practices in an industrialized agricultural landscape. We can put "culture" back in agriculture. We have an irrepressible—or nearly so—interest in procreating. I am not saying that sex is bad or should be stopped, but that sexual relationships are not automatically isolated from the love in relationships and the touching and constructive emotional content. We once lived in humility, and community respect—and it occasionally became unbalanced. I am not saying that there were not outrageous consequences of unbalanced behavior, but to see that there were also times in history, in our culture, in our life, when things are kinder and gentler. We still have the capability and unbroken human traits of kindness, dignity, and lightness within that survived the bad times, and nurtured good behaviors. I am arguing to see these good aspects clearly and to nurture them actively. Indulgence in ignorance is not a positive attribute, although it may accelerate economic exchange. We have gone forth and multiplied and now have dominion over the earth in many ways. Now what? How will the tension between unbridled individual action (e.g., in economic markets), be resolved to make space for the necessary and dominate collective agreements (e.g., for climate change)? It usually seems like addressing that is a huge mystery to me, but that is not the same as saying it is impossible or undoable or unnecessary. It must include nurturing the innately human social abilities we all have. But we need a new cultural narrative to replace existing ones. It means to open some baggage, to know what is inside, and to be willing to be surprised. The dictums of good behavior and attributes still apply—and must apply one by one, every day, by every person, and together.

The Wabash Rule of 22

The movie "Shakespeare in Love" has a scene in which the introduction of Shakespeare's newly completed "Romeo and Juliet" is done by Wabash. Wabash is a nervous character with a terrific stutter and can barely complete a word, much less a sentence. If the play fails this first performance, then the theater is closed. Wabash is the edge of a calamity. Henslowe thrusts Wabash onstage where he gathers himself in a temporary stutter and then … launches into a brilliant and flawless introduction. He speaks clearly and his sentences flow from one to another with pizzazz and energy that reveals a sincere joy for the play. The audience gives boisterous response and the cast is inspired as it explodes into action. William S, however, expected a disaster and is now confused. He asks Henslowe: "What happened? Henslowe pauses, and then looks straight at him: "It is a mystery" and then turns to continue managing the performance.

Entire cultures are now unsure about their human heritage. When we dismiss this heritage intentionally or by omission, then we won't have the sustainable planet we aspire to live in. The reality is that unless we recognize that our common collaborative heritage is sufficient, and that we are whole, then a desirable sustainable system is not possible. Here is one way to look at it. A social worker in Quebec told me that there were 22 people compromised by an addictive behavior. The addiction might be an alcohol, heroin, or sex, and the 22 people were generally those closest to the addicted person who was unwilling to confront either the addiction its consequences. We can use the same empiricism to construct what happens when a positive example ripples through society to reach 22 people in a constructive way. And if one is skillful, then it will be multiples of 22. Our heritage may be a buried mystery, but it can be partially known, appreciated and nurtured. We cannot know everything, but we can be part of the theater, participate with enthusiasm, move beyond our fear of stuttering, and deliver our lines with authenticity and with effect. It may or may not be a disaster, but it is the only 'theater' we have, – our only home, our only life, and sole opportunity.

Acknowledgments I thank Mike Weinstein for the encouragement here and in previous collaborations leading up to this effort. I first explored this topic in Matfield Green, KS, at the 2005 Graduate Student Fellowship meeting of the Land Institute (Wes Jackson, Founder). Any inadequacies found herein, however, are the consequence of being a poor student—a "bug" in the web of this life, who has been incredibly fortunate to have experienced inspiring examples in the natural world, benefited from readings, ordinary life with others, and had precious encounters with profound teachers. Financial support was provided by NSF award DEB-1008184.

References

Abram D (1996) The spell of the sensuous. Random House, New York

Boehm C (1999) Hierarchy in the forest: the evolution of egalitarian behavior. Harvard University Press, Cambridge

Brosnan SF, de Waal FBM (2003) Monkeys reject unequal pay. Nature 425:297–299

De Waal F (2007) Chimpanzee politics: power and sex among apes. Johns Hopkins University Press, Baltimore

Fehr E, Bernhard H, Rockenbach B (2008) Egalitarianism in young children. Nature 454:1079–1083

Gächter S, Renner E, Sefton M (2008) The long-run benefits of punishment. Science 322:1510

Gandhi MK (1909) Hind Swaraj or Indian home rule. Navajivan Publishing House, Ahmedabad (Re-published in 1938)

Henrich J, Ensminger J, McElreath R et al (2010) Markets, religion, community size, and the evolution of fairness and punishment. Science 327:1480–1484

Herrmann B, Thoni C, Gachter S (2008) Antisocial punishment across societies. Science 319:1362–1367

Johansen BE, Mann BA (eds) (2000) Encyclopedia of the Haudenosaunee (Iroquois Confederacy). Greenwood, Westport

Kaplan S (1995) The restorative benefits of nature: toward an integrative framework. J Environ Psych 15:169–182

Kuo FE, Taylor AF (2004) A potential natural treatment for attention-deficit/hyperactivity disorder: evidence from a national study. Am J Public Health 94:1580–1586

Lansing JS (2007) Priests and programmers: technologies of power in the engineered landscape of Bali. Princeton University Press, Princeton

LaPeyre M, Mendelssohn IA, Reams MA et al (2001) Identifying determinants of nation's wetland management programs using structural equation modeling: an exploratory analysis. Environ Manage 27:859–868

Levine RV (2003) The kindness of strangers. Am Scient 91:226–233

Paul DN (2000) We were not the savages. Fernwood, Halifax

Pergams ORW, Zaradic PA (2008) Evidence for a fundamental and pervasive shift away from nature-based recreation. Proc Nat Acad Sci USA 105:2295–2300

Range F, Horn L, Viranyi Z, Huber L (2009) The absence of reward induces inequity aversion in dogs. Proc Nat Acad Sci 106:340–345

Semmann D, Krambeck HJ, Milinski M (2003) Volunteering leads to rock-paper-scissors dynamics in a public goods game. Nature 425:390–393

Shepard P (1996) The only world we've got: a Paul Shepard reader. San Francisco Books, Sierra Club

Sigmund K, De Silaa H, Traulsen A, Hauert C (2010) Social learning promotes institutions for governing the commons. Nature 466:816–863

Taylor AF, Kuo FE, Sullivan WC (2001) Views of nature and self-discipline: evidence from inner city children. J Environ Psychol 21:1–15

Ulrich RS (1984) View through a window may influence recovery from surgery. Science 224: 420–421

Ulrich RS (2007) Visual landscapes and psychological well-being. Landscape Res 4:17–23

Ulrich RS, Simons RF, Losito BD et al (1991) Stress recovery during exposure to natural and urban environments. J Environ Psych 11:201–230

Wilkerson RG, Pickett K (2009a) Income inequality and social dysfunction. Ann Rev Sociol 35:493–511; http://www.equalitytrust.org.uk/

Wilkerson R, Pickett K (2009b) The spirit level: why greater equality makes societies stronger. Bloomsbury Press, London http://www.slideshare.net/equalitytrust/the-spirit-level-slides-from-the-equality-trust

Zinn HA (2003) A people's history of the United States: 1492-present. Harper Collins, New York

Epilogue: The Challenge of Sustainability: Lessons from an Evolutionary Perspective

Simon Levin

Abstract The greatest challenge our generation faces is creating a sustainable future. At the core is maintaining the services ecosystems provide humanity, but our ability to achieve that objective is made more difficult because ecosystems, the biosphere, and the socio-economic system with which they are linked are complex adaptive systems, in which individual agendas translate into global consequences. For management, that introduces problems of the Commons, and of how to achieve cooperation in attaining the best possible solutions for the collective good. At the core are issues of equity, of prosociality, and of the management of public goods and common-pool resources. Progress has been made in addressing these issues, but realism argues that new institutional frameworks will be necessary to create a sustainable future for the global biosphere.

Keywords Sustainability • Equity • Discounting • Prosociality • Public goods • Common pool resources • Commons • Complex adaptive systems

As the chapters in this volume make clear, achieving a sustainable future for our children and their children is the central problem facing our societies. Can we grow economically without unfairly compromising the options for future generations to make choices about their lives (United Nations 1987)? Developing a comprehensive framework for answering that question is the first order of business in assessing and achieving sustainability. How do we measure and aggregate utilities to assess intertemporal social welfare (Arrow et al. 2004)? Are current patterns of consumption consistent with sustainability by this criterion, and if not what must we do differently?

S. Levin (✉)
Department of Ecology and Evolutionary Biology, Guyot Hall, Princeton University, Princeton, NJ 08544-1003, USA
e-mail: slevin60@gmail.com

M.P. Weinstein and R.E. Turner (eds.), *Sustainability Science: The Emerging Paradigm and the Urban Environment*, DOI 10.1007/978-1-4614-3188-6_21,

Sustainability means many things, with different emphases for different people. It includes the stability of financial markets and economic systems, of reliable sources of energy, as well as of biological and cultural diversity. At the core, though, it must mean the preservation of the services that we derive from ecosystems, and this raises a suite of scientific challenges. What are those services? How do they depend upon the features of ecosystems? What aspects of biodiversity and ecosystem organization maintain those features? What are the threats to those aspects, and how do we protect them, within our coupled human–environmental dynamic?

All ecosystems exhibit characteristic regularities in such features as the diversity and distribution of species, the spectrum of sizes of organisms, the balance of nutrients and stoichiometric ratios, and the flow of energy through the levels of the trophic web. If these are preserved, the ecosystem can continue to provide the services on which humanity depends, even if the identities of the component species change; when these are lost, the ecosystem is no longer the same in terms of the services it provides. This has led naturally to a focus on the robustness and resilience of coupled human–environmental systems, and on a search for indicators that systems are nearing qualitative shifts in character; that is, that they are at risk of transition into new basins of attraction, for example, from oligotrophic to eutrophic states (Holling 1973; Levin et al. 1998; Scheffer 2008). Addressing this crucial challenge is only in its infancy, but it represents an extremely promising and exciting area of investigation (Scheffer et al. 2009). The unique features of the challenge arise from the fact that ecosystems and the biosphere, as well as the socioeconomic systems with which they are interlinked, are complex adaptive systems (CAS), made up of individual agents that interact with one another locally, changing behaviors over ecological and evolutionary time, with macroscopic consequences that feed back to influence individual behaviors (Levin 1998). Sustainability science must extract the signal from the noise, focusing on those macroscopic features, and in how they arise and are sustained as the collective consequences of large numbers of interactions at microscopic scales.

The research agenda for sustainability is by nature interdisciplinary and multidisciplinary, leaving no discipline out, from physics and chemistry, biology and mathematics, to psychology, sociology and economics, to the humanities. How do we measure the services we derive from ecosystems, and how do we value those services and aggregate individual utilities to derive measures of social welfare? How do we assess and maintain the robustness of these services? CAS exhibit a range of features that pose special challenges; these have been well studied within individual disciplines, and more recently across disciplines because of the complementarity of insights that can emerge from interdisciplinary studies (Levin 1998, 2003). Independent of the context, CAS exhibit self-organization; the potential for multiple stable basins of attraction, with attendant path dependence and hysteresis; and contagious spread and risk of systemic collapse. Dynamics are played out on multiple scales of space, time and complexity, with the potential for destabilization through the dynamics of slow variables. Robustness in such systems depends on fluctuation and variation, and on a delicate balance of heterogeneity, redundancy, and modularity. For coupled human–environmental systems, therefore, a suite of research challenges present themselves: How do these systems self-organize over time? What features underlie their robustness and resilience, as well as resistance to changes that would improve human welfare? Does robustness normally increase over time, or does system evolution carry with it the seeds of its own collapse? What are appropriate indicators of the erosion of robustness, and increasing vulnerability to shocks? The research agenda to address such questions must include the development of agent-based and hierarchical models of self-organization, the elucidation of a statistical mechanics of ensembles of heterogeneous agents, and the description of emergent macroscopic dynamics.

The challenges already laid out are daunting, but have been at the center of research in ecology for decades. They will surely occupy the attention of ecologists for decades to come; but even if we could resolve them completely tomorrow, our work would have only just begun. Our ability to achieve a sustainable future is limited not primarily by our lack of understanding of biology or physics or chemistry or geology, but rather by a suite of obstacles that relate to psychology, sociology, economics, behavior and culture. They involve issues of intergenerational and intragenerational equity, and of the management of public goods and common pool resources. They inspire a quest to design mechanisms for achieving cooperation in the Global Commons, for example through the establishment and maintenance of social norms and more formal institutions and forms of government (Levin 2009).

At the core of issues of equity is discounting. We discount our own futures, and we discount the interests of others. Much of the inaction on environmental issues like climate change is because of discounting, and associated problems of managing public goods and the Commons. We need to develop theoretical and empirical approaches to problems of public goods and common pool resources, combining game theory and dynamical systems theory on networks to ask how cooperation can arise among independent agents, and how the emergence of groups, norms, customs and traditions depends upon and helps maintain prosocial behavior (Levin 2010). We need also to understand more generally how cooperation arises in natural systems, and to elucidate the role of leadership and the dynamics of collective decision-making (Couzin et al. 2005); and then we need to learn how to apply the insights we derive from other natural systems to human decision-making and the central questions of sustainability.

One of the greatest achievements of human societies has been the capacity to overcome competitive inefficiencies, and to find ways to avoid the "tragedy of the commons" (Hardin 1968). Through these advances, humans have organized themselves into cooperative groups, finding common purpose and laying the foundations for cultures, nations, and to a limited extent a global society. Hardin's solution to the tragedy of the commons was "mutual coercion, mutually agreed upon." Ostrom (1990, 2009) has led in explicating a framework and related theory of self-organization that helps diagnose whether small-scale fisheries will engage in such mutual agreements to improve their long-run capabilities and the sustainability of their fishery, and her work is inspiring other such studies in a wide range of systems. Such solutions involve some degree of prosociality toward other contemporary individuals as well as future generations. How are individual strategies shaped by prosociality, and how does such prosociality arise?

Intergenerational Equity

A fundamental problem in evolutionary theory is to understand how an organism allocates resources over its lifetime, balancing growth and reproduction, and trading off current needs against discounted expected future needs. Since evolution is about genetic combinations that are most successful in reproducing themselves in future generations, the problem of intergenerational allocation of resources is a natural extension—for example how many offspring an individual should have, when she should have them, and what fraction of resources to invest in each. In evolutionary theory, it is natural therefore to derive an implied prosociality, measured in terms of how much an individual values each offspring, and discounted in relation to the growth rate of the population. Other relatives will be valued as well, at lower levels of "prosociality".

This core problem in evolutionary theory has obvious analogues in economics. Parents plan their expenditures over their lifetimes not only to balance their own immediate comforts, but also to leave a legacy for their heirs, as extensions of themselves. This is known in the literature as the "dynasty problem" (Becker 1976), and has been the subject of a broad research literature. For the most part, however, this literature has not dealt adequately with the issue of uncertainty, which is a core topic in the evolutionary literature, or with the implications of individual allocation decisions for the expanding inequity of wealth within and among societies. For example, Arrow and Levin (2009) investigate these issues, determining the optimal (wealth) consumption strategy for an individual in relation to the probability distribution of offspring, the discount rate, and the effective "prosociality" assigned to each offspring. In our simple model, uncertainty can either increase or decrease current consumption, depending upon the shape of the utility curve. Furthermore, once the optimal strategy is implemented, the result will be a logarithmic distribution of wealth, with a variance (inequity) that grows over time at a rate proportional to the variance in the offspring probability distribution.

Work of this sort is a beginning, but much remains to be done. Data show that the distribution of wealth is not lognormal, but has a fatter tail. Possible explanations lie in the increased access wealthy individuals have to high returns on their investments, in the fact that the number of offspring one has is dependent on wealth, and in the shape of the utility curve. Certainly, the next decade must see considerable work in depth on the factors that are contributing to the increasing inequity in the distribution of wealth within and among populations if we are to make progress in achieving a sustainable future.

Intragenerational Equity: Public Goods and Common Pool Resources

Understanding why individuals forgo consumption in order to benefit their children, or more generally their kin, is not difficult. The greater challenge, with hope for humanity, relates to prosociality toward unrelated individuals. What are the consequences of such prosociality, and how has it arisen? Under what circumstances is prosociality sufficient to achieve the collective good, and how may it be enhanced otherwise to avoid defection from socially desirable behaviors? Avinash Dixit and I have explored one approach to these questions, beginning from the basic assumption that every individual can allocate resources selfishly or to a common pool or public good, and receives a payoff that is a nonlinear function of investment in self and the total community investment in the common pool/public good:

$$v_i = y(x_i, Z) - F(x_i + z_i) \qquad (1)$$

Here v is individual utility, x is investment in self, z is investment in the public good, and Z is the total (or average) community investment in the public good. F represents a cost function, dependent on the total investment.

Such a representation treats individuals as totally selfish, whereas there is considerable evidence that humans exhibit prosocial behavior toward other individuals, including nonkin. Therefore, as a second step, we modify (1) to account for that prosociality, replacing (1) by

$$v_i = y(x_i, Z) - F(x_i + z_i) + \gamma \sum_j y(x_j, Z) \qquad (2)$$

where γ is "prosociality," the value an individual places on other individuals in its group. In this simple formulation, all individuals are valued equally, but we modify this to consider a variety of topologies in which an individual i has a specific prosociality γ_{ij} for each other individual j. Of particular interest is the situation in which individuals exhibit prosociality only (or more strongly) toward other individuals within their own groups; in this case, the model is modified yet again to allow leakage of benefits, namely the incidental collateral benefits one might receive from actions intended to benefit others. In this model framework, one then computes the Nash equilibrium, which allows computation of the game-theoretic optimal strategies for all individuals in the population.

This is a powerful theoretical framework, but the test of its usefulness is in the application and testing of it in particular systems. We have therefore begun to apply the approach to the sharing of grazing lands among Maasai herdsmen, in collaboration with Dan Rubenstein. The sharing of grazing land is an effective strategy for dealing with uncorrelated variations in rainfall, and hence uncorrelated variations in land quality. However, the maintenance of sharing arrangements can be difficult to sustain without agreements (or top–down control), and the robustness of those agreements to defection is a topic of central interest.

Evolution and Emergence of Prosociality

The approaches described above all assume that prosociality exists, and ask what its consequences are. This is reasonable, because as already mentioned there is considerable evidence, in human and nonhuman populations alike, of prosocial behavior. It remains a puzzle, however, to understand why prosociality exists. Some of the explanations are undoubtedly rooted in genetics and in kinship, but prosociality also arises culturally, among unrelated individuals. Understanding this phenomenon is a rich area of investigation, including the concomitant emergence and cultural evolution of groups and institutions that foster prosociality (Axelrod and Hamilton 1981; Gintis and Bowles 2004; Nowak et al. 2004; Boyd and Richerson 2009; Levin 2009). Prosociality can emerge as a norm of behavior (Fehr 1999; Durrett and Levin 2005; Ehrlich and Levin 2005; Akçay et al. 2009), for example in which individuals change behavior based on homophilous imitation and other information gained from neighbors on a social network, and in which rewards and punishments coevolve with prosociality to stabilize those behaviors.

Ecological systems and socioeconomic systems alike are CASs, and it is their nature as such that poses unique challenges for management. Just as Adam Smith's Invisible Hand does not guarantee a collectively optimal system for the dynamics of economic resources, nor does a purely free-market approach assure a healthy future for our environmental systems and the services they provide us. Indeed, it is clear that the selfish agendas of individuals and nations too often trump the collective good, leading us to discount disagreeable futures (Levin 1999). Just as for economic systems, sound stewardship requires a mix of free market and top-down regulation. New institutions are needed that are flexible and adaptive, like the human immune system, and polycentric (Ostrom 2009). Finding the pathway to sustainable management of these CAS is the greatest interdisciplinary challenge of our generation.

Acknowledgment This material is based upon work supported by the National Science Foundation under Grant No. 0955699.

References

Akçay E, Van Cleve J, Feldman MW et al (2009) A theory for the evolution of other-regard integrating proximate and ultimate perspectives. PNAS 106:19061–19066

Arrow K, Dasgupta P, Goulder L et al (2004) Are we consuming too much? J Econ Perspect 18:147–172

Arrow K, Levin SA (2009) Intergenerational resource transfers with random offspring. PNAS 106:13702–13706

Axelrod R, Hamilton WD (1981) The evolution of cooperation. Science 211:1390–1396

Becker G (1976) Altruism, egoism and genetic fitness: economics and sociobiology. J Eco Lit 14:817–826

Boyd R, Richerson PJ (2009) Culture and the evolution of human cooperation. Philosoph Trans Royal Soc B 364:3281–3288

Couzin ID, Krause J, Franks NR et al (2005) Effective leadership and decision-making in animal groups on the move. Nature 433:513–516
Durrett R, Levin SA (2005) Can stable social groups be maintained by homophilous imitation alone? J Econ Behav Organ 57:267–286
Ehrlich PR, Levin SA (2005) The evolution of norms. PloS Biol 3:943–948. http://biology.plosjournals.org/perlserv/?request=get-toc&issn=1545-7885&volume=3&issue=6
Fehr E (1999) A theory of fairness, competition and cooperation. Quart J Econ 114:817–868
Gintis H, Bowles S (2004) The evolution of strong reciprocity: cooperation in heterogeneous populations. Theor Popul Biol 61:17–28
Holling CS (1973) Resilience and stability of ecological systems. Annu Rev Ecol Syst 4:1–24
Levin SA (1998) Ecosystems and the biosphere as complex adaptive systems. Ecosystems 1:431–436
Levin SA, Barrett S, Aniyar S et al (1998) Resilience in natural and socioeconomic systems. Environ Devel Econ 3:225–236
Levin SA (1999) Fragile dominion: complexity and the commons. Perseus Books Group, Reading
Levin SA (2003) Complex adaptive systems: exploring the known, the unknown and the unknowable. Bull Am Math Soc 40:3–19
Levin SA (ed) (2009) Games, groups, and the global good. Springer, Berlin
Levin SA (2010) Crossing scales, crossing disciplines: collective motion and collective action in the Global Commons. Philosoph Trans Royal Soc B 365:13–17
Hardin G (1968) The tragedy of the Commons. Science 162:1243–1248
Nowak MA, Sasaki A, Taylor C et al (2004) Emergence of cooperation and evolutionary stability in finite populations. Nature 428:646–650
Ostrom E (1990) Governing the commons: the evolution of institutions for collective action. Cambridge University Press, Cambridge
Ostrom E (2009) A general framework for analyzing sustainability of social-ecological systems. Science 325:419–422
Scheffer M (2008) Critical transitions in nature and society. Princeton University Press, Princeton, NJ
Scheffer M, Bascompte J, Brock WA et al (2009) Early warning signals for critical transitions. Nature 461:53–59
United Nations (1987) Report of the World Commission on Environment and Development A/RES/42/187. United Nations Press, New York

Index

M.P. Weinstein and R.E. Turner (eds.), *Sustainability Science: The Emerging Paradigm and the Urban Environment*, DOI 10.1007/978-1-4614-3188-6,

MIX
Papier aus verantwortungsvollen Quellen
Paper from responsible sources
FSC® C105338

If you have any concerns about our products,
you can contact us on
ProductSafety@springernature.com

In case Publisher is established outside the EU,
the EU authorized representative is:
Springer Nature Customer Service Center GmbH
Europaplatz 3, 69115 Heidelberg, Germany

Printed by Libri Plureos GmbH
in Hamburg, Germany